D0983457

Mass Spectrometry

Volume 8

A Specialist Periodical Report

Mass Spectrometry

Volume 8

A Review of the Recent Literature Published between July 1982 and June 1984

Senior Reporter
M. E. Rose, *Department of Chemistry, Sheffield City Polytechnic*

Reporters
M. A. Baldwin, *The School of Pharmacy, University of London*
J. H. Bowie, *University of Adelaide, Adelaide, South Australia, Australia*
R. M. Caprioli, *University of Texas, Houston, Texas, U.S.A.*
J. R. Chapman, *Kratos Analytical, Manchester*
J. Charalambous, *Polytechnic of North London*
J. Dannacher, *Physikalisch-Chemisches Institut der Universität Basel, Basel, Switzerland*
D. J. Harvey, *University of Oxford*
T. R. Kemp, *Kratos Analytical, Manchester*
N. M. M. Nibbering, *University of Amsterdam, Amsterdam, The Netherlands*
I. Powis, *University of Nottingham*
J.-P. Stadelmann, *Digital Equipment Corporation AG, Basel, Switzerland*

The Royal Society of Chemistry
Burlington House, London W1V 0BN

ISBN 0-85186-328-0
ISSN 0305-9987

Set by Unicus Graphics Ltd, Horsham
and printed in Great Britain
by Whitstable Litho Ltd, Whitstable, Kent

Foreword

In regretting the resignation of Bob Johnstone as Senior Reporter of this series of reports on mass spectrometry, I am sure that I reflect the views of all its regular Reporters. We hope that the commitments that precipitated his resignation bear fruit in fresh fields. He has overseen five fine volumes of the series and, in the process, introduced the occasional specialized chapters that enable the books to respond rapidly to novel developments and growth areas in mass spectometry. It is due not only to the timing of my appointment as Senior Reporter that the format of this volume is unchanged but also to its success. In the future, some innovations will be introduced to try to maintain or improve the high standards set in previous volumes.

The change in Senior Reporter presents an opportunity to ring some other changes. The nomenclature has been brought more into line with IUPAC recommendations so, for example, g.c.-m.s. has been abandoned in favour of GC/MS. The abbreviations used are familiar to practising mass spectometrists, but since they constitute a considerable amendment to SPR terminology a summary of the abbreviations is provided after the list of contents. Another change is more apparent than real. The absence of some regular Reporters from this volume is a temporary situation. The chapter on natural products, so long the domain of Dai Games, does not appear pending its revamping. It will reappear in Volume 9, probably under the authorship of another regular Reporter missing from this volume, Ian Howe. This shift reflects Ian's move into the biochemical and medical sphere. Dai Games will return to report on liquid chromatography/mass spectometry and some related methods.

The specialized chapters in this volume concern photoelectron–photoion coincidence spectroscopy, Fourier-transform ion cyclotron resonance, and fast-atom-bombardment mass spectrometry (FABMS). Co-contributors, Josef Dannacher and Jean-Pierre Stadelmann, and Nico Nibbering provide explanations of the fundamentals of photoelectron–photoion coincidence spectroscopy and Fourier-transform ion cyclotron resonance, respectively, then describe how modern methodologies are improving our understanding of gas-phase ion chemistry. Richard Caprioli's brief was to take the reporting of FABMS out of the realms of the qualitative and into more quantitative and systematic aspects. He fulfils the task by tackling the underlying theme of the relationship between known solution phenomena and observed ionic distributions in the gas phase following FAB of appropriate solutions.

Some new Reporters are welcomed to the SPR series, namely John Chapman, Michael Baldwin, and John Charalambous. I am indebted to these authors as well as to Professors Caprioli and Nibbering for stepping into the breach at short notice. If this series of books has neglected any area of mass spectrometry, it is that of inorganic applications. Mass spectrometry has become a major tool for inorganic chemists. John Charalambous was asked to modify the chapter on inorganic mass spectrometry as a first step in bringing the reporting of the topic into line with current, expanding practice, particularly with regard to the new and varied ion sources designed specifically for inorganic applications. It is intended to take further steps in this direction in future volumes.

A glance at the contributors to this volume indicates the international flavour of the book. I regard this diversity as advantageous. Indeed, in the future the proportion of overseas Reporters may be larger if major funding for British academic mass spectrometrists continues at its present level, which is generally insufficient to finance the instruments that are needed to make an impact on developing methodologies. Partly for the same reason, contributions from staff of instrument manufacturers are likely. If the present volume appears to be biased in favour of one manufacturer, it was not intentional. The planned impartiality was disrupted by Trevor Kemp's recent change of allegiance.

Lastly, to myself. Having compiled the first collective subject index for Volumes 1–5 (published in Volume 5), contributed a chapter to Volumes 7 and 8, and now being Senior Reporter of Volume 8, I feel as though I have served a form of apprenticeship. I hope that my experiences put me in good stead to continue the success of the series. It seems to be traditional for a new Senior Reporter to invite suggestions for improvements and/or specialized topics from readers and reviewers. This seems to me to be unnecessary, not because I believe that the volume could not be improved but because I am sure that you will make your opinions known to me whether I welcome them or not! Nevertheless, as any user of computerized mass spectrometry knows, feedback is key for maintenance of good-quality output.

M. E. ROSE

Contents

Abbreviations

AE	Appearance energy
API	Atmospheric-pressure ionization
CI	Chemical ionization
CID	Collision-induced decomposition (or dissociation)
DCI	Desorption (or direct) chemical ionization
DLI	Direct liquid introduction
EHI	Electrohydrodynamic ionization
EI	Electron impact
FA	Flowing afterglow
FAB(MS)	Fast-atom bombardment (mass spectrometry)
FD	Field desorption
FI	Field ionization
FT-ICR	Fourier-transform ion cyclotron resonance
FTMS	Fourier-transform mass spectrometry
GC/FTIR/MS	Gas chromatography/Fourier-transform infrared spectrometry/mass spectrometry
GC/MS	Gas chromatography/mass spectrometry
(HP)LC/MS	(High-performance) liquid chromatography/mass spectrometry
ICP/MS	Inductively coupled plasma/mass spectrometry
ICR	Ion cyclotron resonance
IE	Ionization energy
IKE(S)	Ion kinetic energy (spectroscopy)
KERD	Kinetic-energy-release distribution
LAMMA	Laser microprobe mass analysis
LD	Laser desorption
MIKE(S)	Mass-analysed ion kinetic energy (spectroscopy)
MQDT	Multi-channel quantum-defect theory
MSM	Multiple-scattering method
MS/MS	Mass spectrometry/mass spectrometry (*i.e.* tandem mass spectrometry)
NICI(MS)	Negative-ion chemical ionization (mass spectrometry)
PEPICO	Photoelectron–photoion coincidence spectroscopy
PES	Photoelectron spectroscopy
QET	Quasi-equilibrium theory
(RE)MPI(MS)	(Resonance-enhanced) multi-photon ionization (mass spectrometry)

RIMS	Resonance-ionization mass spectrometry
RRKM	Ramsberger, Rice, Kassel, and Marcus (theory)
SFC/MS	Supercritical-fluid chromatography/mass spectrometry
SIM	Selected-ion monitoring
SIMS	Secondary-ion mass spectrometry
TGA/MS	Thermogravimetric analysis/mass spectrometry
TLC/MS	Thin-layer chromatography/mass spectrometry
TOF(D)	Time of flight (distribution)

1

Ionization Processes and Ion Dynamics

BY I. POWIS

1 Introduction

Research into the basic processes in which ions are produced and interact continues to be an active and expanding area. Traditional methods for the generation and monitoring of ions are being extended and complemented, most notably by the introduction of new sources of ultraviolet radiation and molecular-beam technology. With such experimental advances new questions can be addressed and old problems tackled in fresh ways.

For the reviewer, however, the old problem of presenting a critical account of developments that are embodied in a large number of publications remains to be tackled, as before, by a selective citation of the literature. My objective, therefore, has been to identify work in which fundamental aspects of the behaviour of molecular ions are featured. Whatever criteria one employs, difficult choices remain to be made, and it is reasonable to acknowledge that in such circumstances the ultimate arbiter is perhaps a personal preference.

2 Ionization Processes

Molecular Photoionization. – Of all the processes leading to positive-ion formation, photoionization is the best able to be characterized at a fundamental level in a general manner applicable to all systems. The pursuance of this aim is a vigorous, developing field at present, as regards both theoretical and experimental advances. Apart from the immediate advantage of being able to model photoionization processes such as are encountered in a wide variety of situations, photoionization represents the simplest electron-scattering process and is of considerable intrinsic interest. Studies of photoionization provide an important means for the investigation of molecular electronic structure and dynamics.

An idealized experimental objective would be the measurement of partial cross-sections for the production of specified quantum states of the molecular-ion products and the electron, as a function of photon energy and relative orientations. Indeed, most of these requirements are currently feasible and have been attained, at least for some specific systems. A full description of the photoelectron includes its spin polarization, and whilst this has previously been considered in atomic systems it is just now receiving attention in connection with molecular systems[1] including, recently, the case of chiral molecules.[2,3]

[1] N. A. Cherepkov, *Adv. At. Mol. Phys.*, 1983, **19**, 395.
[2] N. A. Cherepkov, *Chem. Phys. Lett.*, 1982, **87**, 344.
[3] N. A. Cherepkov, *J. Phys. B*, 1983, **16**, 1543.

Most attention, however, focuses upon the various resonances that appear with changing photon wavelength in measured cross-sections and, often even more markedly, in the photoelectron angular distributions. Although pseudo-photon dipole (*e*, 2*e*) experiments have been used to establish a substantial body of absolute photoionization cross-section data[4,5] and continue to be used for systems such as HF[6] and HCl,[7] the bulk of recently published experimental data has been obtained using tunable synchrotron radiation sources. It is the consequent ability to scan continuously from below threshold up to photon energies of ~100 eV with high intensity which allows the rich dynamical structure, apparent with highly differentiated cross-section measurements, to be uncovered.

Concomitant with the growth in quantity and quality of experimental data is the development of theoretical treatments for the understanding of such processes. The relationship between theory and experiment here appears to be particularly close, with approximately simultaneous treatments for a given system being quite common. Naturally enough, first-row diatomics such as HCl,[7-9] N_2,[10,11] and O_2[12-14] and triatomics such as N_2O,[15,16] CS_2,[17] and COS[17] receive attention as theoretically tractable systems, but increasingly larger polyatomics can be tackled in this manner: examples include acetylene,[18-21] cyanogen,[22,23] ethylene,[24,25] and even benzenoid compounds.[26]

[4] C. E. Brion and J. P. Thomson, *J. Electron Spectrosc. Relat. Phenom.*, 1984, **33**, 287.
[5] C. E. Brion and J. P. Thomson, *J. Electron Spectrosc. Relat. Phenom.*, 1984, **33**, 301.
[6] F. Carnovale and C. E. Brion, *Chem. Phys.*, 1983, **74**, 253.
[7] S. Daviel, Y. Iida, F. Carnovale, and C. Brion, *Chem. Phys.*, 1984, **83**, 391.
[8] T. A. Carlson, M. O. Krause, A. Fahlman, P. R. Keller, J. W. Taylor, T. Whitley, and F. A. Grimm, *J. Chem. Phys.*, 1983, **79**, 2157.
[9] P. Natalis, P. Pennetreau, L. Longton, and J. E. Collin, *Chem. Phys.*, 1982, **73**, 191.
[10] F. A. Grimm and T. A. Carlson, *Chem. Phys.*, 1983, **80**, 389.
[11] R. R. Lucchese, G. Raseev, and V. McKoy, *Phys. Rev. A*, 1982, **25**, 2572.
[12] P. Morin, I. Nenner, P. M. Guyon, L. F. A. Ferreira, and K. Ito, *Chem. Phys. Lett.*, 1982, **92**, 103.
[13] P. Morin, I. Nenner, M. Y. Adam, M. J. Hubin-Franskin, J. Delwiche, H. Lefebvre-Brion, and A. Giusti-Suzor, *Chem. Phys. Lett.*, 1982, **92**, 609.
[14] P. M. Dittman, D. Dill, and J. L. Dehmer, *J. Chem. Phys.*, 1982, **76**, 5703.
[15] C. M. Truesdale, S. Southworth, P. H. Kobrin, D. W. Lindle, and D. A. Shirley, *J. Chem. Phys.*, 1983, **78**, 7117.
[16] T. A. Carlson, P. R. Keller, J. W. Taylor, T. Whitley, and F. A. Grimm, *J. Chem. Phys.*, 1983, **79**, 97.
[17] T. A. Carlson, M. O. Krause, and F. A. Grimm, *J. Chem. Phys.*, 1982, **77**, 1701.
[18] A. C. Parr, D. W. Ederer, J. B. West, D. M. P. Holland, and J. L. Dehmer, *J. Chem. Phys.*, 1982, **76**, 4349.
[19] P. R. Keller, D. Mehaffy, J. W. Taylor, F. A. Grimm, and T. A. Carlson, *J. Electron Spectrosc. Relat. Phenom.*, 1982, **27**, 223.
[20] Z. H. Levine and P. Soven, *Phys. Rev. Lett.*, 1983, **50**, 2074.
[21] D. Lynch, M. T. Lee, R. R. Lucchese, and V. McKoy, *J. Chem. Phys.*, 1984, **80**, 1907.
[22] D. M. P. Holland, A. C. Parr, D. L. Ederer, J. B. West, and J. L. Dehmer, *Int. J. Mass Spectrom. Ion Phys.*, 1983, **52**, 195.
[23] J. Kreile, A. Schweig, and W. Thiel, *Chem. Phys. Lett.*, 1983, **100**, 351.
[24] D. Mehaffy, P. R. Keller, J. W. Taylor, T. A. Carlson, M. O. Krause, F. A. Grimm, and J. D. Allen, jun., *J. Electron Spectrosc. Relat. Phenom.*, 1982, **26**, 213.
[25] F. A. Grimm, *Chem. Phys.*, 1983, **81**, 315.
[26] D. Mehaffy, P. R. Keller, J. W. Taylor, T. A. Carlson, and F. A. Grimm, *J. Electron Spectrosc. Relat. Phenom.*, 1983, **28**, 239.

Photoionization resonances and structures are conveniently considered as either one-electron phenomena, such as shape resonances (*i.e.* quasi-bound states where an electron is temporarily trapped by a potential barrier), or two-electron phenomena such as autoionization (where photoionization takes place *via* excitation into a discrete bound state lying above an ionization threshold followed by decay into the ionization continuum).

As with other photoabsorption processes, photoionization cross-sections depend upon dipole transition amplitudes between initial and final states. The final-state wavefunction describes the continuum electron orbital, and herein lies the computational challenge. Although in principle obtainable from the solution of the Schrödinger equation using a static-exchange potential (*i.e.* at the Hartree–Fock level of approximation, appropriate for the description of one-electron phenomena), significant difficulties arise in the calculation in this manner of the continuum function. This is due primarily to the non-central nature of the potential in a non-spherical molecule and the inapplicability of the algebraic variational methodology commonly applied to bound-state problems.

One of the more generally successful approaches has been the multiple-scattering method (MSM),[27] in which these difficulties are circumvented by representing the molecular potential as a cluster of spherical potentials positioned on the atomic sites. The MSM also often employs a local exchange approximation. A recent extension to this approach covers the evaluation of the interstitial integrals encountered with such a model potential.[28] The MSM potential is clearly very approximated; nevertheless the MSM formalism is routinely capable of producing qualitatively correct predictions of such features as shape resonances. It thus fulfils its aim of providing a valuable tool for exploratory investigation of these phenomena. A particular advantage of MSM at the present time is its applicability to large molecules such as benzene.[26] In this particular example calculated asymmetry parameters were found to agree with experiment for the first two ionizations, $(1e_{1g})^{-1}$ and $(3e_{2g})^{-1}$. A disagreement for the third $(1a_{2u})^{-1}$ channel was ascribed to vibronic mixing effects, which are not reproduced in the calculations.

Thiel[29] has demonstrated how shape resonances may be visualized by means of radial-density plots derived from MSM continuum functions, using the well known σ_u resonances in the N_2^+ $X^2\Sigma_g^+$ and CO_2^+ $\tilde{C}^2\Sigma_g^+$ ionizations as examples. The results for CO_2 (Figure 1) show dramatically enhanced densities in the $l = 3$ or 5 partial-wave components of the σ_u channel at the resonant photon energy, close to the atomic centres. Similar plots have been presented for the ionization of cyanogen.[23] The nature of these radial-density plots at resonance, with clear nodal structure and localized around the atoms, is suggestive of discrete valence-like orbitals. Thiel's work[23, 29] demonstrates strong correlations between the resonant continuum functions and the unoccupied antibonding virtual orbitals that are obtained from minimal-basis-set MO calculations. Other results for ethylene[25] suggest, though, that the correlation may not always be so

[27] D. Dill and J. L. Dehmer, *J. Chem. Phys.*, 1974, **61**, 692.
[28] J. R. Swanson and D. Dill, *J. Chem. Phys.*, 1982, **77**, 2010.
[29] W. Thiel, *J. Electron Spectrosc. Relat. Phenom.*, 1983, **31**, 151.

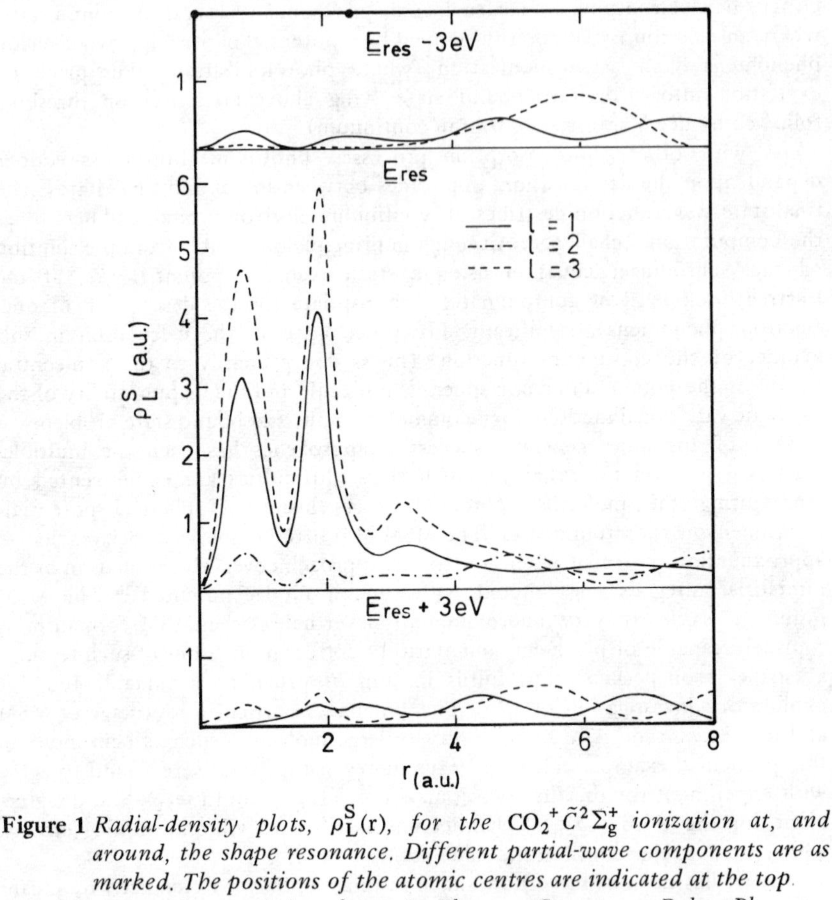

Figure 1 *Radial-density plots,* $\rho_L^S(r)$, *for the* $CO_2^+ \tilde{C}^2\Sigma_g^+$ *ionization at, and around, the shape resonance. Different partial-wave components are as marked. The positions of the atomic centres are indicated at the top.*
(Reproduced with permission from *J. Electron Spectrosc. Relat. Phenom.*, 1983, **31**, 151)

good as to enable simple predictions of the occurrence and relative energy of shape resonances to be made without resort to full scattering calculations.

MSM calculations provide a good qualitative and even semi-quantitative guide to the occurrence of shape resonances. However, they tend to predict exaggerated features with too narrow a width, as in the $(7\sigma)^{-1}$, $(6\sigma)^{-1}$ ionizations of N_2O.[16] The development of Hartree–Fock level computations capable of quantitative agreement with experiment is therefore of some interest. One such procedure is the moment-theory approach in which the electronic Hamiltonian is diagonalized in a large augmented basis set, using standard L^2 methods. The many discrete virtual orbitals lying in the continuum region of the spectrum that are thus obtained are not physical but may be smoothed, typically using Stieltjes or Tchebycheff moment theory, although other procedures are being evaluated.[30] While these calculations may readily be carried out in the static-exchange approximation, they yield no information concerning angular distributions, since the scattering equations are not solved. Moreover, the cross-sections so obtained may be unduly sensitive to the choice of underlying discrete pseudo-spectrum; Tchebycheff results for NO,[31] for example, appear to show spurious resonance structure that is not found in other calculations.[32,33]

Another approach, potentially the most reliable at the Hartree–Fock level of approximation, is to solve for the continuum function using single-centre expansion techniques. Although these methods are at present only practicable for small molecules such as N_2, NO,[32,33,11] and CO,[34] the Schwinger variational treatment of Lucchese and McKoy has been successfully applied to linear molecules like CO_2[35,36] and HCCH.[21] An adaptation of the Schwinger method promises to be particularly useful in the case of highly polar molecules.[37]

A one-electron phenomenon, other than shape resonance, that is well known in atomic photoionization is the Cooper minimum. This arises when the matrix element describing the transition into one of the dominant continuum channels changes sign. A distinct minimum in the net cross-section can result as this element passes through zero. The effect on the asymmetry parameter, β, may be even more striking. Consider ionization of the $3p$ orbital of argon. Dipole selection rules permit s and d photoelectron channels, the latter of which changes sign at the Cooper minimum. The d channel will display anisotropy, and at the minimum only the isotropic s channel contributes to the net ionization, causing the β parameters to fall markedly to zero. Arguing that the lone-pair orbitals of S and Cl should behave similarly, Carlson *et al.* have now obtained experimental evidence of Cooper minima in the molecular photoionization of CS_2,[17] CCl_4,[38] Cl_2,[39] and HCl.[8] Figure 2 shows how the asymmetry parameters for ionization of CCl_4 are affected at the Cooper minimum. For all these molecules MSM-type calculations give a good qualitative account of the phenomenon. It is remarkable that even the non-lone-pair orbitals examined ($5\sigma_g$ of Cl_2, 5σ of HCl, $6t_2$ and $6a_1$ of CCl_4) show partial Cooper minima at the appropriate photoelectron energies.

In general the electric-dipole selection rules operative in photoionization will permit several continuum channels of alternative symmetries to arise in production of a given ion state. These will be degenerate and hence cannot be distinguished by photoelectron spectroscopy alone, although to do so would be

[30] C. E. Woodward, S. Nordholm, and G. Bacskay, *Chem. Phys.*, 1982, **69**, 267.
[31] J. J. Delaney, I. H. Hillier, and V. R. Saunders, *J. Phys. B*, 1982, **15**, 1477.
[32] M. E. Smith, R. R. Lucchese, and V. McKoy, *J. Chem. Phys.*, 1983, **79**, 1360.
[33] L. A. Collins and B. I. Schneider, *Phys. Rev. A*, 1984, **29**, 1695.
[34] R. R. Lucchese and V. McKoy, *Phys. Rev. A*, 1983, **28**, 1382.
[35] R. R. Lucchese and V. McKoy, *Phys. Rev. A*, 1982, **26**, 1406.
[36] R. R. Lucchese and V. McKoy, *Phys. Rev. A*, 1982, **26**, 1992.
[37] M. E. Smith, R. R. Lucchese, and V. McKoy, *Phys. Rev. A*, 1984, **29**, 1857.
[38] T. A. Carlson, M. O. Krause, F. A. Grimm, P. Keller, and J. W. Taylor, *J. Chem. Phys.*, 1982, **77**, 5340.
[39] T. A. Carlson, M. O. Krause, F. A. Grimm, and T. A. Whitley, *J. Chem. Phys.*, 1983, **78**, 638.

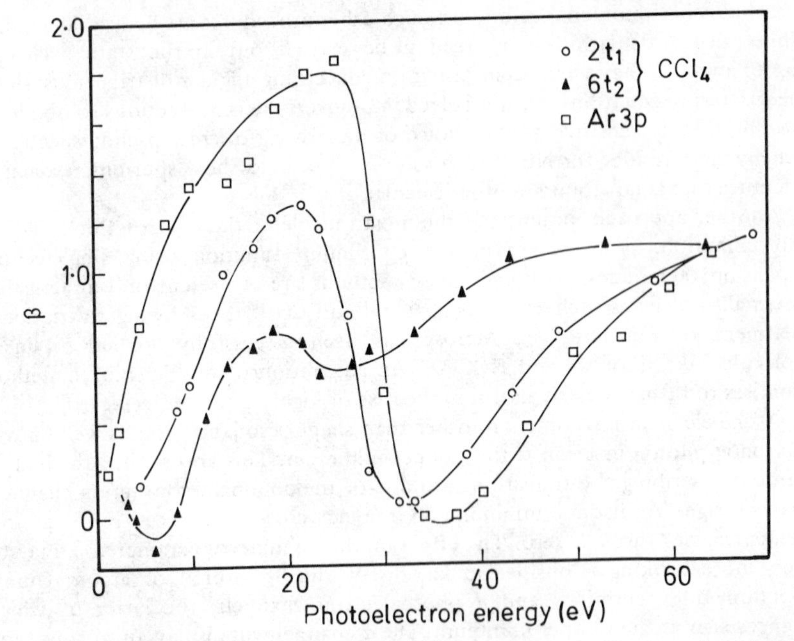

Figure 2 *Photoelectron angular-distribution parameters, β, as a function of photoelectron energy for ionization of the* $2t_1$ *lone-pair orbital and the* $6t_2$ *bonding orbital of* CCl_4. *For comparison the* 3p *orbital of* Ar *is also shown.*
(Reproduced with permission from *J. Chem. Phys.*, 1982, **77**, 5340)

desirable since the differing behaviour of the possible channels is an important facet of theoretical treatments. Nor can photoelectron angular-distribution measurements from a *randomly* oriented sample distinguish the channels; the necessary averaging over all orientations 'washes out' the dynamical detail. A few years ago Dehmer and Dill[40] performed an experiment on H_2 that could distinguish between the allowed σ_g and π_g photoelectron channels, providing a more rigorous test for theory. This was achieved using effectively 'fixed' H_2, the orientation of the molecular axis at the time of photoionization being determined by observing the direction of axial recoil dissociation fragments from the H_2^+ produced initially. An alternative procedure for determining molecular alignment is the measurement of the degree of polarization of fluorescence of the resulting ion target. This approach has been applied to CO $X^1\Sigma^+ \rightarrow CO^+$ $B^2\Sigma^+ + e^-$ $(\epsilon\sigma, \epsilon\pi)$ and N_2 $X^1\Sigma_g^+ \rightarrow N_2^+$ $B^2\Sigma_u^+ + e^-$ $(\epsilon\sigma_g, \epsilon\pi_g)$ photoionization.[41] Although fair agreement is found with theoretical calculations for CO, the N_2 data display quite poor agreement with Hartree–Fock calculations of the channel

[40] J. L. Dehmer and D. Dill, *Phys. Rev. A*, 1978, **18**, 164.
[41] J. A. Guest, K. H. Jackson, and R. N. Zare, *Phys. Rev. A*, 1983, **28**, 2217.

ratio.[11] It has been suggested that one problem with these calculations may be the restriction to a frozen-core hole state to represent the excited ion. A different technique for distinguishing degenerate continua makes use of spin-polarization photoelectron spectroscopy, and its use for the ionization of CH_3Br has been described.[42]

Like the above one-electron effects, autoionization resonances may be manifest as structure in ionization cross-sections as for H_2,[43] H_2S,[44] SO_2,[45] $HgCl_2$,[46] $HgBr_2$,[47] and HgI_2,[47] as well as the pronounced non-Franck–Condon vibrational distributions seen in photoelectron spectra of molecules such as HCl,[9] CO_2,[48] and fluorobenzenes.[49] A theoretical description of autoionization has to address the nature of the discrete-continuum-state interaction. Multichannel quantum defect theory (MQDT) has been combined with *ab initio* determination of electronic parameters to treat autoionization of the Hopfield Ryberg series of N_2 converging to the $B^2\Sigma_u^+$ state of the ion.[50] These calculations are successful in showing the $X^2\Sigma_g^+$ ionization to be more affected by this autoionization than the $A^2\Pi_u$ state.

A complete understanding of ionization resonances requires, as well as a full treatment of electron correlation, a consideration of vibronic interactions and competing neutral predissociation processes.[51] Simultaneous vibrational autoionization and electronic predissociation in NO have been treated[52] using an MQDT approach. It might be expected that $\Delta v > 1$ autoionization would compete less successfully with predissociation than $\Delta v = 1$. However, this is not found to be the case, and the calculation shows that it is the $^2\Pi$ Rydberg-valence-state interactions that both induce $\Delta v > 1$ autoionization and determine the resonance widths. Vibronic interactions play an important role in vibrational autoionization of N_2. Vibrational branching ratios for a number of autoionizing features in the photoion-yield curve of $X^1\Sigma_g^+$ N_2^+ (v) between 800 and 760 Å, using charge exchange to distinguish $v = 0$ and $v = 1$ and 2, have been reported.[53] High-resolution data[54] between the $v = 0$ or 1 thresholds reveal considerable structure corresponding to autoionizing $np\pi_u$ Rydbergs that converge to the $X^2\Sigma_g^+$ $v' = 1$ state of the ion. Ionization from the $v'' = 1$ state of the neutral species gives strong regular structure due to good Franck–Condon overlap, but

[42] F. Schäfers, M. A. Baig, and U. Heinzmann, *J. Phys. B*, 1983, **16**, L1.
[43] P. M. Dehmer and W. A. Chupka, *J. Chem. Phys.*, 1983, **79**, 1569.
[44] H. F. Prest, W. B. Tzeng, J. M. Brom, and C. Y. Ng, *Int. J. Mass Spectrom. Ion Phys.*, 1983, **50**, 315.
[45] C. Y. R. Wu and C. Y. Ng, *J. Chem. Phys.*, 1983, **76**, 4406.
[46] S. H. Linn, J. M. Brom, jun., W. B. Tzeng, and C. Y. Ng, *J. Chem. Phys.*, 1983, **78**, 37.
[47] S. H. Linn, W. B. Tzeng, J. M. Brom, jun., and C. Y. Ng, *J. Chem. Phys.*, 1983, **78**, 50.
[48] A. C. Parr, D. L. Ederer, J. L. Dehmer, and D. M. P. Holland, *J. Chem. Phys.*, 1983, **77**, 111.
[49] G. Dujardin, S. Leach, O. Dutuit, T. Govers, and P. M. Guyon, *J. Chem. Phys.*, 1983, **79**, 644.
[50] M. Raoult, H. Le Rouzo, G. Raseev, and H. Lefebvre-Brion, *J. Phys. B*, 1983, **16**, 4601.
[51] P. M. Guyon, T. Baer, and I. Nenner, *J. Chem. Phys.*, 1983, **78**, 3665.
[52] A. Giusti-Suzor and C. Jungen, *J. Chem. Phys.*, 1984, **80**, 986.
[53] K. Tanaka, T. Kato, and I. Koyano, *Chem. Phys. Lett.*, 1983, **97**, 562.
[54] P. M. Dehmer, P. J. Miller, and W. A. Chupka, *J. Chem. Phys.*, 1984, **80**, 1030.

ionization from $v'' = 0$ gives weaker irregular features as a result of the interactions with Rydberg states converging to N_2^+ $A^2\Pi_u$. Competition between vibrational and electronic autoionization of the series converging to $B^2\Sigma_u^+$ ($v = 1$) has been described.[55]

To distinguish between shape and autoionization resonances is not always easy. New photoelectron data[12] obtained with the ACO synchrotron help to confirm that a double resonance in the $B,b(^{2,4}\Sigma_g^-)$-state photoionization of O_2 at ~21 eV is due to the autoionization of $3s\ \sigma_g\ ^3\Sigma_u^-\ v' = 0$ or 1 Rydbergs converging to the $c^4\Sigma_u^-$ state of the ion. However, the identification[13] at the theoretically expected energy of ~21.5 eV of a shape resonance calls into question the identity of a broad resonance at ~19 eV in the b-state ionization. Vibrationally resolved measurements in fact reveal that the net broad feature arises from narrower features at slightly different photon energies for each vibronic level. MQDT calculations help to show that these are autoionizations of Rydbergs converging to the $B^2\Sigma_u^-$ state, the strong vibrational selectivity arising from similar b- and B-state geometries.[13]

A similar, but as yet incompletely resolved, ambiguity in the ionization of acetylene continues to attract attention.[18,56] This is the broad dip (or double bump) in the $(1\pi_u)^{-1}$ ionization cross-section at ~14 eV. Vibrationally resolved photoelectron studies[18,19] of the $\tilde{X}^2\Pi_u$ band reveal also a sharp dip in the β parameter in this region. Calculations involving channel-coupling effects by Levine and Soven[20] confirm earlier suggestions that autoionization of a $^1\Pi_u$ state ($2\sigma_u \rightarrow 1\pi_g$ transition) accounts for the observed behaviour at around 15 eV. It is further suggested[20] that intensity borrowing by this autoionizing transition from a bound $1\pi_u \rightarrow 1\pi_g$ transition could account for the discrepancies at lower energies (~12.5 eV), rather than a second $3\sigma_g \rightarrow 3\sigma_u$ autoionizing transition.[18] On the other hand Lynch *et al.*[21] believe that the $1\pi_u \rightarrow 1\pi_g$ transition lies above the ionization threshold and in fact leads to a shape-resonance-enhanced π_g continuum channel found in their Hartree–Fock calculations, which underlies the autoionization structures. The involvement of the $2\sigma_u \rightarrow 1\pi_g$ autoionization at least does now seem to be well established.

Inevitably the molecule H_2 has a prominent position in our growing knowledge of the photoionization process. In part this is because as electronically the simplest molecule it is a natural species to treat with newly developed theoretical methods such as the electronic *ab initio* quantum defect theory,[57] or with calculations using novel wavefunctions,[58,59] or in a description of dissociative photoionization.[60] Of equal significance, though, is the fact that H_2 is almost the only molecule for which rotationally resolved photoelectron spectra have been

[55] A. Tabche-Fouhaile, K. Ito, I. Nenner, H. Frohlich, and P. M. Guyon, *J. Chem. Phys.*, 1982, **77**, 182.
[56] D. M. P. Holland, J. B. West, A. C. Parr, D. L. Ederer, R. Stockbauer, R. D. Buff, and J. L. Dehmer, *J. Chem. Phys.*, 1983, **78**, 124.
[57] G. Raseev and H. Le Rouzo, *Phys. Rev. A*, 1983, **27**, 268.
[58] J. A. Richards and F. P. Larkins, *J. Phys. B*, 1984, **17**, 1015.
[59] Y. Itikawa, H. Takagi, H. Nakamura, and H. Sato, *Phys. Rev. A*, 1983, **27**, 1319.
[60] S. Kanfer and M. Shapiro, *J. Phys. B*, 1983, **16**, L655.

obtained.[61-63] Consequently β parameters for the ionization specified as H_2 $(X^1\Sigma_g^+ v'', J'') \rightarrow H_2^+ (^2\Sigma_g^+ v', J') + e^-$ are available. The β parameters associated with *S*-branch ($\Delta J = 2$) transitions have been reported in two experiments with good agreement at 584 Å (0.87 ± 0.19,[61] 0.81 ± 0.17[63]) but with a significant discrepancy at 736 Å (0.08 ± 0.15,[61] 0.54 ± 0.16[63]). For a pure *s*- ($l = 0$) electron wave $\beta = 0.2$ is the result predicted, and so the higher β values indicate the significant role of higher *f* ($l = 3$) waves in the photoelectron continuum.

Multi-photon Ionization. – The use of an intense laser source to bring about ionization by absorption of multiple photons in the visible and near UV, rather than by absorption of a single VUV photon, considerably expands the study and uses of photoionization phenomena. For the chemist the most interesting aspects centre around resonance-enhanced multi-photon ionization (REMPI), in which a neutral species is first excited to a discrete excited neutral state by absorption of one or more photons (in a true multi-photon, simultaneous absorption). This is then subsequently ionized by the absorption of further photons to raise the system into the ionization continua. A convenient short-hand representation to adopt in discussion of REMPI is $(n + m)$, where n photons excite the transition to the resonant intermediate states and m further photons are absorbed to ionize the system.

REMPI offers several distinctive advantages as an ionization source for mass spectrometry.[64,65] These include the ability to regulate the degree of ion fragmentation by control of the laser intensity and pulse duration,[66] and the achievement of isotopically selective ionization of mixtures by tuning the laser to a particular resonant frequency, as recently described for toluene and fluorobenzene.[67] This selectivity, combined with the extreme sensitivity that can be achieved by single-atom counting, is well demonstrated in resonance-ionization mass-spectrometric (RIMS) techniques for elemental analysis.[68-73] Whilst the use of a time-of-flight analyser in multi-photon ionization (MPI) studies is common,

[61] J. E. Pollard, D. J. Trevor, J. E. Reutt, Y. T. Lee, and D. A. Shirley, *Chem. Phys. Lett.*, 1982, **88**, 434.
[62] J. E. Pollard, D. J. Trevor, J. E. Reutt, Y. T. Lee, and D. A. Shirley, *J. Chem. Phys.*, 1982, **77**, 34.
[63] M. W. Ruf, T. Bregel, and H. Hotop, *J. Phys. B*, 1983, **16**, 1549.
[64] E. W. Schlag and H. J. Neusser, *Acc. Chem. Res.*, 1983, **16**, 355.
[65] M. P. Irion, W. D. Bowers, R. L. Hunter, F. S. Rowland, and R. T. McIver, jun., *Chem. Phys. Lett.*, 1982, **93**, 375.
[66] P. Hering, A. G. M. Maaswinkel, and K. L. Kompa, *Int. J. Mass Spectrom. Ion Phys.*, 1983, **46**, 273.
[67] O. Dimopoulo-Rademann, K. Rademann, B. Brutschy, and H. Baumgärtel, *Chem. Phys. Lett.*, 1983, **101**, 485.
[68] D. L. Donohue, J. P. Young, and D. H. Smith, *Int. J. Mass Spectrom. Ion Phys.*, 1982, **43**, 293.
[69] L. J. Moore, J. D. Fassett, and J. C. Travis, *Anal. Chem.*, 1984, **56**, 2770.
[70] M. W. Williams, D. W. Beekman, J. B. Swan, and E. T. Arakawa, *Anal. Chem.*, 1984, **56**, 1384.
[71] J. D. Fassett, J. C. Travis, L. J. Moore, and F. E. Lytle, *Anal. Chem.*, 1982, **55**, 765.
[72] J. P. Young and D. L. Donohue, *Anal. Chem.*, 1983, **55**, 88.
[73] J. P. Young, D. L. Donohue, and D. H. Smith, *Int. J. Mass Spectrom. Ion Processes*, 1984, **56**, 307.

the pulsed nature of the lasers that are used is also ideally suited for use in conjunction with Fourier-transform mass spectrometers,[65,74] offering, for analytical applications, the use of very high mass-resolving power. MPI also has applications for the detection of radical intermediates in reaction studies: MPI spectra (showing ion current *versus* photon wavelength) have been reported for CF_3,[75] Me,[76] and CH_2OH,[77] and detection of both Me and NO_2 by MPI has been utilized in a study of i.r. dissociation of nitromethane.[78] This technique, because of its resonant nature, permits identification of the quantum-state distribution of excited reaction products. An example is the case of NO, produced either by the reaction[79] $O(^1D_2) + N_2O \rightarrow 2NO$ or by photodissociation of NO_2.[80]

Much of the interest in REMPI stems from the power and versatility of the technique for investigating the physics and chemistry of excited-state species, including ionization processes and the subsequent behaviour of ions so prepared. The involvement of more than one photon in the ionization process raises extra complexity in any theoretical treatment, gives additional insights to those gained from studies of one-photon ionizations, and also yields greater flexibility for probing the processes of interest.

In MPI photoelectron spectroscopy the photoelectron-energy distributions will be dependent on the excited neutral rather than on the ground state as in conventional PES studies. This may be utilized to aid identification of the ionizing state, as in studies of H_2S[81,82] where autoionization of a super-excited state as well as direct ionization of the resonant intermediate state have been considered. In other situations variation of the vibronic level of the ionizing intermediate (by tuning the laser) may be exploited to yield a series of photoelectron spectra. The variations in vibrational structure that occur as a consequence of the changing Franck–Condon factors may then be used to obtain additional vibrational information about the ion, as has been demonstrated for the chlorobenzene molecule.[83] Other MPI–PES studies have been used to investigate Jahn–Teller effects[84] in the benzene cation.

In general one would expect the vibrational structure of MPI–PES to be determined by the Franck–Condon factors between the ion and the ionizing neutral state. Furthermore, since the excited neutral is often a Rydberg state with very similar geometry to the ion, a propensity for $\Delta v = 0$ transitions may be anticipated. This is indeed observed for toluene,[85] benzene,[86] and aniline[87]

[74] T. J. Carlin and B. S. Freiser, *Anal. Chem.*, 1983, **55**, 955.
[75] M. T. Duignan, J. W. Hudgens, and J. R. Wyatt, *J. Phys. Chem.*, 1982, **86**, 4156.
[76] J. W. Hudgens, T. G. Di Giuseppe, and M. C. Lin, *J. Chem. Phys.*, 1983, **79**, 571.
[77] C. S. Dulcey and J. W. Hudgens, *J. Phys. Chem.*, 1983, **87**, 2296.
[78] B. H. Rockney and E. R. Grant, *J. Chem. Phys.*, 1983, **79**, 708.
[79] N. Goldstein, G. D. Greenblatt, and J. R. Weisenfeld, *Chem. Phys. Lett.*, 1983, **96**, 410.
[80] R. J. S. Morrison and E. R. Grant, *J. Chem. Phys.*, 1982, **77**, 5994.
[81] J. C. Miller, R. N. Compton, T. E. Carney, and T. Baer, *J. Chem. Phys.*, 1982, **76**, 5648.
[82] Y. Achiba, K. Sato, K. Shobatake, and K. Kimura, *J. Chem. Phys.*, 1982, **77**, 2709.
[83] S. L. Anderson, D. M. Rider, and R. N. Zare, *Chem. Phys. Lett.*, 1982, **93**, 11.
[84] S. R. Long, J. T. Meek, and J. P. Reilly, *J. Chem. Phys.*, 1983, **79**, 3206.
[85] J. T. Meek, S. R. Long, and J. P. Reilly, *J. Phys. Chem.*, 1982, **86**, 2809.
[86] Y. Achiba, K. Sato, K. Shobatake, and K. Kimura, *J. Chem. Phys.*, 1983, **79**, 5213.
[87] M. A. Smith, J. W. Hager, and S. C. Wallace, *J. Chem. Phys.*, 1984, **80**, 3097.

and raises the exciting prospect of one being able to generate vibrationally state-selected ions in those cases where the $\Delta v = 0$ transition dominates the ionization step by simply tuning the laser to excite through a specific vibrational level of the resonant intermediate. Examples of such systems include the (3 + 1) ionization of N_2^+ ($A^2\Pi_u$ $v = 1$ or 2) through the $v' = 1$ or 2 levels of the $^1\Pi_u$ intermediate,[88] (3 + 1) ionization of H_2 through the $C^1\Pi_u$ $v = 0$–4 Rydberg state,[89] the (1 + 1) ionization of H_2 through the $B^1\Sigma_u^+$ $v = 0$ level,[90] as well as some (3 + 1) ionizations of NO and NH_3.[91] Where individual rotational levels of the intermediate are accessible there is also a possibility for rotationally state-selective generation of ions through the intervention of rotational selection rules for the photoionization step,[89] though this has yet to be demonstrated. However, even when the quantum-state distribution of the ions is less well defined than in these examples, it will still be considerably narrower than that obtained by ionization of a thermal neutral species. This can be exploited, for example, to reduce congestion in photodissociation spectra of MeI^+.[92]

Closer examination of MPI–PES vibrational distributions reveals often quite substantial deviations from Franck–Condon distributions, which, as in the case of single-photon ionization, are of considerable interest for the information they reveal concerning photoionization dynamics. Generally speaking, however, the degree of understanding of these features is less well advanced than for the one-photon case. Non-Franck–Condon vibrational distributions have been identified in the MPI–PES of H_2 $C^1\Pi_u$[89] and (tentatively) H_2 $B^1\Sigma_u^+$.[93] In reports of the ionization of the interesting (E,F) double-minimum state of H_2[94,95] non-Franck–Condon distributions have also been noted, although the peak of maximum intensity is explained by good Franck–Condon overlap. Discrepancies in the experimental data have arisen for the MPI–PES of the $\tilde{C}'$ state of ammonia, since Achiba *et al.*[91] do not observe the zero-kinetic-energy photoelectron peaks that were reported by Glownia *et al.*[96] in addition to $\Delta v = 0$ peaks. The latter authors propose that a vibrational autoionization mechanism is operative in the $\tilde{C}'$- and $\tilde{B}$-state ionizations. Further, the $\tilde{D}$ state, which yields exclusively zero-energy electrons, is thought to do so as a result of mixing with the $\tilde{B}$ state and a consequent internal conversion, followed by further excitation and autoionization.

The preceding examples were all $(n + 1)$ REMPI processes. Where more than one photon is absorbed in the ionizing step greater deviations from the distribution predicted by Franck–Condon factors between the resonant intermediate and the final ion state are found, perhaps not surprisingly. An example is CO:

[88] S. T. Pratt, P. M. Dehmer, and J. L. Dehmer, *J. Chem. Phys.*, 1984, **80**, 1706.
[89] S. T. Pratt, P. M. Dehmer, and J. L. Dehmer, *Chem. Phys. Lett.*, 1984, **105**, 28.
[90] H. Rottke and K. H. Welge, *Chem. Phys. Lett.*, 1983, **99**, 456.
[91] Y. Achiba, K. Sato, K. Shobatake, and K. Kimura, *J. Chem. Phys.*, 1983, **78**, 5474.
[92] W. A. Chupka, S. D. Colson, M. S. Seaver, and A. M. Woodward, *Chem. Phys. Lett.*, 1983, **95**, 171.
[93] S. T. Pratt, P. M. Dehmer, and J. L. Dehmer, *J. Chem. Phys.*, 1983, **78**, 4315.
[94] E. E. Marinero, R. Vasudev, and R. N. Zare, *J. Chem. Phys.*, 1983, **78**, 692.
[95] S. L. Anderson, G. D. Kubiak, and R. N. Zare, *Chem. Phys. Lett.*, 1984, **105**, 22.
[96] J. H. Glownia, S. J. Riley, S. D. Colson, J. C. Miller, and R. N. Compton, *J. Chem. Phys.*, 1982, **77**, 68.

$(2 + 1)$,[97] $(2 + 2)$,[98,99] and $(3 + m)$[100] REMPI processes have been observed for this species. For the $(3 + m)$ process the $v = 3$ and $v = 1$ or 2 levels of the $A^1\Pi$ intermediate state require $m = 2$ and $m = 3$ photons, respectively, for ionization. MPI–PES of these levels shows that formation of CO^+ $^2\Sigma^+$ $v = 0$ is strongly preferrred, with modulated intensities for the higher vibrational levels of the ion.[100] These cannot be explained by calculated Franck–Condon factors. Similar observations hold for MPI–PES of the same levels in the $(2 + 2)$ process.[99] In this case, at least, it appears possible to attribute some of the deviations to accidental resonances at higher energies.

Another molecule that has been extensively studied by MPI–PES is NO.[91,101–105] It is interesting to note that it affords the first rotationally resolved photoelectron spectrum of a molecule other than H_2.[101] As previously mentioned, ionization *via* vibrationally selected levels of the F and H Rydberg states is seemingly a direct process with $\Delta v = 0$.[91] In contrast, extensive vibrational excitations are found for the $(3 + 2)$ process *via* the $C^2\Pi$ Rydberg state and for the $(2 + 2)$ ionization *via* the $A^2\Sigma^+$ state.[103] These non-Franck–Condon effects are ascribed to perturbation of the resonant state due to Rydberg-valence interaction[102] and accidental resonances near the three-photon level,[104–106] respectively. Autoionization processes may also be important,[103] but unlike one-photon ionizations there is currently little knowledge of the role of resonances in the actual ionization step. This is quite generally true, although theoretical treatments of multi-photon autoionization of H_2 have been presented.[107–109]

As MPI–PES studies of photoelectron-energy distributions become more commonplace, efforts have been made to measure also angular distributions for NO,[91,105] H_2,[95] and rare gases.[105,110,111] A major distinction as compared to single-photon ionization is that the photoemission is from a well defined state prepared, and consequently aligned, by the initial resonant photon absorption. The resulting angular distributions are expected to contain higher harmonics and

[97] G. W. Loge, J. J. Tiee, and F. B. Wampler, *J. Chem. Phys.*, 1983, **79**, 196.
[98] R. W. Jones, N. Sivakumar, B. H. Rockney, P. L. Houston, and E. R. Grant, *Chem. Phys. Lett.*, 1982, **91**, 271.
[99] S. T. Pratt, P. M. Dehmer, and J. L. Dehmer, *J. Chem. Phys.*, 1983, **79**, 3234.
[100] S. T. Pratt, E. D. Poliakoff, P. M. Dehmer, and J. L. Dehmer, *J. Chem. Phys.*, 1983, **78**, 65.
[101] W. G. Wilson, K. S. Viswanathan, E. Sekreta, and J. P. Reilly, *J. Phys. Chem.*, 1984, **88**, 672.
[102] M. G. White, M. Seaver, W. A. Chupka, and S. D. Colson, *Phys. Rev. Lett.*, 1982, **49**, 28.
[103] J. Kinman, P. Kruit, and M. J. van der Wiel, *Chem. Phys. Lett.*, 1982, **88**, 576.
[104] J. C. Miller and R. N. Compton, *Chem. Phys. Lett.*, 1982, **93**, 453.
[105] M. G. White, W. A. Cupka, M. Seaver, A. Woodward, and S. D. Colson, *J. Chem. Phys.*, 1984, **80**, 678.
[106] T. Ebata, H. Abe, N. Mikami, and M. Ito, *Chem. Phys. Lett.*, 1982, **86**, 445.
[107] K. R. Dastidar and P. Lambropoulos, *Chem. Phys. Lett.*, 1982, **93**, 273.
[108] K. R. Dastidar and P. Lambropoulos, *Phys. Rev. A*, 1984, **29**, 183.
[109] K. R. Dastidar, *Chem. Phys. Lett.*, 1983, **101**, 255.
[110] R. Hippler, H. J. Humpert, H. Schwier, S. Jetzke, and H. O. Lutz, *J. Phys. B*, 1983, **16**, L713.
[111] K. Sato, Y. Achiba, and K. Kimura, *J. Chem. Phys.*, 1984, **80**, 57.

thus to be more informative than those obtained from a statistical, randomly oriented ground-state ionization. It is, however, necessary to recognize complications that arise owing to, for example, the intense laser fields in MPI.[112,113] The theoretical position for MPI of diatomic molecules has been elaborated in a number of papers.[114–116]

Various aspects of the physics of MPI have been discussed[117,118] and include competitive third-harmonic generation of VUV photons.[119–121] This phenomenon is of particular interest to chemists as it may be exploited to obtain tunable VUV by conversion of UV laser output; as an example, this has been used to study two-photon ionization of H_2.[90]

Mass-spectrometric Studies. Mass-spectrometric studies of fragmentation formation (MPI–MS) have been reported for a range of molecules including CS_2,[122] benzene,[123,124] halobenzenes,[123–125] toluene,[123,126] benzylamine,[127] and 3-phenyl-1-(dimethylamino)propane.[128] In many cases fragmentation is found to result from photodissociation of the molecular ion rather than from dissociative autoionization of a super-excited neutral species. However, the prominent $C_6H_6^+$ ion detected in MPI–MS of benzaldehyde[129–131] arises from ionization of the neutral photodissociation product C_6H_6, which is formed in competition with multi-photon ionization; the $C_6H_6^+$ ion precursor is nicely identified by comparing its MPI–PES with the MPI–PES of benzene.[131] Similar photoionization of neutral photodissociation products has been inferred in a study of acetone.[132]

[112] S. N. Dixit and P. Lambropoulos, *Phys. Rev. A*, 1983, **27**, 861.
[113] W. Ohnesorge, F. Diedrich, G. Leuchs, D. S. Elliot, and H. Walther, *Phys. Rev. A*, 1984, **29**, 1181.
[114] B. Ritchie, E. J. McGuire, J. M. Peek, and C. W. Hand, *J. Chem. Phys.*, 1982, **77**, 877.
[115] S. N. Dixit and V. McKoy, *J. Chem. Phys.*, 1984, **80**, 5867.
[116] J. C. Hansen and R. S. Berry, *J. Chem. Phys.*, 1984, **80**, 4078.
[117] B. Dick and R. M. Hochstrasser, *Chem. Phys. Lett.*, 1983, **102**, 484.
[118] M. H. Mittleman, *Phys. Rev. A*, 1984, **29**, 2245.
[119] M. G. Payne and W. R. Garrett, *Phys. Rev. A*, 1982, **26**, 356.
[120] M. G. Payne, W. R. Ferrell, and W. R. Garrett, *Phys. Rev. A*, 1983, **27**, 3053.
[121] M. G. Payne and W. R. Garrett, *Phys. Rev. A*, 1983, **28**, 3409.
[122] M. Seaver, J. W. Hudgens, and J. J. DeCorpo, *Chem. Phys.*, 1982, **70**, 63.
[123] D. W. Squire, M. P. Barbalas, and R. B. Bernstein, *J. Phys. Chem.*, 1983, **87**, 1701.
[124] K. R. Newton and R. B. Bernstein, *J. Phys. Chem.*, 1983, **87**, 2246.
[125] B. D. Koplitz and J. K. McVey, *J. Chem. Phys.*, 1984, **80**, 2271.
[126] V. T. Varoslavtsev, G. A. Abakumov, and A. P. Simonov, *Kantovaya Elektron (Moscow)*, 1984, **11**, 752.
[127] J. H. Catanzarite, Y. Haas, H. Reisler, and C. Wittig, *J. Chem. Phys.*, 1983, **78**, 5506.
[128] D. A. Lichtin, D. W. Squire, M. A. Winnik, and R. B. Bernstein, *J. Am. Chem. Soc.*, 1983, **105**, 2109.
[129] J. J. Young, D. A. Gobeli, R. S. Pandolfi, and M. A. El-Sayed, *J. Phys. Chem.*, 1983, **87**, 2255.
[130] A. V. Polevoi, V. M. Matyuk, G. A. Grigoreva, and V. K. Potapov, *Khim. Vys. Energ.*, 1984, **18**, 195.
[131] S. R. Long, J. T. Meek, P. J. Harrington, and J. P. Reilly, *J. Chem. Phys.*, 1983, **78**, 3341.
[132] M. Baba, H. Shinohara, N. Nishi, and N. Hirota, *Chem. Phys.*, 1984, **83**, 221.

Otner MPI–MS studies have sought to investigate the unimolecular dissociation kinetics of benzene[133,134] and chlorobenzene[135] by observation of metastable ions, or else by measurement of kinetic-energy releases,[136] again for benzene. The effects of IR multiple-photon excitation, either preceding[137] or following[138,139] MPI, have also been examined. Typically, the lasers that are used for MPI work have ns pulse widths, but the use of ps lasers extends the possibilities for investigating dissociative ionization.[66,140] Additionally the use of ps lasers in two-colour experiments is proving to be an impressive technique for measuring very rapid (sub-picosecond) intramolecular relaxation processes affecting the resonant intermediate through competition with the ionization step.[141–143] A better understanding of such relaxation processes of highly excited neutral species is, of course, of direct relevance to similar phenomena affecting the ions themselves.

Electron Collision and Other Ionization Phenomena. – In comparison to the studies of photoionization there is much less work on electron-impact ionization *per se* to report, particularly as regards theoretical understanding of ionization cross-sections.[144] A knowledge of empirical cross-sections is nevertheless of great importance for the modelling of planetary atmospheres and plasmas, and a compilation of such data for light atomic ions has appeared,[145] as have individual studies for SO_2[146] and C_2N_2.[147]

The use of molecular-beam sources allows rotational effects in molecular ionization to be studied. The rotational distribution of N_2^+ $B^2\Sigma_u^+$ ions formed from supersonically cooled N_2 (10 K) has been determined by monitoring the B–X emission.[148,149] At high electron-impact energies electric-dipole selection rules hold good, but below 800 eV first $\Delta J = 3$ and then larger transitions are recorded, indicating a breakdown of the dipole rotational selection rule.[149] Similar deviations may be present below 200 eV impact energy for CO ioniza-

133 U. Boesl, H. J. Neusser, R. Weinkauf, and E. W. Schlag, *J. Phys. Chem.*, 1982, **86**, 4857.
134 H. Kühlewind, H. J. Neusser, and E. W. Schlag, *Int. J. Mass Spectrom. Ion Phys.*, 1983, **51**, 255.
135 J. L. Durant, D. M. Rider, S. L. Anderson, F. D. Proch, and R. N. Zare, *J. Chem. Phys.*, 1984, **80**, 1817.
136 T. E. Carney and T. Baer, *J. Chem. Phys.*, 1982, **76**, 5968.
137 Y. Haas, H. Reisler, and C. Wittig, *Chem. Phys. Lett.*, 1982, **92**, 109.
138 Y. Haas, H. Reisler, and C. Wittig, *J. Chem. Phys.*, 1982, **77**, 5527.
139 J. S. Chou, D. Sumida, and C. Wittig, *Chem. Phys. Lett.*, 1983, **100**, 209.
140 A. L'Huillier, G. Mainfray, and P. M. Johnson, *Chem. Phys. Lett.*, 1984, **103**, 447.
141 B. I. Greene and R. C. Farrow, *J. Chem. Phys.*, 1983, **78**, 3336.
142 J. W. Perry, N. F. Scherer, and A. H. Zewail, *Chem. Phys. Lett.*, 1983, **103**, 1.
143 D. A. Gobeli, J. R. Morgan, R. J. St. Pierre, and M. A. El-Sayed, *J. Phys. Chem.*, 1984, **88**, 178.
144 H. Ehrhardt, *Comments At. Mol. Phys.*, 1983, **13**, 115.
145 K. L. Bell, H. B. Gilbody, J. G. Hughes, A. E. Kingston, and F. J. Smith, *J. Phys. Chem. Ref. Data*, 1983, **12**, 891.
146 O. J. Orient and S. K. Srivastava, *J. Chem. Phys.*, 1984, **80**, 140.
147 O. I. Smith, *Int. J. Mass Spectrom. Ion Processes*, 1983, **54**, 55.
148 S. P. Hernandez, P. J. Dagdigian, and J. P. Doering, *Chem. Phys. Lett.*, 1982, **91**, 409.
149 S. P. Hernandez, P. J. Dagdigian, and J. P. Doering, *J. Chem. Phys.*, 1982, **77**, 6021.

tion.[150] The effects of initial-state rotational energy have been investigated for H_2 and N_2 ionization.[151] Whereas the ionization-efficiency curve for N_2 is unaffected by changes in the initial rotational temperature, that for H_2 shows a marked shift, indicating the efficacy of rotational energy in promoting ionization. This phenomenon is not well understood but correlates with significant changes in rotational constants between ion and neutral species.

Processes other than electron-impact direct ionization have been described, including autoionizing levels of H_2^+ lying above the first excited ion state.[152] Ion-pair formation by methane[153] and the diatomic species N_2 and O_2[154] was investigated, and in both cases multiply charged ions were also reported. Multiply charged ion formation has been examined in detail with electron impact on the rare gases, both single ionization of the ions[155,156] and double ionization[156,157] being considered. In the latter case a comparison of experimental and theoretical cross-sections suggests that indirect inner-shell ionization followed by autoionization is the dominant mechanism.[157] The appearance energies of some molecular doubly charged ions from small organic species have been determined either by electron-impact ionization of the neutral molecule[158] or by charge-stripping collisions of the unipositive ion.[159,160] Dissociation of doubly charged molecular ions is usually dominated by Coulomb repulsions, but at short range intermolecular interactions can produce local wells in the otherwise repulsive curves, allowing long-lived states to exist. Under such circumstances the spectroscopy of the quasi-bound levels can be used to investigate the doubly charged ion structure. An example is N_2^{2+}, for which MCSCF calculations show several states possessing local minima.[161] Photofragment spectroscopy has been successfully used to probe transitions between the $X^1\Sigma_g^+$ $(v = 0–2)$ levels of N_2^{2+} to the predissociated $^1\Pi_u$ quasi-bound levels.[162] Related behaviour has been investigated for CO^{2+} by ion-kinetic-energy spectroscopy.[163]

The formation of molecular ions by field ionization is considered by Block.[164] Although the usual ionization mechanism is thought to be electron tunnelling, surface reactions and field-induced reactions have also to be considered. The H_3^+

[150] P. J. Dagdigian and J. P. Doering, *J. Chem. Phys.*, 1983, **78**, 1846.
[151] S. N. Foner and R. L. Hudson, *Chem. Phys. Lett.*, 1983, **100**, 559.
[152] F. Pichou, R. I. Hall, M. Landau, and C. Schermann, *J. Phys. B*, 1983, **16**, 2445.
[153] P. Plessis, P. Marmet, and R. Dutil, *J. Phys. B*, 1983, **16**, 1283.
[154] F. Feldmeier, H. Durchholz, and A. Hofmann, *J. Chem. Phys.*, 1983, **79**, 3789.
[155] C. Achenbach, A. Mueller, E. Salzborn, and R. Becker, *J. Phys. B*, 1984, **17**, 1405.
[156] D. Mathur and C. Badrinathan, *Int. J. Mass Spectrom. Ion Processes*, 1984, **57**, 167.
[157] M. S. Pindzola, D. C. Griffin, C. Bottcher, D. H. Crandell, R. A. Phaneuf, and D. C. Gregory, *Phys. Rev. A*, 1984, **29**, 1749.
[158] B. Brehm, U. Froebe, and H. P. Neitzke, *Int. J. Mass Spectrom. Ion Processes*, 1984, **57**, 91.
[159] M. Rabrenovic and J. H. Beynon, *Int. J. Mass Spectrom. Ion Processes*, 1984, **56**, 85.
[160] F. Maquin, D. Stahl, A. Sawaryn, P. Schleyer, W. Koch, G. Frenking, and H. Scharz, *J. Chem. Soc., Chem. Commun.*, 1984, 504.
[161] P. R. Taylor, *Mol. Phys.*, 1983, **49**, 1297.
[162] P. C. Crosby, R. Moeller, and H. Helm, *Phys. Rev. A*, 1983, **28**, 766.
[163] J. M. Curtis and R. K. Boyd, *J. Chem. Phys.*, 1984, **80**, 1150.
[164] J. H. Block, *Ber. Bunsenges. Phys. Chem.*, 1982, **86**, 852.

mass peak seen in the field-ionization spectrum of molecular hydrogen shows no tail, unlike the molecular-ion peak, which is ascribed to it being formed at a constant, well defined potential; this is indicative of a reaction taking place only at the surface of the ionizer tip.

Various techniques are available for the formation and characterization of negative ions. Bae *et al.* have employed electron capture *via* two successive collisions of a positive-ion beam with alkali-metal vapour in an unsuccessful search for the previously predicted species H_2^- and H_3^- (also N^- and N_2^-)[165] but found, unexpectedly, the metastable species He_2^-, which has an autodetachment lifetime of at least 10^{-4} s.[166] The same method was used in a study of He^- photodetachment[167] in which a sharp resonant feature at a photon energy of ~1.2 eV, corresponding to the first fully allowed transition of He^-, was confirmed. Resonances are also found in electron-attachment cross-sections of various chloroethylene anions[168,169] at low electron energies and may be associated with dissociative channels.

Photodetachment has been investigated by photoelectron spectroscopic measurements[170,171] and by measurement of the ion/neutral-species ratio, either in an optogalvanic apparatus as demonstrated for I^- [172] or in special laser–ion-beam rigs.[173,174] In an extremely high-resolution study of detachment for rotationally resolved levels of $B^2\Sigma_u^+$ C_2^- ($v = 6$–9)[174] lifetimes ranging from 10^{-7} to 10^{-10} s were found, indicating that vibrational–electronic coupling (and to a lesser extent rotational–electronic coupling) plays a major role in the autodetachment of an electron. Vibrational–electron couplings were also seen to be important in studies of IR-excited acetone enolate anions.[175]

3 Ionization of van der Waals Molecules and Reactions of Ion Clusters

Ionization studies of dimer and larger cluster molecules are considered separately here because of the special interest of these species. Such studies provide excellent opportunities to investigate molecular interactions. The structure of excited Rydberg states can be examined by spectroscopic study of autoionization features. A neutral dimer that may be only very weakly bound by van der Waals forces can, when ionized, possess stable ionic states with an intermolecular potential characteristic of weak chemical bonding. Ion/molecule interactions

[165] Y. K. Bae, M. J. Coggiola, and J. R. Peterson, *Phys. Rev. A*, 1984, **29**, 2888.
[166] Y. K. Bae, M. J. Coggiola, and J. R. Peterson, *Phys. Rev. Lett.*, 1984, **52**, 747.
[167] J. R. Peterson, M. J. Coggiola, and Y. K. Bae, *Phys. Rev. Lett.*, 1983, **50**, 664.
[168] E. Illenberger, H. Baumgärtel, and S. Süzer, *J. Electron Spectrosc. Relat. Phenom.*, 1984, **33**, 123.
[169] R. Kaufel, E. Illenberger, and H. Baumgärtel, *Chem. Phys. Lett.*, 1984, **106**, 342.
[170] H. B. Ellis, jun. and G. B. Ellison, *J. Chem. Phys.*, 1983, **78**, 6541.
[171] G. B. Ellison, P. G. Engelking, and W. C. Lineberger, *J. Phys. Chem.*, 1982, **86**, 4873.
[172] C. R. Webster, I. S. McDermid, and C. T. Rettner, *J. Chem. Phys.*, 1983, **78**, 646.
[173] P. A. Schulz, R. D. Mead, P. L. Jones, and W. C. Lineberger, *J. Chem. Phys.*, 1982, **77**, 1153.
[174] U. Hefter, R. D. Mead, P. A. Schulz, and W. C. Lineberger, *Phys. Rev. A*, 1983, **28**, 1429.
[175] R. F. Foster, W. Tumas, and J. Brauman, *J. Chem. Phys.*, 1983, **79**, 4644.

and reactive processes may be investigated with the 'collision complexes' that a dimeric ion resembles. From a technological standpoint ionization of excimers is important in modelling gas lasers. The higher clusters are significant as they represent a stage mid-way between gas and condensed phases; one may then seek to understand phase transitions, nucleation, or solvation phenomena by considering the properties of these clusters. Again from the point of view of chemical dynamics, the heat-bath effect of increased cluster size (increased number of degrees of freedom) on reactivity may be investigated.

As the supersonic molecular-beam technology required to generate significant concentrations of neutral dimers and higher clusters becomes more widespread, so the experimental literature on this topic grows. Ng[176] has reviewed many photoionization studies of these species. Several recent papers[177–179] have continued the investigations of photoionization structure of rare-gas dimers. Many features are assigned to autoionizing levels. Between the atomic fine-structure thresholds, the autoionization structure is suggestive of the monomer autoionization,[178] but elsewhere is more complex,[178,179] indicating that there are risks in identifying the molecular spectra as simply perturbed atomic-ionization spectra. Other reports dealing with rare-gas dimers and trimers have included photoionization of excited-state dimers,[180] photoelectron spectra of Xe complexes,[181] and double ionization (by electron impact) of NeKr.[182]

The structure and properties of large clusters can be probed in studies of their ionization and are of wide-ranging interest, for both fundamental and applied reasons. Argon *n*-mers up to $n = 60$ have been observed.[183] The binding energy released by formation of chemical bonds on ionization of the van der Waals bound cluster ($\sim$1.2 eV for $Ar_2 \rightarrow Ar_2^+$) is thought to 'boil off' other atoms in the pure clusters. However, no such loss of Ar from the cluster is observed when Ar_nEt_2O ($n = 1$–22) is generated; rather it appears to be the ether molecule that is fragmented.[184] Assuming that the ether receives the excess of energy on ionization, this result indicates weak coupling of its vibrations to those of the Ar atoms in the cluster with some fascinating implications regarding the structure of the cluster species.

Supersonic beam techniques have been used for the generation and subsequent ionization of many other cluster species ranging from large sodium-metal clusters[185,186] through aromatic rare-gas van der Waals complexes[187–190] to

176 C. Y. Ng, *Adv. Chem. Phys.*, 1983, **52**, 263.
177 M. G. White and J. R. Grover, *J. Chem. Phys.*, 1983, **79**, 4124.
178 P. M. Dehmer and S. T. Pratt, *J. Chem. Phys.*, 1982, **77**, 4804.
179 S. T. Pratt and P. M. Dehmer, *J. Chem. Phys.*, 1982, **76**, 4865.
180 A. W. McCown, M. N. Ediger, and J. G. Eden, *Phys. Rev. A*, 1984, **29**, 2611.
181 E. D. Poliakoff, P. M. Dehmer, J. L. Dehmer, and R. Stockbauer, *J. Chem. Phys.*, 1982, **76**, 5214.
182 K. Stephan, T. D. Märk, and H. Helm, *Phys. Rev. A*, 1982, **26**, 2981.
183 H. P. Birkhofer, H. Haberland, M. Winterer, and D. R. Worsnop, *Ber. Bunsenges. Phys. Chem.*, 1984, **88**, 207.
184 A. J. Stace, *J. Phys. Chem.*, 1983, **87**, 2286.
185 M. M. Kappes, R. W. Kunz, and E. Schumacher, *Chem. Phys. Lett.*, 1982, **91**, 413.
186 K. I. Peterson, P. D. Dao, R. W. Farley, and A. W. Castleman, jun., *J. Chem. Phys.*, 1984, **80**, 1780.

hydrogen-bonded species such as formic acid[191] and methanol[192] dimers. Ammonia clusters have been widely investigated with VUV photoionization,[193] multi-photon ionization,[194,195] and electron impact.[196] Although under some conditions the species $(NH_3)_n^+$ are detected,[193,196] it is protonated $(NH_3)_nH^+$ that dominate the mass spectra. This phenomenon can be attributed to the energetically and kinetically favoured process

$$NH_3^+ + NH_3 \rightarrow NH_4^+ + NH_2 \qquad (1)$$

occurring in an intracluster ion/molecule reaction. Protonated species dominate also the mass spectra of mixed ammonia/water clusters[197] and again are attributed to proton transfer *via* the intracluster reaction (1) in preference to mechanisms involving the H_2O species. Protonation of mixed water/alcohol dimers has been experimentally investigated,[198] and proton transfer in pure water dimer has been examined theoretically[199] as part of a wider *ab initio* investigation.[199,200]

The ammonia dimer is but one example of a van der Waals species that is ionized and seen to dissociate. Recently results for $(N_2)_2$,[201] $(H_2S)_2$,[202,203] $(H_2S)_3$,[203] $(C_2H_4)_2$,[204] $(C_2H_2)_2$,[205] and $(C_2H_2)_3$[206] have been presented. These fragmentations too are discussed in terms of ion/molecule reactions between the constituent moieties. In this sense a preliminary collisional association step is circumvented by the ionization of a van der Waals species, and the experimental observations relate directly to the behaviour of the ion/molecule complex. The appearance energies and relative fragment abundances reported in the above studies provide data on the energetics and reaction propensities of different ion states, whereas information about the dynamics of the decomposition of these

187 N. Gonohe, A. Shimizu, H. Abe, N. Mikani, and M. Ito, *Chem. Phys. Lett.*, 1984, **107**, 22.
188 L. F. Dimauro, M. Heaven, and T. A. Miller, *Chem. Phys. Lett.*, 1984, **104**, 526.
189 S. Leutwyler, U. Even, and J. Jortner, *J. Chem. Phys.*, 1983, **79**, 5769.
190 J. Jortner, U. Even, S. Leutwyler, and Z. Berkovitch-Yellin, *J. Chem. Phys.*, 1983, **78**, 309.
191 S. Tomoda, Y. Achiba, K. Nomoto, K. Sato, and K. Kimura, *Chem. Phys.*, 1983, **74**, 113.
192 S. Tomoda and K. Kimura, *Chem. Phys.*, 1983, **74**, 121.
193 H. Shinohara, N. Nishi, and N. Washida, *Chem. Phys. Lett.*, 1984, **106**, 302.
194 H. Shinohara, *J. Chem. Phys.*, 1983, **79**, 1732.
195 O. Echt, S. Morgan, P. D. Dao, R. J. Stanley, and A. W. Castleman, jun., *Ber. Bunsenges. Phys. Chem.*, 1984, **88**, 217.
196 K. Stephan, J. H. Futrell, K. I. Peterson, A. W. Castleman, jun., H. E. Wagner, D. Djuric, and T. D. Märk, *Int. J. Mass Spectrom. Ion Phys.*, 1982, **44**, 167.
197 K. Y. Choo, H. Shinohara, and N. Nishi, *Chem. Phys. Lett.*, 1983, **95**, 102.
198 A. J. Stace and A. K. Shukla, *J. Am. Chem. Soc.*, 1982, **104**, 5314.
199 S. Tomoda and K. Kimura, *Chem. Phys.*, 1983, **82**, 215.
200 S. Tomoda and K. Kimura, *Chem. Phys. Lett.*, 1983, **102**, 560.
201 K. Stephan, T. D. Märk, J. H. Futrell, and H. Helm, *J. Chem. Phys.*, 1984, **80**, 3185.
202 E. A. Walters, and N. C. Blais, *J. Chem. Phys.*, 1984, **80**, 3501.
203 H. F. Prest, W. B. Tzeng, J. M. Brom, and C. Y. Ng, *J. Am. Chem. Soc.*, 1983, **105**, 7531.
204 Y. Ono, S. H. Lin, W. B. Tzeng, and C. Y. Ng, *J. Chem. Phys.*, 1984, **80**, 1482.
205 Y. Ono and C. Y. Ng, *J. Chem. Phys.*, 1982, **77**, 2947.
206 Y. Ono and C. Y. Ng, *J. Am. Chem. Soc.*, 1982, **104**, 4752.

pseudo-collision complexes may be obtained by photodissociation,[207–209] metastable-ion,[210–215] and MIKES[211] experiments.

Futrell *et al.*[210] have examined both true, unimolecular metastable-ion dissociations and collision-induced dissociations of propane, CO_2, and NH_3 cluster ions generated by electron-impact ionization of molecular beams. The failure to observe reactions (2) and (3) as metastable-ion processes was suggested

$$(CO_2)_2^+ \rightarrow CO_2^+ + CO_2 \tag{2}$$

$$(NH_3)_2H^+ \rightarrow NH_4^+ + NH_3 \tag{3}$$

to indicate weak coupling in the complex since collisions do effectively initiate these reactions. However, Bowers and co-workers[211] have seen these dissociations as metastable-ion processes, and both sets of authors are agreed that reaction (4) and others do occur as metastable-ion dissociations. The metastable-

$$(NH_3)_2^+ \rightarrow NH_4^+ + NH_2 \tag{4}$$

ion abundances and kinetic-energy releases reported by Bowers are well fitted by statistical phase-space theory. The reason for the discrepancies between these two groups is unclear. The experiments of Bowers produce the complexes by association reactions in a high-pressure ion source, and this may lead to significant differences in the initial phase-space distribution prior to dissociation, a point emphasized in recent trajectory calculations for the ionization/dissociation of ArH_2^+ complexes.[216] Similar discrepancies are found for dissociations of the molecular-beam-generated ions Ar_2^+,[212] Ar_3^+,[213] and N_4^+ [214] as compared to the same species generated by association reactions,[215] except that here metastable-ion decompositions are detected only by the former technique.

The study of complexes formed in a high-pressure mass source has been reviewed.[209] In general, of course, the formation of ionic clusters under these conditions will be a more complex process, but this may be advantageous as, for example, when a sequence of hydrogen association[217] and dissociation[218] processes leads to the production of H_4^+. This was previously predicted to be stable but was not observed by photoionization of $(H_2)_n$ clusters.

[207] A. McCown, M. N. Ediger, S. M. Stazak, and J. G. Eden, *Phys. Rev. A*, 1983, **28**, 1440.
[208] M. F. Jarrold, A. J. Illies, and M. T. Bowers, *J. Chem. Phys.*, 1983, **79**, 6086.
[209] A. W. Castleman, P. M. Holland, D. E. Hunton, R. G. Keesee, T. G. Lindeman, K. I. Peterson, F. J. Schelling, and B. L. Upschutte, *Ber. Bunsenges. Phys. Chem.*, 1982, **86**, 866.
[210] J. H. Futrell, K. Stephan, and T. D. Märk, *J. Chem. Phys.*, 1982, **76**, 5893.
[211] A. J. Illies, M. F. Jarrold, L. M. Bass, and M. T. Bowers, *J. Am. Chem. Soc.*, 1983, **105**, 5775.
[212] K. Stephan, A. Stamatovic, and T. D. Märk, *Phys. Rev. A*, 1983, **28**, 3105.
[213] K. Stephan and T. D. Märk, *Chem. Phys. Lett.*, 1982, **90**, 51.
[214] K. Stephan and T. D. Märk, *J. Chem. Phys.*, 1983, **78**, 2953.
[215] A. J. Illies and M. T. Bowers, *Org. Mass Spectrom.*, 1983, **18**, 553.
[216] A. Hashim, J. S. Hutchinson, and E. R. Weiner, *J. Chem. Phys.*, 1983, **79**, 736.
[217] R. J. Beuhler, S. Ehrenson, and L. Friedman, *J. Chem. Phys.*, 1983, **79**, 5982.
[218] N. J. Kirchner, J. R. Gilbert, and M. T. Bowers, *Chem. Phys. Lett.*, 1984, **106**, 7.

4 Spectroscopy and Structure of Ions

Molecular-ion spectroscopy, despite having only developed, on the whole, as an active research area within the last decade, has inevitably come to exert a profound influence on our knowledge of ion structures and relaxation dynamics. Some indications of the current status in this field, though by no means complete, for it is a rapidly growing area, can be found in recent contributed volumes.[219,220]

One of the notable aspects of this subject is the extremely high resolution that can be obtained using coaxial ion- and laser-beam techniques. Amongst the advantages accruing from such methods are sub-Doppler linewidths, due to so-called kinematic compression of fast ion beams, and the ability to Doppler-tune the ions into resonance with single mode, but fixed frequency lasers where necessary. A good illustration is to be found in recent work on N_2O^+,[221,222] where the transition $\tilde{X}^2\Pi_{3/2}(0, 0, 0) \rightarrow \tilde{A}^2\Sigma^+(1, 0, 0)$ has been investigated. Although the upper level is in fact detected by monitoring the NO^+ predissociation product, it is relatively long-lived, which allows sharp (<100 MHz) lines to be recorded. Hyperfine structure, attributed mainly to upper-state Fermi-contact interactions on the outer N nucleus, is reported.[222] Moreover, isotopically substituted samples help to reveal that it is the outer atom that is cleaved in the predissociation of the (1, 0, 0) level; this may possibly conflict with earlier assumptions about N_2O^+ fragmentation. In a subsequent experiment[223] kinetic-energy releases accompanying fragmentation of selected levels of N_2O^+ were determined. It is of interest to consider the role of the ν_2 bending mode, since it is thought to exert a significant effect on the predissociation lifetimes of the $\tilde{A}$ state,[224] but there is apparently little effect on the product NO^+ vibrational distributions that are inferred.[223]

There is another reason for interest in the observation of a $\nu'_2 = \nu''_2 = 1$ transition[221] in that it permits the vibronic Renner–Teller interaction of the linear $^2\Pi$ ground state to be examined. Although this transition was only tentatively assigned in the ion-beam work, it appears to be corroborated by a new analysis of the N_2O^+ emission spectrum,[225] for in both cases it is suggested that the splitting parameter should be revised, with the new suggestions being in exact agreement with one another.

[219] 'Molecular Ions: Spectroscopy, Structure and Chemistry', ed. T. A. Miller and V. E. Bondybey, North-Holland, Amsterdam, 1983.
[220] 'Ions and Light', Gas Phase Ion Chemistry, Vol. 3, ed. M. T. Bowers, Academic Press, London, 1984.
[221] S. Abed, M. Broyer, M. Carré, M. L. Gaillard, and M. Larzillière, *Chem. Phys.*, 1983, **74**, 97.
[222] S. Abed, M. Broyer, M. Carré, M. L. Gaillard, and M. Larzillière, *Phys. Rev. Lett.*, 1982, **49**, 120.
[223] J. Lerme, S. Abed, R. A. Holt, M. Larzillière, and M. Carré, *Chem. Phys. Lett.*, 1983, **96**, 403.
[224] S. Miret-Arles, G. Celgado-Barrio, O. Atabek, and J. A. Beswick, *Chem. Phys. Lett.*, 1983, **98**, 554.
[225] J. F. M. Aarts and J. H. Callomon, *Chem. Phys. Lett.*, 1982, **91**, 419.

These kinds of data on the potential-energy surfaces and dissociative behaviour of N_2O^+ are of more than purely theoretical interest because of the role of N_2O^+ as an intermediary in the atmospherically important reaction

$$O^+(^4S_u) + N_2(X^1\Sigma_g^+) \rightarrow NO^+(X^1\Sigma^+) + N(^4S_u).$$

Indeed, the utility of ion-beam photopredissociation studies for preparing 'half-collision' complexes representing significant ion/molecule reactions has been highlighted,[226] with particular reference to O_2^+ and CH^+. Predissociation of rotationally quasi-bound levels of the latter ion (by tunnelling through the centrifugal barrier on the $A^1\Pi$ potential curve) is well established by photofragment spectra of the ion.[227] It is believed that this phenomenon is associated with an enhancement of radiative association of C^+ and H. However, in the interstellar medium where this is considered to be an important reaction, C^+ is predominantly in the lower $^2P_{1/2}$ state yet the CH^+ $A^2\Pi$ surface correlates to C^+ $^2P_{3/2}$. Very recently various calculations[228–230] of the $A^2\Pi$ predissociation have concluded that non-adiabatic interactions leading to predissociation to the lower C^+ state are significant. This would in turn have implications for the predictions of the radiative association rate. It will now be of interest to see whether any of the newly predicted predissociating levels can be assigned in experimental spectra of CH^+.

Various other electronic photopredissociation studies of HeH^+,[231] Cs_2^+,[232] O_2^+,[233] N_2^+,[234] O_3^+,[235,236] and SO_2^+[237] have been reported. High-resolution spectra of the non-predissociated N_2^+ $X \rightarrow A$ transition have also been recorded using charge-exchange detection of the excited state.[238] In an important paper Chupka and co-workers[92] have demonstrated how it is possible to relieve congestion in MeI^+ predissociation spectra by using a more selective method for ion generation (*i.e.* MPI), and this raises the enticing prospect of more highly resolved data on larger ions to come in the near future.

Much important high-resolution ion-beam spectroscopy is performed in the IR region, and because of its simplicity the hydrogen molecular ion attracts the attention of theoreticians and experimentalists alike.[239] Carrington *et al.* have extended their earlier measurements of HD^+ levels close to the dissociation limit

[226] J. T. Moseley, *J. Phys. Chem.*, 1982, **86**, 3282.
[227] H. Helm, P. C. Cosby, M. M. Graff, and J. T. Moseley, *Phys. Rev. A*, 1984, **25**, 304.
[228] M. M. Graff, J. T. Moseley, J. Durup, and E. Roueff, *J. Chem. Phys.*, 1983, **78**, 2355.
[229] M. M. Graff and J. T. Moseley, *Chem. Phys. Lett.*, 1984, **105**, 163.
[230] S. J. Singer, K. F. Freed, and Y. B. Band, *Chem. Phys. Lett.*, 1984, **105**, 158.
[231] D. Basu and A. K. Barua, *J. Phys. B*, 1984, **17**, 1537.
[232] H. Helm, P. C. Cosby, and D. L. Huestis, *J. Chem. Phys.*, 1983, **78**, 6451.
[233] J. C. Hansen, J. T. Moseley, A. L. Roche, and P. C. Cosby, *J. Chem. Phys.*, 1982, **77**, 1206.
[234] H. Helm and P. C. Cosby, *J. Chem. Phys.*, 1982, **77**, 5396.
[235] J. F. Hiller and M. L. Vestal, *J. Chem. Phys.*, 1982, **77**, 1248.
[236] S. P. Goss and J. D. Morrison, *J. Chem. Phys.*, 1982, **76**, 5175.
[237] T. F. Thomas, F. Dale, and J. F. Paulson, *J. Chem. Phys.*, 1983, **79**, 4078.
[238] J. C. Hansen, C. H. Kuo, F. J. Grieman, and J. T. Moseley, *J. Chem. Phys.*, 1983, **79**, 1111.
[239] A. Carrington and R. A. Kennedy in ref. 220, p. 393.

with the publication of a number of rotational components of the v 17–14 band.[240] For the lower N components excellent agreement is found (to a few thousandths of a wavenumber) with the non-adiabatic calculations of Wolniewicz and Poll,[241] but the discrepancies increase for higher N levels. It further appears from photoelectron spectra[62] that the calculations may diverge from experiment for higher v values as the dissociation threshold is approached. Further investigation of this point poses a challenge for both theoretical and experimental procedures of the highest accuracy. Similar possibilities for interplay of theory and experiments are to be found in IR studies of HeH^+.[242]

The monitoring of fluorescence following the initial ionization of a gaseous sample allows various features of the ionic states to be investigated.[243,244] When used in conjunction with supersonically cooled molecular beams spectral congestion is minimized and good-quality spectra are obtained; this technique has been applied to $A \rightarrow X$ emissions of F_2^+ [245] and Cl_2^+ [246] as well as to larger species such as substituted diacetylenes.[247–249] As well as providing spectroscopic constants and structural information, the state distribution of the ions can be inferred, as successfully demonstrated for the rotational distributions of N_2^+ [148] and H_2O^+.[250] Fluorescence of the first and second electronically excited states of CO_2^+ has been widely studied,[243,251–253] with particular attention paid to the nature of the suggested coupling between $\tilde{A}$ and $\tilde{B}$ states, as discussed by Johnson *et al.*[254] These authors have identified those levels of the $\tilde{A}$ state responsible for perturbation of the $\tilde{B}(0, 0, 0)$ state.[254] A consequence of the coupling is that a proportion of the $\tilde{B}$–$\tilde{X}$ oscillator strength is re-distributed to the red.

Investigations of the Jahn–Teller effect feature prominently in the development of molecular-ion spectroscopy, particularly for benzenoid cations. There is a continuing interest in this class of ions, since, with a combination of techniques such as laser-induced fluorescence and emission studies,[255] vibrationally resolved features may be observed with good sensitivity in the gas phase. They are thus

[240] A. Carrington, J. Buttenshaw, and R. A. Kennedy, *Mol. Phys.*, 1983, **48**, 775.
[241] L. Wolniewicz and J. D. Poll, *J. Chem. Phys.*, 1980, **73**, 6225.
[242] A.Carrington, R. A. Kennedy, T. P. Softley, P. G. Fournier, and E. G. Richard, *Chem. Phys.*, 1983, **81**, 251.
[243] J. P. Maier, M. Ochsner, and F. Thommen, *Faraday Discuss. Chem. Soc.*, 1983, **75**, 77.
[244] T. Ibuki, N. Sato, and S. Iwata, *J. Chem. Phys.*, 1983, **79**, 4805.
[245] R. P. Tuckett, A. R. Dale, D. M. Jaffey, P. S. Jarrett, and T. Kelly, *Mol. Phys.*, 1983, **49**, 475.
[246] R. P. Tuckett and S. D. Peyerimhoff, *Chem. Phys.*, 1984, **83**, 203.
[247] D. Klapstein, J. P. Maier, L. Misev, F. Thommen, and W. Zambach, *J. Electron Spectrosc. Relat. Phenom.*, 1983, **31**, 283.
[248] S. Leutwyler, D. Klapstein, and J. P. Maier, *Chem. Phys.*, 1983, **78**, 151.
[249] D. Klapstein, R. Kuhn, S. Leutwyler, and J. P. Maier, *Chem. Phys.*, 1983, **78**, 167.
[250] S. Leutwyler, D. Klapstein, and J. P. Maier, *Chem. Phys.*, 1983, **74**, 441.
[251] J. Rostas and R. P. Tuckett, *J. Mol. Spectrosc.*, 1982, **96**, 77.
[252] M. Endoh, M. Tsuji, and Y. Nishimura, *J. Chem. Phys.*, 1982, **77**, 4027.
[253] M. A. Johnson, J. Rostas, and R. N. Zare, *Chem. Phys. Lett.*, 1982, **92**, 225.
[254] M. A. Johnson, R. N. Zare, J. Rostas, and S. Leach, *J. Chem. Phys.*, 1984, **80**, 2407.
[255] C. Cossart-Magos, D. Cossart, S. Leach, J. P. Maier, and L. Misev, *J. Chem. Phys.*, 1983, **78**, 3673.

free of perturbations attributable to the lowered site symmetry encountered in matrix-isolation studies. The data gained empirically are complemented by theoretical calculations of the distorted structures, most recently using *ab initio* methods to investigate the $C_6F_6^+$ ion.[256] Benzene[257] and fluorinated benzene[258,259] cations have also been shown to be fertile ground for the investigation, by fluorescence measurements, of radiationless transitions. Both Jahn–Teller effects and these unimolecular relaxation processes can be viewed as non-adiabatic vibronic coupling phenomena, which are discussed further in the next section.

5 Intramolecular Relaxation and Decay Processes

Intramolecular energy redistribution in polyatomic ions has long been central to considerations of dissociation processes (see for example the review by Lifshitz[260]). Recent emphasis and development in this area, however, underline the fact that the really fundamental questions, that is those concerning the dynamics of an excited state immediately following its preparation, have relevance to both spectroscopic and kinetic studies.

One category of experimental observation of interest in this context is so-called 'isolated-state' behaviour. Generally this terminology is used to imply that a system does not appear to undergo the kind of complete statistical degradation of energy envisaged in the formulation of Quasi-Equilibrium Theory (QET), but rather shows evidence of non-statistical behaviour that can be correlated with a specific electronic state(s); the state is thus 'isolated' and, in this interpretation, is not subject to internal conversion.

An example, with many interesting but not yet understood facets, is the dissociation of the methyl nitrite/nitromethane cation system.[261-263] There is evidence that the ground state of $MeNO_2^+$ isomerizes to $MeONO^+$ prior to dissociation, yet the excited states of the isomers show very different behaviour in their breakdown curves,[261,263] in itself an indication of non-statistical behaviour. Most striking, however, is the formation of MeO^+ near the threshold for the $MeONO^+$ dissociation. This has a rate that is appreciably slower than the concurrent dissociation to NO^+;[262] neither the rate nor the observed isotope effects for CD_3O^+ are in accordance with expectations based on QET. It was suggested that this is a non-competing electronic predissociation of an isolated state that therefore does not interconvert with the surface leading to the faster NO^+ dissociation.[262]

Other examples of isolated-state behaviour are known from earlier studies. Galloy *et al.*[264] have studied one, the dissociation of $\tilde{B}$, $\tilde{C}$, and $\tilde{D}$ states of

256 K. Raghavachari, R. C. Haddon, T. A. Miller, and V. E. Bondybey, *J. Chem. Phys.*, 1983, **79**, 1387.
257 O. Braitbart, E. Castellucci, G. Dujardin, and S. Leach, *J. Phys. Chem.*, 1983, **87**, 4799.
258 G. Dujardin and S. Leach, *Faraday Discuss. Chem. Soc.*, 1983, **75**, 23.
259 G. Dujardin and S. Leach, *J. Chem. Phys.*, 1983, **79**, 658.
260 C. Lifshitz, *J. Phys. Chem.*, 1983, **87**, 2304.
261 J. P. Gilman, T. Hsieh, and G. G. Meisels, *J. Chem. Phys.*, 1983, **78**, 1174.
262 J. P. Gilman, T. Hsieh, and G. G. Meisels, *J. Chem. Phys.*, 1983, **78**, 3767.
263 I. K. Ogden, N. Shaw, C. J. Danby, and I. Powis, *Int. J. Mass Spectrom. Ion Processes*, 1983, **54**, 41.
264 C. Galloy, C. Lecomte, and J. C. Lorquet, *J. Chem. Phys.*, 1982, **77**, 4522.

$MeOH^+$ to Me^+, using *ab initio* methods. Although this group of excited states appears to interconvert readily, the Me^+ fragment, unlike the other products, is formed in a non-statistical manner. What the calculations reveal is that there exists a localized region of non-adiabatic interaction between the $\tilde{B}$ and $\tilde{A}$ states that efficiently converts $\tilde{B}$ to $\tilde{A}$. Moreover, this interaction zone acts as a 'funnel' producing a very specific population of $MeOH^+$ in the $\tilde{A}$-state configuration space. From here the trajectories are assumed either to run directly to Me^+ or to branch, become complex, and yield the other 'statistical' products. Thus it is argued[264] that here the 'isolated-state' behaviour is accounted for by the branching of the reactive flux into two distinct regions of phase space, which perhaps ironically is consequent upon efficient non-adiabatic interactions of the electronic states and the very specific population that results. How general an explanation of isolated-state behaviour this will prove to be is not yet clear. Yet the general notion, that as the trajectories run through the total phase space (which includes all electronic hypersurfaces) they branch, some possibly becoming quasi-ergodic, others not, does seem to have wider validity. Phase-space branching has certainly been postulated to explain the bimodal energy distributions recorded for the dissociation of enol ions,[265-267] though the mechanism for the binary branching is not clear.

Other experimentally studied examples are to be found in the dissociations of MeI^+,[268] most notably for the $\tilde{B}$ state. Here a direct dissociation to CH_2I^+ is observed simultaneously with more or less 'statistical' dissociation to Me^+ and I^+. It seems plausible that in this case there is competition (and hence phase-space branching) between the direct dissociation and internal conversion to lower surfaces, which subsequently undergo vibrationally ergodic dissociation.

The importance of internal conversions brought about by non-adiabatic interactions (*i.e.* breakdown of the Born–Oppenheimer approximation) has been highlighted by a number of theoretical studies of molecular ions.[264,269-272] When a single co-ordinate is considered, the region of interaction may occur at an avoided crossing between two adiabatic electronic potential-energy surfaces, as for $MeOH^+$.[264] Even more significant are conical intersections. These arise at an intersection between two non-degenerate surfaces of different symmetry. Another co-ordinate that removes an element of symmetry can cause the two states to reduce to the same representation in the lowered molecular symmetry, so that they now repel one another. The resulting surfaces in a two-dimensional co-ordinate space resemble two cones touching at their apex. The Jahn–Teller interaction is a special case where the initial degeneracy is not accidental but a consequence of the electronic-state symmetry. The multi-mode nature of the vibronic interactions that result at conical intersections has been stressed.[269] Even in the simplest two-mode case the locus of the intersection will be determined by a mode that preserves the symmetry element, giving rise to a seam of interaction. This can equivalently be viewed as a modulation of the energy

[265] C. Lifshitz, *Int. J. Mass Spectrom. Ion Phys.*, 1982, **43**, 179.
[266] C. Lifshitz, P. Berger, and E. Tzidony, *Chem. Phys. Lett.*, 1983, **95**, 109.
[267] F. Tureček and F. W. McLafferty, *J. Am. Chem. Soc.*, 1984, **106**, 2525.
[268] I. Powis, *Chem. Phys.*, 1983, **74**, 421.
[269] H. Köppel, L. S. Cederbaum, and W. Domcke, *J. Chem. Phys.*, 1982, **77**, 2014.

gap between the surfaces by a totally symmetric tuning mode, which is thus coupled to the non-totally symmetric mode forming the conical intersection.

It is to be expected that such non-adiabatic interactions will become more probable as molecular size goes up, for, as the number of vibrational modes increases, so also do the chances for two surfaces to approach in energy or to intersect. For ions which typically may have a 'floppier' structure than neutral species and so explore more of their configuration space (witness the important role of isomerization in ion decomposition) these effects may be even more likely. There is, however, a more fundamental reason for the non-adiabatic effects to be pronounced in ions. In open-shell molecules, adjacent electronic-state configurations usually differ in only one electron, and the coupling matrix elements $\langle l|\partial/\partial q|u\rangle$ can be large because the coupling is a single particle effect.[270] In contrast, neighbouring states in closed-shell species may differ in more than one electron, and vibronic coupling occurs only through configuration interaction.

The ubiquity of these interactions has implications for both spectroscopy and kinetics. Conical intersections in particular acquire prominence because their multi-dimensional nature leads to a dramatic enhancement of the non-adiabatic effects.[269] This is evident in an investigation of the $C_2H_4^+$ ion, where it is found that the second PES band, even though well separated from neighbouring bands, has totally non-adiabatic vibronic structure.[269] Further calculations on this ion show that conical intersections lead to ultrafast ($\sim 10^{-14}$ s)[270,271] non-radiative decay. Relaxations at this rate, and with the efficiency inferred, would readily account for the absence of detectable emissions from many organic cations.

The time evolution of a system on these extremely short femtosecond time-scales can be investigated by forming an autocorrelation function by Fourier transform of the electronic spectrum.[270,273] This represents the manner in which the initially generated wave packet propagates and evolves. Lorquet and co-workers[274] have shown how for HBr^+ the experimentally determined autocorrelation function reveals the wave packet splitting into two after crossing through the region of non-adiabatic interaction. The same approach can be applied to $C_2H_4^+$ [270,275] to estimate the probability of remaining on the upper surface.

In terms of dissociation dynamics there are two aspects to consider. The first is the specific manner in which non-adiabatic interactions can 'funnel' population onto a localized region of the lower surface.[264,272,276] Again in the case of

[270] H. Köppel, *Chem. Phys.*, 1983, **77**, 359.
[271] H. D. Meyer, *Chem. Phys.*, 1983, **82**, 199,
[272] D. Dehareng, X. Chapuisat, J. C. Lorquet, C. Galloy, and G. Raseev, *J. Chem. Phys.*, 1983, **78**, 1246.
[273] A. J. Lorquet, J. C. Lorquet, J. Delwiche, and M. J. Hubin-Franskin, *J. Chem. Phys.*, 1982, **76**, 4692.
[274] D. Dehareng, B. Leyh, M. Desouter-Lecomte, J. C. Lorquet, J. Delwiche, and M. J. Hubin-Franskin, *J. Chem. Phys.*, 1983, **79**, 3719.
[275] J. C. Lorquet, A. J. Lorquet, D. Dehareng, and B. Leyh, *Bull. Soc. Chim. Belg.*, 1983, **92**, 609.
[276] M. Desouter-Lecomte, C. Sannen, and J. C. Lorquet, *J. Chem. Phys.*, 1983, **79**, 894.

$C_2H_4^+$ this is thought to be a key feature controlling energy redistribution in the dissociation to $C_2H_2^+$.[276] It would also explain the lack of any significant barrier to dissociation that would be expected for reaction on a single electronic surface, a feature confirmed again by energy-release measurements for this dissociation.[277] The second aspect is the effect that the internal-conversion step exerts on the overall rate of dissociation. Whilst this could be approached by running classical or quasi-classical trajectories through the interaction region, as for $C_2H_4^+$ [271] or H_2O^+,[272] a more general method is to use a transition-state theory that has been modified to allow for the interaction. This has been done for a rate-determining spin–orbit interaction in CO_2^+ dissociation[278] as well as the aforementioned conical intersections of $C_2H_4^+$.[276]

Transition-state and Related Dissociation Treatments. – There is, inevitably, a continuing wider interest in statistical or transition-state treatments. A statistical treatment has been considered[279] for the product-energy distributions to be expected when simultaneous dissociation to more than two fragments occurs: evidence is starting to emerge of a number of ionic systems where this may in fact happen. Troe[280] has presented a simplified form of the Statistical Adiabatic Channel Model suitable for the calculation of specific unimolecular rate constants, $k(E, J)$. The 'tightness' or 'looseness' of the transition state is here controlled by a single parameter. This touches upon one of the currently more controversial issues in the area of ionic-system reaction dynamics. It has been observed that, whereas 'tight' transition states are often invoked for the understanding of unimolecular-ion dissociation rates, the 'loose' orbiting transition state has proved to be a highly successful device for the treatment of product-energy distributions (and the reverse ion/molecule bimolecular collisions). This may suggest that there are two dynamically important regions to be considered. Variational rate theory treatments[281] aim to locate the transition state, or region of minimum reactive flux, by locating the minimum in the sum-of-states function. For an ion-dissociation system it is easy to see that as the product separation increases more energy becomes converted to fixed potential energy, which, by the above criterion, places a transition state at the top of the centrifugal barrier in the effective potential function. However, a second effect occurs: as the reactant vibrations evolve into product rotations the density of states associated with these degrees of freedom increases. This counters the effect of the decreasing sum of states described above and can cause there to be a second minimum in the flux, this time located within the orbiting transition state;[282] this has been referred to[283] as the locked-rotor transition state from consideration of the locking of free rotations that occurs in the reverse bimolecular collision process.

[277] P. D. Lightfoot, C. J. Danby, and I. Powis, *Chem. Phys. Lett.*, 1983, **96**, 232.
[278] M. Th. Praet, J. C. Lorquet, and G. Raseev, *J. Chem. Phys.*, 1982, **77**, 4611.
[279] T. Baer, A. E. DePristo, and J. J. Hermans, *J. Chem. Phys.*, 1982, **76**, 5917.
[280] J. Troe, *J. Chem. Phys.*, 1983, **79**, 6017.
[281] W. L. Hase, *Acc. Chem. Res.*, 1983, **16**, 258.
[282] M. T. Bowers, M. F. Jarrold, W. Wagner-Redeker, P. R. Kemper, and L. M. Bass, *Faraday Discuss. Chem. Soc.*, 1983, **75**, 57.
[283] J. A. Dodd, D. M. Golden, and J. I. Brauman, *J. Chem. Phys.*, 1984, **80**, 1894.

Bowers and co-workers[282,284] have suggested that close to threshold a transition-state switching occurs, as dynamical control of the dissociation rate passes from the loose orbiting transition state to the tight inner transition state (or entropy bottleneck), although this has been disputed by others. Using a theoretical study of the $Me + Me^+$ system, it has been argued[283] that entropy bottlenecks associated with the locked-rotor transition state are not of kinetic significance because of the strong ion/molecule attractions. It was contended that if such switching does occur it must be associated with a double-well potential curve. Although the transition-state-switching model has been successfully applied to experimental data for $C_4H_8^+$, $C_4H_6^+$, and $C_6H_6^+$,[282,284] a more definitive test of these ideas must await more detailed data.

Brand and Baer[285] have reported an energy-selected study of the dissociation of six $C_5H_{10}^+$ isomers. An analysis of the asymmetric broadening of the fragment time-of-flight peak shapes reveals that all isomers except the lowest-energy 2-methylbut-2-ene isomer exhibit two-component decay-rate behaviour, and it is suggested that the slower rate channel results from isomerization to 2-methylbut-2-ene ions prior to dissociation. In many other cases complete isomerization of several precursors to common dissociating forms has been inferred from indistinguishable breakdown curves and other data; examples are various $C_3H_6O^+$ isomers,[296–288] ethylene oxide and acetaldehyde,[289] and various $C_4H_6^+$ isomers.[290,291] What is noteworthy in the present context, however, is that an RRKM-calculated rate for the 2-methylbut-2-ene dissociation is too slow. As the authors comment,[285] this is unlikely to reflect incomplete energy randomization given the isomerization that has occurred (a configurational randomization!), and, although no such calculations were performed, this may be a case to which the transition-state-switching mechanism can be applied, with the effective activation energy above threshold lying lower than predicted on purely energetic grounds.

In a number of other large molecular-ion systems, experimentally determined rate data can be compared successfully with RRKM/QET calculations.[292–294] This would also seem to be essentially true in the case of buta-1,3-diene,[295,296] although the $C_2H_4^+$-fragment channel may not be fully competitive.[291,296] A

[284] M. F. Jarrold, L. M. Bass, P. R. Kemper, P. A. M. van Koppen, and M. T. Bowers, *J. Chem. Phys.*, 1983, **78**, 3756.
[285] W. A. Brand and T. Baer, *J. Am. Chem. Soc.*, 1984, **106**, 3154.
[286] R. Bombach, J. P. Stadelmann, and J. Vogt, *Chem. Phys.*, 1982, **72**, 259.
[287] R. Bombach, J. Dannacher, E. Honegger, J. P. Stadelmann, and R. Neier, *Chem. Phys.*, 1983, **82**, 459.
[288] F. Tureček and F. W. McLafferty, *J. Am. Chem. Soc.*, 1984, **106**, 2528.
[289] K. Johnson, I. Powis, and C. J. Danby, *Chem. Phys.*, 1982, **70**, 329.
[290] F. N. Preuninger and J. M. Farrar, *J. Chem. Phys.*, 1982, **77**, 263.
[291] R. Bombach, J. Dannacher, and J. P. Stadelmann, *Helv. Chim. Acta*, 1983, **66**, 701.
[292] W. A. Brand and T. Baer, *Int. J. Mass Spectrom. Ion Phys.*, 1983, **49**, 103.
[293] L. Szepes and T. Baer, *J. Am. Chem. Soc.*, 1984, **106**, 273.
[294] R. Bombach, J. Dannacher, and J. P. Stadelmann, *J. Am. Chem. Soc.*, 1983, **105**, 4205.
[295] R. Arakawa, M. Arimura, and Y. Yoshikawa, *Int. J. Mass Spectrom. Ion Phys.*, 1982, **44**, 257.
[296] R. Bombach, J. Dannacher, and J. P. Stadelmann, *J. Am. Chem. Soc.*, 1983, **105**, 1824.

particularly appropriate system for the investigation of the energy-randomization hypothesis of the RRKM/QET approaches is bromobenzene and its deuteriated analogue. The aromatic ring's structure and vibrational frequencies are effectively unchanged in the dissociation, and so its influence is mainly as an energy sink. Since the C—H vibrations are not therefore participating directly in dissociation, any isotope effects will reflect only this role as an energy sink. In fact excellent agreement was found between experimental rate/energy-dependence curves and RRKM calculations for both isotopes,[297] supporting assumptions that all vibrational modes participate in energy randomization.

Other relaxation processes of the halobenzene ions have received much attention. Extremely reliable rate/energy-dependence curves for the corresponding dissociation of iodobenzene have been reported[298] and checked for internal consistency by repeating measurements for various instrumental residence times. This work is discussed in Chapter 3. Ion-trapping techniques may be used to extend ion observation times into the millisecond or even second time regimes. Photodissociation of iodobenzene[299,300] and bromobenzene[299,301] ions trapped in an ion-cyclotron resonance cell has been examined. Of interest in these studies is the determination of IR radiative relaxation times since these will set limits on the lifetimes of metastable-ion dissociations through their competition.[302] Although electronic radiative lifetimes of trapped ions such as HCl^+ and HBr^+ are readily determined by direct measurement,[303] the much longer IR-emission lifetimes pose a greater challenge. By kinetic modelling of multiple-photon dissociation processes, emission lifetimes of several hundred milliseconds were deduced for iodo- and bromo-benzenes.[300,301] More generally, trapped-ion fragmentation studies may be used to investigate extremely slow dissociations and related problems of the 'kinetic shift'.[304–306]

An alternative approach to the investigation of dissociation dynamics is the measurement of fragment kinetic energies, particularly where full kinetic-energy-release distributions (KERDs) can be obtained. In very many cases the form of the KERDs indicates vibrationally quasi-ergodic behaviour.[263,268,277,289,307–310]

[297] T. Baer and R. Kury, *Chem. Phys. Lett.*, 1982, **92**, 659.
[298] J. Dannacher, H. M. Rosenstock, R. Buff, A. C. Parr, R. L. Stockbauer, R. Bombach, and J. P. Stadelmann, *Chem. Phys.*, 1983, **75**, 23.
[299] J. P. Honovich and R. C. Dunbar, *J. Phys. Chem.*, 1983, **87**, 3755.
[300] N. B. Lev and R. C. Dunbar, *Chem. Phys.*, 1983, **80**, 367.
[301] R. C. Dunbar, *J. Phys. Chem.*, 1983, **87**, 3105.
[302] R. C. Dunbar, *Int. J. Mass Spectrom. Ion Processes*, 1983, **54**, 109.
[303] C. C Martner, J. Pfaff, N. M. Rosenbaum, A. O'Keefe, and R. J. Saykally, *J. Chem. Phys.*, 1983, **78**, 7073.
[304] C. Lifshitz and P. E. Eaton, *Int. J. Mass Spectrom. Ion Phys.*, 1983, **49**, 337.
[305] C. Lifshitz, P. Gotchiguian, and R. Roller, *Chem. Phys. Lett.*, 1983, **95**, 106.
[306] C. Lifshitz, M. Goldenberg, Y. Malinovich, and M. Peres, *Int. J. Mass Spectrom. Ion Phys.*, 1983, **46**, 269.
[307] W. A. Brand, T. Baer, and C. E. Klots, *Chem. Phys.*, 1983, **76**, 111.
[308] B. E. Miller and T. Baer, *Chem. Phys.*, 1984, **85**, 39.
[309] M. Tadjeddine, G. Bouchoux, L. Malegat, J. Durup, C. Pernot, and J. Weiner, *Chem. Phys.*, 1982, **69**, 229.
[310] M. F. Jarrold, A. J. Illies, N. J. Kirchner, W. Wagner-Redeker, M. T. Bowers, M. L. Mandich, and J. L. Beauchamp, *J. Phys. Chem.*, 1983, **87**, 2213.

Most of these data were obtained by photoelectron–photoion coincidence (PEPICO) methods (as indeed were many of the rate/energy data discussed previously – for a comprehensive tabulation of systems studied by PEPICO see Chapter 3 and ref. 311), and the KERDs are available as a function of a defined initial state and vibrational energy. It has been proved particularly convenient for the purposes of interpretation and analysis of such data to plot scaled or reduced forms of the KERDs for different excitation energies on a common diagram; changes in the characteristic dynamics of the dissociation are then readily discerned in the scaled data.[262,268,289,307,308]

Dissociations of various alkyl halides have provided rich detail when studied in this manner.[268,307,308] In these species excited spin–orbit states of the halogen fragments may be produced non-statistically, and both vibrationally ergodic and non-ergodic behaviour can be observed in electronically specific fragmentations. Clearly a detailed knowledge of the couplings and interconversions between the electronic states will be required for a full understanding of the experimental data. In iodopropane[307] $>60\%$ I $^2P_{1/2}$ formation from $\tilde{A}$- and $\tilde{B}$-state fragmentations can be inferred from an energetic argument (that the accompanying Pr^+ fragment would be more completely fragmented itself were the extra 0.94 eV excitation of the I $^2P_{1/2}$ available to it). For iodomethane[268] analysis of the KERDs is taken to indicate preferential formation of I $^2P_{1/2}$ from $\tilde{B}$-state dissociation; however, as discussed earlier in this section, it would seem that this must follow an internal conversion, presumably to the $\tilde{A}$ state. A strong $\tilde{A}$ to $\tilde{B}$ coupling has in fact been proposed elsewhere to explain anisotropy parameters measured in photopredissociation experiments with this ion.[309]

The quantitative and, indeed, qualitative forms of the KERDs that are found for the quasi-ergodic fragmentations are very sensitive to molecular-rotation effects, and especially to the constraints required to conserve angular momentum at the centrifugal barrier. Appropriate phase-space calculations that assume an ion-induced dipole potential in order to locate the centrifugal barrier and that properly conserve angular momentum have a high degree of success in comparisons with the experimental data. A more direct confirmation or test of these models, however, requires greater characterization of the rotational states of the molecular species. Although rotational distributions of products may sometimes be available (for example the CH^+ product from CH_4^+ studied by laser-induced fluorescence[312]), little is known about dissociations of polyatomic reactants whose rotation is characterized other than by an assumed temperature. Such data can be revealing. Powis[313] has investigated rotational predissociation of deuteriated ammonia below its 0 K threshold. In this region only the rotationally hot members of the thermal-ion population are able to dissociate, and so reducing the ionization energy increases the mean rotational state and angular momentum of the dissociating sample. Above threshold the energy releases for the loss of D are statistical;[313,314] remarkably, however,

[311] J. Dannacher, *Org. Mass Spectrom.*, 1984, **19**, 253.
[312] B. H. Mahan and A. O'Keefe, *Chem. Phys.*, 1982, **69**, 35.
[313] I. Powis, *Chem. Phys.*, 1982, **68**, 251.
[314] R. E. Kutina, A. K. Edwards, R. S. Pandolfi, and J. Berkowitz, *J. Chem. Phys.*, 1984, **80**, 4112.

below the 0 K threshold the energy release does not extrapolate to zero but achieves a limiting value. Actually this was shown to be in accord with the phase-space calculations[313] and indicates that states of high angular momentum have a propensity for high energy release. This is attributed to angular-momentum conservation for this system requiring much of the initial angular momentum to appear as orbital motion of the products at the centrifugal barrier; this is then converted to relative translational energy at infinite separation.

Centrifugal-barrier effects in this system are observed in another manner, since tunnelling through the barrier gives rise to metastable ions.[315] In this, and in a companion study of H loss from CH_4^+ and its deuteriated analogues,[316] the tunnelling mechanism was confirmed by examining abundances of metastable ions and energy releases as a function of assumed barrier height, this latter being effectively controlled by variation of the thermal rotational energy. Again, the ion-induced dipole attractive potential/centrifugal-barrier model is demonstrated to be appropriate, although for the specific cases of NH_3^+ and ND_2H^+ the agreement is less satisfactory than for ND_3^+ and the methanes. It is tentatively suggested that this may reflect an isotope dependence of the required $\tilde{A}$–$\tilde{X}$ internal conversion prior to dissociation,[315] which may in fact be brought about by a conical intersection.[269]

Rotational effects in the simpler H_2^+ rotational predissociation have been demonstrated. By examining fragmentations of H_2^+ prepared from different precursor molecules, the rotational distribution of the H_2^+ ion could be varied.[317] Structure observed in the H^+ translational-energy spectra can be correlated with expected predissociations of quasi-bound levels (v, J) of H_2^+. Electronic predissociation of O_2^+ [318] and of the much more complicated system of N_2O^+ [319,320] has likewise been investigated by determination of the partitioning of the excess energy into translation, as has predissociation of CO_2^+ $\tilde{C}^2\Sigma_g^+$.[321] *Ab initio* calculations[278] suggested that predissociation takes place *via* a rate-determining inter-system crossing to $\tilde{a}^4B_1$, which may then either dissociate to O^+ or rapidly interconvert with $\tilde{X}^2\Pi_g$, which in turn correlates with the CO^+ product. However, when CO^+ becomes energetically possible above the $\tilde{C}(0, 0, 0)$ level, the O^+ product is seen to vanish. Bombach *et al.*[321] propose that this is due to an alternative predissociation mechanism *via* a $^2B_1$ ($^2\Pi_u$) state coupled by Coriolis interaction. The measured $v = 0:1$ population ratio of 2:1 for the CO fragment from the $\tilde{C}(0, 0, 0)$ level was nevertheless shown to be compatible with a simple application of statistical adiabatic-channel theory to dissociation on the calculated $\tilde{a}^4B_1$ surface. For all the dissociation channels observed there was apparently a high degree of excitation of the diatomic fragment.

[315] M. F. Jarrold, A. J. Illies, and M. T. Bowers, *Chem. Phys. Lett.*, 1982, **92**, 653.
[316] A. J. Illies, M. F. Jarrold, and M. T. Bowers, *J. Am. Chem. Soc.*, 1982, **104**, 3587.
[317] A. G. Brenton, J. H. Beynon, E. G. Richard, and P. G. Fournier, *J. Chem. Phys.*, 1983, **79**, 1834.
[318] R. Bombach, A. Schmelzer, and J. P. Stadelmann, *Int. J. Mass Spectrom. Ion Phys.*, 1982, **43**, 211.
[319] J. L. Olivier, R. Locht, and J. Momigny, *Chem. Phys.*, 1982, **68**, 201.
[320] J. L. Olivier, R. Locht, and J. Momigny, *Chem. Phys.*, 1984, **84**, 295.
[321] R. Bombach, J. Dannacher, J. P. Stadelmann, and J. C. Lorquet, *J. Chem. Phys.*, 1983, **79**, 4214.

6 Bimolecular Reactions

For some time ion/molecule systems represented arguably the most closely investigated category of bimolecular reactive cross-section studies. This circumstance arose not just because of the relatively large cross-sections exhibited by such reactions but also because of the possibilities for control, selection, and determination of the translational energies of the ionic species. These experimental features are well exemplified in recent merged-beam studies of the reactions of D_2^+ with C[322] and F[323] atoms, in which collision energies are varied over four orders of magnitude and distributions of product kinetic energy are determined. As is now well known, reactive cross-sections can frequently be modelled in terms of capture cross-sections dependent upon long-range ion/induced-dipole, ion/permanent-dipole, and, as are shown to be required for the reaction with C (^3P),[322] ion/quadrupole interactions. Quasi-classical trajectory calculations permit the dynamical implications of model[324,325] and qualitatively realistic[326] ion/molecule potential surfaces to be investigated. However, for open-shell reactants such as those mentioned above, full electronic-state correlation diagrams can be exceedingly complex with many close-lying electronic surfaces,[322,323] and in general much work applicable to the full understanding of state-specific reactions remains to be done.

In the meantime there have been dramatic increases in the availability of experimental state-resolved data for neutral reaction systems due, for example, to the introduction of laser techniques such as laser-induced fluorescence (LIF) for the probing of product-state distributions. From the perspective of this review, perhaps the most significant recent development in ion/molecule studies has been the successful introduction and application of state-selective experimental techniques for ionic species, with the accompanying prospect of much more detailed experimental data to come. Accordingly this section focuses upon such developments.

Despite difficulties associated with relaxation of nascent distributions, LIF has been successfully employed to probe the populations of CO^+ levels formed in charge-exchange experiments.[327–329] An important finding for the thermal reaction

$$N^+\,(^3P) + CO\,(X^1\Sigma^+) \rightarrow N\,(^4S) + CO^+\,(X^2\Sigma^+\ v, J)$$

was that predominantly $v = 0$ is formed with a rotational temperature of just 410 K.[329] Consequently nearly all the excess energy of the reaction must be released into product translation. This runs counter to received wisdom that a small energy defect (*i.e.* quasi-resonance between *internal* energy levels of

[322] G. F. Schuette and W. R. Gentry, *J. Chem. Phys.*, 1983, **78**, 1777.
[323] G. F. Schuette and W. R. Gentry, *J. Chem. Phys.*, 1983, **78**, 1786.
[324] L. M. Babcock and D. L. Thompson, *J. Chem. Phys.*, 1983, **78**, 2394.
[325] L. M. Babcock and D. L. Thompson, *J. Chem. Phys.*, 1983, **79**, 4193.
[326] D. G. Hopper, *J. Chem. Phys.*, 1982, **77**, 314.
[327] J. Danon and R. Marx, *Chem. Phys.*, 1982, **68**, 255.
[328] C. E. Hamilton, M. A. Duncan, T. S. Zwier, J. C. Weisshaar, G. B. Ellison, V. M. Bierbaum, and S. R. Leone, *Chem. Phys. Lett.*, 1983, **94**, 4.
[329] D. R. Guyer, L. Hüwel, and S. R. Leone, *J. Chem. Phys.*, 1983, **79**, 1259.

reactants and products) is favoured by enhanced charge-transfer cross-sections and that conversion to translational energy is correspondingly inefficient. However, recent experiments in which internal energy levels of reactant ions are varied to allow state-resolved cross-sections to be determined allow a systematic investigation of the applicability of this idea and the related assumption that good Franck–Condon overlap factors favour charge transfer. A threshold-electron–secondary-ion coincidence method was used to prepare state-selected Ar^+ ($^2P_{1/2,3/2}$),[330] N_2^+ ($X^2\Sigma_g^+$, v),[331] and NO^+ ($a^2\Sigma^+$ v; $b^3\Pi$ v)[332] for subsequent charge-transfer reaction with N_2 and CO for the former and Ar for the latter two reactants. The cross-sections so obtained are discussed and compared with a simple model of the energy-gap and Franck–Condon-factor dependences. Close examination reveals only a limited applicability for this approach, although a similar treatment relating to the charge transfer of vibrationally selected H_2^+ with Ar,[333] N_2, CO, and O_2 [334] found that the cross-sections could be quite successfully modelled, at least in terms of an energy-gap dependence, once additional allowance for the difference between endo- and exo-ergic channels was made.

These rather mixed results are not unexpected, since such charge transfers intrinsically involve more than one electronic surface and are thus likely to be sensitive to the details of the potentials involved. The $H_2^+ + Ar$ system[333] is particularly intereresting, as there is evidence at higher collisional energies for a coupling between the nominally adiabatic and non-adiabatic charge- and proton-transfer channels. This is attributed to interactions in the region of an avoided crossing.

Full quantal[335] and semi-classical[336, 337] treatments (with quantal treatment of internal states) of symmetric charge transfer, referring to $O_2^+ + O_2$ as a representative system, have been published, with good agreement between the methods. In comparison with experimental data for the system, qualitative agreement was found,[337] which reflects the use of appropriate Franck–Condon factors and energy gaps in the model. Quantitative agreement between the semi-classical method and experimental data has been claimed for the symmetric H_2^+ (v) + H_2 charge-transfer system.[338]

Despite the above demonstration that the simple concept of resonant enhancement of charge-transfer cross-sections cannot be assumed to have universal validity, ion generation by charge transfer from reagents of known recombination energy nevertheless remains a potentially useful means for the preparation of reactant ions having well characterized internal energies. A good

[330] T. Kato, K. Tanaka, and I. Koyano, *J. Chem. Phys.*, 1982, **77**, 337.
[331] T. Kato, K. Tanaka, and I. Koyano, *J. Chem. Phys.*, 1982, **77**, 834.
[332] T. Kato, K. Tanaka, and I. Koyano, *J. Chem. Phys.*, 1983, **79**, 5969.
[333] F. A. Houle, S. L. Anderson, D. Gerlich, T. Turner, and Y. T. Lee, *J. Chem. Phys.*, 1982, **77**, 748.
[334] S. L. Anderson, T. Turner, B. H. Mahan, and Y. T. Lee, *J. Chem. Phys.*, 1982, **77**, 1842.
[335] C. H. Becker, *J. Chem. Phys.*, 1982, **76**, 5928.
[336] A. E. DePristo, *J. Chem. Phys.*, 1983, **78**, 1237.
[337] A. E. DePristo, *J. Chem. Phys.*, 1983, **79**, 1741.
[338] C. L. Liao, C. X. Liao, and C. Y. Ng, *Chem. Phys. Lett.*, 1984, **103**, 418.

example is to be found in studies of the product distributions of the reactions of charge-exchange-produced NH_3^+ with H_2O[339] and H_2S,[340] where data are obtained as a function of NH_3^+ internal energy. This latter quantity was, however, carefully determined by a prior initial calibration in which the product-energy partitioning in the charge-transfer process was directly determined.[339]

The electron/ion coincidence technique provides an alternative method for generation of reactant ions in specific internal quantum states. The previously mentioned charge-transfer results[330–332] in fact form part of a continuing and fruitful series of papers in which the potential of this technique for state-selective ion/molecule chemistry is explored.[330–332, 341, 342] Cross-sections for the vibronically selected reactions (5) and (6) were determined.[341] As before, the

$$O_2^+ (X^2\Pi_g\ v = 19 \text{ or } 20; a^4\Pi_u\ v = 0\text{–}8) + H_2 \rightarrow O_2H^+ + H \quad (5)$$

$$\rightarrow O_2^+ + H_2 \quad (6)$$

ideas of quasi-resonant cross-section enhancement were only partially successful in describing the charge-transfer reaction (6). For both reactions (5) and (6) the cross-sections of the *a*-state reactant are considerably larger than those of the *X*-state O_2^+ and show a maximum at the $v = 2$ level.[341, 342] Isotope effects observed for reaction (5) are consistent with a direct mechanism for this reaction.[342] In this instance, as in many others where experiment now provides state-selected data, a full understanding of the reactions awaits more detailed knowledge of the potential-energy surfaces involved and appropriate theoretical treatments of such information. The new developments in experimental techniques are thus setting a clear challenge for the theoretician.

[339] P. R. Kemper, M. T. Bowers, D. C. Parent, G. Mauclaire, R. Derai, and R. Marx, *J. Chem. Phys.*, 1983, **79**, 160.
[340] W. Wagner-Redeker, P. R. Kemper, M. T. Bowers, and K. R. Jennings, *J. Chem. Phys.*, 1984, **80**, 3606.
[341] K. Tanaka, T. Kato, P. M. Guyon, and I. Koyano, *J. Chem. Phys.*, 1982, **77**, 4441.
[342] K. Tanaka, T. Kato, P. M. Guyon, and I. Koyano, *J. Chem. Phys.*, 1983, **79**, 4302.

2
Structures and Reactions of Gas-phase Organic Ions

BY M. A. BALDWIN

1 Introduction

In mass spectrometry structures and reactions are inextricably linked. All spectroscopic techniques depend on the interactions of matter and energy, but in most cases the matter remains in the same chemical form. This is not so in mass spectrometry, in which the energy brings about chemical changes that are used to characterize the structures. Whether the 70 eV electron-impact spectrum or some more sophisticated technique is being used, the objective is to produce reactions unique to the ion structures.

Structures and reactions are at the heart of mass spectrometry, and the title of this chapter is sufficiently broad to embrace most of the material contained in this book. However, the scope must necessarily be restricted to avoid overlaps, and to make it possible to review two years' research in 20 pages or so. Before selecting work to be included, the Reporter resolved to concentrate on electron-impact positive-ion studies that are essentially experimental rather than theoretical (fundamental aspects being covered in Chapter 1 and negative ions in Chapter 7). Also, because Fourier-transform mass spectrometry, photoelectron–photoion coincidence, and various other techniques are covered in other chapters, it is appropriate to cover mainly work carried out on magnetic-sector instruments. Furthermore, in the writing of any review, arbitrary restrictions will inevitably be introduced by the Reporter in choosing work he believes to be of particular interest.

2 Literature Sources

The continuing healthy interest in gas-phase ion chemistry is shown by the well-being of the two primary journals *Organic Mass Spectrometry* and *International Journal of Mass Spectrometry and Ion Processes* (previously *Physics*), which continue as the major literature sources in the field. The latter journal also published the proceedings of the 9th International Mass Spectrometry Conference held in Vienna in 1982.[1] Publishing these as a two-volume set in a journal (four

[1] 'Mass Spectrometry Advances 1982. Parts A, B and C, D', ed. E. R. Schmidt, K. Varzuma, and I. Fogy, *Int. J. Mass Spectrom. Ion Phys.*, 1983, **45–48**: (*a*) J. H. Beynon, A. G. Brenton, and F. M. Harris, *ibid.*, 1983, **45**, 5; (*b*) T. D. Märk, *ibid.*, 1983, **45**, 125; (*c*) J. Seibl, *ibid.*, 1983, **45**, 147.

volumes of the journal) was a break with the previous policy of publishing them as a book in the series 'Advances in Mass Spectrometry', and this has met with some criticism[2] because the normal refereeing standards of the journal were not maintained, the quality of the contributions being somewhat mixed. The first volume of this compilation (journal Volume 45) contains the plenary and keynote lectures. Plenary lectures of particular relevance to this chapter are reviews of metastable ions and photon-induced reactions,[1a] fundamental aspects of electron-impact ionization,[1b] and structure elucidation of organic compounds.[1c]

After a mixed start the quarterly journal *Mass Spectrometry Reviews* is establishing itself, with some of the contributions being timely, comprehensive, and of high quality.[3] Of interest here are reviews on time-resolved appearance energies,[3a] chemistry of collisionally activated ions,[3b] stereochemical effects in mass spectrometry,[3c] isotope effects in fragmentation,[3d] and the retro-Diels–Alder reaction.[3e] An innovation is the incorporation of selected reviews on mass-spectrometric topics by Budzikiewicz, in which he provides leading references on a different topic each quarter.

Burlingame *et al.* continue to provide an excellent service to the mass-spectrometric community with the latest *Analytical Chemistry Review*.[4] Recent books of interest include 'Ionic Processes in the Gas Phase' edited by Almoster Ferreira,[5] 'Current Topics in Mass Spectrometry and Chemical Kinetics' edited by Beynon and McGlashan,[6] 'Tandem Mass Spectrometry' edited by McLafferty,[7] and 'Chemical Ionisation Mass Spectrometry' by Harrison.[8]

3 Theory and Experiment

Having earlier stated the intention to concentrate on experiment rather than theory, it must now be said that the most significant recent developments in ion-structure studies have come from the marriage of theory and experiment, the major theoretical contribution coming from Radom's group in Canberra. Using *ab initio* calculations the group has studied the cations or radical cations of the C_1 to C_4 alkanes, including cyclopropane and cyclobutane,[9] CH_3O,[10]

[2] A. Maccoll, *Org. Mass Spectrom.*, 1983, **18**, 412.

[3] *Mass Spectrom. Rev.*, ed. M. L. Gross, Wiley, New York: (*a*) C. Lifshitz, *ibid.*, 1982, **1**, 309; (*b*) K. Levsen and H. Schwarz, *ibid.*, 1983, **2**, 76; (*c*) A. Mandelbaum, *ibid.*, 1983, **2**, 223; (*d*) P. J. Derrick, *ibid.*, 1983, **2**, 285; (*e*) F. Turecek and V. Hanus, *ibid.*, 1984, **3**, 85.

[4] A. L. Burlingame, J. O. Whitney, and D. H. Russell, *Anal. Chem. Rev.*, 1984, **56**, 417R.

[5] 'Ionic Processes in the Gas Phase', ed. M. A. Almoster Ferreira, D. Reidel, 1984: (*a*) H. Schwarz, *ibid.*, p. 267.

[6] 'Current Topics in Mass Spectrometry and Chemical Kinetics', ed. J. H. Beynon and M. L. McGlashan, Heyden, 1982.

[7] 'Tandem Mass Spectrometry', ed. F. W. McLafferty, Wiley, New York, 1983: (*a*) P. H. Dawson and D. J. Douglas, *ibid.*, p. 125; (*b*) P. J. Todd and F. W. McLafferty, *ibid.*, p. 149; (*c*) S. A. McLuckey and R. G. Cooks, *ibid.*, p. 303.

[8] 'Chemical Ionization Mass Spectrometry', A. G. Harrison, CRC Press, 1983.

[9] W. J. Bouma, D. Poppinger, and L. Radom, *Isr. J. Chem.*, 1983, **23**, 21.

[10] W. J. Bouma, R. H. Nobes, and L. Radom, *Org. Mass Spectrom.*, 1982, **17**, 315.

CH_3S,[11] CH_3F,[12] CH_4O,[13] CH_4S,[11] CH_5O,[14] CH_5N,[15] CH_3HX (X = OH, SH, NH_2, or F),[16] C_2H_3O,[17] C_2H_4O,[18] C_2H_5O,[19] C_2H_6O,[20] and C_2H_6S,[11] and the dications of CH_2O and CH_4O.[21] Together with Frisch and Raghavachari the same group has also reported on the radical cations of some of the above and CH_2O, CH_3N, H_2O_2, N_2H_2, N_2H_4, HNO, and H_3NO.[22] Using the most advanced computational techniques they have calculated the energies and structures of the various reacting configurations and the energy barriers between them, thereby explaining the course of many of the experimentally observed unimolecular-ion fragmentations. For many of these entities, unusual structures corresponding to unstable neutral species have been predicted to be more stable than the 'expected' classical structures. Indeed, for many small molecules the most stable radical cations are predicted to have non-classical structures for which there are no stable neutral counterparts.

$CH_4O^{+\cdot}$. – The attainment of the most stable structure frequently involves a formal 1,2 hydrogen shift, *e.g.* a more stable isomer of the methanol radical cation was predicted to have the methylenoxonium structure, $CH_2OH_2^{+\cdot}$, this being more stable than the ionized methanol structure by 45 kJ mol^{-1}. However, the barrier to formation of this ion from the methanol radical cation was calculated as 112 kJ mol^{-1}, whereas the barrier to hydrogen loss was only 72 kJ mol^{-1}.[13] In order to detect the methylenoxonium species (1) it was necessary to form it indirectly by the expulsion of neutral methanal from the cation of ethylene glycol (Scheme 1). The use of labelled precursors and collisional activation enabled ion (1) to be identified.[23] The heat of formation of ion (1) was also determined from appearance-energy measurements and was indeed

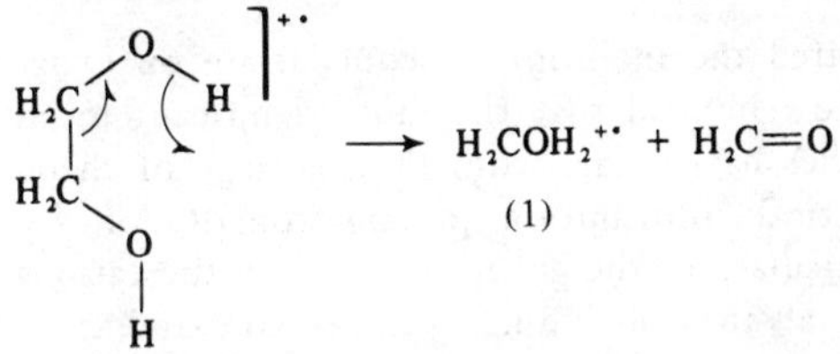

$$[HOCH_2CH_2OH]^{+\cdot} \longrightarrow H_2COH_2^{+\cdot} + H_2C{=}O$$

(1)

Scheme 1

[11] R. H. Nobes, W. J. Bouma, and L. Radom, *J. Am. Chem. Soc.*, 1984, **106**, 2774.
[12] W. J. Bouma, B. F. Yates, and L. Radom, *J. Phys. Lett.*, 1982, **92**, 620.
[13] W. J. Bouma, R. H. Nobes, and L. Radom, *J. Am. Chem. Soc.*, 1982, **104**, 2929.
[14] R. H. Nobes and L. Radom, *Org. Mass Spectrom.*, 1982, **17**, 340.
[15] W. J. Bouma, J. M. Dawes, and L. Radom, *Org. Mass Spectrom.*, 1983, **18**, 12.
[16] R. H. Nobes and L. Radom, *Chem. Phys.*, 1983, **74**, 163.
[17] R. H. Nobes, W. J. Bouma, and L. Radom, *J. Am. Chem. Soc.*, 1983, **105**, 309.
[18] W. J. Bouma, D. Poppinger, S. Saebo, J. K. MacLeod, and L. Radom, *Chem. Phys. Lett.*, 1984, **104**, 198.
[19] R. H. Nobes and L. Radom, *Chem. Phys. Lett.*, 1983, **99**, 107.
[20] W. J. Bouma, R. H. Nobes, and L. Radom, *J. Am. Chem. Soc.*, 1983, **105**, 1743.
[21] W. J. Bouma and L. Radom, *J. Am. Chem. Soc.*, 1983, **105**, 5484.
[22] M. J. Frisch, K. Raghavachari, J. A. Pople, W. J. Bouma, and L. Radom, *Chem. Phys.*, 1983, **75**, 323.
[23] W. J. Bouma, J. K. MacLeod, and L. Radom, *J. Am. Chem. Soc.*, 1982, **104**, 2930.

shown to be less than that of the methanol radical cation, though by a smaller amount than had been predicted (29 ± 8 kJ mol^{-1}).[24] Another characteristic feature of ion (1) was the sharp doubly charged ion peak at m/z 16 formed by charge stripping in the collisional-activation spectrum using helium as collision gas, although this peak was absent from the corresponding methanol spectrum.[24] Bouma and Radom had calculated that the doubly charged ion of methanol was not a stable species,[21] although it was demonstrated later that ionized methanol could give a doubly charged ion peak when it collided with molecular oxygen.[25] Calculation suggested that sufficient energy was being transferred by collision with O_2 for the charge-stripping process to be accompanied by a 1,2 hydrogen rearrangement, to give the methylenoxonium dication.[25] Holmes *et al.* also studied hydrogen loss from $CH_4O^{+\bullet}$ species by deuterium labelling and measurement of kinetic-energy release associated with metastable-ion decompositions, and showed that $H^\bullet$ loss from $CH_2OH_2^{+\bullet}$ was accompanied by rearrangement to the methanol structure.[26]

$CH_2OH_2^{+\bullet}$ Analogues. – Methylenoxonium is a radical cation that can be described as an ionized ylide or an ion–dipole complex (of an ionized carbene and a polar neutral molecule) of the general form $(R^1CHXR^2)^{+\bullet}$ (R^1, R^2 = H or higher homologues, X = F, Cl, Br, I, OH, SH, or NH_2). Many of these have been investigated theoretically and experimentally.[11–13, 15, 19, 22–32] The ionized non-classical forms occupy wells in the potential-energy surfaces and can be detected as stable structures. Their heats of formation are generally less than or similar to those of the ionized isomeric stable neutral molecules, with the possible exception of $CH_2NH_3^{+\bullet}$, which was calculated to be more stable than ionized methylamine by 8 kJ mol^{-1} [15] but was found by experiment to be considerably less stable.[26] In general, there are large energy barriers to interconversion of the two forms. The radical cations of CH_3Cl and CH_2HCl have also been studied by the promising new technique of neutralization–reionization mass spectrometry, clearly demonstrating the relative instability of the neutral form of CH_2HCl, the ratio of

[24] J. L. Holmes, F. P. Lossing, J. K. Terlouw, and P. C. Burgers, *J. Am. Chem. Soc.*, 1982, **104**, 2931.

[25] F. Maquin, D. Stahl, A. Sawaryn, P. von R. Schleyer, W. Koch, G. Frenking, and H. Schwarz, *J. Chem. Soc., Chem. Commun.*, 1984, 504.

[26] J. L. Holmes, F. P. Lossing, J. K. Terlouw, and P. C. Burgers, *Can. J. Chem.*, 1983, **61**, 2305.

[27] Y. Apeloig, B. Ciommer, G. Frenking, M. Karni, A. Mandelbaum, H. Schwarz, and A. Weisz, *J. Am. Chem. Soc.*, 1983, **105**, 2186.

[28] Y. Apeloig, M. Karni, B. Ciommer, G. Frenking, and H. Schwarz, *Int. J. Mass Spectrom. Ion Processes*, 1983, **55**, 319.

[29] J. L. Holmes, P. C. Burgers, J. K. Terlouw, H. Schwarz, B. Ciommer, and H. Halim, *Org. Mass Spectrom.*, 1983, **18**, 208.

[30] J. K. Terlouw, W. Heerma, G. Dijkstra, J. L. Holmes, and P. C. Burgers, *Int. J. Mass Spectrom. Ion Phys.*, 1983, **47**, 147.

[31] B. Ciommer, G. Frenking, and H. Schwarz, *Int. J. Mass Spectrom. Ion Processes*, 1984, **57**, 135.

[32] P. O. Danis, C. Wesdemiotis, and F. W. McLafferty, *J. Am. Chem. Soc.*, 1983, **105**, 7454.

molecular-ion to $HCl^{+\bullet}$ abundance in the spectra obtained by reionization of the neutral species being 10–20 times greater for the methyl chloride structure.[32]

Oxycarbenes. – Various oxycarbenes have been identified, including $HCOH^{+\bullet}$ and $C(OH)_2^{+\bullet}$ [33] and the higher homologues $CH_3COH^{+\bullet}$ and $CH_3OCOH^{+\bullet}$, for which collisional-activation spectra and heats of formation have been reported.[34] Energy barriers of 20–100 kJ mol^{-1} prevent interconversion with the isomeric forms stable as neutral molecules, *i.e.* methanal, methanoic acid, *etc.* The stability of these species had been predicted by earlier theoretical calculations, and more recent calculations predicted the stability of the nitrogen-containing analogue $HCNH_2^{+\bullet}$.[22] This was confirmed experimentally, and its heat of formation and collisional-activation spectrum, and those of the corresponding imine, have been reported.[35] The experimentally determined heat of formation of the carbene radical cation was some 50 kJ mol^{-1} greater than that of the ionized imine, although *ab initio* calculations had predicted the opposite.[22] Burgers *et al.* have compared kinetic-energy releases for $H^\bullet$ loss from the carbonium-ion site in the conventional structures, with the loss of $H^\bullet$ from the heteroatom site in the alternative carbene structures. In each case a high reverse activation energy was found to be associated with the latter process, giving rise to substantial kinetic-energy release for the carbene reactions, as was evidenced by dished or flat-topped metastable peaks. Calculations on the experimental data showed the bond energy between a radical and the heteroatom to be approximately 1 eV higher than that between a radical and the carbon, and this was attributed to higher electron density at the heteroatom site in these carbene-radical cations.[36]

Other Unusual Radical Cations. – Some related unusual structures have been identified, some of which are described here, but this list is not comprehensive. The loss of C_2H_4 from ionized butan-1,4-diol gives a complex between ionized vinyl alcohol and water that may correspond to one or more of the structures (2)–(5), and a similar complex occurs with ionized vinyl alcohol and methanol.[37] The vinylidene cation $H_2C{=}C^{+\bullet}$ has been formed by collisional charge reversal of the corresponding anion and distinguished from the acetylene radical cation by its collisional-activation mass spectrum.[38] A new isomer of $C_2H_4O_2^{+\bullet}$ was formed by dissociative ionization of 1,3-dihydroxyacetone and identified as

$$\mathrm{HO{-}\dot{C}H{-}CH_2{-}\overset{+}{O}H_2}\quad(2)\qquad \mathrm{HO{-}CH(\dot{C}H_2){-}\overset{+}{O}H_2}\quad(3)\qquad \mathrm{CH_2{=}CH\overset{+\bullet}{O}H\cdots HOH}\quad(4)\qquad \mathrm{\dot{C}H_2{-}CHO\cdots\overset{+}{H}\cdots OH_2}\quad(5)$$

[33] P. C. Burgers, A. A. Mommers, and J. L. Holmes, *J. Am. Chem. Soc.*, 1983, **105**, 5976.
[34] J. K. Terlouw, J. Wezenberg, P. C. Burgers, and J. L. Holmes, *J. Chem. Soc., Chem. Commun.*, 1983, 1121.
[35] P. C. Burgers, J. L. Holmes, and J. K. Terlouw, *J. Am. Chem. Soc.*, 1984, **106**, 2762.
[36] P. C. Burgers, J. L. Holmes, and J. K. Terlouw, *J. Chem. Soc., Chem. Commun.*, 1984, 642.
[37] J. K. Terlouw, W. Heerma, P. C. Burgers, and J. L. Holmes, *Can. J. Chem.*, 1984, **62**, 289.
[38] J. L. Holmes and J. E. Szulejko, *Chem. Phys. Lett.*, 1984, **107**, 301.

a radical-ion complex of ketene and water that readily ejects water in both metastable-ion and collisionally activated decompositions.[39] Two new oxonium-ion isomers of $C_3H_8O^{+\bullet}$ have been identified as (6) and (7), and their heats of formation and metastable-ion and collisional-activation spectra have been reported; a proton-bound structure (8) has been suggested as another possibility.[40] Similar oxonium-ion isomers of ethanol and dimethyl ether, having structures (9) and (10), have been predicted to be the most stable radical cations of C_2H_6O, and a third oxonium ion (11) is of comparable stability to ionized dimethyl ether.[20] Ammonium ions of formula $C_2H_7N^{+\bullet}$ that are analogous to structures (9) and (10) have been shown to exist, by means of metstable ions, collisional activation, and charge stripping, *i.e.* $^{\bullet}CH_2CH_2\overset{+}{N}H_3$ and $CH_3\dot{C}H\overset{+}{N}H_3$,[26,41] but the nitrogen analogue of (11), $CH_3\overset{+}{N}H_2\dot{C}H_2$, could not be separately identified.

$\dot{C}H_2{-}CH_2{-}CH_2{-}\overset{+}{O}H_2$ (6)

$CH_3{-}\dot{C}H{-}CH_2{-}\overset{+}{O}H_2$ (7)

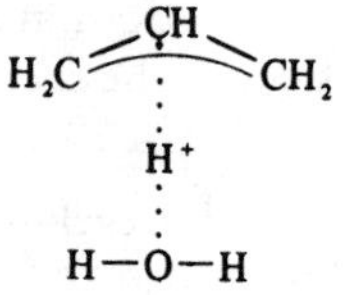

$H_2C{=}\dot{C}H{=}CH_2 \cdots H^+ \cdots H{-}O{-}H$ (8)

$\dot{C}H_2{-}CH_2{-}\overset{+}{O}H_2$ (9)

$CH_3{-}\dot{C}H{-}\overset{+}{O}H_2$ (10)

$CH_3{-}\overset{+}{O}H{-}\dot{C}H_2$ (11)

Non-classical Even-electron Ions. – All the species discussed above are radical cations corresponding to neutral molecules that are unstable, often because of hypervalency, *e.g.* the methylenoxonium neutral species involves three bonds to oxygen. Many even-electron ions correspond to ionized radicals that may also involve hypervalency, or they may be ion–dipole complexes. Before the period under review the well known CH_5^+ ion had already been shown to correspond to a complex between the methyl cation and H_2. However, theoretical studies on protonated methanol have shown the most stable CH_5O^+ isomer to be the methyloxonium ion $CH_3OH_2^+$, with a pyramidal bond arrangement at the oxygen atom.[14] Neutralization–reionization showed that the major decomposition pathway of its neutral hypervalent counterpart was to $CH_3^{\bullet}$ and H_2O.[32] Protonated methanol is one of a series of ions CH_3XH^+ ($X = OH, SH, F,$ or NH_2). Such ions are formed as intermediates in the condensation of the appropriate polar molecule with methyl cation ($CH_3^+ + HX \rightarrow CH_3XH^+ \rightarrow CH_2X^+ + H_2$).[16] This condensation reaction shows an appreciable rate when X is SH or NH_2 but not when X is OH or F. This has been the subject of a theoretical study which shows that the transition structure for H_2 elimination is more stable than the

[39] J. K. Terlouw, C. G. de Koster, W. Heerma, J. L. Holmes, and P. C. Burgers, *Org. Mass Spectrom.*, 1983, **18**, 222.

[40] J. L. Holmes, A. A. Mommers, J. E. Szulejko, and J. K. Terlouw, *J. Chem. Soc., Chem. Commun.*, 1984, 165.

[41] S. Hammerum, D. Kuck, and P. J. Derrick, *Tetrahedron Lett.*, 1984, **25**, 893.

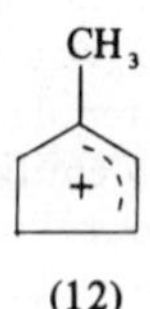

(12)

condensation reactants, $CH_3^+ + HX$, when X is SH or NH_2, but not when X is OH or F.[16]

Many hydrocarbon rearrangements involve pyramidal carbocations, which were discussed in the previous volume in this series.[42] Schwarz has since reviewed this field,[5a] and a further example involving $C_6H_9^+$ has also been reported.[43] Both unimolecular loss of water from $C_6H_{11}O^+$ and protonation of cyclohexa-1,3- or -1,4-diene give the 1-methylcyclopentenylium ion (12), according to the collisional-activation spectra (losses of $CH_3^{\bullet}$ and C_2H_4). This is shown by MNDO calculation to be the most stable isomer.[43]

4 Development of Techniques

The methylenoxonium ion discussed above is characteristic of many ions of unusual structure that have come to light in recent years, partly as a result of theoretical predictions that have stimulated experimentalists to search for and identify these species, but also because of the wide range of new techniques that have been developed to produce new structures and to characterize them: in this case initial generation of an ionic species having no neutral counterpart, isotopic labelling of the precursor molecules, energy-loss measurements for the decomposition of metastable ions, further energization of the ions to produce a structurally characteristic spectrum (collisional activation and charge stripping), neutralization and reionization of the neutral species, and thermochemical measurements with energy-selected electrons.

Advances in mass spectrometry rely heavily on the development of new techniques. The basic requirements for a mass spectrometer are the means of ionizing the sample and separating and detecting the ions. However, an array of alternative methods is available for each of these requirements and for a wide range of additional procedures. Many of the most sophisticated variants to have come from purely academic research are now routinely incorporated in commercially available analytical mass spectrometers. This is very much to the advantage of studies of the chemistry of gas-phase ions as it allows fundamental research to be carried out on instruments purchased primarily for analytical applications.

An example of the rapid acceptance of new techniques is provided by the widespread use of the multiple-sector instruments required for the range of techniques classified as tandem mass spectrometry (or MS/MS). Such techniques allow the study of sequential events, ranging from simple observation of decom-

[42] I. Howe in 'Mass Spectrometry', ed. R. A. W. Johnstone (Specialist Periodical Reports), The Royal Society of Chemistry, London, 1984, Vol. 7, p. 119.
[43] T. Gaumann, R. Houriet, D. Stahl, J.-C. Tabet, N. Heinrich, and H. Schwarz, *Org. Mass Spectrom.*, 1983, **18**, 215.

positions of metastable ions to a variety of unimolecular and bimolecular processes following energy and/or charge transfer. Instrument manufacturers now offer complex combinations of magnetic sectors, electric sectors, and quadrupole mass filters as standard items, and these are equipped with the appropriate reaction chambers (collision cells) to carry out sophisticated chemical experiments to determine structure and reactivity of gas-phase ions.

In the next few sections the applications of a number of techniques will be outlined. Very few studies of structures and reactions of gas-phase ions use only a single technique, and in most cases several sources of evidence are exploited. This should be borne in mind when considering the specific examples presented here.

5 Molecular-ion Formation

Despite the development of a number of new ionization techniques, electron impact remains the most commonly used method for obtaining the molecular ion of a stable molecule. It suffers from the disadvantage of transferring a wide range of energies, the distribution of which can be difficult to determine, though the use of low electron energy and low ion-source temperatures reduces this problem and makes the spectra simpler and more structure-specific, as is shown in a study of n-octane derivatives.[44]

Field Ionization. — A further disadvantage of electron impact is the relatively long time the ions spend in the ion source ($\sim 10^{-6}$ s), making it impossible to study the kinetics of many rapid isomerizations and dissociations. Field ionization provides an alternative that has been exploited in the field-ionization-kinetics technique, but the limited fragmentation obtained following field ionization and difficulties associated with the practicalities of the technique and interpretation of the results have restricted its use. Nibbering has reviewed recent applications of the technique[45] and used field-ionization kinetics to distinguish two mechanisms for loss of propyl radical from the molecular ions of methoxycyclohexane.[46] D- and ^{13}C-labelling showed that the 'classic' mechanism as illustrated in Scheme 2 occurred at the shortest times ($< 10^{-10}$ s) whereas longer-lived ions underwent the complex mechanism shown in Scheme 3. It should be noted that the same carbon atoms are lost in both cases, *i.e.* C-2–C-4, and the mechanisms were only revealed by studying the time dependence of the H/D losses from various labelled precursors. Alkanoic acids,[47,48] and hydrogen migrations in oct-4-ene,[49] have also been studied by this technique. Brand and Levsen have discussed the deter-

[44] R. D. Bowen and A. Maccoll, *Org. Mass Spectrom.*, 1983, **18**, 576.
[45] N. M. M. Nibbering, *Int. J. Mass Spectrom. Ion Phys.*, 1982, **45**, 343.
[46] T. A. Molenaar-Langeveld and N. M. M. Nibbering, *Org. Mass Spectrom.*, 1983, **18**, 426.
[47] R. Weber, K. Levsen, C. Wesdemiotis, T. Weiske, and H. Schwarz, *Int. J. Mass Spectrom. Ion Phys.*, 1982, **43**, 131.
[48] J. J. Zwinselman, N. M. M. Nibbering, C. E. Hudson, and D. J. McAdoo, *Int. J. Mass Spectrom. Ion Phys.*, 1983, **47**, 129.
[49] J. Stocklov, K. Levsen, and G. J. Shaw, *Org. Mass Spectrom.*, 1983, **18**, 444.

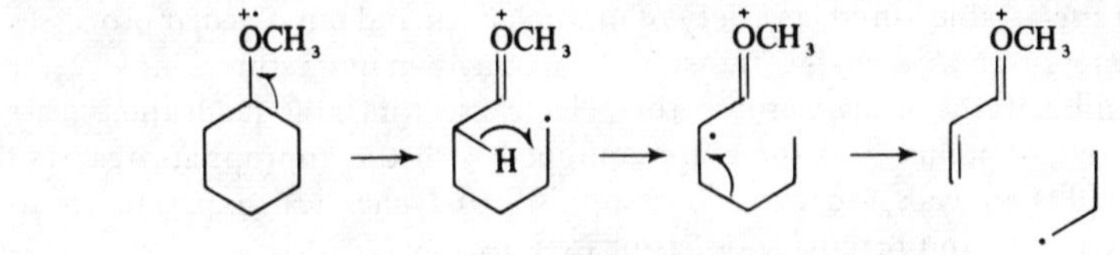

Scheme 2

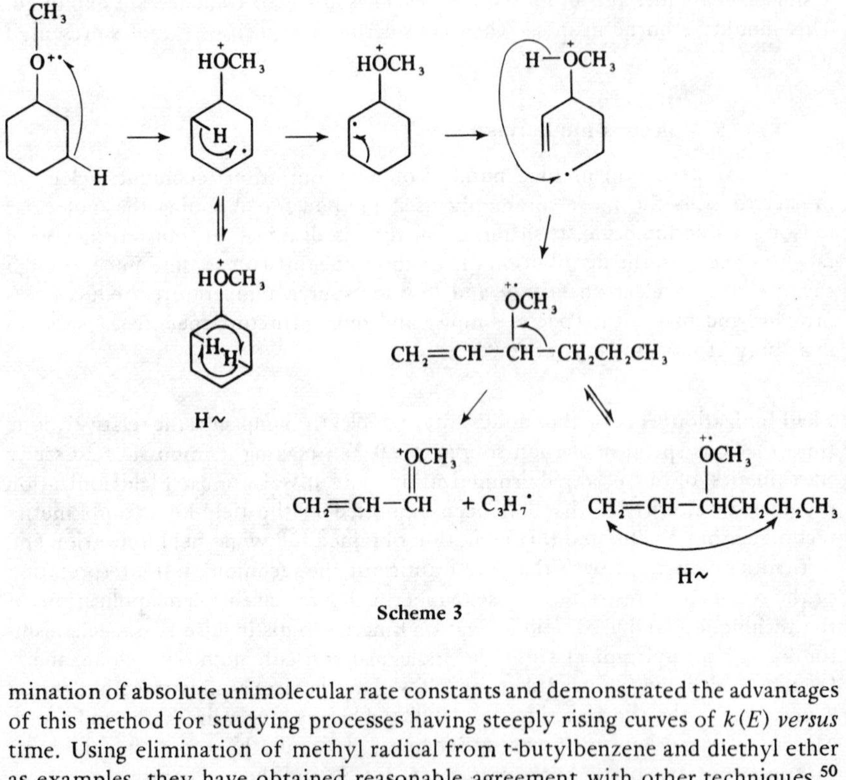

Scheme 3

mination of absolute unimolecular rate constants and demonstrated the advantages of this method for studying processes having steeply rising curves of $k(E)$ *versus* time. Using elimination of methyl radical from t-butylbenzene and diethyl ether as examples, they have obtained reasonable agreement with other techniques,[50] and rate constants have been reported for halobenzenes.[51] Field ionization was also able to distinguish a series of $C_9H_{10}^{+\cdot}$ isomeric ions that give identical 70 eV electron-impact mass spectra.[52]

Photoionization. – Historically, little interest has been shown in photoionization for analytical applications so the instrumentation has not been widely available;

[50] W. A. Brand and K. Levsen, *Int. J. Mass Spectrom. Ion Phys.*, 1983, **51**, 135.
[51] W. A. Brand, J. Stocklov, and H. J. Walther, *Int. J. Mass Spectrom. Ion Processes*, 1984, **59**, 1.
[52] J. O. Lay, jun., M. L. Gross, J. J. Zwinselman, and N. M. M. Nibbering, *Org. Mass Spectrom.*, 1983, **18**, 16.

recent developments in the use of laser-induced desorption techniques may change this situation. The development of laser technology has greatly increased the activity in the field of ionic spectroscopy, much of which is of direct relevance to mass spectrometry, which is frequently employed to detect the decay products following photoionization. The use of multi-photon ionization has been reviewed by Schlag and Neusser[53] (see also Chapter 1). Some examples of photoionization and time-dependent fragmentation studies using multi-photon ionization concern alkyl iodides[54] and substituted benzenes.[55, 56] Photoionization is also fundamental to photoelectron spectroscopy and photoionization–photoelectron-coincidence studies, but they are outside the immediate scope of this chapter (see the following chapter). Accurate thermochemical studies by photoionization have a long history. Traeger has reported 298 K heats of formation for the allyl cation (946.9 ± 1.4 kJ mol^{-1})[57] and the acetyl cation (657.0 ± 1.5 kJ mol^{-1}).[58] Lifshitz has developed a new technique of time-resolved photoionization mass spectrometry,[59] as an extension of her previous work on time-resolved appearance energies by ion trapping.[3a] Results from this will be described in a later section in this chapter.

Charge Exchange. – Ionization by charge exchange can be achieved in a conventional chemical-ionization source. The use of a number of reagent gases having different recombination energies enables breakdown graphs to be constructed that show the energy dependence of various fragmentations. A study of stereoisomeric methylcyclohexanols has revealed dramatic stereochemical differences in water elimination from the *cis*- and *trans*-2-methyl and, in particular, the 4-methyl stereoisomers at low energy. The highly favoured *cis*-1,4 elimination of H_2O from a boat-like conformation can occur for the *trans* isomer (13) but not the *cis* isomer (14), but at higher energies ring opening and 1,3 elimination are favoured. The abundance ratio, $(M - H_2O)^{+\bullet}/M^{+\bullet}$, is 23 times greater for (13) than for (14) at low energy.[60] The energy dependence has been studied for the loss of methyl radical from haloanisoles, leading to resonance-stabilized ions for the *ortho* and *para* isomers but not for the *meta* isomer, which undergoes loss

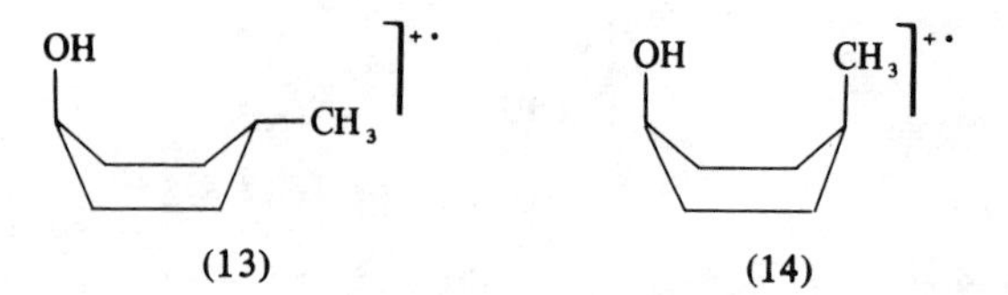

[53] E. W. Schlag and H. J. Neusser, *Acc. Chem. Res.*, 1983, **16**, 355.
[54] D. H. Parker and R. B. Bernstein, *J. Phys. Chem.*, 1982, **86**, 60.
[55] D. W. Squire, M. P. Barbalas, and R. B. Bernstein, *J. Phys. Chem.*, 1983, **87**, 1701.
[56] K. R. Newton and R. B. Bernstein, *J. Phys. Chem.*, 1983, **87**, 2246.
[57] J. C. Traeger, *Int. J. Mass Spectrom. Ion Processes*, 1984, **58**, 259.
[58] J. C. Traeger, R. G. McLoughlin, and A. J. C. Nicholson, *J. Am. Chem. Soc.*, 1982, **104**, 5318.
[59] C. Lifshitz, M. Goldenberg, Y. Malinovich, and M. Peres, *Org. Mass Spectrom.*, 1982, **17**, 453.
[60] A. G. Harrison and M. S. Lin, *Org. Mass Spectrom.*, 1984, **19**, 67.

of CH_2O.[61] Other species studied by this technique include n-butylbenzene[62] and $C_3H_6^{+\bullet}$.[63] Charge exchange with $CS_2^{+\bullet}$ was used to study the gas-phase Diels–Alder reaction between low-energy *o*-quinodimethane ions and neutral styrene[64] and the [1 + 2] cycloaddition reaction between ionized and neutral styrene, showing that the cyclic ion of 1,2-diphenylcyclobutane was formed at low energies only, and then probably by a stepwise process.[65]

6 Formation of Ions with Unstable Neutral Counterparts

Many stable odd-electron ions have no neutral counterparts and must be formed indirectly. They may be formed by isomerization of ions of stable neutral molecules, but if large barriers prevent isomerization they must be formed by dissociative ionization of larger species (*e.g.* formation of the methylenoxonium radical cation, Scheme 1).

Even-electron ions may correspond to ionized radicals that normally have only transient existence, but the ions may readily be formed by decomposition of odd-electron ions. Studies of reactive transient species have been carried out by direct ionization of such species produced in reactors immediately adjacent to the ion source (*e.g.* the heats of formation of 18 organic radicals have recently been obtained from appearance-energy measurements using energy-selected electrons[66]).

Chemical Ionization. – Even-electron ions are readily produced by a number of soft-ionization techniques such as chemical ionization and various desorption methods. Protonation of saturated species can give hypervalent ions, such as protonated methanol discussed above, that are not otherwise accessible. Protonation can also give unique structures not obtainable by decomposition of larger species, *e.g.* the protonation of oxirane gives a cyclic ion (15) distinguishable from isomeric $C_2H_5O^+$ ions by its collisional-activation and charge-stripping spectra.[67] Rearrangements of protonated molecules in chemical ionization have been reviewed.[68]

H
|
O⁺
H_2C—CH_2

(15)

[61] E. J. Reiner and A. G. Harrison, *Int. J. Mass Spectrom. Ion Processes*, 1984, **58**, 97.
[62] A. G. Harrison and M. S. Lin, *Int. J. Mass Spectrom. Ion Phys.*, 1983, **51**, 353.
[63] S. G. Lias and T. J. Buckley, *Int. J. Mass Spectrom. Ion Processes*, 1984, **56**, 123.
[64] E. K. Chess, P.-H. Lin, and M. L. Gross, *J. Org. Chem.*, 1983, **48**, 1522.
[65] G. S. Groenewold, E. K. Chess, and M. L. Gross, *J. Am. Chem. Soc.*, 1983, **106**, 539.
[66] J. L. Holmes and F. P. Lossing, *Int. J. Mass Spectrom. Ion Processes*, 1984, **58**, 113.
[67] P. C. Burgers, J. K. Terlouw, and J. L. Holmes, *Org. Mass Spectrom.*, 1982, **17**, 369.
[58] E. E. Kingston, J. S. Shannon, and M. J. Lacey, *Org. Mass Spectrom.*, 1983, **18**, 183.

Chemical-ionization sources can be used to produce alternative ion types by mechanisms other than protonation, a few examples of which are given below. Several groups have studied cluster ions formed in high-pressure ion sources, *e.g.* Bowers *et al.* have investigated the formation and stability of a number of cluster-ion species, including those of water, ammonia, and carbon dioxide,[69] proton-bound dimers of methanol,[70] and H_5^+, for which collisional activation gives the novel species $H_4^{+\bullet}$.[71] The formation and unimolecular decompositions of the clusters were modelled using statistical phase-space theory with some success, kinetic-energy-release distributions have been reported, and collisionally activated decomposition mechanisms have been discussed. Other commonly used chemical-ionization reagents include NH_4^+, for which the adduct ions can isomerize to incorporate the ammonia into the molecule with expulsion of another group.[72] Reactions of NO^+ with alkenes have been studied by deuterium labelling to rationalize the hydrogen rearrangements associated with the cleavage of the double bond.[73]

Desorption Techniques. – Several desorption techniques have been developed for the ionization of polar and thermally labile species. These frequently form even-electron ions by protonation or hydride abstraction, and they also facilitate gas-phase production of ions from pre-ionized solid- or liquid-phase materials. In general, these techniques have been used for analytical purposes rather than for studies of ion structure or reactions. Secondary-ion mass spectrometry is probably the longest-established of these techniques, and a recent study used collisional activation to probe ion structures for a number of intact cations of quaternary ammonium chlorides and, in some cases, glycerol adducts, when a liquid matrix was used.[74] In the past, secondary-ion mass spectrometry has been used to study some very high-mass clusters of alkali halide ions such as $Cs_{n+1}I_n^+$. More recently the unimolecular and collision-induced decompositions of these species produced by fast-atom bombardment have been used to investigate the stabilities and structures of such clusters.[75,76]

E- and *Z*-isomeric dicarboxylic acids, such as maleic and fumaric acids, protonated by fast-atom bombardment show a strong dependence of fragmentation on geometrical isomerism. The *Z*-isomer, maleic acid, can form an internally proton-bound ion (16) that is much more stable than protonated fumaric acid and exhibits less fragmentation. The major fragmentation of the *Z*-acids is the

[69] A. J. Illies, M. F. Jarrold, L. M. Bass, and M. T. Bowers, *J. Am. Chem. Soc.*, 1983, **105**, 5775.
[70] L. M. Bass, R. D. Cates, M. F. Jarrold, N. J. Kirchner, and M. T. Bowers, *J. Am. Chem. Soc.*, 1983, **105**, 7024.
[71] N. J. Kirchner, J. R. Gilbert, and M. T. Bowers, *Chem. Phys. Lett.*, 1984, **106**, 7.
[72] K. G. Das, R. A. Swamy, J. van Thuijl, W. Okenhout, D. Fraise, and J.-C. Tabet, *Org. Mass Spectrom.*, 1983, **18**, 34.
[73] G. J. Bukovits and H. Budzikiewicz, *Org. Mass Spectrom.*, 1984, **19**, 23.
[74] G. L. Glish, P. J. Todd, K. L. Busch, and R. G. Cooks, *Int. J. Mass Spectrom. Ion Processes*, 1984, **56**, 177.
[75] M. A. Baldwin, C. J. Proctor, I. J. Amster, and F. W. McLafferty, *Int. J. Mass Spectrom. Ion Processes*, 1983, **54**, 97.
[76] T. G. Morgan, M. Rabrenovic, F. M. Harris, and J. H. Beynon, *Org. Mass Spectrom.*, 1984, **19**, 315.

loss of water, which gives another stable ion (17). It is proposed that protonation of the *E*-acids is at the double bond rather than at oxygen.[77] The same study showed that formation of the same protonated molecules by field desorption gave no fragmentation at all. The formation of protonated arginine by field desorption from a heated emitter gave fragmentation by loss of ammonia. ^{15}N-Labelling showed one of the terminal guanidyl nitrogen atoms to be involved, and this was rationalized by invoking the cyclic ion (18) in Scheme 4.[78]

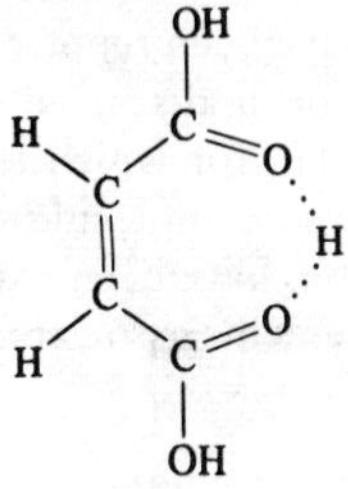

(16) 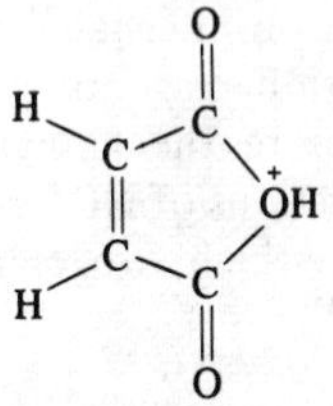 (17)

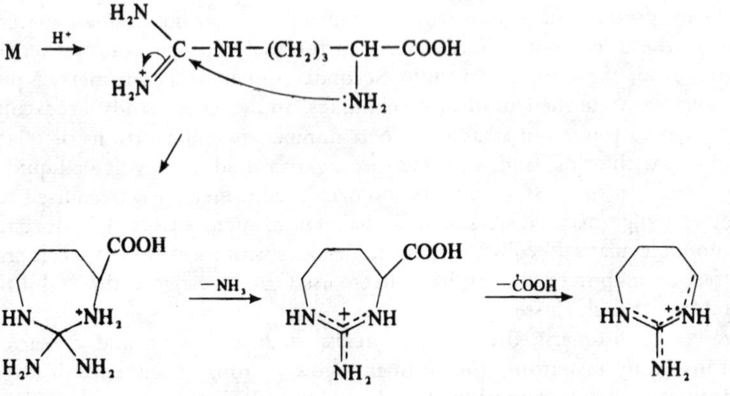

M $\xrightarrow{H^+}$ H_2N—C(=N^+H_2)—NH—$(CH_2)_3$—CH(:NH_2)—COOH

(18) $\xrightarrow{-NH_3}$ $\xrightarrow{-\dot{C}OOH}$

Scheme 4

7 Thermochemistry

The use of ion thermochemistry to characterize ion structures was dealt with in some detail in the previous volume of this title.[42] Thermochemical data are almost never used as the sole means of identifying structures, but they can provide confirmatory evidence. The difficulties associated with measurements of ionization and appearance energies by electron impact are well known and will only be outlined here. One of the major sources of error arising from the

[77] J. W. Dallinga, N. M. M. Nibbering, J. van der Greef, and M. C. Ten Noever de Brauw, *Org. Mass Spectrom.*, 1984, **19**, 11.

[78] J. J. Zwinselman, N. M. M. Nibbering, J. van der Greef, and M. C. Ten Noever de Brauw, *Org. Mass Spectrom.*, 1983, **18**, 525.

thermal-energy spread in the ionizing electrons can be eliminated by using energy-selected electrons. Many of the new values for heats of formation reported by Holmes *et al.* have been measured in this way by Lossing, and much greater confidence can be attached to these data than to earlier values.[66] Another potential source of error is the kinetic shift, an interesting brief discussion of which has recently been presented.[79] This topic has also been reviewed by Lifshitz,[3a] who pioneered electron-impact ion-trapping measurements of appearance energies and has now extended the technique to photoionization using a cylindrical ion trap.[59] This has since been described in more detail, and time-resolved appearance energies are reported for decompositions of iodobenzene, pyridine, aniline, phenol, cyclo-octatetraene, and ethylbenzene at delay times up to 500 μs.[80] Burgers and Holmes have recently compared appearance energies measured by electron impact using energy-selected electrons and measurements on metastable ions.[81] Two of the processes they describe have also been studied by Lifshitz, namely $C_6H_5I^{+\bullet} \rightarrow |C_6H_5^+ + I^\bullet$ (reaction 1) and $C_5H_5N^{+\bullet} \rightarrow |C_4H_4^{+\bullet} +$ HCN (reaction 2). The following appearance energies (eV) obtained from (i) energy-selected electrons, (ii) metastable peaks, and (iii) time-resolved photoionization are as follows. Reaction 1: (i) 11.32 ± 0.05, (ii) 11.4 ± 0.1, and (iii) 10.55 ± 0.1. Reaction 2: (i) 12.34 ± 0.05, (ii) 12.55 ± 0.1, and (iii) 11.84 ± 0.05. Another recent value for reaction 1 from time-resolved electron-impact studies is 10.95 ± 0.1 eV at 10 μs, falling to 10.55 eV at 700 μs.[82] For reaction 2 the time-resolved electron-impact studies gave 11.95 ± 0.2 eV at the longest times,[83] and another recent measurement on second field-free-region metastable ions gave 12.22 ± 0.2 eV[84] [(AE − IE) was measured as 2.57 ± 0.1 eV, the electron-impact IE being taken as 9.65 ± 0.1 eV[83]]. These results show that for processes known to have large kinetic shifts ion trapping gives the lowest measured ionization energies, whether photoionization or electron-impact ionization is employed. Thermochemical calculations for species such as these using data obtained without trapping will be in error, though not necessarily by enough to affect conclusions about ion structures.

Any reverse activation energy for ion decomposition, causing an elevation of appearance energies above the thermochemical minimum, may be reflected in kinetic-energy release causing broadening of metastable peaks. The methods for establishing the magnitude and distributions of such energy release from peak widths and shapes are well established, but the approach to correcting measured appearance energies is less well defined. The well known empirical equation due to Haney and Franklin and the theoretical approach by Klots both date from more than 10 years ago, but significant improvements have not yet been achieved. Derrick and Donchi have discussed the so-called dynamical theory of energy partitioning,[79] and Lifshitz has reviewed intramolecular-energy redistribution in

[79] P. J. Derrick and K. F. Donchi in 'Comprehensive Chemical Kinetics, Vol. 24: Modern Methods in Kinetics', ed. C. H. Bamford and C. H. F. Tipper, Elsevier, 1983.
[80] C. Lifshitz and Y. Malinovich, *Int. J. Mass Spectrom. Ion Processes*, 1984, **60**, 99.
[81] P. C. Burgers and J. L. Holmes, *Int. J. Mass Spectrom. Ion Processes*, 1984, **58**, 1582.
[82] S. Gefen and C. Lifshitz, *Int. J. Mass Spectrom. Ion Processes*, 1984, **58**, 251.
[83] C. Lifshitz, *J. Phys. Chem.*, 1982, **86**, 606.
[84] M. A. Baldwin, J. Gilmore, and M. N. Mruzek, *Org. Mass Spectrom.*, 1983, **18**, 127.

polyatomic ions with particular reference to non-statistical behaviour leading to bimodal energy-release distributions.[85] Lifshitz has also studied the time dependence of kinetic-energy releases.[3a] Derrick and Donchi have recently produced the first comprehensive compilation of kinetic-energy-release data from the literature.[79]

Measurements of ionization and appearance energies and a knowledge of heats of formation, either measured or calculated from Benson's rules, allow calculation of the heats of formation of ionic species. Many such heats of formation have now been established, and the corresponding value for an unknown ion will help to characterize its structure. Holmes *et al.* are working to devise a scheme to estimate ionic heats of formation, analogous to Benson's additivity rules for neutral compounds. In their second report on this subject they show that, for homologous series of ions substituted at the charge-bearing site, the heat of formation is linearly dependent on the logarithm of the number of atoms in the ion. Using this relationship they have predicted the heat of formation of a number of species not accessible to direct measurement, *e.g.* straight-line plots for the effects of $-OCH_3$ substitution and a knowledge of the thermochemistry for $CH_2{=}CH_2^{+\bullet}$ and $CH_2{=}CHOCH_3^{+\bullet}$ allow the heat of formation of $CH_2{=}C(OCH_3)_2^{+\bullet}$ to be predicted to be 530 kJ mol^{-1}.[86]

A new compilation of ionization and appearance-energy measurements has been published to supplement the previous National Bureau of Standards data.[87]

Thermochemical data have been used to assist in the characterization of many of the species referred to earlier in this review. The methylenoxonium ion was prepared by dissociative ionization of $HOCH_2CH_2OH$ and $HOCH_2CHO$, and the appearance energies were measured as 11.42 ± 0.05 and 10.42 ± 0.05 eV, respectively, using energy-selected electrons. Known heats of formation for the neutral species involved allowed calculation of the heat of formation as 824 and 808 kJ mol^{-1}, whereas the heat of formation of the ionized methanol isomer is 845 kJ mol^{-1}, thus allowing the differentiation of the two structures.[24]

The relative importance of two competitive reactions in 2-substituted 2,3-dihydro-4*H*-pyrans was explained on the basis of thermochemistry and charge localization. The reactions illustrated in Scheme 5 both occur when the 2-substituent, X, is OR, but when X = COR α-cleavage occurs to the exclusion of the retro-Diels–Alder (RDA) reaction. Ion enthalpies showed the RDA products to be destabilized with respect to the α-cleavage products by about 60 kJ mol^{-1} when the 2-substituent was an electron-withdrawing group such as acetyl. With electron-donating alkoxy substituents such as OC_2H_5 the RDA fragments were more stable than the α-cleavage products by about 10 kJ mol^{-1}. These measurements were also supported by theoretical calculations.[88]

A completely different approach to thermochemistry is possible by studying clustering reactions and acid/base phenomena. This is somewhat outside the stated scope of this review; as an example, however, proton transfer from

[85] C. Lifshitz, *J. Phys. Chem.*, 1983, **87**, 2304.
[86] J. L. Holmes and F. P. Lossing, *Can. J. Chem.*, 1982, **60**, 2365.
[87] R. D. Levin and S. Lias, 'Ionisation and Appearance Energy Measurements', NSRDS-NBS 71, 1982.
[88] J.-P. Morizur, J. Mercier, and M. Sarraf, *Org. Mass Spectrom.*, 1982, **17**, 327.

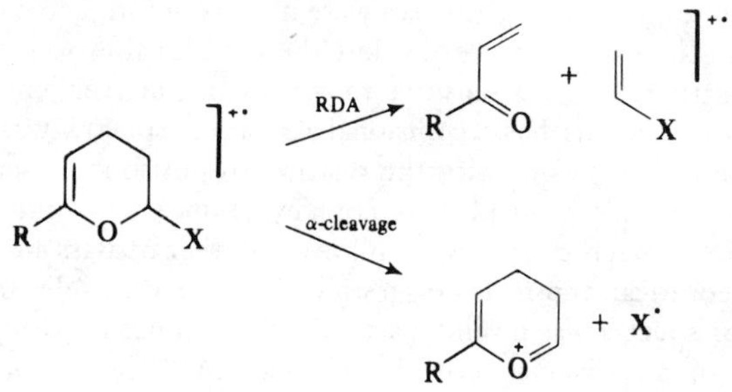

RDA
α-cleavage
R O X
R O
X
+ X˙

Scheme 5

CH_3S^+ species to bases of known proton affinity in an ion-cyclotron-resonance spectrometer enabled the stable isomer to be identified as CH_2SH^+, in agreement with *ab initio* calculations and earlier collisional-activation data.[89]

8 Tandem Mass Spectrometry

At its simplest, tandem mass spectrometry (sometimes called MS/MS) involves ionization and perhaps fragmentation of either a pure compound or a mixture, isolation of ions of a particular m/z value, and observation of the subsequent reactions of the selected ions. This embraces metastable-ion studies where the secondary fragmentations are unimolecular processes, and reactions following energization by photons or by low- or high-energy collisions with neutral molecules. Such collisions can cause decomposition, charge stripping or charge reversal, derivatization, and, most recently, neutralization and reionization. With multiple-sector instruments different types of processes can be observed sequentially. Tandem mass spectrometry has achieved rapid acceptance and is now the most frequently used technique in ion-structure studies. The recent book edited by McLafferty presents reviews of theory, instrumentation, and applications ranging from ion-structure studies to analytical problems from a wide range of backgrounds.[7]

Metastable ions give rise to diffuse peaks in mass spectra, and there was a time when instrument manufacturers fitted metastable suppressors to prevent interference from such peaks. In the 1960s the use of metastable peaks to identify reaction pathways became established, and various workers devised methods to obtain pure metastable-ion spectra, free of normal peaks. It was then appreciated that the relative sizes of metastable peaks could be characteristic of ion structure, but when the phenomenon of collisionally activated dissociation was observed it was realized that the increase in number and size of the peaks would make the collisional-activation spectra far more valuable for ion-structure studies. Metastable-peak sizes were known to be dependent on the internal energy of the ions sampled, but McLafferty proposed that collisional-activation spectra were largely independent of internal energy as long as peaks that also

[89] M. Roy and T. B. McMahon, *Org. Mass Spectrom.*, 1982, 17, 392.

corresponded to unimolecular processes were not taken into account. Metastable peaks arise from low-energy processes, but the collisional processes involve high energies and are therefore less sensitive to the original internal energy of the ion before excitation. On this basis collisional-activation spectra were taken to be additive and were used to quantify the relative proportions present in isomeric mixtures. However, this principle has not always met with universal support. Earlier criticism by Beynon *et al.* was followed by that of Bass and Bowers, who presented a theoretical argument suggesting that the principle of additivity or superposition of spectra was invalid, particularly when peaks arising in part from processes other than collisional activation are ignored.[90] They reported significant effects of internal energy and angular momentum on collisional-activation spectra, particularly for charge-stripping processes.[91,92] In response, Proctor and McLafferty reinvestigated the systems criticized and found no energy dependence outside experimental error for the benzyl/tropylium-ion system or isomeric $C_2H_4O^{+\bullet}$ structures.[93]

One thing to emerge from this controversy was the recognition of the existence of a tacit assumption that collision cross-sections were independent of ion structure. McLafferty and Proctor showed that this was not necessarily true, comparing the collisional-activation yields of the two very different $C_3H_7O_2^+$ isomers (19) and (20). The proton-bound structure was less stable to collisions and yielded approximately twice the total fragment-ion current.[94] It has recently been shown that the variation in collision cross-section is dependent on ion shape, *e.g.* linear $C_6H_6^{+\bullet}$ isomers have higher cross-sections than ionized benzene,[95] and such data can be used as further evidence for structural analysis, as is shown for $C_4H_9NO^{+\bullet}$ isomers.[96]

$$CH_3{-}O{-}CH{=}\overset{+}{O}{-}CH_3 \quad \textbf{(19)} \qquad\qquad CH_3{-}CH{=}O\cdots\overset{+}{H}\cdots O{=}CH_2 \quad \textbf{(20)}$$

Instrumentation. – It is not intended to present a detailed review of instrumentation for tandem mass spectrometry but to indicate the trends. Reference 7 is recommended for comprehensive coverage of this subject. The realization that pure metastable-ion spectra could be obtained from conventional-geometry double-focusing sector instruments by decoupling the accelerating and electric-sector potentials, and subsequent development of reversed-geometry instruments and linked-scan procedures, resulted in these being the first instruments used

[90] L. M. Bass and M. T. Bowers, *Org. Mass Spectrom.*, 1982, **17**, 229.
[91] P. A. M. van Koppen, A. J. Illies, S. Liu, and M. T. Bowers, *Org. Mass Spectrom.*, 1982, **17**, 399.
[92] M. F. Jarrold, A. J. Illies, N. J. Kirchner, and M. T. Bowers, *Org. Mass Spectrom.*, 1983, **18**, 388.
[93] C. J. Proctor and F. W. McLafferty, *Org. Mass Spectrom.*, 1983, **18**, 193.
[94] F. W. McLafferty and C. J. Proctor, *Org. Mass Spectrom.*, 1983, **18**, 272.
[95] M. W. E. M. van Tilborg and J. van Thuijl, *Org. Mass Spectrom.*, 1984, **19**, 217.
[96] J. van Thuijl, *Org. Mass Spectrom.*, 1984, **19**, 243.

extensively for tandem studies, and they continue to be the most widely used because of their general-purpose nature. The development of three-dimensional mapping techniques combines the various scan modes and allows the maximum amount of information to be derived.[97,98] The major limitation of two-sector instruments is mass resolution of both the primary- and secondary-ion beams, so multi-sector tandem instruments have been developed for high-resolution applications. At least one manufacturer offers a four-sector high-resolution instrument of the highest available mass range (15 000 daltons for 8 keV ions).

Sector instruments are ideal for high-energy collision processes, but simultaneous development of the triple-quadrupole mass spectrometer has produced an instrument that does not possess the high-resolution capabilities yet has its own unique advantages. These include the ability to study low-energy collisions, ease of computer control, relatively simple design and construction, and relatively low maintenance costs compared with large sector instruments. The logical outcome of these two lines of development was to combine the two types of analyser, sector and quadrupole, in hybrid instruments that could incorporate the advantages of both types, *e.g.* the ability to study both high- and low-energy collisions with high resolution.[99] The relative merits of the various possible combinations of analysers have been reviewed.[100] Another advantage of multiple analysers is the ability to study processes occurring sequentially in two or more field-free regions, although relatively few reports of such studies have appeared as yet, despite pioneering studies by Beynon *et al.* in 1980. They used a two-sector reversed-geometry instrument to study photodissociation in the second field-free region of the products of metastable ions decomposing in the first field-free region.[101] They have since described metastable-ion studies following charge inversion in a triple-sector instrument (BEE).[102] Quadrupole instruments can also be used to study sequential processes.[103]

Collision Processes. – In ref. 7, collision processes are discussed for low-energy ions[7a] and high-energy ions.[7b] Typical ion energies are a few tens of eVs for triple-quadrupole systems and several keVs for magnetic-sector instruments, leading to essentially different mechanisms for collisional excitation, although in both cases translational energy is converted into internal energy. For low-energy collisions the excitation is essentially vibrational, and the collisional-activation spectra are dependent on the precursor-ion internal energy and translational energy. Although the collision products experience scattering, the wide acceptance

97 M. J. Farncombe, K. R. Jennings, R. S. Mason, and J. Scrivens, *Int. J. Mass Spectrom. Phys.*, 1982, **44**, 91.
98 C. G. MacDonald and M. J. Lacey, *Org. Mass Spectrom.*, 1984, **19**, 55.
99 G. L. Glish, S. A. McLuckey, T. Y. Ridley, and R. G. Cooks, *Int. J. Mass Spectrom. Ion Phys.*, 1982, **41**, 157.
100 J. H. Beynon, F. M. Harris, B. N. Green, and R. H. Bateman, *Org. Mass Spectrom.*, 1982, **17**, 55.
101 C. J. Proctor, B. Krajl, A. G. Brenton, and J. H. Beynon, *Org. Mass Spectrom.*, 1980, **15**, 619.
102 M. Rabrenovic, A. G. Brenton, and J. H. Beynon, *Int. J. Mass Spectrom. Ion Phys.*, 1983, **52**, 175.
103 D. J. Burinsky, R. G. Cooks, E. K. Chess, and M. L. Gross, *Anal. Chem.*, 1982, **54**, 295.

angle of the quadrupole analyser makes the spectra independent of the angular distribution. High-energy collisions bring about electronic excitation that is relatively independent of internal and translational energy, but the narrow acceptance angle of most sector instruments makes the spectra sensitive to scattering. As a result of these instrumental differences the energy dependence is best studied in quadrupole experiments, which has led to the development of energy-resolved mass spectrometry, whereas the angular dependence is best studied in sector experiments by angle-resolved mass spectrometry. As might be anticipated, the scattering experienced by the fragments is related to the energy transferred, and the two techniques can give similar results. The energy transfer is likely to depend on ion structure, and angular and energy dependence can be used to differentiate isomeric structures.[7c, 104, 105]

A somewhat different aspect to fragment-ion abundances in collision-induced spectra comes from a Hammett correlation for fragmentations of $(M-CO)^{+\bullet}$ from tetraphenylcyclopentadienones, and Hammett σ^+ constants, for various substituents on the phenyl rings. The correlation shows the effects of electron density on the competition for the charge on substituted and unsubstituted $PhCCPh^{+\bullet}$.[106]

Typical processes that can occur on collision of a singly charged positive ion are fragmentation to a smaller ion and a neutral species, charge stripping to give a doubly charged ion, neutralization by charge exchange, and derivatization by the neutral target gas through reactive collisions. Similar processes occur for negative ions, and there is the possibility of charge reversal in which a negative ion loses two electrons to form a positive ion.

The first of these processes, fragmentation, gives rise to the normal collisional-activation spectrum that is now widely used to characterize ion structures. Very many of the studies already discussed have used this technique, and no further examples will be given at this stage. Doubly charged ions formed by charge stripping will also occur in the same spectrum, *e.g.* the m/z 16 peak discussed earlier in the $CH_4O^{+\bullet}$ spectrum. Doubly charged ions formed in the ion source can also be studied by collisions with neutral gases to form singly charged ions of twice the normal translational energy, which are transmitted by operating the electric sector at twice the normal potential ($2E$ spectra). An example comes from a study of the doubly charged ions of acetylenes, which includes measurements of appearance energies and molecular-orbital calculations of potential energies.[107] A similar technique allowed the study of the decompositions of metastable doubly charged ions.[102] Doubly charged ions can also be studied in triple-quadrupole instruments.[108]

[104] J. D. Ciupek, D. Zakelt, R. G. Cooks, and K. V. Wood, *Anal. Chem.*, 1982, **54**, 2215.
[105] P. J. Todd, R. J. Warmack, and E. H. McBay, *Int. J. Mass Spectrom. Ion Phys.*, 1983, **50**, 299.
[106] M. M. Bursey, D. J. Harvan, J. R. Hass, E. I. Becker, and B. H. Arison, *Org. Mass Spectrom.*, 1984, **19**, 160.
[107] J. R. Appling, B. E. Jones, L. E. Abbey, D. E. Bostwick, and T. F. Moran, *Org. Mass Spectrom.*, 1983, **18**, 282.
[108] H. I. Kenttamaa, K. V. Wood, K. L. Busch, and R. G. Cooks, *Org. Mass Spectrom.*, 1983, **18**, 561.

Charge reversal of negative ions to form positive ions can be used to produce unique structures not otherwise accessible, as they do not reside in an energy well. The ion CH_2CN^+ can be formed directly by $H^•$ loss from $CH_3CH^{+•}$ or indirectly by charge reversal of CH_2CN^-. Ions formed by the latter process give a unique collisional spectrum that cannot be related directly to any stable structure, although there are indications of partial isomerization to the isocyanide.[109]

The study of reactive collisions by ion-cyclotron-resonance spectroscopy has a long history. Similar studies can now be carried out by tandem mass spectrometry, particularly with the low-energy domain of the triple-quadrupole system. The ion $C_3H_3^+$ formed by decomposition of propargyl halide molecular ions can have either a cyclic structure as given by the chloride and bromide or a linear structure as given by the iodide. Collision of these species with neutral C_2H_2 in a triple-quadrupole instrument gave a collision complex $C_5H_5^+$, and the abundance ratio, $C_5H_5^+/C_3H_3^+$, showed that the linear form produced by the decomposition of propargyl iodide was significantly more reactive.[110]

Neutralization by charge transfer to the target gas molecules can only be useful for mass-spectrometric studies if it is followed by deflection of any remaining ions and then by reionization of the neutral molecules by a subsequent collision process. McLafferty *et al.* have developed the instrumentation for this in a large-sector tandem mass spectrometer. Following work by Porter *et al.*,[111] they use not only stable gases for neutralization but also various metal vapours. The difference between the ionization energy of the metal and the electron affinity of the ion will determine the energy transferred and hence the internal energy of the neutral species. Neutralization may not occur if the ionization energy exceeds the electron affinity. In this way, the t-butyl ion could be differentiated from its isomers.[32] The same apparatus can be used to study the neutral fragments lost on fragmentation of ions, either metastable or collision induced. On the basis of m/z 26:27 abundance ratios in the spectra of the reionized neutral species, it has been reported that metastable aniline ions lose HNC rather than HCN.[32] Burgers *et al.* subsequently compared m/z 12–15 for the neutral species from aniline and pyridine, confirming the earlier report and also showing that the neutral fragment lost from pyridine was HCN.[112] These authors have since reported further studies on the ionization of neutral fragments using a commercial double-focusing mass spectrometer with only minor modifications.[113] One unexpected observation was that methyl ethanoate ions lose $CH_2OH^•$ whereas methyl propanoate ions lose $CH_3O^•$.[114]

[109] M. M. Bursey, D. J. Harvan, C. E. Parker, and J. R. Hass, *J. Am. Chem. Soc.*, 1983, **105**, 6801.

[110] D. D. Fetterholf, R. A. Yost, and J. R. Eyler, *Org. Mass Spectrom.*, 1984, **19**, 105.

[111] G. I. Gellene and R. F. Porter, *Acc. Chem. Res.*, 1983, **16**, 200.

[112] P. C. Burgers, J. L. Holmes, A. A. Mommers, and J. K. Terlouw, *Chem. Phys. Lett.*, 1983, **102**, 1.

[113] P. C. Burgers and J. L. Holmes, *Org. Mass Spectrom.*, 1984, **19**, 452.

[114] P. C. Burgers, J. L. Holmes, A. A. Mommers, J. E. Szulejko, and J. K. Terlouw, *Org. Mass Spectrom.*, 1984, **19**, 442.

Photodissociation. – Collision processes produce energized ions with a distribution of internal energies. Precise control of the energy transferred can be achieved by using monochromatic photon sources. The majority of photodissociation studies have been carried out in ion-cyclotron-resonance spectrometers, but recent development of high-power lasers has made photodissociation possible in the high-velocity ion beams of sector instruments. Beynon *et al.* have published a series of papers on experiments in which an argon-ion laser was used to irradiate the ion beam along the axis of the beam between the two sectors of a double-focusing mass spectrometer. Data from this technique have recently been used to rationalize isomerizations of $C_8H_{10}^{+\bullet}$ in terms of the energy dependence of $CH_3^\bullet$ loss and the associated kinetic-energy release.[115] A similar experimental arrangement has been used to demonstrate that photodissociation can distinguish different isomeric structures that give virtually identical collision spectra. Of the $C_4H_5N^{+\bullet}$ ions formed by electron impact on compounds (21)–(25), ionized pyrrole (21) does not give any photoinduced decomposition and so stands out from the other four isomers, all of which give photodissociation spectra that are virtually identical at any one photon wavelength. However, the cross-sections for fragmentation following photoexcitation are substantially different for all except compounds (24) and (25), the $C_4H_5N^{+\bullet}$ ions from which probably have a common structure.[116] The use of photodissociation in the ion source of a tandem mass spectrometer has also been described.[117]

Unimolecular Reactions. – The emphasis on collisional activation may appear to have diminished the role of metastable-ion studies, but there are still many situations where such studies can be highly informative. Collisional activation samples low-energy ions that would not otherwise have fragmented, and the structures of such ions may differ from those of the reacting species. Metastable ions of a series of deuterium- and ^{13}C-labelled C_8H_8 isomers were compared

N–H

(21)

H$_2$C—CH$_2$ / CH / CN

$CH_3-CH=CH-CN$ (22)

$CH_2=CH-CH_2-CN$ (23)

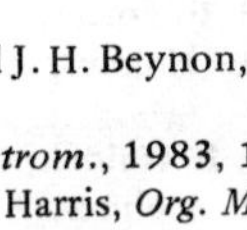

(24)

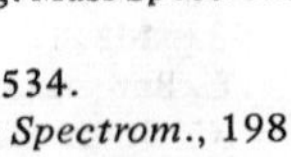

$CH_2=C(CH_3)-CN$ (25)

[115] J. M. Curtis, R. K. Boyd, B. Shushan, T. G. Morgan, and J. H. Beynon, *Org. Mass Spectrom.*, 1984, **19**, 207.

[116] E. Weger, W. A. Brand, and K. Levsen, *Org. Mass Spectrom.*, 1983, **18**, 534.

[117] M. S. Kim, T. G. Morgan, E. E. Kingston, and F. M. Harris, *Org. Mass Spectrom.*, 1983, **18**, 582.

with ions decomposing in the ion source by loss of $CH_3^{\bullet}$ to study interconversion of ethylbenzene, 7-methylcycloheptatriene, and *p*-xylene ions. The data were fitted to a dynamic model of interconversion and decomposition calculated from RRKM theory. For high-energy ions, fast hydrogen rearrangements were postulated to occur independently of skeletal isomerizations,[118] but there has been some criticism of the statistical calculations used in this work.[115] Unimolecular decompositions of some crowded triarylethenol radical cations were shown to be preceded by conformationally controlled CH_3/H transfers between *ortho* and *ipso* positions of opposing rings.[119]

Because metastable-ion fragmentations are low-energy processes their rates are more sensitive to substitution with heavy isotopes. Isotope effects can reveal which atoms are involved in rate-limiting steps and have been used to detect hidden hydrogen rearrangements. They have been reviewed by Derrick.[3d] An example of the use of isotope effects comes from controversy over the stepwise or concerted nature of the well known McLafferty rearrangement. The molecular ion of benzyl ethyl ether undergoes CH_3CHO elimination by the mechanism depicted in Scheme 6, *i.e.* by breaking a C–O bond and by breaking a C–H bond. If the mechanism was concerted an isotope effect would be anticipated for both bond cleavages, whereas a step-wise mechanism would only show an isotope effect for the rate-determining step. Labelling with ^{18}O and deuterium showed significant isotope effects on metastable-peak sizes for both cleavages for competitive losses from doubly substituted species (26) and (27), supporting the concerted mechanism.[120]

A more dramatic effect was observed for O loss from $COOH^+$ when the H was replaced by D. The mechanism involves a rate-determining 1,2 hydrogen transfer, and the isotope effect is large enough to make an alternative reaction pathway

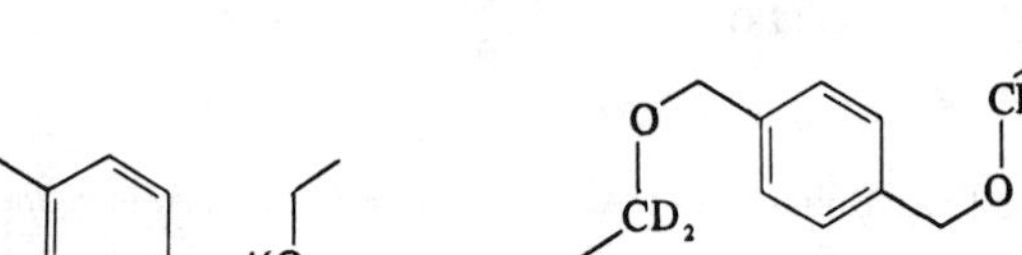

H CHCH3 O C H2 → H H ĊH2 + CHCH3 O

Scheme 6

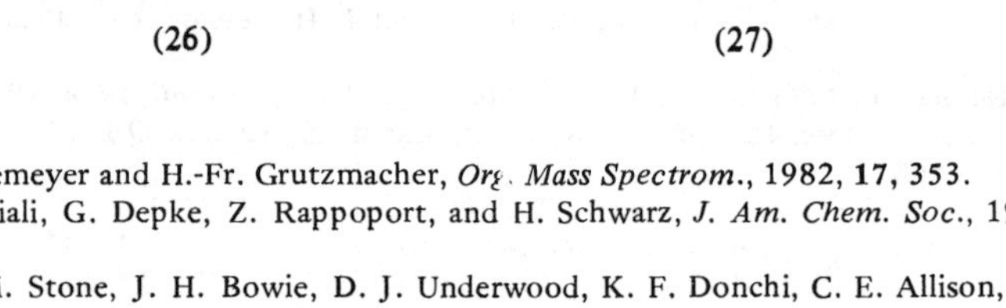

18O 16O

(26)

O CD2 CH3 CH2 CH3 O

(27)

[118] J. Grotemeyer and H.-Fr. Grutzmacher, *Org. Mass Spectrom.*, 1982, **17**, 353.
[119] S. E. Biali, G. Depke, Z. Rappoport, and H. Schwarz, *J. Am. Chem. Soc.*, 1984, **106**, 496.
[120] D. J. M. Stone, J. H. Bowie, D. J. Underwood, K. F. Donchi, C. E. Allison, and P. J. Derrick, *J. Am. Chem. Soc.*, 1983, **105**, 1688.

competitive for the deuterium analogue. Loss of O from $COOH^+$ results in the formation of HCO^+, with a characteristic gaussian metastable peak. Loss of O from $COOD^+$ results in the formation of both DCO^+ and COD^+, the latter process having a dish-shaped metastable peak, thereby giving rise to a composite metastable peak.[121]

9 Proximity Effects

Proximity effects in mass spectrometry continue to receive considerable attention and will be discussed here as the final example of the way structure determines the course of unimolecular-ion reactions in the gas phase, *ortho* substitution of benzene rings providing the best known example. The classic case of $OH^\bullet$ loss from *o*-nitrotoluene continues to generate interest. Loss of $OH^\bullet$ also occurs from *meta* and *para* isomers to a limited extent, and the resulting ions have been studied by collisional activation[122] and photodissociation[123] to demonstrate different structures for the two isomers. Metastable peaks for the loss of CO from $(M - OH^\bullet)^+$ ions of *o*-nitrotoluene have been compared with those of isomeric ions produced from a number of different sources. The most likely structure for metastable $(M - OH^\bullet)^+$ ions proved to be that of 1,2-benzisoxazolenium (**28**), whereas ions fragmenting in the source probably have the isomeric structure (**29**).[124] Schwarz has studied a similar *ortho* effect: loss of $OH^\bullet$ from *o*-nitrostyrene[125] and loss of ethene (alkene) from cyclopropylnitrobenzene and related cycloalkyl analogues.[126] Ramana has continued his studies of *ortho* effects with an investigation of elimination of $SH^\bullet$ from *o*-methoxy aromatic thioamides,[127] and Luijten and van Thuijl have reported on $CHO^\bullet$ and CH_2O loss from nitrodiazoles.[128] An interesting *ortho* effect is the loss of $CH_3^\bullet$ from the molecular ions of *N*-(2′-hydroxyethyl)-2-piperidone. Kinetic-energy-release data, deuterium labelling, and kinetic isotope effects suggest that a rate-determining isomerization precedes the fragmentation, as shown in Scheme 7.[129]

N $\overset{+}{O}$

(28)

O^+ N

(29)

[121] P. C. Burgers, J. L. Holmes, and A. A. Mommers, *Int. J. Mass Spectrom. Ion Phys.*, 1983, **54**, 283.
[122] M. A. Baldwin and H. J. Bowley, *Org. Mass Spectrom.*, 1982, **17**, 580.
[123] T. G. Morgan, E. E. Kingston, F. M. Harris, and J. H. Beynon, *Org. Mass Spectrom.*, 1982, **17**, 594.
[124] C. G. Herbert, E. A. Larka, and J. H. Beynon, *Org. Mass Spectrom.*, 1984, **19**, 306.
[125] G. Depke, W. Klose, H. Schwarz, W. Blum, and W. J. Richter, *Org. Mass Spectrom.*, 1983, **18**, 568.
[126] G. Depke, W. Klose, and H. Schwarz, *Org. Mass Spectrom.*, 1983, **18**, 496.
[127] D. V. Ramana and S. K. Viswanadham, *Org. Mass Spectrom.*, 1983, **18**, 418.
[128] W. C. M. M. Luijten and J. van Thuijl, *Org. Mass Spectrom.*, 1982, **17**, 304.
[129] B. Steiner, D. Schumann, and H. Hoffmann, *Org. Mass Spectrom.*, 1983, **18**, 130.

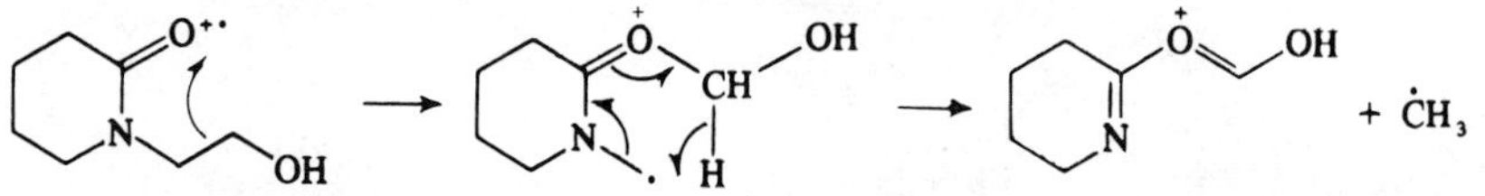

Scheme 7

Proximity effects other than *ortho* effects are well known, many stereochemically controlled reactions falling into this category.[3c] The example discussed earlier of protonated *E*- and *Z*-unsaturated dicarboxylic acids is characteristic of many such studies.[60] A similar study of the protonated esters of related acids has been made by chemical-ionization and triple-quadrupole experiments,[130] and stereospecific reactions of negative ions of cyclohexan-1,4-diols have been studied by collisional activation in a reversed-geometry sector instrument.[131] Many of the fragment ions in the mass spectra of steroids have been examined by tandem mass spectrometry. A study of 3-hydroxy steroids having substituents on the D-ring showed that the collisional-activation spectrum of $C_{16}H_{26}O^{+\bullet}$ could indicate the absolute stereochemistry of the hydroxyl group and the A/B ring junction.[132]

10 Conclusion

The majority of mass spectrometers are bought for analytical applications and the majority of mass spectrometrists are engaged in analytical work. However, academic studies into structures and reactions of gas-phase ions have also produced tremendous advances in mass-spectrometric techniques, which unquestionably have been to the benefit of analytical chemistry. Tandem mass spectrometry provides an excellent example of this. The techniques that were developed for distinguishing different ion structures are finding increasing application in analytical studies, for both structural and mixture analysis. Furthermore, a greater understanding of the complex chemistry that gives rise to the mass spectra of known compounds will in turn facilitate the structural analysis of new compounds by mass spectrometry.

[130] A. Weisz, A. Mandelbaum, J. Shabanowitz, and D. F. Hunt, *Org. Mass Spectrom.*, 1984, **19**, 238.
[131] T. Gauman, D. Stahl, and J. C. Tabet, *Org. Mass Spectrom.*, 1983, **18**, 263.
[132] M. T. Cheng, M. P. Barbalas, R. F. Pegues, and F. W. McLafferty, *J. Am. Chem. Soc.*, 1983, **105**, 1510.

3

Photoelectron – Photoion Coincidence Spectroscopy

BY J. DANNACHER AND J.-P. STADELMANN

1 Introduction

The introduction of mass spectrometry as an analytical method has revolutionized the domain of chemical elementary analysis. No other technique can match the accuracy, sensitivity, and speed of mass-spectrometric determination of the elemental composition of an organic molecule. Generally, the mass spectrum leaves little room for speculation on the formula of an unknown compound, provided that the measurement is not impaired by, for example, impurities, chemical transformations of the sample before ionization, or lack of stability of even the least excited molecular ions formed. With increasing molecular size, the number of isomers compatible with a given elemental composition increases considerably. Deducing the structure – or at least some elements of structure – has therefore always been attempted by observation of fragmentation behaviour under typical experimental conditions, *i.e.* 70 eV electron bombardment, double-focusing magnetic-sector instruments. An impressively large number of mass-spectrometric studies has shown unambiguously that the structure of the initially ionized molecule is represented by the corresponding mass spectrum. However, for a long time the important factors that govern this transformation of chemical structure into spectral information concerning relative abundance *versus* m/z ratio have at best been qualitatively understood. With reference to well known thermal and some photochemical types of dissociation reactions of organic molecules in solution, empirical rules have been formulated for deriving chemical structures from mass-spectrometric data. Bearing in mind that nothing more than a set of pairs of m/z values and relative abundances is measured, detailed ionic structures, reaction pathways, and dissociation mechanisms frequently proposed to vindicate the assignment of a mass spectrum to a particular molecule are better considered as possible rationalizations than as experimentally established facts. Although such empiricism has proved very useful for interpreting mass spectra, the behaviour of highly excited positive ions cannot be comprehensively described within the framework of this approach. As an illustration, it should be recalled that the absence of a particular signal in a mass spectrum may originate from the initial excitation-energy distribution or from rapid non-dissociative relaxation processes that precede the formation of fragment ions. Clearly, an empirical 'chemical model' does not allow for either of these two phenomena, therefore preventing any fundamentally correct conclusions in such cases.

In recent years a whole series of novel experimental techniques has been developed to investigate thoroughly the formation, excitation, relaxation, characterization, and detection of positive ions in the gas phase.[1] Owing to these efforts, mass spectra brought about by photon impact can now be interpreted in terms of the physical and chemical properties of the ionized molecules. For simplicity, ionizing and exciting particles are restricted to quanta in the far-UV region. The advantages of photon-impact techniques are essentially based on the very simple threshold law for photoionization of gaseous molecules and the availability of sufficiently monochromatic far-UV radiation sources. The study of absorption and emission of electromagnetic radiation by positive ions and, more especially, photoelectron spectroscopy have furnished much valuable information on the nature, energy, and accessibility of a great many of the electronic states of singly charged cations. When the original population of the accessible vibronic states of the molecular ion is known, the initial distribution of the excitation energy connected with the process of photoionization can be defined. This distribution, commonly referred to in mass spectrometry as the energy-deposition function, largely determines the type and abundance of the product ions formed. Once generated, the excited molecular cations can diminish or redistribute their energy content in several distinct ways. Conceivable relaxation processes may either occur on a single electronic hypersurface or involve radiative or radiationless transitions between different electronic states. Vibrational predissociation and isomerization to a more stable species are examples of the former case; the emission of electromagnetic radiation and internal conversion belong to the second group of relaxation processes.

The quoted development of novel experimental probes has stimulated the elaboration of new theoretical approaches and the formulation of more tractable versions of older methods.[2] Nowadays, the dissociation of very small cations consisting of up to four atoms has become more accessible to sophisticated *ab initio* calculations, but even for the smallest systems relevant to organic mass spectrometry the use of such *ab initio* calculations is still out of the question.[2] At present we must content ourselves with statistical models to describe the dissociative processes of such larger ionic species. While the fundamental aspects of these rather old statistical theories have remained unchanged,[3,4] a more straightforward connection between theory and experiment has recently been arrived at. The more detailed and accurate experimental data define a few quantities that earlier used to be free parameters. In this respect photoelectron–photoion coincidence (PEPICO) spectroscopy is of prime importance for testing any theoretical model for ionic-decomposition phenomena and for incorporating the large body of mass-spectrometric information into the sphere of gas-phase ion chemistry.

Studying ionic-decay processes by means of internal-energy-selective measurements has always been a goal for mass spectrometrists. The first noteworthy

[1] 'Gas Phase Ion Chemistry', ed. M. T. Bowers, Academic Press, New York, 1979.
[2] For a recent review of this field see: J. C. Lorquet, *Org. Mass Spectrom.*, 1981, **16**, 469.
[3] W. Forst, 'Theory of Unimolecular Reactions', Academic Press, New York, 1973.
[4] P. J. Robinson and K. A. Holbrook, 'Unimolecular Reactions', Wiley-Interscience, New York, 1972.

attempts to realize such experiments involved charge-exchange mass spectrometry[5] and a preliminary form of PEPICO spectroscopy developed in the late sixties.[6] The major breakthrough in this area was due to J. H. D. Eland who, in 1972, described the design and successful operation of the first PEPICO spectrometer incorporating differential photoelectron-energy analysis.[7] Shortly after the publication of this remarkable piece of scientific work, an important variant of the original experiment, termed threshold PEPICO spectroscopy, was developed.[8] In the following years a few more coincidence spectrometers of both types were built in other laboratories. The capacity of the instruments was gradually improved, and coincidence data on some 150 molecular cations were reported in about as many research papers during the past 15 years. Individual research groups developed specific preferences for studying particular sorts of molecules and/or selected kinds of processes. Among the more extensively investigated topics are the predissociation of very small molecular cations consisting of up to five atoms, the breakdown diagrams of small organic molecular cations, the average kinetic energy or, in particularly favourable cases, the distribution of the kinetic energy of fragment ions as a function of the internal energy of the decaying parent ions, the determination of rate/energy functions, the effect of competition between all sorts of isomerization reactions and dissociative processes, and the depletion of excited-electronic-state populations by ion fluorescence. Readers interested in further details are referred to the original literature, which, to the best of our knowledge, is comprehensively compiled in Table 1.

So far the number of PEPICO studies has remained remarkably small, in striking contrast to the great importance of the method in several fields of chemical physics. This is because the technique is extremely time-consuming, which results in a very low rate of data production. An extensive coincidence study of the radical cation of, say, a 12-atom organic molecule may easily exceed 1000 hours of measuring time, if high mass-, energy-, and time-resolution and an adequate signal-to-noise ratio are required. Such a slow and experimentally demanding method seems unattractive to many scientists, and the number of research groups who really master the technique has therefore remained notably small. As a closer inspection of the references reveals, the vast majority of studies (*cf.* Table 1) are due to (i) J. H. D. Eland's group, partly in collaboration with C. J. Danby in Oxford (U.K.) and with B. Brehm and co-workers in Freiburg (West Germany), (ii) R. Stockbauer at Argonne (U.S.A.) and in co-operation with H. M. Rosenstock at the National Bureau of Standards (Washington D.C., U.S.A.), (iii) T. Baer and his co-workers in Chapel Hill (U.S.A.), (iv) G. G. Meisels *et al.* in Lincoln (U.S.A.), (v) I. Powis and co-workers originally in Oxford (U.K.) and now in Nottingham (U.K.), (vi) Y. Niwa and T. Tsuchiya in Tokyo (Japan), and (vii) the Reporters' group in Basel (Switzerland).

In writing this review article on PEPICO spectroscopy the authors wish to give a comprehensive survey of the method, with particular emphasis on recent progress in our knowledge of the decay processes of 'larger' molecular cations,

[5] E. Lindholm, *Z. Naturforsch., Teil A*, 1954, **9**, 535.
[6] B. Brehm and E. von Puttkamer, *Z. Naturforsch., Teil A*, 1970, **25**, 1062.
[7] J. H. D. Eland, *Int. J. Mass Spectrom. Ion Phys.*, 1972, **8**, 143.
[8] R. Stockbauer, *J. Chem. Phys.*, 1973, **58**, 3800.

Table 1 *Molecules whose cations have been studied by photoelectron–photoion coincidence spectroscopy*

Number of atoms	*Molecule*	*References*
2	O_2	9,[a,b,c] 10,[d,e] 11,[d,f] 12,[a,b,g,h] 13,[a,b,g] 14,[a,b,g] 15[a,b,g]
	$^{18}O_2$	15,[a,b,g]
	NO	12,[a,b,g,h] 16[d,e]
	ICl	12[a,b,g,h]
	IBr	12[a,b,g,h]
	I_2	12[a,b,g,h]
	H_2	17[d,e]
	Xe_2	18[a]
3	CO_2	19,[a,b,g] 20,[b,d,g,l] 21[a,b,g]
	SO_2	22,[a,b,g] 23[b,d]
	CS_2	24[a,b,g]
	N_2O	25,[a,b,i] 26,[a,b,g] 27,[b,d,f] 28[b,d,h,l]
	COS	25,[a,b,i] 26[a,b,g]
	H_2O	29[b,d]
	D_2O	29,[b,d] 30[a,b,c]

[9] C. J. Danby and J. H. D. Eland, *Int. J. Mass Spectrom. Ion Phys.*, 1972, **8**, 153.
[10] T. Baer, P. T. Murray, and L. Squires, *J. Chem. Phys.*, 1978, **68**, 4901.
[11] P. M. Guyon, T. Baer, L. F. A. Ferreira, I. Nenner, A. Tabché-Fouhaillé, R. Botter, and T. R. Govers, *J. Phys. B*, 1978, **11**, L141.
[12] J. H. D. Eland, *J. Chem. Phys.*, 1979, **70**, 2926.
[13] R. G. C. Blyth, I. Powis, and C. J. Danby, *Chem. Phys. Lett.*, 1981, **84**, 272.
[14] R. Bombach, A. Schmelzer, and J.-P. Stadelmann, *Chem. Phys.*, 1981, **61**, 215.
[15] R. Bombach, A. Schmelzer, and J.-P. Stadelmann, *Int. J. Mass Spectrom. Ion Phys.*, 1982, **43**, 211.
[16] L. Squires and T. Baer, *J. Chem. Phys.*, 1976, **65**, 4001.
[17] K. Tanaka and I. Koyano, *J. Chem. Phys.*, 1978, **69**, 3422.
[18] E. D. Poliakoff, P. M. Dehmer, J. L. Dehmer, and R. Stockbauer, *J. Chem. Phys.*, 1982, **76**, 5214.
[19] J. H. D. Eland, *Int. J. Mass Spectrom. Ion Phys.*, 1972, **9**, 397.
[20] R. Frey, B. Gotchev, O. F. Kalman, W. B. Peatman, H. Pollak, and E. W. Schlag, *Chem. Phys.*, 1977, **21**, 89.
[21] R. Bombach, J. Dannacher, J.-P. Stadelmann, and J. C. Lorquet, *J. Chem. Phys.*, 1983, **79**, 4214.
[22] B. Brehm, J. H. D. Eland, R. Frey, and A. Kuestler, *Int. J. Mass Spectrom. Ion Phys.*, 1973, **12**, 197.
[23] M. J. Weiss, T.-C. Hsieh, and G. G. Meisels, *J. Chem. Phys.*, 1979, **71**, 567.
[24] B. Brehm, J. H. D. Eland, R. Frey, and A. Kuestler, *Int. J. Mass Spectrom. Ion Phys.*, 1973, **12**, 213.
[25] J. H. D. Eland, *Int. J. Mass Spectrom. Ion Phys.*, 1973, **12**, 389.
[26] B. Brehm, R. Frey, A. Kuestler, and J. H. D. Eland, *Int. J. Mass Spectrom. Ion Phys.*, 1974, **13**, 251.
[27] T. Baer, P. M. Guyon, I. Nenner, A. Tabché-Fouhaillé, R. Botter, L. F. A. Ferreira, and T. R. Govers, *J. Chem. Phys.*, 1979, **70**, 1585.
[28] I. Nenner, P. M. Guyon, T. Baer, and T. R. Govers, *J. Chem. Phys.*, 1980, **72**, 6587.
[29] R. Stockbauer, *J. Chem. Phys.*, 1980, **72**, 5277.
[30] J. H. D. Eland, *Chem. Phys.*, 1975, **11**, 41.

Table 1 *(cont.)*

Number of atoms	*Molecule*	*References*
	H_2S	31[a,b,g]
	D_2S	31[a,b,g]
	Xe_3	18,[a] 32[a]
4	BF_3	33[b,d]
	C_2H_2	31[a,b,g]
	NH_3	34,[d,e] 35[a,g,f]
	COF_2	36[a,b,g,j]
	$COCl_2$	36[a,b,g,j]
	ND_3	37[a,g,j]
	NF_3	38[a,b,k]
	H_2CO	39,[a,b,c,k] 40,[a,b,c] 41[a,b,c,k]
	D_2CO	39,[a,b,c,k] 40,[a,b,c] 41,[a,b,c,k] 42[a,b,k,l,m]
5	CH_4	8,[b,c,d] 43,[c,d] 44,[b,d,k,l] 45[a,g,j]
	CD_4	8[b,c,d]
	MeF	46[a,b,c,k]
	MeCl	46[a,b,c,k]
	MeBr	46,[a,b,c,k] 47[b,d]
	MeI	46,[a,b,c,k] 47,[b,d] 48,[d,g] 49[a,b,g]
	CD_3I	48[d,g]
	CH_2Cl_2	9[a,b,c]
	CH_2Br_2	47,[b,d] 50[d,e]

[31] J. H. D. Eland, *Int. J. Mass Spectrom. Ion Phys.*, 1979, **31**, 161.
[32] E. D. Poliakoff, P. M. Dehmer, J. L. Dehmer, and R. Stockbauer, *J. Chem. Phys.*, 1981, **75**, 1568.
[33] C. F. Batten, J. A. Taylor, B. P. Tsai, and G. G. Meisels, *J. Chem. Phys.*, 1978, **69**, 2547.
[34] T. Baer and P. T. Murray, *J. Chem. Phys.*, 1981, **75**, 4477.
[35] I. Powis, *J. Chem. Soc., Faraday Trans. 2*, 1981, **77**, 1433.
[36] K. M. Johnson, I. Powis, and C. J. Danby, *Int. J. Mass Spectrom. Ion Phys.*, 1979, **32**, 1.
[37] I. Powis, *Chem. Phys.*, 1982, **68**, 251.
[38] P. I. Mansell, C. J. Danby, and I. Powis, *J. Chem. Soc., Faraday Trans. 2*, 1981, **77**, 1449.
[39] R. Bombach, J. Dannacher, J.-P. Stadelmann, and J. Vogt, *Chem. Phys. Lett.*, 1980, **76**, 429.
[40] R. Bombach, J. Dannacher, J.-P. Stadelmann, and J. Vogt, *Int. J. Mass Spectrom. Ion Phys.*, 1981, **40**, 275.
[41] R. Bombach, J. Dannacher, J.-P. Stadelmann, and J. Vogt, '29. Annual Conference on Mass Spectrom. and Allied Topics', Minneapolis, 1981.
[42] R. Bombach, J. Dannacher, J.-P. Stadelmann, and J. Vogt, *Chem. Phys. Lett.*, 1981, **77**, 399.
[43] R. Stockbauer, *Int. J. Mass Spectrom. Ion Phys.*, 1977, **25**, 401.
[44] R. Stockbauer and H. M. Rosenstock, *Int. J. Mass Spectrom. Ion Phys.*, 1978, **27**, 185.
[45] I. Powis, *J. Chem. Soc., Faraday Trans. 2*, 1979, **75**, 1294.
[46] J. H. D. Eland, R. Frey, A. Kuestler, H. Schulte, and B. Brehm, *Int. J. Mass Spectrom. Ion Phys.*, 1976, **22**, 155.
[47] B. P. Tsai, T. Baer, A. S. Werner, and S. F. Stephen, *J. Phys. Chem.*, 1975, **79**, 570.
[48] D. M. Mintz and T. Baer, *J. Chem. Phys.*, 1976, **65**, 2407.
[49] I. Powis, *Chem. Phys.*, 1983, **74**, 421.
[50] T. Baer, L. Squires, and A. S. Werner, *Chem. Phys.*, 1974, **6**, 325.

Table 1 *(cont.)*

Number of atoms	*Molecule*	*References*
	CH_2I_2	47[b,d]
	$CHBr_3$	47[b,d]
	CHI_3	47[b,d]
	CF_4	26,[a,b,g] 51,[a,b,g,k] 52[a,b,g,j,k]
	CF_3Cl	52,[a,b,g,j,k] 53[a,g,j]
	CF_3Br	52,[a,b,g,j,k] 53[a,g,j]
	CF_3I	54[a,b,c]
	CH_2CO	43[c,d]
	HCOOH	55[a,b,c]
6	C_2H_4	56,[b,c,d] 57,[a,g] 58[j]
	C_2D_4	56,[b,c,d] 57[a,g]
	C_2H_3F	59[a,b,c,m]
	C_2H_3Br	60[d,g]
	cis-$C_2H_2F_2$	61,[a,b,c,i,m] 62[a,b,c,i,m]
	trans-$C_2H_2F_2$	61,[a,b,c,i,m] 62[a,b,c,i,m]
	C_4H_2	63[a,b]
	MeOH	64[a,b,c]
	CD_3OD	64,[a,b,c] 65[a,b]
7	SF_6	51[a,b,g,k]
	Allene	44,[b,d,k,l] 66,[b,d] 67[a,b,c]
	Propyne	68[b,d]

[51] I. G. Simm, C. J. Danby, J. H. D. Eland, and P. I. Mansell, *J. Chem. Soc., Faraday Trans. 2*, 1976, **72**, 426.
[52] I. Powis, *Mol. Phys.*, 1980, **39**, 311.
[53] I. Powis and C. J. Danby, *Chem. Phys. Lett.*, 1979, **65**, 390.
[54] R. Bombach, J. Dannacher, J.-P. Stadelmann, J. Vogt, L. R. Thorne, and J. L. Beauchamp, *Chem. Phys.*, 1982, **66**, 403.
[55] Y. Niwa, T. Nishimura, F. Isogai, and T. Tsuchiya, *Chem. Phys. Lett.*, 1980, **74**, 40.
[56] R. Stockbauer and M. G. Inghram, *J. Chem. Phys.*, 1975, **62**, 4862.
[57] P. D. Lightfoot, C. J. Danby, and I. Powis, *Chem. Phys. Lett.*, 1983, **96**, 232.
[58] R. Bombach, J. Dannacher, and J.-P. Stadelmann, *Int. J. Mass Spectrom. Ion Processes*, 1984, **58**, 217.
[59] J. Dannacher, A. Schmelzer, J.-P. Stadelmann, and J. Vogt, *Int. J. Mass Spectrom. Ion Phys.*, 1979, **31**, 175.
[60] B. E. Miller and T. Baer, *Chem. Phys.*, 1984, **85**, 39.
[61] J.-P. Stadelmann and J. Vogt, *Adv. Mass Spectrom.*, 1980, **8**, 47.
[62] J.-P. Stadelmann and J. Vogt, *Int. J. Mass Spectrom. Ion Phys.*, 1980, **35**, 83.
[63] J. Dannacher, E. Heilbronner, J.-P. Stadelmann, and J. Vogt, *Helv. Chim. Acta*, 1979, **62**, 2186.
[64] T. Nishimura, Y. Niwa, T. Tsuchiya, and H. Nozoye, *J. Chem. Phys.*, 1980, **72**, 2222.
[65] Y. Niwa, T. Nishimura, H. Nozoye, and T. Tsuchiya, *Int. J. Mass Spectrom. Ion Phys.*, 1979, **30**, 63.
[66] A. C. Parr, A. J. Jason, and R. Stockbauer, *Int. J. Mass Spectrom. Ion Phys.*, 1978, **26**, 23.
[67] J. Dannacher and J. Vogt, *Helv. Chim. Acta*, 1978, **61**, 361.
[68] A. C. Parr, A. J. Jason, R. Stockbauer, and K. E. McCulloh, *Int. J. Mass Spectrom. Ion Phys.*, 1979, **30**, 319.

Table 1 *(cont.)*

Number of atoms	*Molecule*	*References*
	Cyclopropene	69,[b,d] 70[b,d]
	3-Chloropropyne	71,[b,d,m] 72[b,d,k,m]
	3-Bromopropyne	72[b,d,k,m]
	1-Chloropropyne	73[a,b,c,i,m]
	MeONO	74,[b,d,k,m] 75,[b,d,k] 76[b,d,k]
	$MeNO_2$	75,[b,d,k] 77[a,b,g]
	CD_3NO_2	76,[b,d,k] 77[a,b,g]
	MeCHO	78,[a,b,g,j] 79[a,b,c]
	CH_2—O—CH_2 (ring)	78[a,b,g,j]
	Thiirane	80[c,d,j]
8	C_2H_6	8,[b,c,d] 58[j]
	C_2D_6	8[b,c,d]
	EtCl	71,[b,d,m] 72[b,d,k,m]
	EtBr	60[d,g]
	EtI	81,[d,g,j] 82,[b,d] 83[b,d,l]
	CH_2ClCH_2Cl	71,[b,d,m] 72[b,d,k,m]
	C_2F_6	51,[a,b,g,k] 84,[a,b,k] 85,[a,b,k] 86,[b,c,d,l] 87[a,b,g,k]
9	MeHgMe	88[a,b,g]
	Penta-1,3-diyne	63[a,b,i,m]

69 A. C. Parr, A. J. Jason, and R. Stockbauer, *Int. J. Mass Spectrom. Ion Phys.*, 1980, **33**, 243.
70 A. C. Parr, A. J. Jason, R. Stockbauer, and K. E. McCulloh, *Adv. Mass Spectrom.*, 1980, **8**, 62.
71 T. Baer, A. S. Werner, and B. P. Tsai, *J. Chem. Phys.*, 1975, **62**, 2497.
72 B. P. Tsai, A. S. Werner, and T. Baer, *J. Chem. Phys.*, 1975, **63**, 4384.
73 J. Dannacher and J.-P. Stadelmann, *Chem. Phys.*, 1980, **48**, 79.
74 G. G. Meisels, T. Hsieh, and P. Gilman, *J. Chem. Phys.*, 1980, **73**, 4126.
75 J. P. Gilman, T. Hsieh, and G. G. Meisels, *J. Chem. Phys.*, 1983, **78**, 1174.
76 J. P. Gilman, T. Hsieh, and G. G. Meisels, *J. Chem. Phys.*, 1983, **78**, 3767.
77 I. K. Ogden, N. Shaw, C. J. Danby, and I. Powis, *Int. J. Mass Spectrom. Ion Processes*, 1983, **54**, 41.
78 K. Johnson, I. Powis, and C. J. Danby, *Chem. Phys.*, 1982, **70**, 329.
79 R. Bombach, J.-P. Stadelmann, and J. Vogt, *Chem. Phys.*, 1981, **60**, 293.
80 J. J. Butler and T. Baer, *J. Am. Chem. Soc.*, 1982, **104**, 5016.
81 T. Baer, U. Buechler, and C. E. Klots, *J. Chim. Phys.*, 1980, **77**, 739.
82 T. Baer, *J. Am. Chem. Soc.*, 1980, **102**, 2482.
83 H. M. Rosenstock, R. Buff, M. A. A. Ferreira, S. G. Lias, A. C. Parr, R. L. Stockbauer, and J. L. Holmes, *J. Am. Chem. Soc.*, 1982, **104**, 2337.
84 I. G. Simm, C. J. Danby, and J. H. D. Eland, *J. Chem. Soc., Chem. Commun.*, 1973, 832.
85 I. G. Simm, C. J. Danby, and J. H. D. Eland, *Int. J. Mass Spectrom. Ion Phys.*, 1974, **14**, 285.
86 M. G. Inghram, G. R. Hanson, and R. Stockbauer, *Int. J. Mass Spectrom. Ion Phys.*, 1980, **33**, 253.
87 I. G. Simm and C. J. Danby, *J. Chem. Soc., Faraday Trans. 2*, 1976, **72**, 860.
88 C. S. T. Cant, C. J. Danby, and J. H. D. Eland, *J. Chem. Soc., Faraday Trans. 2*, 1975, **71**, 1015.

Table 1 *(cont.)*

Number of atoms	*Molecule*	*References*
	[1-^{2}H]Penta-1,3-diyne	63^{a,b,i,m}
	[5-$^2H_3$]Penta-1,3-diyne	63^{a,b,i,m}
	[$^2H_4$]Penta-1,3-diyne	63^{a,b,i,m}
	EtOH	89^{a,b,c}
	C_2D_5OD	89^{a,b,c}
	Thiophene	90^{b,c,d,j,m}
	Furan	91^{b,c,d,j,m}
	3-Butyne-2-one	91^{b,c,d,j,m}
	Pyrrole	92^{b,d,j,m}
10	Methacrylonitrile	92^{b,d,j,m}
	Allyl cyanide	92^{b,d,j,m}
	Cyclopropyl cyanide	92^{b,d,j,m}
	Acetone	88^{a,b,g} 93,^{c,d} 94,^{d,g,j} 95,^{a,b,c,g} 96,^{a,g,j} 97,^{a,b} 98,^{a,b,c} 99^{a,b,c,k}
	Cyclopropanol	100^{a,b,c}
	Allyl alcohol	100^{a,b,c}
	Propionaldehyde	101^{a,b,c}
	1,2-Epoxypropane	98,^{a,b,c} 99^{a,b,c,k}
	Buta-1,3-diene	102,^{b,d,j,m} 103,^{d,g,j} 104,^{b,d,m} 105,^{a,b} 106^{a,b,j,l,m}
	Buta-1,2-diene	102^{b,d,j,m}
	1-Butyne	102^{b,d,j,m}
	2-Butyne	102^{b,d,j,m}
	Cyclobutene	102^{b,d,j,m}
	2-Methylcyclopropene	107^{b,d,j}
	Bicyclobutane	108^{a,b,j,m}

[89] Y. Niwa, T. Nishimura, and T. Tsuchiya, *Int. J. Mass Spectrom. Ion Phys.*, 1982, **42**, 91.
[90] J. J. Butler and T. Baer, *J. Am. Chem. Soc.*, 1980, **102**, 6764.
[91] G. D. Willett and T. Baer, *J. Am. Chem. Soc.*, 1980, **102**, 6769.
[92] G. D. Willett and T. Baer, *J. Am. Chem. Soc.*, 1980, **102**, 6774.
[93] R. Stockbauer, *Int. J. Mass Spectrom. Ion Phys.*, 1977, **25**, 89.
[94] D. M. Mintz and T. Baer, *Int. J. Mass Spectrom. Ion Phys.*, 1977, **25**, 39.
[95] I. Powis and C. J. Danby, *Int. J. Mass Spectrom. Ion Phys.*, 1979, **32**, 27.
[96] K. Johnson, I. Powis, and C. J. Danby, *Chem. Phys.*, 1981, **63**, 1.
[97] K. Johnson, I. Powis, and C. J. Danby, *Chem. Phys. Lett.*, 1982, **89**, 177.
[98] J.-P. Stadelmann, *Chem. Phys. Lett.*, 1982, **89**, 174.
[99] R. Bombach, J.-P. Stadelmann, and J. Vogt, *Chem. Phys.*, 1982, **72**, 259.
[100] R. Bombach, J. Dannacher, E. Honegger, J.-P. Stadelmann, and R. Neier, *Chem. Phys.*, 1983, **82**, 459.
[101] J. Dannacher and J.-P. Stadelmann, to be published.
[102] A. S. Werner and T. Baer, *J. Chem. Phys.*, 1975, **62**, 2900.
[103] C. E. Klots, D. Mintz, and T. Baer, *J. Chem. Phys.*, 1977, **66**, 5100.
[104] T. Baer, D. Smith, B. P. Tsai, and A. S. Werner, *Adv. Mass Spectrom.*, 1978, **7**, 56.
[105] J. Dannacher, J.-P. Flamme, J.-P. Stadelmann, and J. Vogt, *Chem. Phys.*, 1980, **51**, 189.
[106] R. Bombach, J. Dannacher, and J.-P. Stadelmann, *J. Am. Chem. Soc.*, 1983, **105**, 1824.
[107] T. Baer, *J. Electron Spectrosc. Relat. Phenom.*, 1979, **15**, 225.
[108] R. Bombach, J. Dannacher, J.-P. Stadelmann, and R. Neier, *Helv. Chim. Acta*, 1983, **66**, 701.

Table 1 *(cont.)*

Number of atoms	*Molecule*	*References*
	Thietane	80[c,d,j]
	Pyridazine	109[b,d,j,l,m]
	Pyrimidine	109[b,d,j,l,m]
	Pyrazine	109[b,d,j,l,m]
11	C_3H_8	110,[a,b,c] 111,[b,d,j] 112[b,d]
	$MeCD_2Me$	111[b,d,j]
	$MeCH_2CD_3$	111[b,d,j]
	C_3D_8	111[b,d,j]
	$Me^{13}CH_2Me$	112[b,d]
	CH_2ICH_2Me	82,[b,d] 83,[b,d,j,l] 113[c,d,g,j]
	CH_2BrCH_2Me	83[b,d,j,l]
	MeCHBrMe	83[b,d,j,l]
	MeCHIMe	83,[b,d,j,l] 113[c,d,g,j]
	Pyridine	114,[a,b,j,m] 115[b,d,j,l,m]
12	Benzene	116,[a,b,k,m] 117,[a,b] 118,[a,b,c] 119[b,d,j,m]
	Hexa-1,5-diyne	119[b,d,j,m]
	Hexa-2,4-diyne	119,[b,d,j,m] 120,[a,b,c,i] 121[a,b,c,i,m]
	$[^2H_6]$Hexa-2,4-diyne	122[a,b,c,i,m]
	Hexa-1,3-diyne	123[a,b,c,i,m]
	But-1-ene	104,[b,d] 124,[b,d] 125[b,d]
	cis-But-2-ene	104,[b,d] 124,[b,d] 125,[b,d] 126[b,d]
	trans-But-2-ene	104,[b,d] 124,[b,d] 125,[b,d] 126[b,d]
	Isobutene	104,[b,d] 124,[b,d] 125[b,d]

[109] R. Buff and J. Dannacher, *Int. J. Mass Spectrom. Ion Processes*, 1984, **62**, 1.
[110] B. Brehm, J. H. D. Eland, R. Frey, and H. Schulte, *Int. J. Mass Spectrom. Ion Phys.*, 1976, **21**, 373.
[111] R. Stockbauer and M. G. Inghram, *J. Chem. Phys.*, 1976, **65**, 4081.
[112] J. P. Gilman, T. Hsieh, and G. G. Meisels, *J. Chem. Phys.*, 1982, **76**, 3497.
[113] W. A. Brand, T. Baer, and C. E. Klots, *Chem. Phys.*, 1983, **76**, 111.
[114] J. H. D. Eland, J. Berkowitz, H. Schulte, and R. Frey, *Int. J. Mass Spectrom. Ion Phys.*, 1978, **28**, 298.
[115] H. M. Rosenstock, R. Stockbauer, and A. C. Parr, *Int. J. Mass Spectrom. Ion Phys.*, 1981, **38**, 323.
[116] J. H. D. Eland and H. Schulte, *J. Chem. Phys.*, 1975, **62**, 3835.
[117] J. H. D. Eland, *Int. J. Mass Spectrom. Ion Phys.*, 1974, **13**, 457.
[118] J. H. D. Eland, R. Frey, H. Schulte, and B. Brehm, *Int. J. Mass Spectrom. Ion Phys.*, 1976, **21**, 209.
[119] T. Baer, G. D. Willett, D. Smith, and J. S. Phillips, *J. Chem. Phys.*, 1979, **70**, 4076.
[120] J. Dannacher, *Chem. Phys.*, 1978, **29**, 339.
[121] J. Dannacher, *Adv. Mass Spectrom.*, 1980, **8**, 37.
[122] J. Dannacher, J.-P. Stadelmann, and J. Vogt, *J. Chem. Phys.*, 1980, **74**, 2094.
[123] J. Dannacher, J.-P. Stadelmann, and J. Vogt, *Int. J. Mass Spectrom. Ion Phys.*, 1981, **38**, 69.
[124] G. G. Meisels, M. J. Weiss, T. Hsieh, and G. L. M. Verboom, 'Proc. 26th Annu. Conf. Mass Spectrom. All. Topics', St. Louis, 1978.
[125] T. Hsieh, J. P. Gilman, M. J. Weiss, and G. G. Meisels, *J. Phys. Chem.*, 1981, **85**, 2722.

Table 1 *(cont.)*

Number of atoms	*Molecule*	*References*
	Cyclobutane	124,[b,d] 125[b,d]
	Methylcyclopropane	104,[b,d] 124,[b,d] 125[b,d]
	Chlorobenzene	104,[b,d] 127,[b,d,j,m] 128[b,d,j,l,m]
	Bromobenzene	104,[b,d] 127,[b,d,j,m] 129,[b,d,j,l,m] 130[d,j,m]
	[$^2H_5$]Bromobenzene	130[d,j,m]
	Iodobenzene	104,[b,d] 127,[b,d,j,m] 131[a,b,d,j,l,m]
13	Benzonitrile	116,[a,b,k,m] 132[b,d,j,l,m]
	Tetrahydrothiophene	80,[c,d,j] 133[d,j,m]
	$Cr(CO)_6$	65[a,b]
14	*cis*-Hexa-1,3,5-triene	134[a,b,c,i,m]
	trans-Hexa-1,3,5-triene	134,[a,b,c,i,m] 135[b,d,j,l,m]
	Cyclohexa-1,3-diene	135[b,d,j,l,m]
	Cyclohexa-1,4-diene	135[b,d,j,l,m]
	1,4-Dioxane	136[b,d,j,m]
	Butanoic acid	137[b,d,j,m]
	Ethyl acetate	138[b,d,j,m]
	Aniline	139[b,d,j,m]
	1,4-Oxathian	140[d,j,m]
	1,4-Dithian	140[d,j,m]
	Me_3SiCl	141[b,d,j,m]
	Me_3SiBr	141[b,d,j,m]
	Me_3SiI	141[b,d,j,m]

[126] G. G. Meisels, G. M. L. Verboom, M. J. Weiss, and T. Hsieh, *J. Am. Chem. Soc.*, 1979, **101**, 7189.
[127] T. Baer, B. P. Tsai, D. Smith, and P. T. Murray, *J. Chem. Phys.*, 1976, **64**, 2460.
[128] H. M. Rosenstock, R. Stockbauer, and A. C. Parr, *J. Chem. Phys.*, 1979, **71**, 3708.
[129] H. M. Rosenstock, R. Stockbauer, and A. C. Parr, *J. Chem. Phys.*, 1980, **73**, 773.
[130] T. Baer and R. Kury, *Chem. Phys. Lett.*, 1983, **92**, 659.
[131] J. Dannacher, H. M. Rosenstock, R. Buff, A. C. Parr, R. L. Stockbauer, R. Bombach, and J.-P. Stadelmann, *Chem. Phys.*, 1983, **75**, 23.
[132] H. M. Rosenstock, R. Stockbauer, and A. C. Parr, *J. Chim. Phys.*, 1980, **77**, 745.
[133] J. J. Butler and T. Baer, *Org. Mass Spectrom.*, 1983, **18**, 248.
[134] M. Allan, J. Dannacher, and J. P. Maier, *J. Chem. Phys.*, 1980, **73**, 3114.
[135] S. G. Lias and P. Ausloos, to be published.
[136] M. L. Freiser-Monteiro, L. Freiser-Monteiro, J. J. Butler, T. Baer, and J. R. Hass, *J. Phys. Chem.*, 1982, **86**, 739.
[137] L. Freiser-Monteiro, M. L. Freiser-Monteiro, J. J. Butler, and T. Baer, *J. Phys. Chem.*, 1982, **86**, 747.
[138] J. J. Butler, M. Freiser-Monteiro, M. L. Freiser-Monteiro, T. Baer, and J. R. Hass, *J. Phys. Chem.*, 1982, **86**, 753.
[139] T. Baer and T. E. Carney, *J. Chem. Phys.*, 1982, **76**, 1304.
[140] T. Baer, W. A. Brand, T. L. Bunn, and J. J. Butler, *Faraday Discuss. Chem. Soc.*, 1983, **75**, 45.
[141] L. Szepes and T. Baer, *J. Am. Chem. Soc.*, 1984, **106**, 273.

Table 1 *(cont.)*

Number of atoms	*Molecule*	*References*
15	Toluene	142,[a,b,j,l,m] 143[a,b,j,l,m]
	Cyclopentane	140,[d,j,m] 144[b,d,j,m]
	Pent-1-ene	144[b,d,j,m]
	trans-Pent-2-ene	140,[d,j,m] 144[b,d,j,m]
	3-Methylbut-1-ene	144[b,d,j,m]
	2-Methylbut-1-ene	144[b,d,j,m]
	2-Methylbut-2-ene	140,[d,j,m] 144[b,d,j,m]
16	Styrene	145[b,d,j,m]
	Cyclo-octatetraene	145[b,d,j,m]
17	Neopentane	110[a,b,c]
	all-*trans*-Hepta-1,3,5-triene	134[a,b,c,i,m]
	Me_4Si	143[b,d,j,m]
24	t-Butylbenzene	146[b,d,j,m]
26	Me_6Si_2	141[b,d,j,m]

[a] Fixed-wavelength technique. [b] Branching ratios, breakdown curves. [c] Average kinetic-energy release. [d] Variable-wavelength technique. [e] Application to bimolecular processes. [f] Synchrotron radiation used. [g] Distribution of the kinetic energy released. [h] He-II radiation used. [i] Competition between emission and fragmentation. [j] Analysis of the experimental data by means of statistical theory. [k] State-specific decay processes. [l] Time-resolved experiments. [m] Determination of ion lifetimes and rate/energy functions

and to familiarize mass spectrometrists with the fundamentals of and the results obtained from this unusually powerful technique. To this end, the basic aspects of the experiment and the essence of the evaluation of the coincidence data are described first. Although purposely kept as simple as possible, this part of the article necessarily contains a few rather abstract concepts and definitions that need to be clarified by means of a suitable example: for scientific as well as didactic reasons, the cation of molecular oxygen is the species chosen. The subsequent sections are concerned with our understanding of the decomposition phenomena of 'larger' and thus mass-spectrometrically more relevant cations. After a brief overview of the principal aspects of the statistical theory of mass spectra, there follows a critical discussion of the pertinent literature. At the outset investigations of the least endothermic fragmentation pathway are reviewed, and later on the complexities arising from competing and/or consecutive frag-

[142] R. Bombach, J. Dannacher, and J.-P. Stadelmann, *Chem. Phys. Lett.*, 1983, **95**, 259.
[143] R. Bombach, J. Dannacher, and J.-P. Stadelmann, *J. Am. Chem. Soc.*, 1983, **105**, 4205.
[144] W. A. Brand and T. Baer, *J. Am. Chem. Soc.*, 1984, **106**, 3154.
[145] D. Smith, T. Baer, G. D. Willett, and R. C. Ormerod, *Int. J. Mass Spectrom. Ion Phys.*, 1979, **30**, 155.
[146] W. A. Brand and T. Baer, *Int. J. Mass Spectrom. Ion Phys.*, **1983**, 49, 103.

mentations are summarized. A further section deals with the measurement and interpretation of fragment-ion translational energies, and is followed by a discussion of the importance of radiative relaxation and isomerization reactions. The present article concludes with a few remarks on non-statistical phenomena as reported in a number of coincidence studies.

2 The Essence of Photoelectron–Photoion Coincidence Experiments

The hitherto developed and currently used PEPICO spectrometers can be grouped into so-called fixed- and variable-wavelength instruments, respectively. The following comparison of the main features of both sorts of apparatus reveals the complementary character of the two techniques. To illustrate these considerations, typical values for important experimental quantities are quoted. Unless otherwise stated, these values refer to the Reporters' He-Iα fixed-wavelength instrument in Basel (*cf.* Figure 1) and to the variable-wavelength spectrometer at the National Bureau of Standards in Washington D.C. (*cf.* Figure 2), respectively.

Fixed-wavelength Instruments. – This version of the experiment owes its name to the fact that the wavelength of the ionizing photons is kept constant. The standard photon source used in all the mass-spectrometrically relevant fixed-wavelength coincidence studies consists of a d.c. discharge in pure helium. When suitable conditions are selected, more than 99% of the emitted far-UV photons are He-Iα quanta with an energy of 21.22 eV. The only significant 'spectral impurity' is the He-Iβ line, equivalent to a photon energy of 23.09 eV. So far the presence of these somewhat more energetic quanta has mostly been neglected in evaluating the data, since more significant sources of error were usually involved. Apart from being adequately monoenergetic, the far-UV radiation emitted by such a helium-resonance lamp is also of remarkably high intensity.[147] Therefore, extremely narrow ($\theta \approx 0.1$ mm) but still sufficiently intense photon beams can be produced. In combination with a suitable inlet system, *e.g.* a gas-jet, an essentially point-shaped ionization zone is obtained. The importance of keeping this ionization zone as tiny as possible can hardly be overestimated, inasmuch as deficiencies here – in the heart of a coincidence spectrometer – lead inevitably to data of inferior quality. This becomes immediately evident when a few aspects of the experiment are recalled. First of all, ionization of the molecules takes place in an electric field. Moreover, the whole experiment depends to a very large extent on accurate measurement of the times that it takes for the ions to travel the distance between the point of their formation and the ion detector. In general, ions generated at different positions gain different energies in the electric field, have unequal distances to travel, and may be subject to unequal discrimination effects. In other words, two ions possessing identical properties at the moment of their formation may nevertheless give rise to a different response if they were formed at distinct positions. Hence the ionization zone must be kept small to minimize the corresponding distortion of the measured time-of-flight distributions. Similar arguments apply to the energy

[147] J. R. Samson, 'Techniques of Vacuum Ultraviolet Spectroscopy', J. Wiley and Sons, New York, 1967.

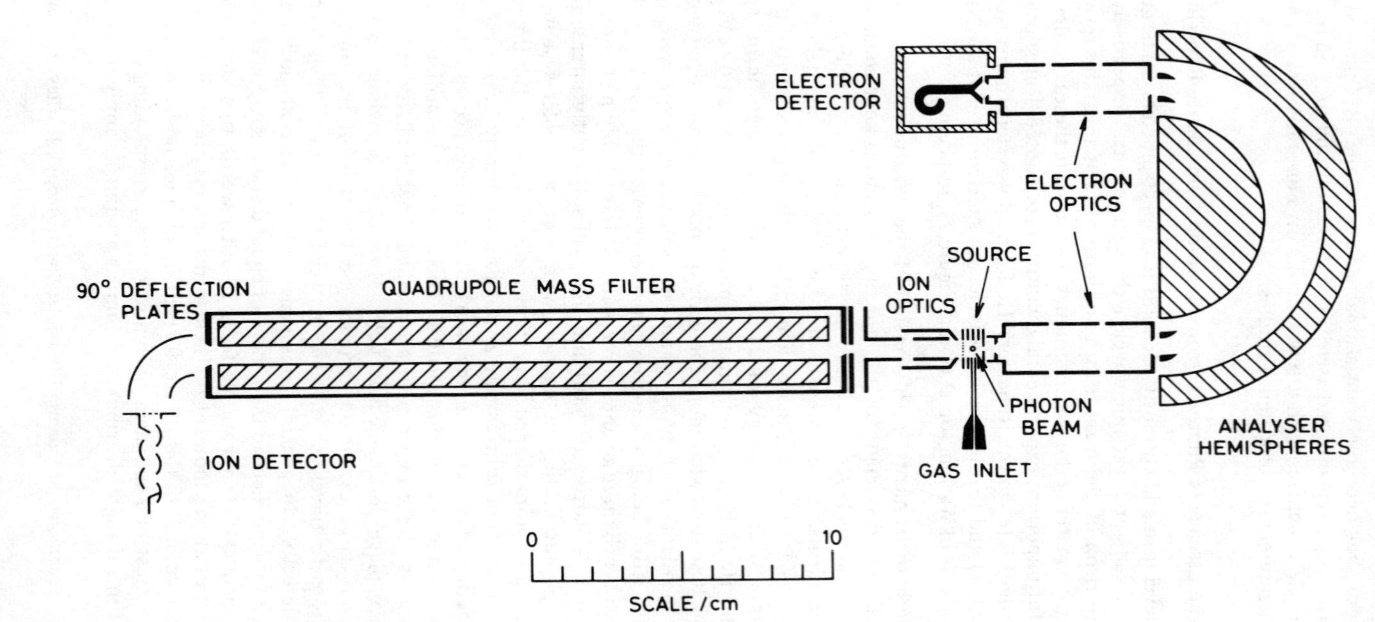

Figure 1 *Fixed-wavelength coincidence spectrometer, Basel, Switzerland*
(Reproduced with permission from *Int. J. Mass Spectrom. Ion Phys.*, 1982, 43, 211)

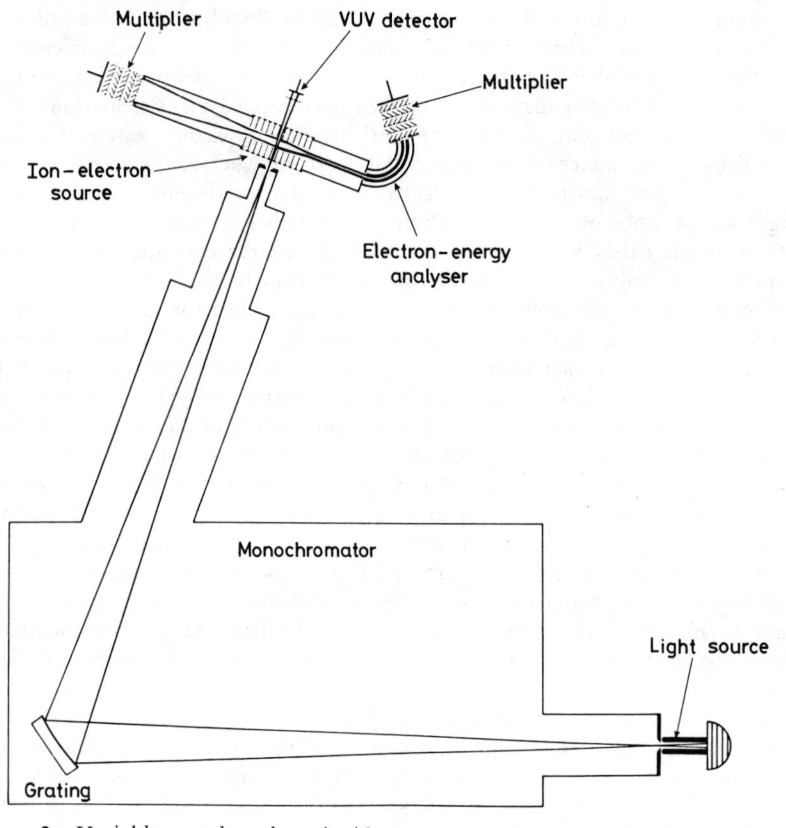

Figure 2 *Variable-wavelength coincidence spectrometer, National Bureau of Standards, Washington D.C., U.S.A.*
(Reproduced with permission from *Int. J. Mass Spectrom. Ion Phys.*, 1977, **25**, 89)

analysis and collection efficiency of the photoelectrons. In addition, confining ion formation to a very small volume is the key to ensuring that the electron and ion analysers not only share a common source but also sample the same ionization events. Note that a serious reduction of the attainable signal-to-noise ratio occurs when the solid angles accessible for ion and electron analysis, respectively, only partially overlap. Lastly, a small ionization region makes it possible to extract most of the charged particles from the source and to transfer them efficiently into the respective analysers by means of suitable ion and electron optics.

Referring to the spectrometer depicted in Figure 1, a constant electric field of typically 2 V cm^{-1} is applied to the source. This value is sufficient for effective extraction of most of the ions, without significantly degrading the photoelectron-energy resolution. A hemispherical analyser equipped with suitable entrance and exit lens systems is used for the energy analysis of the photoelectrons. The transmission function of this analyser is a Gaussian distribution with a full width at half-maximum (fwhm) of 50 meV. The electron transmission coefficient, f_e, *i.e.* the relative probability that a photoelectron with an energy corresponding to the pass energy of the analyser is detected, amounts to 1%. Calculations of the relevant potentials and trajectories reveal that this f_e value is essentially determined by the diameter of the source exit aperture, whereas the actual focusing/analyser system transmits photoelectrons of appropriate energy and entrance angle with a probability of unity. Therefore a classical intensity/resolution trade-off situation exists with the diameter of the source exit aperture as crucial parameter. Possibly less obvious is the connection between the magnitude of this diameter and the background count rate in the electron channel. When the source exit aperture is continuously increased there is at first a gain in intensity proportional to the increased solid angle and a concomitant reduction of the energy resolution. However, as soon as the diameter of the exit aperture exceeds a certain value, the resulting width of the photoelectron beam becomes larger than the ideal acceptance angle of the analyser. Thus, the analyser itself begins to discriminate against electrons of appropriate energy but with adverse entrance conditions. These electrons are scattered and reflected on the analyser surfaces and contribute significantly to the background count rate. Therefore, the diameter of the source exit aperture of the described instrument has been calculated to match the acceptance angle of the analyser, and only virtual slits have been used to focus the photoelectrons. This has resulted in a remarkably low background count rate of only a few tenths per second at typical signal intensities of about 10^2–10^3 counts per second. Although a value of $f_e \approx 1\%$ at a resolution of 50 meV is an unusually large efficiency relative to other fixed-wavelength coincidence instruments and in particular compared with conventional photoelectron spectrometers, collecting only 1% of all of the generated photoelectrons is the main reason for the slowness of the method. While optimizing the described experimental arrangement may result in at best a five-fold increase of f_e without significantly degrading the energy resolution, achieving substantially larger f_e values requires a vastly different design of the electron-energy analyser system. In this context it is recalled that photoionizing gaseous molecules with He-Iα resonance radiation generates photoelectrons of various energies. By relying on conventional means, only a very small fraction of these photoelectrons possesses appropriate energy to be transmitted by the analyser, and of these, as described above, only one in a hundred will reach the detector. To measure the kinetic-energy distribution of these photoelectrons the pass energy of the analyser must therefore be scanned, typically over a range of about 10 eV. An elegant and very promising alternative to make the fixed-wavelength experiment considerably faster is based on the use of position-sensitive detectors. The fundamental idea of this approach is to disperse the multi-energetic photoelectron beam in the electrostatic field of a preferentially second-order focusing deflection-type analyser and to detect simultaneously n of the beam's components with particular kinetic energies and thus at geometrically distinct positions. Whereas this method does not alter f_e, it reduces by a factor of n the time necessary for an energy sweep. A conservative estimate for n implies that in this way the time requirements of the experiment could be reduced by at least one order of magnitude. Since the energy of the ionizing quanta is kept constant, the partial-photoionization cross-sections remain constant and correspond essentially to direct-photoionization processes only. Moreover, when the measurements are made at the magic angle, the photoelectron intensity becomes independent of the anisotropy parameter, β. Therefore, by accounting for the energy dependence of f_e, the photoelectron-energy distribution directly reflects the energy-deposition function for He-Iα photoionization.

In coincidence spectroscopy the mass and energy analysis of the photoions is founded, exclusively or at least in part, on an exact measurement of the ions' times-of-flight. Pure time-of-flight methods are appropriate where low mass resolution is acceptable and where a particularly simple design is considered to be important. Furthermore, it is worth pointing out that relying solely on time-of-flight techniques makes it possible to monitor simultaneously all types of fragment ions characterized by their respective mass. For most purposes, in particular when organic radical cations are to be studied, the use of an additional mass selector is necessary, not only for achieving sufficient mass resolution but also for attaining a significantly better signal-to-noise ratio of the coincidence count rate. For a number of reasons, quadrupole mass filters are especially suitable for realizing such an improved mass analysis in a coincidence spectrometer. Quadrupole mass filters have very high inherent transmission, are unusually compact mass selectors, and are capable of analysing ions of various energies. The latter property can be used to carry out time-resolved coincidence measurements by varying the residence time of the internal-energy- and mass-selected photoions in the mass filter. Extensive work in the Reporters' laboratory has demonstrated that the use of a quadrupole system does not significantly broaden the time-of-flight distributions and in fact increases the ion transmission coefficient, provided that the ionization zone is compact and that perfect alignment of the quadrupole axis is achieved. Note that using a quadrupole mass filter does not exclude using pure time-of-flight techniques. The quadrupole system can function either as a conventional drift tube with a constant potential on all four rods or as a classical filter, when operated in its integral mode, *i.e.* transmitting only ions with a mass that exceeds a predetermined value. In addition, since the resolution/transmission trade-off is particularly easily adjusted by changing the ratio of two voltages, there is also a high degree of flexiblity when the quadrupole is used to select a single mass or a small range of masses.

Another experimental parameter, at least as important as mass and time resolution in the ion channel, is the ion-transmission coefficient, $f_i(m_k, \vec{v}, \vec{r})$. This analogue to f_e is defined as the relative probability that a photoion of mass m_k and initial velocity vector $\vec{v}$ and with an original place for formation characterized by a position vector $\vec{r}$ is detected. Although the importance of $f_i(m_k, \vec{v}, \vec{r})$ has been pointed out in the very first coincidence studies, a number of discrepancies in the literature can be traced back to invalid assumptions regarding this important quantity. At present, generally valid calculations of $f_i(m_k, \vec{v}, \vec{r})$ are not feasible, and thus approximate methods must be used to estimate $f_i(m_k, \vec{v}, \vec{r})$. The significance of $f_i(m_k, \vec{v}, \vec{r})$ is discussed further in the section on coincidence-data evaluation. However, to give the reader an idea of the orders of magnitude involved, the apparatus depicted in Figure 1 typically operates with $f_i(m_K, \vec{v}, \vec{r})$ values between 0.25 and 0.60 for thermal molecular ions of mass not exceeding 300. The key role of a highly compact ionization zone for achieving such high $f_i(m_K, \vec{v}, \vec{r})$ values has already been mentioned.

Variable-wavelength Instruments. – As its name implies, this version of the experiment is based on photoionizing the sample molecules by means of far-UV quanta of variable energy. Again, there exists a standard laboratory 'light source' used in all those studies that are significant to the field of organic mass spectrometry. A low-pressure capillary spark discharge in hydrogen provides quanta with energies of up to ~13 eV, and the more energetic photons are obtained from the (experimentally more exacting) Hopfield continuum of helium, which extends essentially up to the energy of the He-Iα resonance line. The emitted radiation is passed through a conventional monochromator. As compared with the fixed-wavelength approach, the potentially accessible energy range is thus the same. However, the two methods differ significantly in the intensity and the narrowness of the principal photon beam, as well as in complications due to scattered photons of various energies and the nature of the ionization processes. Relative to the He-Iα resonance line, the intensity of these polychromatic photon sources is a number of orders of magnitude lower even before monochromator losses. At typical monochromator slit widths, photon beams with fwhm between 1 and 2 Å, equivalent to an energy resolution of between 5 and 50 meV, are produced. Serious intensity problems arise particularly at those wavelengths where the 'light source' emits only very few quanta. To achieve a sufficient ionization rate, the photon/molecule collision region must be enlarged and/or higher sample pressures used. Being concerned with unimolecular reactions, bimolecular processes must, of course, be avoided, so increasing the pressure is not generally practicable. Therefore, a photon beam of large cross-section is used. Typically, the ionization region of a variable-wavelength instrument extends several millimetres along the source drawout field and a couple of centimetres in the other two directions. As a consequence, the above-mentioned advantages of keeping the ionization region as compact as possible are lost.

A significant, but nevertheless rarely discussed, problem inherent in the variable-wavelength technique concerns the presence of scattered photons. Owing to non-ideal effects of the monochromator, photons other than those of the selected wavelength may reach the source region. The intensity and spectral distribution of this so-called scattered light[147] is hard to ascertain. The number of scattered photons depends on the nature and operating conditions of the light source, on the properties of the monochromator, and on the intensity and wavelength of the selected main photon beam. Clearly, the 'spectral purity' of the ionizing photon beam becomes a function of the photon energy. Furthermore, the importance of scattered light for coincidence measurements will also depend on the properties of the ionized molecules. In other words, when the monochromator is set to a particular photon energy, scattered light of given intensity and spectral distribution may seriously distort the data for one compound and essentially not affect the data for another, simply because of the different partial-photoionization cross-sections of the two species.

The major mechanisms for photoionizing gaseous molecules are direct ionization and autoionization. In the former process a sufficiently energetic photon imparts its entire energy to a molecule, which then rapidly ionizes by ejecting a photoelectron. Depending on the relevant partial-photoionization cross-sections, the generated molecular cation is formed in a more or less excited state and the photoelectron is ejected with a correspondingly smaller or larger translational energy. Direct photoionization is not a resonance process, since the free photoelectron can carry whatever kinetic energy is needed to account for the conservation of energy. On the other hand, autoionization is a two-step mechanism initiated by a transition to a superexcited state of the neutral molecule, whose energy exceeds the ionization limit. This is a resonance process that can only be brought about by photons of a particular energy and that is governed by the normal optical-selection rules. In the second step the superexcited neutral molecule spontaneously autoionizes by emitting a photoelectron. Whereas the same ionic states as in direct ionization are involved, the relative probabilities for their population through autoionization are generally distinctly different depending on the lifetime and the geometry of the superexcited neutral molecule. Since the energy of the ionizing photons is kept constant and because it is larger than the energy of most of the relevant Rydberg states, the He-Iα fixed-wavelength instruments are 'blind' at those energies where the photoelectron intensity is essentially zero owing to a vanishing Franck–Condon factor for the corresponding direct-ionization process. In the variable-wavelength experiment all autoionizing states are usually encountered and hence both ionization mechanisms can contribute to the total ionization rate at any photon energy. Numerous photoionization efficiency curves have revealed that the photoionization cross-section can fluctuate widely owing to autoionization phenomena. Autoionization may affect the variable-wavelength coincidence data in two ways. First of all, it makes the experiment especially vulnerable to scattered-light effects because even a relatively small number of scattered photons of a particular energy may cause about as many ionization events as the selected main beam. Secondly, the original thermal-energy distribution of the molecules may be distorted on autoionization. On the other hand, high vibrational levels of the electronic states of the ions, which are mostly inaccessible to direct ionization, can frequently be populated through autoionization mechanisms. In general, variable-wavelength experiments are also feasible where the He-Iα photoelectron spectrum shows a Franck–Condon gap. The variable-wavelength experiment is thus capable of making any energy input to the ion between zero and the maximum quantum energy of the 'light source'. However, at those energies where the electron count rate of the He-Iα photoelectron spectrum is essentially zero, the variable-wavelength data are significantly less accurate because of the reasons given above and incomplete discrimination of the steradiancy analyser towards energetic electrons (see below).

The principal component of the electron-energy analyser of a variable-wavelength instrument is a so-called steradiancy analyser,[148] almost completely pervious to the generated threshold photoelectrons and strongly discriminating against energetic ones. At first glance it is tempting to rate the variable-wavelength

[148] T. Baer, W. B. Peatman, and E. W. Schlag, *Chem. Phys. Lett.*, 1969, **4**, 243; R. Spohr, P. M. Guyon, W. A. Chupka, and J. Berkowitz, *Rev. Sci. Instrum.*, 1971, **43**, 1872.

technique as the superior variant in view of its potentially higher electron-energy resolution and its ability for selecting threshold photoelectrons with essentially 100% efficiency. However, the major problem with a steradiancy analyser is that its transmission function never vanishes as the electron energy increases because there is no discrimination against those electrons whose initial velocity vectors are parallel to the source field. This hot-electron tail can be reduced significantly if an additional deflection-type analyser is used, as shown in Figure 2.[8] Although this deflection-type analyser may be operated at moderate resolving power, the quoted modification inevitably reduces the electron transmission coefficient significantly. Usually the shape of the $^2P_{1/2}$ threshold-photoelectron band of krypton or xenon is used as approximate transmission function of the electron-energy analyser system. Transmission functions defined in this way are strongly peaked at very low electron energies, slightly broadened owing to finite photon- and electron-energy resolution, and markedly tailed towards higher acceptance energy as a result of the detection of some of these hot electrons. The reported variable-wavelength coincidence data on positive ions consisting of at least six atoms have been measured typically with fwhm $\approx$ 30 meV. Identifying a particular threshold-photoelectron rare-gas signal with the electron-energy transmission function of a pure steradiancy analyser is acceptable as far as the fwhm is concerned, but it is clearly inadequate to characterize the high-energy tail. More recently, an alternative method that is purely computational (and consequently preserves high transmission) has therefore been devised[149] to determine the entire transmission function and subsequently to use it to deconvolute the coincidence data. The proposed approach involves extensive calculations and a variety of calibration experiments on CO/Ar/Xe gas mixtures of known compositions and photoionization cross-sections. Since the He-Iα photoelectron spectrum is used as an approximate energy-deposition function for photoionization, the method is expected to be the more successful the closer the He-Iα and threshold-photoelectron spectra agree, as in this case the implicitly assumed step-function threshold law appears to be justified. As an illustration, the technique was applied to correct the effect of the 'hot-electron tail' on the breakdown diagrams of the propane cation, diverse $C_4H_8^{+\bullet}$ isomers, and the $CF_3I^{+\bullet}$ ion.[149] Where possible, comparison with the He-Iα coincidence data was made. This study has revealed that incomplete discrimination towards energetic photoelectrons may dramatically distort ionic-breakdown diagrams. Most notably, the extent of distortion depends in general on the relevant partial-photoionization cross-sections for population of the various electronic states accessible. Therefore it is recommended that those variable-wavelength data that have not been corrected for this 'hot-electron effect' are interpreted with caution.

Despite their very large transmission factors, the signal-to-noise ratio in the photoelectron channel of variable-wavelength instruments rarely exceeds a value of ten, with the exception of some of the most intense photoelectron bands

[149] T. Hsieh, J. P. Gilman, M. J. Weiss, G. G. Meisels, and P. M. Hierl, *Int. J. Mass Spectrom. Ion Phys.*, 1980, **36**, 317; J. P. Gilman, T. Hsieh, and G. G. Meisels, *Int. J. Mass Spectrom. Ion Phys.*, 1983, **51**, 513.

known. On the one hand, this is a direct consequence of the limited number of photoionization events produced by too few photon/molecule collisions (see above). On the other, working with photoelectrons of very small kinetic energy is particularly susceptible to distortion by various scattering processes producing electrons of low kinetic energy and thereby an almost irreducible background count rate.

The mass and energy analysis of the photoions is normally based exclusively on flight-time measurements in variable-wavelength coincidence experiments. The very low mass-resolving power ($m/\Delta m$) inherent in this technique is unacceptable when organic radical cations are to be studied. To some extent, $m/\Delta m$ can be increased by extracting the photoions by means of a high-voltage pulse triggered by the photoelectron signal.[8] However, with this modification, $m/\Delta m$ rarely exceeds ~60 for thermal ions and is much smaller when fragment ions with even moderately larger kinetic energies are involved. Note that pulsed-ion extraction leads to a time correlation of otherwise accidental coincidences, so two separate measurements are necessary in order to determine both the total and the accidental coincidence rates. It is difficult to design appropriate ion optics to focus the entire extended ionization region of variable-wavelength instruments onto the entrance aperture of the mass filter, and efficient use of a quadrupole system is thus impeded. Therefore, it is not surprising that distortions of the time-of-flight distributions and decreased ion transmisson coefficients have been observed when quadrupole mass filters have been incorporated into variable-wavelength instruments.[127]

3 Production, Evaluation, and Presentation of Coincidence Data

Generally, the photoionization of gaseous molecules by means of monochromatic radiation generates molecular cations in various vibronic states and an equivalent number of photoelectrons. The kinetic-energy distribution of these photoelectrons contains information on the electronic structure of the photoionized molecules and, most notably in the present context, represents the distribution of the internal energy (E^*) of the initially formed ensemble of positive molecular ions. Since an ionizing photon imparts its entire energy ($h\nu$) to a molecule, detecting a photoelectron with a particular kinetic energy $E_{KIN}(e^-)$ indicates that a molecular cation with

$$E^* = h\nu - E_{KIN}(e^-) - I_1^a \quad (1)$$

has been formed. With I_1^a being the first adiabatic ionization energy of the molecule under consideration, E^* is measured relative to the zeroth vibrational level of the electronic ground state of the molecular cation. In order to identify corresponding electron and ion signals (a prerequisite for making internal-energy-selective measurements feasible) PEPICO spectroscopy exploits the fact that a molecular cation and its conjugate photoelectron are formed at exactly the same time. In principle this is accomplished in the following way. In detecting a photoelectron of given kinetic energy we know that t_e seconds earlier a molecular cation with internal energy E^* (*cf.* equation 1) had been formed. The time of flight of the photoelectron (t_e) can be calculated from its kinetic energy and the geometry and potentials of the apparatus. The much longer time of flight of the corresponding molecular ion (t_i) or its ionic-decay product is a function of the m/z value of the ion, its initial velocity component along the spectrometer axis ($v_{\parallel}$), and its original place of formation, as well as of some specific constants of the spectrometer used. Thus, by combining photoelectron detection with measurement of t_i, it can be determined whether molecular cations possessing a particular internal energy E^* survive the sampling time (t_s) of the experiment or whether they fragment within t_s seconds. When dissociation does take place, the m/z values of the generated fragment ions, the relative probabilities for their formation (termed branching ratios $b[E^*, m_k]_{t_s}$), the kinetic energy released on fragmentation, and, in favourable cases, the decay rates can be measured. Either by varying $E_{KIN}(e^-)$ while keeping $h\nu$ constant (fixed-wavelength technique) or by collecting nominally zero-kinetic-energy electrons and scanning the wavelength of the ionizing photons (variable-wavelength technique), more or less excited molecular ions can be selected and their subsequent fate can be studied. When an extended internal-energy range is probed and all types of fragment ion are considered, the breakdown diagram of the investigated molecular cation (*i.e.* the branching ratios $b[E^*, m_k]_{t_s}$ for the molecular ions and the various types of fragment ion as a function of E^*) is obtained.

Unfortunately, coincidence experiments are much more involved in practice. Some of the complicating factors have already been hinted at in the experimental section, and the origin of further difficulties will be outlined here. A vital aspect, particularly when mass-spectrometrically relevant systems are studied, concerns the attainable resolution in selecting a particular internal energy, E^*. As a consequence of the limited photoelectron-energy resolution and of the initial thermal-energy content of the ionized molecules, a more or less extended distribution of internal energies and *not* a discrete value of E^* is actually involved. Therefore, coincidence data obtained at a nominal internal energy E^* are in fact average values weighted by the instrument – as well as by the molecule-dependent, effective sampling function of the experiment. For obvious reasons, referring to such coincidence experiments as a 'state-selective' technique is an exaggeration that should be avoided. More recent coincidence studies, in particular H. M. Rosenstock's work, have provided convincing evidence that this spread in E^* should not be neglected if reliable data are to be obtained. To account for the spread in E^* it has been proposed[44] that an effective sampling function be constructed from the transmission function of the electron-energy analyser and the thermal-energy distribution of the parent molecules. This is an approximate solution only, as it is affected by the already mentioned uncertainties in the transmission function of the photoelectron-energy analyser and as it implies that the thermal-energy distribution of the molecules is preserved on photoionization. However, recently published results[131] have clearly shown that approximating the distribution of E^* in this way is a reasonable working hypothesis, certainly superior to ignoring the spread in E^* or to trying to account for it by simply adding the average thermal energy of the molecules as a fourth term in equation 1. Effective sampling functions constructed in the described way have been used to obtain more accurate breakdown and rate/energy data, while at best the average thermal energy of the molecules has been considered in evaluating the kinetic-energy-release data. The limited value of these latter data is apparent because, even for a relatively small organic cation consisting of, say, a dozen atoms, a typical internal-energy spread of the order of a few tenths of an eV is involved.

Commonly, the molecular ion and the various sorts of fragment ion are characterized exclusively by their mass-to-charge ratio (*i.e.* $m_k = m/z$) in coincidence spectroscopy. Since the coincidence experiment is also sensitive to the ions' translational energies, a more detailed characterization by m_k and the kinetic energy is sometimes possible. However, it is worth pointing out that neither the electronic and/or geometric structure nor the internal and external rotational energy of the fragments can be measured in coincidence spectroscopy. When such quantities are inferred, additional information from independent sources has been utilized.

It is well known that mass spectrometers possess different sensitivities for detecting different ions. This is also true in coincidence spectroscopy, where the already noted ion transmission coefficient $f_i(m_k, \vec{v}, \vec{r})$ represents the transmission function of the ion channel. Clearly, $f_i(m_k, \vec{v}, \vec{r})$ must be taken into account in evaluating the data. In the present context it is appropriate to confine the discussion initially to a point-like ionization zone and to thermal molecular cations. When all the ions are formed at exactly the same position, the dependence of f_i on $\vec{r}$ becomes a constant. Moreover, undissociated molecular cations possess a well defined distribution of thermal translational energy, which is independent of their internal energy, E^*. For a given temperature (T) and molecular-ion mass (m_p), the dependence of f_i on $\vec{v}$ can hence be worked out and included implicitly in f_i. Therefore, in this case f_i depends solely on m_k, and this dependence, $f_i(m_k)$, can be established precisely in the following manner. To determine a particular branching ratio $b[E^*, m_k]_{t_s}$ [*i.e.* the relative probability that a molecular ion originally prepared with nominal energy E^* dissociates within the sampling time of the experiment (t_s) into a fragment ion with mass m_k] the true photoelectron count rate, $E[E^*]$, and the true coincidence count rate, $C[E^*, m_k]$, must be measured. Bearing in mind that average quantities are actually involved, the generally valid relation between the quantities in question is given in equation 2.[24]

$$b[E^*, m_k]_{t_s} \cdot f_i(m_k, \vec{v}, \vec{r}) = C[E^*, m_k]/E[E^*] \qquad (2)$$

Suppose that E^* is chosen in such a way that fragmentation cannot take place for energetic reasons. Therefore, the molecular-ion branching ratio is by necessity equal to unity and equation 2 reduces to equation 3.

$$f_i(m_p) = C[E^*, m_p]/E[E^*] \qquad (3)$$

Since $f_i(m_p)$ is a true constant respecting E^*, the molecular-ion branching ratio can be determined precisely for any value of E^*.

When fragment ions are considered, the situation becomes much more complex, even if the assumption of a point-like ionization zone is at first retained. The origin of the difficulties is the kinetic energy imparted to the fragments in the course of dissociative processes and the dependence of this kinetic energy on the internal energy, E^*, of the decaying molecular ion. In brief, there is no longer a comparably simple calibration procedure to establish $f_i(m_k, \vec{v}, \vec{r})$ as outlined above for molecular ions, and moreover $\vec{v}$ and consequently also $f_i(m_k, \vec{v}, \vec{r})$ depend on E^*. It is therefore necessary to use approximate methods for estimating $f_i(m_k, \vec{v}, \vec{r})$, a quantity that is required in order to evaluate the data (*cf.* equation 2). Relative to molecular ions with the same mass, fewer fragment ions are normally transmitted because their larger kinetic energy, equivalent to larger velocity vectors $\vec{v}$, leads to increased discrimination. Hence, in fixed-wavelength coincidence spectroscopy the well known $f_i(m_k)$ values for a thermal ion of the same mass are used at first for approximating the essentially unknown fragment-ion transmission coefficient. Relying on equation 2, lower limits to the true breakdown curves can be obtained in this way. These preliminary individual breakdown curves are then added up. The deviation of the resulting 'sum curve' from its theoretical value of unity immediately reveals whether or not fragment-ion discrimination effects are important. If needed, a second more reliable set of breakdown curves can then be computed by partitioning the established total deviation of the 'sum curve' from unity according to the average kinetic energies of the various fragment ions. It must be emphasized that underestimating the fragment-ion discrimination is a frequent source of error in coincidence spectroscopy. Even the release of relatively small amounts of kinetic energy may result in a serious reduction of the corresponding ion transmission coefficient. This is clearly evidenced by the results[8] collected in Table 2 and by the notable deviation of the sum curve from unity as reported frequently in the literature. Even nowadays, the relative areas of the time-of-flight signals are very often directly equated to the branching ratios, and thus the problem of having m_k- as well as E^*-dependent fragment-ion transmission coefficients is simply ignored. Such breakdown data evaluated on the basis of a constant ion transmission coefficient are therefore only approximate. In order to assess the reliability of these data, at least the behaviour of the sum curve, which unfortunately is often missing in important studies, ought to be checked.

Table 2 *Ion transmission coefficients for ethane cation and its fragment ions as a function of the average kinetic energy imparted to the fragment ions in the course of the respective dissociative processes. A Maxwellian distribution is assumed for the released kinetic energy. Data have been drawn from a more extended table given in reference 8*

	Average kinetic enervy/eV *imparted to the fragment ions on dissociation*			
Ion	0	0.0045	0.015	0.045
$C_2H_6^{+\bullet}$	0.240			
$C_2H_5^+$	0.246	0.226	0.190	0.129
$C_2H_4^{+\bullet}$	0.251	0.230	0.192	0.130
$C_2H_3^+$	0.256	0.235	0.195	0.131
$C_2H_2^{+\bullet}$	0.262	0.239	0.199	0.132
CH_3^+	0.359	0.316	0.246	0.149

The primary source of information in PEPICO spectroscopy is a time-of-flight distribution (TOFD) for ions of the selected mass, m_K, originally stemming from molecular cations with known internal energy, E^*. When mass analysis of the photoions is based on pure time-of-flight techniques, a whole group of such signals, each associated with a particular value of m_K, can be considered simultaneously. On the other hand, only a 'single TOFD' is recorded if an additional independent mass selector is utilized. Presupposing sufficient mass-resolving power, the following discussion of a 'single TOFD' is independent of these instrumental details. Again, it is useful to begin with some comments on the signal shape for thermal molecular ions all produced at exactly the same location in the ion source. TOFDs for molecular ions closely resemble a Gaussian distribution, owing to their initial thermal energies. Spectrometer-specific constants, the mass number, and $v_{\parallel}$ determine the time of flight of a particular molecular ion (t_i). Although t_i is independent of any initial velocity component perpendicular to the spectrometer axis ($v_{\perp}$), $v_{\perp}$ does nevertheless influence the shape of the TOFD inasmuch as the magnitude of $v_{\perp}$ essentially determines whether or not a particular ion will reach the ion detector. A detailed analysis of the TOFDs for thermal molecular cations by means of extensive ion-trajectory calculations makes it possible to determine the ion-channel transmission function that transforms the well known original thermal-velocity distribution into the experimentally observed signal. In such an analysis the finite width of the ionization region may also be included. The apparatus function established in this way can be utilized to address the problem of deriving from the observed TOFD the average value or even the distribution (KERD) of the kinetic-energy release associated with a particular dissociation of internal-energy-selected molecular ions. While in most cases average kinetic energies may now be reasonably estimated, the determination of KERDs is an intricate task that still lacks a rigorous solution in the study of mass-spectrometrically relevant systems. KERDs for the decay of up to ten-atom ions have already been reported. However, a closer inspection of these data reveals that the published KERDs are often no more than one possible rationalization of experimental results (see below).

To summarize, breakdown curves can now be obtained with typical errors of the order of 5%. When properly evaluated, breakdown data reported by different laboratories agree quite well in general. Note that these data are essentially derived from a measurement of $C[E^*, m_K]$ corresponding to the area of a coincidence signal. On the other hand, kinetic-energy data on one process still show large discrepancies.[150] As has been expounded, deriving kinetic energies and in particular KERDs requires that the shape of the coincidence signal is subject to a detailed analysis. Obviously this has not yet been fully realized, and in view of the exemplified complexities involved the accuracy of kinetic-energy data is generally hard to ascertain. In this context some comments on 'slowly' fragmenting (*i.e.* metastable) ions must be made here. When the molecular-ion lifetime (τ) is comparable to the sampling time of the experiment (t_s), fragmentation occurs far away from the position of ionization. Under these circumstances the

[150] Compare, for example, the kinetic-energy-release data on acetone and methyl iodide as reported by different research groups.

dependence of $f_i(m_K, \vec{v}, \vec{r})$ on $\vec{r}$ becomes most important. Phenomenologically, slow fragmentation reactions ($\tau \approx t_s$) cause asymmetrically broadened daughter-ion signals with more or less pronounced tailing towards higher mass. Relatively early on,[102] simple methods were developed to determine absolute fragmentation rates by modelling this asymmetry. Qualitatively, this method is certainly correct, but comparison of the derived rate/energy functions with corresponding data obtained from the more recently developed time-resolved coincidence experiments has revealed that the earlier approach consistently leads to an overestimation of the fragmentation rates. A refined definition of the ion transmission function that accounts properly for the spread in E^* appears to be mandatory for reliable results.

4 A Selection of Important Results

Predissociation of Molecular-oxygen Cation. – An ideal example to illustrate the above considerations and the remarkable experimental progress achieved in recent years is the decay of the molecular-oxygen cation. Although the present review is devoted to coincidence results on systems consisting of at least half a dozen atoms, a brief introductory discussion of the coincidence results on this diatomic cation will facilitate readers' appreciation of data available through coincidence experiments.

When molecular oxygen is photoionized by means of He-Iα quanta, five distinct electronic states of the molecular radical cation ($\tilde{X}\,^2\Pi_g$, $\tilde{a}\,^4\Pi_u$, $\tilde{A}\,^2\Pi_u$, $\tilde{b}\,^4\Sigma_g^-$, $\tilde{B}\,^2\Sigma_g^-$) are populated. The relative initial population of the accessible vibronic levels and thus the energy-deposition function can be inferred from the photoelectron spectrum (*cf.* Figure 3). Within the energy range of interest, only two sets of dissociation products are possible:

$$O^+(^4S) + O(^3P) \qquad (18.7334 \text{ eV})$$

$$O^+(^4S) + O(^1D) \qquad (20.7008 \text{ eV})$$

with precisely known[151] dissociation limits as indicated in parentheses. The relevant Franck–Condon factors are such that, for energetic reasons, molecular ions originally formed in the lowest three electronic states cannot fragment. Stable molecular ions are also produced when the lowest three vibrational levels ($v = 0$, 1, or 2) of the $\tilde{b}\,^4\Sigma_g^-$ state are initially populated. Although the $v = 3$ or 4 levels of the $\tilde{b}\,^4\Sigma_g^-$ state lie well below the lowest dissociation limit, some of these molecular ions fragment owing to sufficient thermal rotational excitation[15] at ambient temperatures. Molecular ions prepared with an energy in between the two dissociation limits may either fragment to ground-state products or relax by emission of electromagnetic radiation. The situation is no longer trivial above the second dissociation limit because two fragmentation channels are energetically feasible and radiative relaxation is still a possibility. Thus, it is important to know whether all the molecular ions whose initial energy exceeds the second-dissociation-limit fragment and to what extent the two accessible decay channels

[151] C. Pernot, J. Durup, J.-P. Ozenne, J. A. Beswick, P. C. Cosby, and J. T. Mosely, *J. Chem. Phys.*, 1979, **71**, 2387.

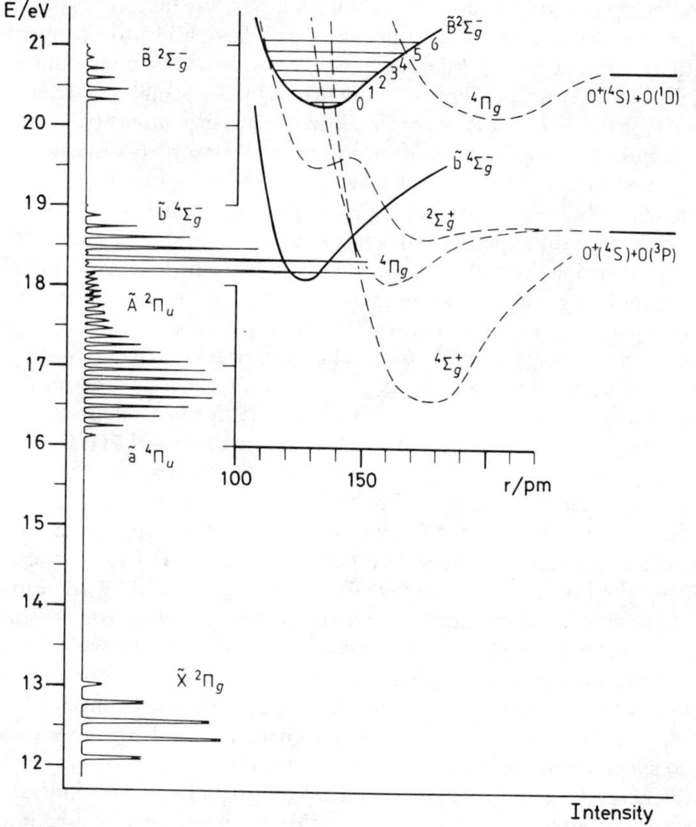

Figure 3 *The photoelectron spectrum of O_2 and the relevant potential curves for the fragmentation of O_2^+ (r = internuclear separation)*

are involved. Although the mass of the charged fragment is the same for both processes, the relative importance of the two dissociation pathways can nevertheless be determined by PEPICO spectroscopy. Since the dissociation products are monoatomic, any excess of energy relative to the respective dissociation limits must be converted into translational energy of the separating fragments. The formation of an excited oxygen atom is accompanied by a small kinetic-energy release of the order of 0.1 eV, whereas ground-state products separate with a substantially higher relative translational energy of ~2 eV. Therefore, an analysis of the kinetic-energy distribution of the O^+ fragment ions is likely to furnish information about the relative importance of the two fragmentation pathways.

As noted above, thermal molecular ions give rise to Gaussian-like coincidence signals under typical experimental conditions. When ions with quasi-thermal kinetic-energy distributions are investigated, the signals are more or less broadened depending on the magnitude of the formal temperature involved. The dissociation of internal-energy-selected diatomic cations often leads to fragments having essentially a single-valued kinetic energy. As long as discrimination effects are negligible, the corresponding coincidence signal resembles a trapezoid with relatively sharp flanks and a continuously decreasing intensity towards longer flight times. The signal is transformed into a bimodal distribution when the single-valued kinetic energy and hence also $\vec{v}$ are progressively increased, as then particularly those ions with mean flight times (*i.e.* small $v_{\parallel}$ and large $v_{\perp}$) will escape detection for experimental reasons.

The principal aspects of the coincidence data on the decay of excited O_2^+ and the corresponding analysis can be exemplified by reviewing the reported results for molecular ions initially generated in the $v = 4$ level of the electronic $\tilde{B}\,^2\Sigma_g^-$ state, *i.e.* at an energy of 20.816 eV. The charged fragments carry either ~1 eV, when the neutral fragment is in its ground state, or merely ~60 meV, when the neutral fragment is electronically excited. The former relatively large single kinetic-energy release is expected to bring about a bimodal TOFD, with a short- and a long-time component due, respectively, to 'forward flyers' (*i.e.* O^+ ions whose initial velocity vector is directed essentially parallel to the source field) and 'backward flyers' (*i.e.* O^+ ions whose initial velocity vector is directed essentially antiparallel to the source field). On the other hand, the coincidence signal for the low-kinetic-energy O^+ fragments ought to be of trapezoidal form with a possible local minimum at its centre as a consequence of some minor discrimination effects. For a given geometry, this minimum will be more pronounced the smaller is the intensity of the source field.

The results of three independent studies, Eland's pioneering work[9] and two more recent investigations,[13,15] are depicted in Figure 4. In the earliest study[9] only a small hump, correctly attributed to the 'forward flyers' of the high-kinetic-energy O^+ fragments, was detected (*cf.* Figure 4a). These measurements[9] indicated that only the lower fragmentation channel was effective in depleting the initially populated state. The more recent studies,[13–15] however, have shown that both reactions occur. Referring to Figure 4b, the O^+ TOFD consists of three well separated parts. The two outer components correspond to the 'forward flyers' and 'backward flyers' of the high-kinetic-energy O^+ fragments. The trapezoidal centre part is due to the low-kinetic-energy O^+ fragments. In the conditions used in these experiments,[13] the centre part shows no resolved structure. The TOFD obtained in the Reporters' own laboratory is presented in Figure 4c. Again, three local maxima are observed. However, a detailed analysis shows clearly that now the component at the shortest flight times must be ascribed to the 'forward flyers' of the high-kinetic-energy O^+ fragments, while the other two adjacent parts of the signal are both due to low-kinetic-energy O^+ fragments. To preserve high photoelectron-energy resolution only a weak source field was applied in this study,[15] and thus the signal component corresponding to the 'backward flyers' of the high-kinetic-energy O^+ fragments was not detected. The weakness of the source field is also responsible for discrimination towards some of the low-kinetic-energy O^+ fragments, as evidenced by the significant minimum of the corresponding TOFD. For obvious reasons, the respective areas of the measured TOFDs are not a direct measure for the branch-

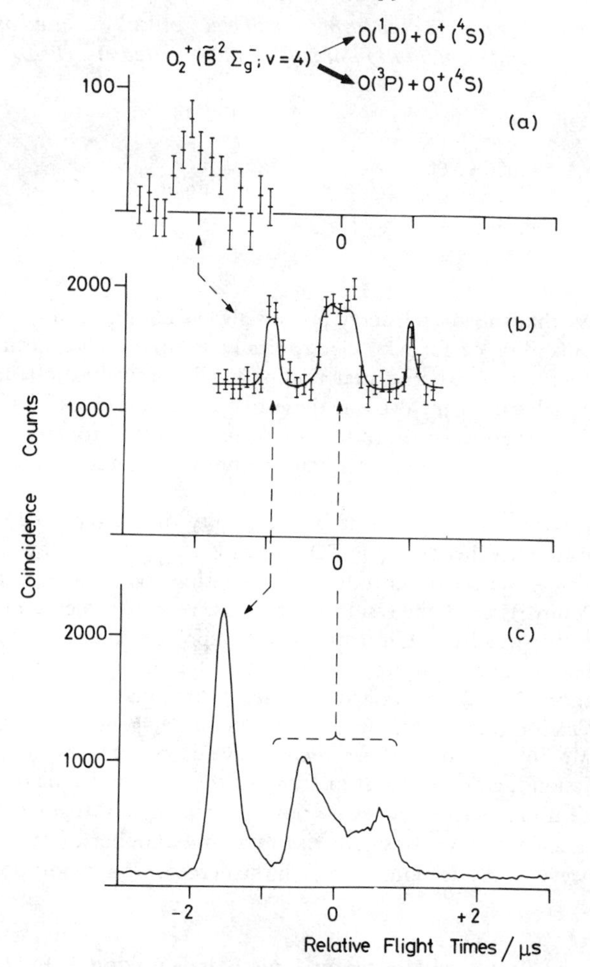

Figure 4 *Competitive predissociations of* O_2^+ $(B^2\Sigma_g^-,\ v = 4)$. *To facilitate comparison of the individual data a time-scale relative to the most probable flight time of thermal* O^+ *ions is used to represent the originally reported* O^+ *TOFs. The solid line in* (b) *is calculated, while all the other data are experimental*

(Reproduced with permission from (a) *Int. J. Mass Spectrom. Ion Phys.*, 1972, **8**, 153, (b) *Chem. Phys. Lett.*, 1981, **84**, 272, and (c) *Int. J. Mass Spectrom. Ion Phys.*, 1982, **43**, 211)

ing ratios into the two decay channels. Their importance can only be quantified when the observed kinetic-energy distribution is analysed in detail by explicitly accounting for the different ion transmission factors. The results of such an analysis for the $\tilde{B}\,^2\Sigma_g^-$, $v = 4$, and its two neighbouring vibrational levels are collected in Table 3. Although some discrepancies are obvious, the data consis-

Table 3 *Branching ratios for the formation of electronically excited oxygen atoms formed by fragmentation of sufficiently excited* O_2^+ $(\tilde{B}^2\Sigma_g^-, v)$ *ions*

v	*Ref.* 15	*Ref.* 14	*Ref.* 13
3	0.006 ± 0.001	–	0.005 ± 0.002
4	0.07 ± 0.02	0.25 ± 0.05	0.13 ± 0.02
5	0.03 ± 0.01	–	0.05 ± 0.02

tently suggest that the lower-energy process always dominates and, most notably, that the probability of forming electronically excited oxygen atoms is highest for $v = 4$. This latter experimental finding has been well rationalized in terms of the potential curves involved in these two predissociations.[13] The transition probabilities to the respective states are indeed expected to vary little (or considerably) with increasing energy when ground- (or excited-) state O atoms are produced (*cf.* Figure 3).

It is instructive to compare briefly the quality of the reported experimental data. The difference between the TOFD in Figure 4a on one hand and that in Figures 4b and 4c on the other reflects the experimental progress achieved within a decade. Comparison of the results of the more recent studies demonstrates the importance of a quadrupole instrument for improving the signal-to-noise ratio. The partial-photoionization cross-sections for the initial population of the five electronic states of O_2^+ are about the same.[152] As noted above, only the $\tilde{B}$ state is fully predissociated, *i.e.* only about 20% of all generated molecular ions fragment. Moreover, most O^+ fragments have large kinetic energies and thus low transmission factors. The vast majority of the ions impinging on the detector are therefore undissociated molecular ions if only time-of-flight mass analysis is utilized. These molecular ions give rise to a large number of accidental coincidences. However, only O^+ ions can reach the detector if a quadrupole mass filter is used.

The Essence of Statistical Models for Ionic Fragmentation. – In studying larger positive ions, as opposed to diatomic cations, a significantly different situation obtains. Clearly resolved vibrational fine structure in photoelectron spectra is scarce, and even bands corresponding to distinct electronic states often overlap appreciably. The thermal vibrational energy of larger cations gives rise to further difficulties (see above). In brief, the initial energy input to the molecular ion is less well defined. Moreover, the originally generated molecular ions may now isomerize in competition with other relaxation pathways. At the same time, the pattern of products becomes more complex because even for a moderately large organic radical cation the number of decay channels increases considerably with increasing internal energy of the reactant. This is true when the fragment ions involved are characterized solely by their mass, let alone when the variety of conceivable and mostly unknown geometric and electronic structures of neutral

[152] O. Edqvist, E. Lindholm, L. E. Selin, and L. Åsbrink, *Phys. Scr.*, 1970, **1**, 25.

and charged fragments are taken into account. Furthermore, the polyatomic fragments can be vibrationally and, in principle, also rotationally excited, and several generations of daughter ions are usually involved. Experimental methods for unravelling such a complex situation are just beginning to emerge. Theoretical treatments are still not developed sufficiently to interpret the fragmentations of a 'large' organic cation in terms of potential-energy surfaces. As with diatomics, the relevant energy-deposition function and ion transmission coefficients also influence the measured fragmentation patterns of polyatomic cations. However, the decay processes of mass-spectrometrically interesting systems can only be rationalized when the dissociative and non-dissociative relaxation of the excited molecular cations is treated within the framework of a suitable model.

The current theories for describing the dissociative ionization of organic molecules are statistical in nature. The application of RRKM calculations[3,4] to the decay processes of excited cations – often referred to as quasi-equilibrium theory (QET) – was originally put forward by Rosenstock *et al.*[153] In the present context only the most important aspects of this model will be recapitulated, while readers interested in more detailed explanations are encouraged to consult the original literature. The relevant Franck–Condon factors for the ionization of organic molecules are such that the attainable vibrational excitation is typically of the order of 1–2 eV. Referring to the electronic ground state, this is often insufficient to effect even the least endothermic dissociation. As a consequence, most fragmentations can only be induced through initial population of an excited electronic state. For a statistical treatment of ionic fragmentation, electronic-excitation energy is postulated to be converted rapidly into vibrational and internal rotational energy of a highly excited ground-state molecular ion prior to the subsequent, much slower dissociative steps taking place on this ground-state hypersurface. Furthermore, random energy flow is presupposed, so that any isoenergetic set of quantum numbers is equally probable and energy is transported back and forth within this highly excited molecular cation. The fragmentation rate of a particular process can then be calculated from the one-dimensional passage of a system point over a barrier of height E_0 along the reaction co-ordinate q (*cf.* Figure 5). E_0 denotes the critical energy of the process in question, *i.e.* the true energy difference between the vibronic ground states of reactant and transition state. 'Local accumulation' of the available energy and thus bond breaking will be less likely the larger is the number of isoenergetic sets of quantum numbers that compete. The rate is therefore inversely proportional to the density of states of the reactant at internal energy E^*, *i.e.* $\rho(E^*)$. On the other hand, the rate is expected to increase in proportion to the total number of levels of the transition state between zero and $(E^* - E_0)$, *i.e.* the integrated density of states of the transition state $W^{\neq}(E^* - E_0)$, because each of these levels represents an independent pathway to the product space. Moreover, a statistical factor, σ, accounting for the possibility of having several equivalent saddlepoints due to symmetry, must be included. For a single fragmentation this yields equation 4,[3,153]

[153] H. M. Rosenstock, M. B. Wallenstein, A. L. Wahrhaftig, and J. Eyring, *Proc. Natl. Acad. Sci.*, 1952, **38**, 667.

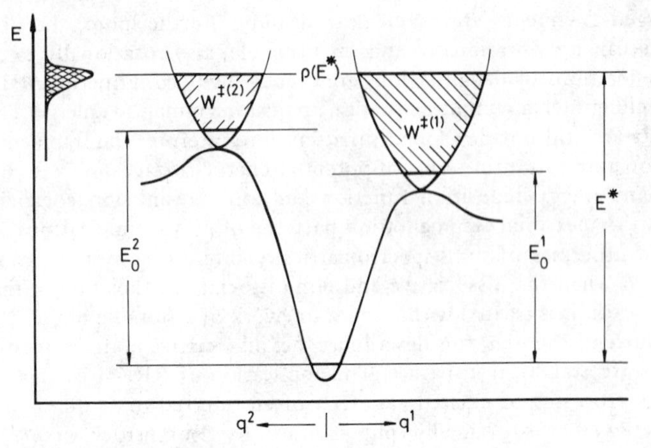

Figure 5 *The essence of statistical models for ionic fragmentation. The reactants are prepared with nominal internal energy,* E*. *The actual distribution of internal energies is symbolized in the upper left-hand corner. The figure refers to two competitive dissociations. When only a single decay channel is considered, the left-hand side of the figure can be disregarded. See text for the definition of the symbols*

$$k(E^*-E_0)=\begin{cases}0 & \text{for } E^*<E_0\\ \dfrac{\sigma}{h}\cdot\dfrac{W^{\neq}(E^*-E_0)}{\rho(E^*)} & \text{for } E^*\geqslant E_0\end{cases}\tag{4}$$

where h denotes Planck's constant. When several fragmentation channels are accessible, each individual rate $k^{(i)}(E^*-E_0^{(i)})$ can be formulated in this way, and the total depletion rate of the molecular-ion population $[k_{tot}(E^*)]$ is simply the sum of these individual rate constants (equations 5 and 6).

$$k^{(i)}(E^*-E_0^{(i)})=\begin{cases}0 & \text{for } E^*<E_0^{(i)}\\ \dfrac{\sigma^{(i)}}{h}\cdot\dfrac{W^{\neq(i)}(E^*-E_0^{(i)})}{\rho(E^*)} & \text{for } E^*\geqslant E_0^{(i)}\end{cases}\tag{5}$$

$$k_{tot}(E^*)=\sum_{i=1}^{k} k^{(i)}(E^*-E_0^{(i)})\tag{6}$$

The situation of two competing fragmentations is schematically depicted in Figure 5. To make use of these equations in practice, $E_0^{(i)}$ and $\sigma^{(i)}$, as well as the distribution of the rovibrational levels of the reactant and transition states in the relevant energy range, must be known. Whereas reasonable estimates for the former two quantities may usually be obtained from independent sources, knowledge of the rovibrational levels of a highly excited molecular cation and of the transition state is practically nil. So far, whenever calculations have been carried out, they have been based on a harmonic oscillator and rigid rotor level spacing, respectively. Despite these shortcomings a rigorous test of the usefulness of statistical models for ionic fragmentation is desirable, as now and in coming years they are likely to be the only tools at hand.

Coincidence Studies on the Least Endothermic Fragmentation Reactions. – Testing the QET is particularly simple when only a single fragmentation pathway is involved. Therefore, diverse molecular cations whose least endothermic dissociation is sufficiently separated from higher-energy decay channels have been investigated. According to QET, the rate of a particular fragmentation (*cf.* equation 4) is a continuously increasing function of the available excess of energy $(E^* - E_0)$, assuming its minimum value (equation 7)

$$k_{min}(E^* - E_0) = \frac{\sigma}{h} \cdot \frac{1}{\rho(E^*)} \tag{7}$$

at the reaction threshold, *i.e.* at $E^* = E_0$. Note that k_{min} is independent of the properties of the transition state. When $1/k_{min}$, equivalent to the maximum molecular-ion lifetime τ_{max}, is significantly smaller than t_s, fragmentation is fast compared to the time-scale of the experiment, and an absolute measurement of the decay rates involved as a function of $(E^* - E_0)$, *i.e.* the rate/energy function is not feasible. However, when $1/k_{min}$ is comparable to or larger than t_s, a more or less extended segment of the rate/energy function can be quantified by means of coincidence experiments. As $\rho(E^*)$ is an exponentially increasing function of the number of degrees of freedom as well as of E^*, 'larger' cations and/or sufficiently large critical energies are required for achieving sufficiently small threshold rates. Bearing in mind that t_s is of the order of 1–100 μs, suitably low threshold rates occur for the fragmentation of, say, a ten-atom cation, with a critical energy of some 2 eV.

Deriving rate/energy functions from coincidence data was pioneered by Baer's group.[102] The corresponding measurements were carried out on a variable-wavelength instrument with pure time-of-flight ion analysis. If, for a particular excess of energy, $1/k(E^* - E_0)$ is comparable to t_s, dissociation occurs during ion acceleration. Therefore, the resulting fragment ions enter the drift tube with a velocity that depends on the lifetime of the respective precursor. As a consequence, the fragment ion TOFD is tailed towards the molecular-ion signal. The extent of this asymmetry as a function of E^* was exploited to determine rate/energy functions. By making plausible assumptions about the quantities required rate/energy functions were then computed (*cf.* equation 4) and compared with the experimental results. The degree of accord between theory and experiment was variable. However, commenting on these results appears to be of little use as more recent investigations (see below), in particular the work of Rosenstock have revealed that a refined analysis of data is required in order to deduce rate/energy functions from asymmetrically broadened daughter-ion signals. A key factor is once again the distribution of the internal energy of the reactants. Presupposing a discrete internal energy for the precursor ions, or simply shifting the energy axis by a constant energy increment, *e.g.* by the average thermal energy of the ionized molecules, is clearly inadequate. No matter which method of data handling is employed, consistent and reliable data can only be obtained if the detailed distribution of E^* is properly accounted for. Fortunately, this has now been realized, and a corresponding improvement in analysis of asymmetric signal shape has been successfully tested on the lowest-energy bromobenzene cation fragmentation.[130] A further potential source of error involved in deriving rate/energy data from the extent of tailing of the daughter-ion signals is the dependence of $f_i(m_k, \vec{v}, \vec{r})$ on $\vec{r}$. There is no doubt that the detection probability for fragment ions formed close to or far away from the ionization region will be different. Without detailed ion-trajectory calculations, it is difficult to estimate the importance of this effect in general. However, the approximate nature of data evaluated on the basis of a constant f_i value must be borne in mind, in particular when instruments with rather low ion transmission coefficients are used. In this respect a slow fragmentation with a large activation-energy barrier for the reverse reaction is about the worst situation conceivable, because large fluctuations of $f_i(m_k, \vec{v}, \vec{r})$ are likely to prevent any reliable analysis of the daughter-ion signal.

In recent years, a more dependable approach for determining rate/energy functions – often referred to as time-resolved coincidence spectroscopy – has been devised.[44] For a single fragmentation channel, the molecular- and fragment-ion branching ratios are related to the corresponding rate constant by equations 8 and 9, respectively.

$$b[m_p, E^*]_{t_s} = \exp[-k(E^* - E_0) \cdot t_s] \qquad (8)$$

$$b[m_f, E^*]_{t_s} = 1 - \exp[-k(E^* - E_0) \cdot t_s] \qquad (9)$$

When the lifetime, τ, of the molecular ions is comparable to the time-scale, t_s, of the experiment, the degree of fragmentation at a particular energy increases with increasing t_s. Several studies have shown that this effect can be used to determine dependable rate/energy functions (*cf.* Table 1, time-resolved studies). Lately, the approach has been rigorously tested on an extensive data set for fragmentation of iodobenzene cation into phenyl cation and iodine radical.[131] In order to exclude conceivable instrumental artifacts and to be able to vary t_s substantially, three distinct instruments have been used for this investigation. The phenyl-cation breakdown curves for $t_s = 1$, 5.9, 21, or 58 μs are shown in Figure 6. The energy where 50% of the initially generated iodobenzene cations have decayed (often termed crossover energy) decreases by ~0.4 eV when t_s is increased from 1 μs to 58 μs. It is worth pointing out that these results are not distorted by ion discrimination effects. We have already noted that an exact determination of the molecular-ion breakdown curve is always possible (*cf.* equations 2 and 3), and as long as only a single fragmentation is involved this applies, though somewhat less directly, to the fragment-ion breakdown curve as well. One major source of error is thus *a priori* excluded. As a working hypothesis a rate/energy dependence as defined by equation 4 is then postulated for the process in question. With the emphasis being on developing a consistent, straightforward, and generally applicable strategy, the quantities required to calculate

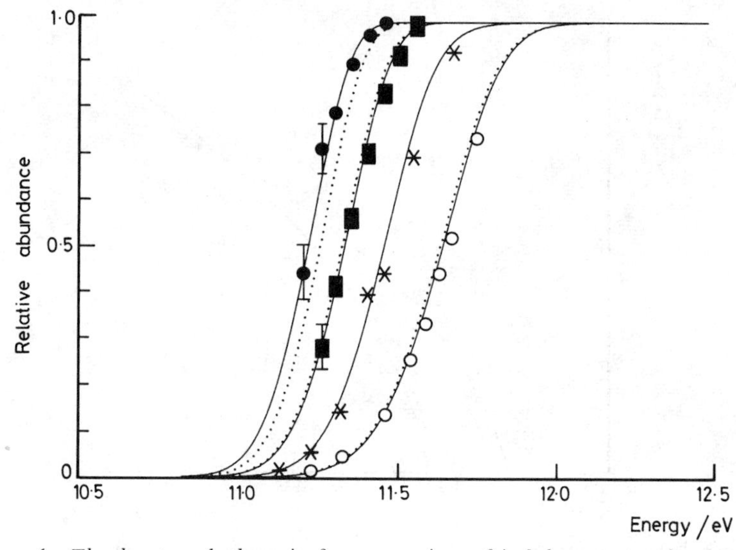

Figure 6 *The least endothermic fragmentation of iodobenzene cation yielding phenyl cation plus iodine radical. Sampling-time dependence of the* Ph^+ *breakdown curve (● 57 μs, ■ 21 μs, * 5.9 μs, ○ 1 μs). Solid lines represent individual best fits to the four data sets; dotted lines represent best average fit to all data sets*
(Reproduced with permission from *Chem. Phys.*, 1983, **75**, 23)

$k(E^* - E_0)$ are obtained as follows. The vibrational frequencies of the parent molecule are utilized for characterizing the ionic reactant. One of these fundamentals is selected as reaction co-ordinate. The remaining set of frequencies is multiplied by a constant factor, F, and then used to characterize the transition state. The reaction-path degeneracy, σ, is evaluated. Estimating starting values for F and the critical energy E_0, a first rate/energy function is computed based on an exact-count algorithm of a harmonic oscillator level spacing. This rate/energy function is converted into a breakdown curve according to equations 8 and 9, respectively. To account for the spread in E^*, this breakdown curve is convoluted with the effective sampling function of the experiment, which in turn consists of the thermal vibrational- and rotational-energy distribution of the ionized molecules and the transmission function of the electron-energy analyser. The resulting breakdown curve is fitted to the experimental data of the corresponding sampling time, by systematically varying the only two adjustable parameters (*i.e.* E_0 and F) by means of a non-linear multi-dimensional regression technique.

Remarkably, a single pair of parameter values was found to account essentially for the entire experimental data set on the lowest-energy decay channel of iodobenzene cation. Some marginal discrepancies with the longest sampling-time data have been traced back to a very low Franck–Condon factor for He-Iα photoionization.[131] The deduced rate/energy function (*cf.* Figure 7) can be

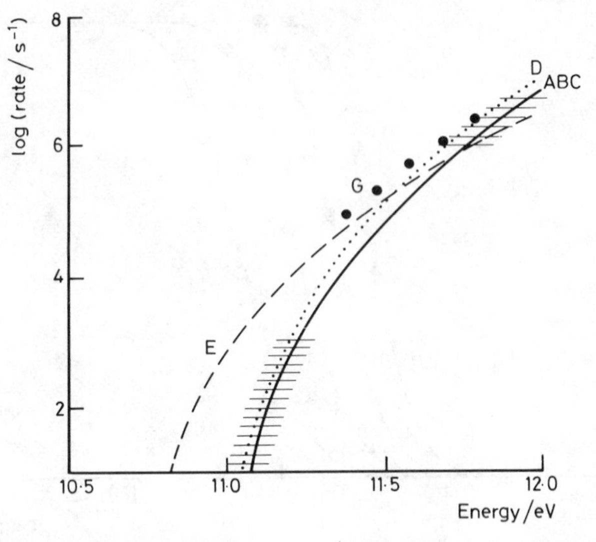

Figure 7 *Rate/energy functions for* Ph^+ *formation by iodine loss from iodobenzene cation. Solid line* ABC *represents best average fit, dotted line* D *is derived from the longest sampling-time data set, and points* G^{127} *and* E *are obtained when the effective sampling function is ignored. The dashed regions represent extrapolated values*
(Reproduced with permission from *Chem. Phys.*, 1983, **75**, 23)

employed to extrapolate the experimental data to lower excitation energies. Most notably, this makes it possible to define consistent threshold energies, even when considerable kinetic shifts are involved. So far, the derived F values have been interpreted in terms of loose ($F < 1$) and tight ($F > 1$) transition states. Sometimes, formal activation entropies have been calculated and compared with the equivalent quantities of the corresponding neutral process. Whether these F values can be used for characterizing consistently various types of unimolecular reactions has not yet been shown.

Rate/Energy Functions of Competitive Dissociations. – To be of any relevance to mass spectrometry, models for ionic fragmentation must be applicable to several competing dissociations. In the present context, this raises the question of whether the approach outlined in the preceding section can be suitably extended (*cf.* equations 5 and 6) to establish the required input parameters. According to the most recent investigations, this seems to be feasible in principle. To begin with, it is useful to examine competing dissociations of sufficiently long-lived molecular cations. This makes it possible to measure the energy dependence of the total decay rate with essentially the same precision as in the case of a single fragmentation reaction, by relying on time-resolved coincidence experiments or on an analysis of the daughter-ion signal shapes. With a view to deriving rate/energy functions for the individual dissociative processes, the branching ratios for the respective product ions must be ascertained. No matter which method of data analysis is used, all the aforesaid potential sources of error connected with the determination of branching ratios will now affect the precision of the deduced rate data. However, if a derived set of rate/energy functions is consistent with the measured molecular-ion decay rate and accounts also for its diverse fragment-ion breakdown curves, a rationalization of the experimental data based on the model used seems to be reasonable.

Relying on a daughter-ion peak-shape analysis, the energy dependence of the total decay rate of excited furan[91] and hexamethyldisilane cation[141] was determined. These total rate/energy functions were then partitioned into contributions from three[91] or two[141] competing dissociations by approximating the required branching ratios by the relative signal areas. The reported results support the concept of reaction competition in the sense of the QET for the low-energy fragmentations of both molecular cations. The energetically accessible low-energy fragmentations of buta-1,3-diene,[106] bicyclobutane,[108] and toluene cations[142,143] have been investigated by means of time-resolved experiments. The studies on the two $C_4H_6^{+\bullet}$ isomers have furnished accurate total rate/energy functions and independently confirmed that the low-energy dissociation behaviour of excited $C_4H_6^{+\bullet}$ cations is independent of the structure of the originally ionized molecule, as had been proposed earlier.[102] The detailed QET analysis of the time-resolved data has identified buta-1,3-diene cation as the common reactant structure. Particularly extensive results have been presented for the dissociative processes of bicyclobutane cation.[108] In view of the remarkable accord between computed and measured data for quite an energy range, the most important conclusions of this study deserve a few more words of comment. Four processes, *i.e.* loss of $CH_3^{\bullet}$, $H^{\bullet}$, H_2, and $C_2H_3^{\bullet}$, were found to compete, and the corresponding rate/energy functions were determined. The critical energies led to consistent thermochemical data and revealed that the apparent threshold energies of the two higher-energy reactions were subject to competitive shifts of the order of 2 eV. (Recall that a competitive shift originates in too low a rate constant of a particular dissociation relative to the total depletion rate of the reactant population.) A detailed QET analysis of coincidence data is a unique method for quantifying competitive shifts. This is of utmost importance, since in principle all but the least endothermic fragmentations are subject to the effect and since the corresponding experimental threshold energies are thus only upper limits for the true values. Evidently, rate/energy functions are the kind of information required to assess this effect and its importance on the mass-spectrometric fragmentation pattern. Surprisingly, the expected competitive shift for a fifth reaction channel – involving the loss of an acetylene molecule – did not appear in the case of the decay of bicyclobutane molecular cations. The experimental onset energy practically coincides with the well known dissociation limit of this reaction. Since the corresponding breakdown curve could not be rationalized, the assumption of reaction competition had to be abandoned for this decay channel.

A time-resolved coincidence study on the toluene cation[142,143] has provided rate/energy functions for tropylium- and benzylium-cation formation despite the fact that both fragment ions have the same mass. The situation resembles the two-channel fragmentation of excited oxygen molecular cations. However, now two isomeric ionic structures and not a ground state and an electronically excited neutral species, respectively, are involved. Also, in the case of toluene cation it is a precise measurement of the energy dependence of the molecular-ion decay rate and not an accurate measurement of fragment-ion kinetic energies that makes such an analysis feasible. The successful partitioning of the toluene-cation decay rate is to some extent fortuitous, as the rate/energy functions for the two reactions are dramatically different and happen to cross each other at a rate falling within the absolutely measurable range of the instrument used.

Most recently, the breakdown diagrams of ethylene and ethane cations have been subjected to a similar analysis.[58] The longest molecular-ion lifetimes of both species are much shorter than t_s. Therefore, a direct measurement of absolute rates is not possible. Nevertheless, the breakdown diagrams can still be analysed and individual rate/energy functions can still be derived. Gratifying agreement between calculated and measured data was noted for loss of $H^{\cdot}$ and H_2 from both molecular ions. Whether or not a third decay channel (*i.e.* loss of $CH_3^{\cdot}$), which is accessible to excited ethane cations, does indeed compete with the former two processes could not be definitely decided as only fragmentary experimental data are available.

Consecutive Fragmentations. – So far only very few attempts have been made to analyse coincidence data on consecutive fragmentations in a comparably detailed manner as exemplified for a single dissociation and for several competing dissociations. While the QET model applies equally to dissociations of molecular and fragment ions, the decay processes of the latter species are experimentally less well characterized because their internal-energy content is usually incompletely known. This results from a lack of knowledge of the partitioning of the excess of energy in the primary reactions. One notable exception concerns first-generation daughter ions that differ from the molecular ion by only one atom. Neglecting external rotational excitation, any excess of energy involved must then be converted into translational energy of the separating fragments or into internal energy of the ionic-decay product. Hence, the internal energy of the subsequently decaying daughter ion can be ascertained, provided that the kinetic energy released can be determined. The only secondary fragmentation that has hitherto been investigated by time-resolved coincidence spectroscopy is acetylene loss from $C_7H_7^+$, in turn formed by $H^{\cdot}$ abstraction from excited toluene cations. This particular sequence of two consecutive fragmentations is the only noteworthy low-energy dissociation pathway of excited toluene cations and thus an obvious choice for the first studies. The experiments have shown unambiguously that the declining section of the $C_7H_7^+$ primary fragment ion is displaced towards higher energies when the period of time between ion formation and ion detection is shortened. However, the rate/energy function derived in the course of this study[143] is only approximate because the kinetic energy released in the primary process has only been estimated.

Another particularly interesting class of consecutive fragmentations involves first-generation daughter ions that can also be formed directly by photoionizing a stable molecule. Once the rate/energy functions of the dissociations connected with the latter method of ion production have been determined, the resulting product-ion distribution can be calculated for any internal-energy distribution of this species. Conversely, a comparison of the product-ion distribution arising from the two methods of ion production may yield the internal-energy distribution in question. Realistically, this can only be done if the secondary fragments involved are not produced through diverse parallel/consecutive dissociation pathways. Qualitatively, such an analysis has recently been presented for the subsequent decay processes of ethylene cations formed as primary fragment ions by acetylene loss from the bicyclobutane cation.[108]

If the currently valid concept of a unimolecular reaction is at least basically correct, the rate/energy function is the most important piece of information for quantifying such a process kinetically. Therefore, the more recently devised methods of deducing rate/energy functions for unimolecular fragmentations of excited cations have purposely been discussed quite extensively. Once complete breakdown diagrams can be analysed in a similar way, our understanding of ionic fragmentation patterns will increase considerably.

The Release of Kinetic Energy in Ionic Fragmentations. – Coincidence spectroscopy has been used frequently for determining the kinetic-energy release that accompanies the fragmentations of internal-energy-selected cations. The dependence of the 'average' translational energy released on the internal energy of the decaying reactant has been reported in various studies. Referring to mass-spectrometrically relevant systems, these data are usually inferred from the fwhm of the corresponding daughter-ion signal, by implicitly assuming a Boltzmann distribution for the released kinetic energy. In most of these studies neither the spread in E^* nor the crucial role of $f_i(m_K, \vec{v}, \vec{r})$ has been considered in detail. It is therefore not surprising that kinetic-energy-release data on one process reported by different laboratories sometimes differ significantly.[150] Average translational energies derived in this way usually reveal little of the nature of the reactions involved and are inadequate for defining the internal-energy content of the generated fragments.

In recent years, workers like Baer[48] and Powis[36] have begun to deduce the distribution of the released kinetic energy (KERD) from a detailed analysis of the fragment-ion coincidence signal. Currently, their work represents our most reliable source of information concerning the transformation of excess of energy into translational degrees of freedom of the separating fragments. Apart from making use of different experimental variants, both research groups rely essentially on the same approach for determining KERDs. The fragment-ion TOFDs to be analysed are expressed as a linear combination of n basis functions. These basis functions are calculated TOFDs for fragment ions of the appropriate mass that possess a particular fictitious single-valued kinetic energy. In order to minimize their linear dependence, the n basis functions are spaced linearly in velocity and thus according to $n^2\Delta E$ in energy. Typically some 10 basis functions and ΔE values between 5 and 10 meV are used to approximate the measured coincidence signals, which in turn are defined by 50–100 individual data points. Fitting the experimental TOFDs by means of a linear-regression technique yields the optimum values for the coefficients of the n basis functions. As an illustration, some results on Et^+ formed by dissociative photoionization of EtBr[60] are shown in Figure 8. The KERDs connected with this dissociation have been studied as a

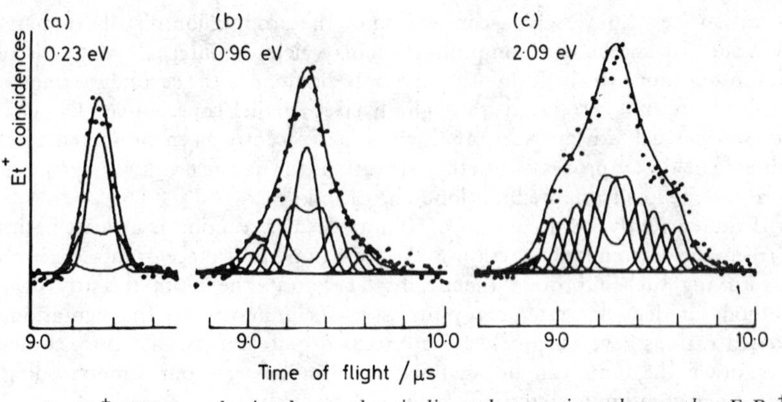

Figure 8 Et^+ *TOFs obtained at the indicated energies above the* $EtBr^{+\bullet}$ *dissociation limit. The solid line through the experimental points is a least-squares fit resulting from a weighted sum of the single-energy-release basis functions corresponding to* 12, 48, 108, 192, 300, 432, 588, 768, *and* 962 meV *energy release. The basis functions are depicted below the TOFDs*
(Reproduced with permission from *Chem. Phys.*, 1984, **85**, 39)

function of the internal energy of the reactant.[60] For excesses of energy of up to ~1 eV, the derived KERDs are in accordance with a statistical model, whereas the fragmentation of more highly excited bromoethane cations gives rise to non-statistical KERDs. Since this change of character coincides energetically with the onset energy of the first excited electronic state of bromoethane cation, direct dissociation on the $\tilde{A}$-state manifold was proposed as an explanation.[60] Sudden changes of the shape of KERDs, and/or of the fraction of the available excess of energy converted into kinetic energy of the fragments, have been interpreted in terms of different electronic and/or geometric product structures.[81,113]

These studies have made it clear that the distribution of the translational energy of the fragments reflects important details of the decay processes of internal-energy-selected cations. Therefore, it is hoped that these promising investigations are continued and that the spread of E^* will henceforth be included in the KERD analysis. However, it must be pointed out that KERDs derived in this way are one possible rationalization of the experimental data. Coincidence experiments on triatomic cations[31] have shown that different sorts of KERDs are frequently compatible with the measured TOFDs.

Competition Between Radiative and Dissociative Relaxation Pathways. – In recent years gas-phase emission spectra of about 100 organic radical cations have been published.[154] In some instances the radiatively depopulated excited electronic state is energetically located above one or even above several experimentally (EI, PI) determined fragmentation thresholds. The method of choice for proving that

[154] J. P. Maier, *Chimia*, 1980, 34, 219.

radiative/ and non-radiative/dissociative decay processes are indeed competitively depopulating the same ionic state is fixed-wavelength PEPICO spectroscopy. Variable-wavelength experiments are not unambiguous, as fragmentation might originate from autoionization, while the directly populated ionic state might be depleted purely radiatively. Fixed-wavelength experiments on a number of such cations (*cf.* Table 1, entry i) have indeed revealed that part of the initially formed excited molecular ions radiate their excess of energy whereas the rest undergo a radiationless transition, presumably to the electronic ground state from where fragmentation takes place. This rather unique situation makes it possible to determine the depletion rate of the initially populated excited electronic state by experiments relying on time-resolved photon detection and to establish the rate of fragmentation by means of electron–ion coincidence measurements. One of the fundamental assumptions of the QET, *i.e.* the depopulation of the excited electronic states by rapid radiationless transitions, can thus be tested directly. The quoted measurements have revealed that these radiationless processes are indeed not always fast enough for quenching the radiative decay completely.

So far, competition between radiative/ and non-radiative/dissociative depopulation processes has only been established for the first excited electronic state of some 10 organic radical cations. In all these cases the radiative transition to the ground-state manifold was dipole-allowed. However, it has not yet been consistently explained why in some instances the first excited electronic state is depleted both by radiative transition to the ground state and by internal conversion followed by fragmentation. As an illustration, it is still recondite that buta-1,3-diene cations prepared in their $\tilde{A}$-state do not fluoresce at all ($\Phi \leqslant 10^{-4}$), while radiative and radiationless processes effectively compete in depleting the analogous state of the *trans*- as well as *cis*-hexa-1,3,5-triene cations.

When excited molecular cations relax radiatively instead of fragmenting, rather puzzling mass spectra may result. In the cases investigated up to now, the $\tilde{A} \rightarrow \tilde{X}$ electronic transition converts molecular ions possessing sufficient energy for fragmentation into stable ones, with a concomitant increase in the relative abundance of molecular ions in the mass spectrum. In general, radiative depletion of the excited-state population may also distort the fragment-ion abundances. Commonly, large signals may not appear in the mass spectrum or may have only small sizes because the non-dissociative pathways quench the corresponding fragmentation reactions. However, ion fluorescence occurs only infrequently and is thus not an important phenomenon for the domain of organic mass spectrometry.

Competition between Isomerization Reactions and Dissociative Processes. – It is well known that the mass spectrum of an organic compound cannot be completely rationalized by simple cleavages. Often, the formation of particular fragment ions can only be explained if extensive rearrangements prior to dissociation are postulated. ^{13}C and more especially ^{2}H labelling experiments have frequently failed to furnish mechanistic details on specific fragmentation reactions, since the rearrangements give rise to complete scrambling of carbon and hydrogen atoms. This complicates the interpretation of ionic fragmentation considerably, because the structure of the dissociating reactant may have little in common with the structure of the initially ionized molecule. The vast majority of the coincidence data on unsaturated hydrocarbon cations suggest that the energy barriers for various isomerizations are usually lower than the critical energies of even the least endothermic fragmentations. Therefore, isomerization to the most stable ionic structure is likely to precede the dissociations of unsaturated hydrocarbon cations. This may hold true for a more or less extended energy range. The simplest way to establish the extent of such isomerizations consists in comparing the breakdown diagrams of the molecular ions in question. Isomerization to a common precursor structure is probable if the breakdown diagram (relative to a common enthalpy of formation scale) is independent of the structure of the initially ionized species. The identical breakdown diagrams of allene, propyne, and cyclopropene cation[44, 66, 68–70] have been interpreted in this way, and similar results have been reported for other isomeric hydrocarbon cations, *e.g.* $C_4H_8^{+\cdot}$ [124–126] and $C_6H_6^{+\cdot}$.[119,155] In other instances, the same conclusion was drawn based on rate/energy functions for a particular dissociation of originally different isomeric molecules, *e.g.* $C_4H_6^{+\cdot} \rightarrow C_3H_3^+ + CH_3^\cdot$.[102,104] It is worth noting that identical breakdown diagrams in general mean distinct mass spectra, owing to the different energy-deposition functions connected with the photoionization of different isomeric molecules.

The currently available coincidence data imply that isomerization to the most stable ionic structure is considerably less pronounced for hetero compounds. An extensive study on the $C_3H_6O^{+\cdot}$ system has begun.[98–101,156] The data already published have revealed that the breakdown diagrams of most of these radical cations are different. A notable exception concerns ionized cyclopropanol and allyl alcohol, whose breakdown diagrams[100] concur for an energy range of several eV. Obviously, the presence of a particular functional group and hence of a particular electronic structure largely determines the relative heights of the barriers towards isomerization as well as fragmentation. In addition, the work on $C_3H_6O^{+\cdot}$ has also shown that gaps in the relevant energy-deposition functions can be responsible for the lack of certain fragments.[156] Structure-specific breakdown diagrams have also been reported for the radical cations of diverse molecules containing heteroatoms, *e.g.* references 136–138 and the three diazines in reference 109. The latter work is especially interesting when compared with fragmentation of the benzene cation, which is known to be preceded by complete scrambling of the carbon and hydrogen atoms.[157] In contrast, extensive isomerization of the ionized diazines does not take place, since only the 1,2-isomer dissociates by loss of an N_2 molecule, which is by far the least endothermic fragmentation channel for all the three isomers. On the other hand, the lowest-energy fragmentation pathway detected for the 1,3- and the 1,4-diazine cations is loss of HCN, *i.e.* a decay channel that lies about 2 eV above the lowest dissociation limit. The rate/energy functions obtained from time-resolved coincidence experiments imply that a cleavage of the aromatic ring system involving a

[155] H. M. Rosenstock, J. Dannacher, and J. F. Liebman, *Radiat. Phys. Chem.*, 1982, **20**, 7.
[156] J. Dannacher and J.-P. Stadelmann, unpublished results on various C_3H_6O isomers.
[157] K. R. Jennings, *Z. Naturforsch., Teil A*, 1967, **22**, 454; J. Horman, A. N. Yeo, and D. H. Williams, *J. Am. Chem. Soc.*, 1970, **92**, 2131.

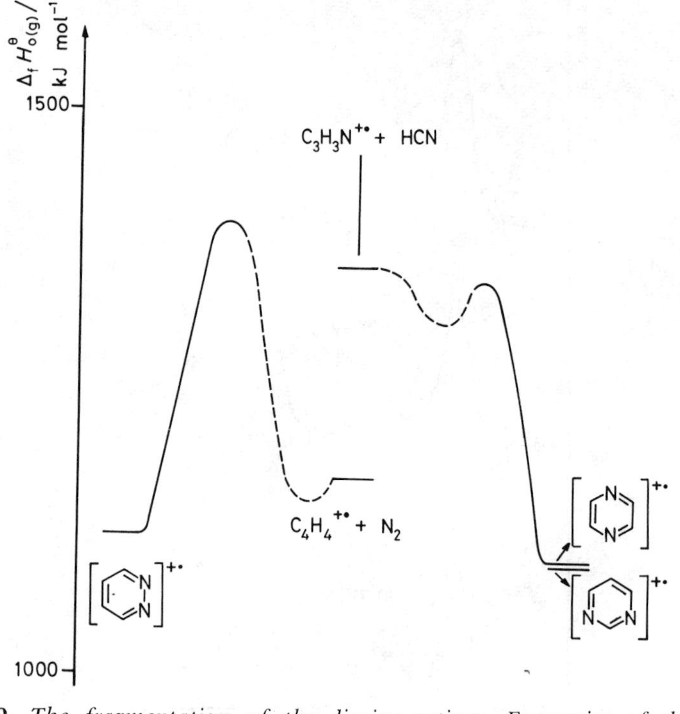

Figure 9 *The fragmentation of the diazine cations. Energetics of the least endothermic decay channels*

critical energy of some 2 eV is the rate-determining step in the lowest-energy fragmentation of the three diazine cations. This initial bond-breaking may occur everywhere in the ring system except for the N–N bond in the 1,2-isomer. A brief summary of these results is presented in Figure 9.

Recently, the problem of isomerization prior to or in competition with fragmentation has been studied thoroughly.[140,144] This work was instigated by experimental findings that had been interpreted in terms of isolated electronic-state dissociations. Baer and co-workers observed strongly tailed TOF signals for loss of $CH_3^{\bullet}$ and essentially symmetrical coincidence peak shapes for the loss of C_2H_4 from diverse $C_5H_{10}^{+\bullet}$ isomers. This indicates a slow formation rate for $C_4H_7^+$ fragment ions and a significantly larger formation rate for the $C_3H_6^{+\bullet}$ fragment ions (*cf.* Figure 10). Based on a rigorous mathematical analysis,[140] this behaviour was traced back to competition between isomerization and dissociation. This is a remarkable result, as evidently even such a two-component decay can be interpreted by statistical models for ionic fragmentation (Figure 11).

State-specific Fragmentations. – For radical cations consisting of at least six atoms, convincing evidence for the existence of state-specific fragmentation behaviour has been presented only rarely. Possibly the least unambiguous

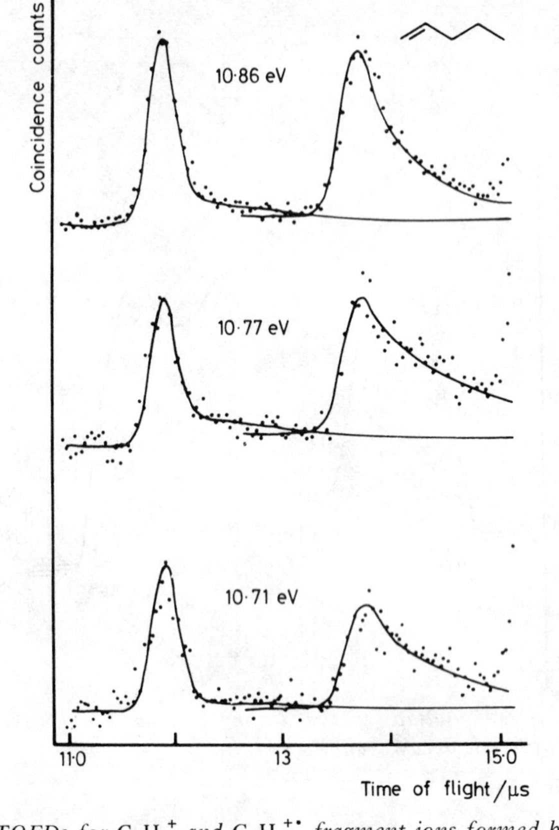

Figure 10 *TOFDs for* $C_4H_7^+$ *and* $C_3H_6^{+\bullet}$ *fragment ions formed by dissociative photoionization of pent-1-ene. The strongly tailed signal of the former fragment ion implies molecular-ion lifetimes that are comparable to the sampling time of the experiment. On the other hand, the essentially symmetrical TOF of the* $C_3H_6^{+\bullet}$ *ion signal indicates fast dissociations of the molecular ions relative to the time-scale of the experiment*

(Reproduced with permission from *J. Am. Chem. Soc.*, 1984, **106**, 3154)

examples reported so far concern the fluorinated ethylene cations[59,61,62] and especially $C_2F_6^{+\bullet}$.[51,84–87] In these cases loss of fluorine becomes the preferred fragmentation pathway for those molecular ions originally formed by ejection of a photoelectron from an orbital of essentially fluorine lone-pair character. The corresponding breakdown diagrams show significant local increases of the $(M-F)^+$ fragment-ion breakdown curves at the respective internal energies. Other examples of state-specific dissociations are significantly less reliable. It can be surmised that most of the reported disagreements between a statistical model

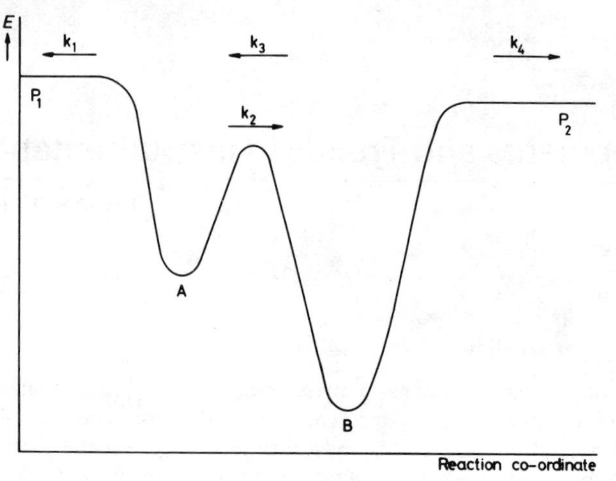

Figure 11 *Model for a competitive isomerization/fragmentation mechanism giving rise to two-component decay. Referring to the data given for Figure* 10, *isomer* A *may be pent*-1-*ene cation and* 2-*methylbut*-2-*ene cation,* i.e. *the most stable* $C_5H_{10}^{+\bullet}$ *structure may be isomer* B. *The initially generated cation* A *can either fragment with rate* k_1 *or isomerize to* B *with rate* k_2. B *can either isomerize back to* A *with rate* k_3 *or dissociate with rate* k_4. *Note that the dissociation products need not be different. The relative magnitudes of the various* k_i *determine the experimentally observed rates*
(Reproduced with permission from *J. Am. Chem. Soc.*, 1984, **106**, 3154)

and coincidence data can be settled when modern, refined experimental techniques are used and comparison between theory and experiment is put on a firm and consistent basis. This applies, for example, to the buta-1,3-diene cation,[102,104] where earlier coincidence measurements implied that the QET could not account for the measured rate/energy data. However, more recent coincidence experiments[106,108] have revealed that practically quantitative agreement can be achieved, as long as some aspects that had been neglected in the earlier study are rigorously taken into account.

4

Developments and Trends in Instrumentation

BY T. R. KEMP

1 Introduction

Following the recent upheaval in mass spectrometry brought about by the introduction of fast-atom bombardment (FAB), the dust is now settling, leaving a legacy of general interest in this and other techniques for the analysis of large molecules. One important by-product has been the improvement in performance of commercial mass spectrometers to cope with the extra demands. Instrumental developments, though not racing forward at their former breathtaking pace, are continuing steadily at both ends of the scale: the ion-trap detector, which will be covered in a separate Report in Volume 9 of this series, provides an example of one end of the scale, whilst at the other end are the magnetic-sector instruments capable of the analysis of molecules with molecular weights of 50 000 daltons and greater. Parallel developments of other types of analyser are taking place. The time-of-flight mass spectrometer, with its virtually unlimited mass range, has a promising future. Fourier-transform mass spectrometry is covered separately in Chapter 6.

2 Ionization Techniques

The ionization of polar molecules of all sizes has continued to attract a great deal of attention after the upsurge of interest generated by the development of FABMS, and several review articles have appeared covering the many methods of ionization that are now being applied to this problem. The general principles, applications, and results of most of the desorption techniques in current use are covered in a review of the mass spectrometry of large, fragile, and involatile molecules.[1] The newest of all the techniques, FAB, has seen the most activity, and the instrumentation and, more particularly, applications of the technique have been summarized.[2] Such is the interest in FAB that several workers have described simple, inexpensive modifications to existing (and often old) instruments to enable FAB operation. The conversion of existing ion sources to operate under FAB conditions has often been described, with field-ionization (FI) and field-desorption (FD) sources being particularly adaptable. The VG ZAB combined FI/FD source can be modified on a non-permanent basis and FAB conditions achieved by using a commercially available fast-atom gun from

[1] K. L. Busch and R. G. Cooks, *Science*, 1982, **218**, 247.
[2] K. L. Rinehart, jun., *Science*, 1982, **218**, 254.

Ion Tech mounted on one of the standard inlet flanges.[3] By using a solid top-hat cathode on the gun instead of the normal hollow one, the resulting reduced gas flow gives a more economical operation, lower source pressure, and less erosion of the atom-beam exit aperture; this reduces the distortion of the atom beam and results in a longer lifetime of the atom gun.[4] The VG ZAB FD source has also been modified by mounting a small capillaritron atom gun directly on to the ion-source block.[5] Operation of the capillaritron gun at voltages lower than the mass-spectrometer accelerating potential provides atom bombardment (since ions are prevented from entering the source and striking the sample), whilst operation of the gun at voltages higher than the accelerating voltage provides ion bombardment. The secondary-ion beam has been shown to have excellent stability, better than 0.7% instability having been recorded using the *m/z* 93 ion from glycerol. By a simple conversion requiring only a couple of hours work, a capillaritron-type fast-atom gun in which the ratio of ions to atoms in the emitted beam is variable can be fitted to a quadrupole instrument without interfering with the instrument's normal CI/EI functions.[6] Such a conversion allows rapid switchover between the EI/CI and FAB conditions.[7] The capillaritron source used in this modification comprises a very fine bore capillary and a concentric accelerating electrode.[8] Ions are formed in the microdischarge inside the capillary and/or near the exit orifice; the gas flow and the ion-beam energy may be varied so that the contribution of fast atoms in the beam can be varied between 40% and 80%. Experiments with a similar capillaritron source using beam energies between 2 and 6 keV indicate that the higher beam energies yield FAB spectra that favour higher-mass ions.[9]

The modification of a standard, commercially available ion gun to provide a diffuse beam of fast neutral particles suitable for FABMS has been achieved by addition of a metal-ion-beam neutralizer and an ion-deflector grid at the end.[10] The advantage of this system is that it does not require the addition of a gas-collision chamber to produce the static neutral beam. A clean and stable beam and elimination of jet-orifice wear are also features of a small, rugged, and reliable gun that produces a high-intensity beam of mixed ions and neutral particles.[11] This gun may be fitted as an accessory to any mass spectrometer having a source-pumping speed of better than 0.5 ml min^{-1}.

[3] M. A. Baldwin, D. M. Carter, and K. J. Welham, *Org. Mass Spectrom.*, 1983, **18**, 176.
[4] D. Jerolamon and R. D. Sedgwick, presented at the 31st Annual Conference on Mass Spectrometry and Allied Topics, Boston, 8–13 May, 1983.
[5] M. A. Rudat, *Anal. Chem.*, 1982, **54**, 1917.
[6] J. F. Mahoney, J. Perel, P. C. Goodley, C. N. Kenyon, and K. Faull, *Int. J. Mass Spectrom. Ion Phys.*, 1983, **48**, 419.
[7] K. F. Faull, A. N. Tyler, H. Sim, J. D. Barchov, I. J. Massey, C. N. Kenyon, P. C. Goodley, J. F. Mahoney, and J. Perel, *Anal. Chem.*, 1984, **56**, 308.
[8] J. F. Mahoney, D. M. Goebel, J. Perel, and A. T. Forrester, *Biomed. Mass Spectrom.*, 1983, **10**, 61.
[9] In ref. 5.
[10] M. M. Ross, J. R. Wyatt, R. J. Colton, and J. E. Campana, *Int. J. Mass Spectrom. Ion Processes*, 1983, **54**, 237.
[11] R. A. McDowell and H. R. Morris, *Int. J. Mass Spectrom. Ion Phys.*, 1983, **46**, 443.

Conversion of AEI/Kratos instruments (MS9, MS12, and MS50) with the Ion Tech fast-atom gun, sample probe, and a simple ion source (which is also fitted with a filament and electron trap) enables alternate or simultaneous FAB/EI, the EI mode being particularly useful for calibration purposes.[12] Some difficulties have been experienced in the calibration of mass spectrometers for accurate mass measurement under FAB conditions, due primarily to the lack of suitable reference materials and the selective desorption of either reference material or sample. It is possible to use ions from the reference material (*e.g.* glycerol) as reference peaks, and this technique has been used successfully, by interpolation between the matrix ions using an accelerating voltage scan.[13] However, under ideal FAB conditions, the abundance of the matrix ions in the region of the interesting sample ions (*i.e.* those at high mass) is too low for them to be useful as references. It is possible to add a suitable internal standard to the matrix along with the sample material; a surfactant, which has useful ions between m/z 280 and m/z 780, has been used successfully for this purpose.[14] An interesting alternative is the use of a sample probe with a rotating target comprising two separate surfaces upon which the sample and calibrant are independently mounted.[15] As the probe is rotated, the sample and calibrant are presented alternately to the ion optics of the mass spectrometer. Although this technique has the advantage of enabling a wide range of reference compounds to be used, giving an almost unlimited reference mass range, the results achieved are not as good as those obtained using the internal reference method. A systematic study has been undertaken of the experimental variables in FAB with the aim of optimizing the conditions for accurate mass measurement at high resolution.[16] During the investigation of such parameters as the shape and material of the target, the nature of the ionizing gas, the preparation of the sample, its concentration in the matrix, and the composition of the matrix itself, it was concluded that the sample preparation had the most influence on the quality of the spectrum obtained.

The effects of the matrix material on the success of the FAB experiment have also been examined in some detail.[17] Several liquid organic substances have been demonstrated as possible matrix materials. The solubility of the sample in the matrix has been shown to be very important, and the sensitivity of the technique can be improved by spiking the sample solution with acids or salts. However, use of a matrix material is not always desirable or, indeed, possible. For example, glasses and glazes from archeological artifacts have been examined directly under FAB conditions by attaching the sample to the target with a small piece of double-sided cellophane tape.[18] In a similar study, bombarding a calibrated glass sample with xenon atoms of 6 keV energy during a depth-profiling experiment,

[12] A. M. Hogg, *Int. J. Mass Spectrom. Ion Phys.*, 1983, **49**, 25.
[13] U. Rapp, H. Kaufmann, M. Hohn, and R. Pesch, *Int. J. Mass Spectrom. Ion Phys.*, 1983, **46**, 371.
[14] J. M. Gilliam, P. W. Landis, and J. Occolowitz, *Anal. Chem.*, 1983, **55**, 1531.
[15] In ref. 14.
[16] S. A. Martin, C. E. Costello, and K. Biemann, *Anal. Chem.*, 1982, **54**, 2362.
[17] J. Meili and J. Siebl, *Int. J. Mass Spectrom. Ion Phys.*, 1983, **46**, 367.
[18] G. D. Dolnikowski, J. Throck Watson, and J. Allison, *Anal. Chem.*, 1984, **56**, 197.

a burn rate of 0.8 nm min^{-1} was indicated.[19] The sputter rate in FAB has also been closely examined using a film of Ta_2O_5, where the colour of the surface is dependent on the thickness of the Ta_2O_5 film.[20] After sputtering a sample of Ta_2O_5 of known thickness for a given length of time, the sputter rate can be calculated by comparing the colour of the used sample surface with a calibration strip.

The preparation of samples as targets for secondary-ion mass spectrometry (SIMS) has also been examined. One group of workers, working on the principle that a porous target, with its greater surface area, performs better than a smooth surface, has devised a method of producing surfaces with the required porosity by the thermal decomposition of metal salts on a metal foil.[21] Of the three types of metal surface produced (silver, gold, and tantalum), the porous silver surface produced the best results. In some cases, the addition of glycerol to the surface aided secondary-ion emission. A successful method of sample deposition on to the SIMS target involves the use of an electrospray, which has recently been modified to enable the charged droplets to be focused over an area as small as 1 mm in diameter, thus allowing very small samples to be examined.[22]

Of great interest to all workers in the FAB and SIMS fields is obviously the importance of the size and energy of the primary impacting particle, since this, more than any other parameter, gives an insight into the mechanism of the ionizing process. A review of all aspects of SIMS and FAB places particular emphasis on the study of the mechanisms involved and on an examination of the atomic structure of the surface at the point of impact.[23] Another review of organic SIMS also discusses the mechanism of the ionization process.[24] An examination of the secondary-ion yield in SIMS, using standardized sample-handling techniques and indium ions from a liquid-metal-ion source as the primary beam, has led to a greater understanding of the mechanisms of secondary-ion emission.[25] Several workers have made important comparisons between the results obtained in SIMS and those obtained in FABMS. In one group of experiments, which compares the results of FABMS using argon atoms with SIMS using liquid-metal-ion sources of gallium, indium, and bismuth, a 10- to 50-fold increase was observed in molecular-ion abundance under SIMS conditions, although no essential qualitative differences were noticed between the liquid-metal-ion SIMS and other soft-ionization techniques, including FAB.[26] In

[19] C. J. Wakefield, D. Hazelby, L. C. E. Taylor, and S. Evans, *Int. J. Mass Spectrom. Ion Phys.*, 1983, **46**, 491.
[20] B. L. Bentz and P. J. Gale, *Anal. Chem.*, 1983, **55**, 1434.
[21] K. D. Kloppel, K. Weyer, and G. von Bunau, *Int. J. Mass Spectrom. Ion Phys.*, 1983, **51**, 47.
[22] K. G. Standing, R. Beavis, W. Ens, and B. Schueler, *Int. J. Mass Spectrom. Ion Phys.*, 1983, **53**, 125.
[23] B. J. Garrison and N. Winograd, *Science*, 1982, **216**, 805.
[24] A. Benninghoven, *Int. J. Mass Spectrom. Ion Phys.*, 1983, **53**, 85.
[25] D. F. Barofsky, A. M. Ilias, E. Barofsky, and J. H. Murphy, presented at the 32nd Annual Conference on Mass Spectrometry and Allied Topics, San Antonio, 27 May to 1 June, 1984.
[26] D. F. Barofsky, U. Giessmann, A. E. Bell, and L. W. Swanson, *Anal. Chem.*, 1983, **55**, 1318.

an experiment providing simultaneous operation of a xenon-atom gun directed at 40° to the target surface and a caesium-ion gun directed at 20° to the surface, a three-fold increase in sensitivity was achieved with the caesium ions.[27] A similar result was observed using the source-mounted capillaritron fast-atom gun already described to compare the results of argon-atom and xenon-atom FAB with caesium-ion SIMS.[28] The relative abundances of the higher-mass ions are greater than the low-mass ions when the heavier particles are used. The caesium-ion gun used in these experiments is also source-mounted close to the sample, and this mounting procedure eliminates the need for a special line-of-sight inlet port.[29] Since it has a low gas load, it is available for installation on instruments with a limited source-pumping capacity. The gun also features simple construction from low-cost materials. An enhancement factor of approximately 10 has been achieved in the FAB mode by using mercury atoms instead of argon atoms, although the spectra show no significant differences in fragmentation pattern.[30] It has been noted in another series of experiments that for a sample of vitamin B_{12} the yield of secondary ions is approximately in the ratio of the atomic weights of the primary beam for xenon, argon, and neon (*i.e.* xenon gives a higher yield than argon, which gives a higher yield than neon).[31] Unfortunately, this relationship could not be shown to hold true for other sample materials. Krypton has recently been used successfully to study oligosaccharides.[32]

One method of ionizing such large polar molecules intact, which has a rapidly growing following, is ^{252}Cf plasma desorption. Torgerson's review illustrates the long history of the technique;[33] Macfarlane's vigorous article examines all aspects of the method.[34] Another similar method, which has also continued to attract experimental work, uses lasers to desorb ions from a target. The range of analytical techniques that make use of lasers has been described, together with some of the history of the use of lasers in mass spectrometry.[35] A special double-focusing mass spectrometer has been designed primarily to measure the kinetic-energy distributions of the ions produced by laser desorption particularly (but also other soft-ionization techniques like FAB and SIMS).[36] The instrument is characterized by a readily interchangeable source so that not only can the ionization method be switched easily but also an EI source can be fitted for calibration purposes. The detection system incorporates a channel-plate system with vidicon camera and multichannel analyser. For inorganic analysis, resonance-ionization mass spectrometry using laser ionization offers the ad-

[27] W. Aberth, K. L. Straub, and A. L. Burlingame, *Anal. Chem.*, 1982, **54**, 2029.
[28] M. A. Rudat and C. N. McEwen, *Int. J. Mass Spectrom. Ion Phys.*, 1983, **46**, 351.
[29] C. N. McEwen, *Anal. Chem.*, 1983, **55**, 967.
[30] R. Stoll, U. Schade, F. W. Rollgen, U. Giessmann, and D. F. Barofsky, *Int. J. Mass Spectrom. Ion Phys.*, 1982, **43**, 227.
[31] H. R. Morris, M. Panico, and N. J. Haskins, *Int. J. Mass Spectrom. Ion Phys.*, 1983, **46**, 363.
[32] J.-L. Aubagnac, F. M. Devienne, and R. Combabieu, *Org. Mass Spectrom.*, 1983, **18**, 173.
[33] D. F. Torgerson, *Int. J. Mass Spectrom. Ion Phys.*, 1983, **53**, 27.
[34] R. D. Macfarlane, *Anal. Chem.*, 1983, **55**, 1247A.
[35] R. J. Cotter, *Anal. Chem.*, 1984, **56**, 485A.
[36] G. J. Q. van der Peyl, W. J. van der Zande, K. Bederski, A. J. H. Boerboom, and P. G. Kistemaker, *Int. J. Mass Spectrom. Ion Phys.*, 1983, **47**, 7.

vantage over thermal ionization of high elemental selectivity, but with a precision equal to that of single-filament thermal ionization.[37]

Perhaps the ultimate particle-desorption ion source uses fast iron dust particles (0.2–1.5 μm in diameter) in the velocity range 1–60 km s^{-1}.[38] These particles are normally used to simulate cosmic dust impinging on spacecraft. In mass spectrometry experiments they have been used to desorb organic materials that have been electrosprayed on to metal matrices (adenosine on silver foil and tetrabutyl ammonium iodide on iron are cited examples). The velocity of the particles over the stated range has little effect on the relative abundances of the ions in the spectra, but the highest ion yields are obtained using the slowest (*i.e.* largest) particles.

Although probably the most commonly used techniques for producing ions from large, polar molecules, the particle-desorption techniques are not the only methods available. Field ionization and field desorption continue to attract experimentation. In a recent study[39] the temperature distribution across FI/FD emitter wires has been accurately predicted by computer. The findings in some respects run contrary to popular opinion: long emitters produce hotter central temperatures than short ones; at a given heating current, a pure tungsten wire is hotter than a carbon emitter, which is hotter than a cobalt emitter, all of a similar size and shape; long dendrites (which have long been the aims of all field-desorption researchers) cool the wire quicker than short ones, although cooling along the axis of the dentrite is only slight.[39] The conclusions to be drawn from this study are that, for accurate comparisons of emitter temperatures from literature references, it is important to know the emitter length and heating current and the dendrite length, density, and composition.

Organic ions may also be desorbed from hot filaments by surface ionization, although this technique is usually more suitable for inorganic materials. However, several organic compounds produced $(M-1)^+$ ions when desorbed by surface ionization from a hot rhenium wire inserted through the insertion lock of a Finnigan quadrupole instrument.[40] In an extension of the work, a variation of sensitivity with the temperature of the emitter was noted; the emitter temperature was measured with an optical pyrometer.[41] The home-built combination ion source used in this work allowed an interchange between surface ionization and EI. In a comparison of the results achieved by the two ionization methods, it was noted that some compounds produced a base peak of significantly higher abundance (greater than 13 times in one case) in surface ionization than in EI, which indicates a potential use for the technique for specific organic compounds. A special fixture has been described that both forms a V-shaped filament from 0.03 mm thick rhenium ribbon and holds it in place for spot-

[37] D. L. Donohue, D. H. Smith, J. P. Young, H. S. McKown, and C. A. Pritchard, *Anal. Chem.*, 1984, **56**, 379.

[38] F. R. Krueger and W. Knabe, *Org. Mass Spectrom.*, 1983, **18**, 83.

[39] D. F. Fraley, L. G. Pedeven, and M. M. Bursey, *Int. J. Mass Spectrom. Ion Phys.*, 1982, **43**, 99.

[40] T. Fujii, *Int. J. Mass Spectrom. Ion Processes*, 1984, **57**, 63.

[41] T. Fujii, presented at the 31st Annual Conference on Mass Spectrometry and Allied Topics, Boston, 8–13 May, 1983.

welding on to support posts.[42] The entire operation takes only 30 s and provides a convenient method for the manufacture of filaments for surface ionization. A technique that provides a pulse of 100 A at peak current intensity and 5 ns duration to a tungsten filament upon which the sample is coated allows a time-of-flight instrument to measure the masses of the desorbed ions.[43] The preparation of a 'thermionic potassium-ion glass' ($K_2O:Al_2O_3:SiO_2$, 1:1:2) was described as long ago as 1936 as a source of gaseous potassium ions.[44] The glass is heated by passing a current through its rhenium wire support; non-volatile samples (*e.g.* peptides deposited on the thermionic emitter) are desorbed as $(M+K)^+$ ions from the surface of the glass. Molecular species can also be 'desorbed' from a conventional FD emitter coated with nitrocellulose (typically from a 2.5% solution), then with the sample.[45] The loaded (pun intended!) emitter is then inserted into the FD/EI source in the filament-on mode; when the current is applied to the emitter, the nitrocellulose explodes and propels the sample into the electron beam. Only low-mass ions (due to H_2O, CO, NO, and NO_2, the decomposition products) are detected from the nitrocellulose.

Of the methods available for direct ionization of samples from the liquid phase, by far the most closely studied uses the technique of atmospheric-pressure ionization (API). A review article discusses the principles and applications of API.[46] It is a 'soft' ionization technique, which produces mainly quasi-molecular ions; fragment ions can be obtained by collision-induced dissociation (CID) in a high-pressure region, then analysed by a second mass spectrometer using the technique of MS/MS.[47] An API source coupled with a triple-quadrupole MS/MS instrument in this way is claimed to require less maintenance than other ion sources such as EI or CI.[48] A modification of the API technique uses metastable argon atoms produced in a corona discharge in the argon gas at or near to atmospheric pressure.[49] The sample is injected into the corona discharge in a matrix, such as glycerol or nujol. Alternatively, the sample (in its matrix) can be smeared on to a heated needle tip situated in the corona discharge and close to the sampling orifice of the mass spectrometer.[50] By suitable choice of matrix, a useful spectrum can be obtained from 10^{-7}–10^{-9} g of sample, giving a detection limit in the single-ion monitoring mode of about 10^{-13} g. By injecting a small amount (about 100 p.p.m.) of xenon or krypton into the corona discharge, the ionization of compounds having a higher ionization energy than the xenon or krypton can be quenched;[51] in this way, ions having the same nominal

[42] J. J. Stoffels and C. R. Lagergren, *Int. J. Mass Spectrom. Ion Processes*, 1983/4, **55**, 217.
[43] P. K. D. Feigl, F. R. Krueger, and B. Schueler, *Org. Mass Spectrom.*, 1983, **18**, 442.
[44] D. Bombick, J. D. Pinkston, and J. Allison, *Anal. Chem.*, 1984, **56**, 396.
[45] R. S. Gohlke and L. S. Wakeman, *Anal. Chem.*, 1982, **54**, 2114.
[46] R. K. Mitchum and W. A. Korfmacher, *Anal. Chem.*, 1983, **55**, 1485A.
[47] B. A. Thompson, J. V. Iribarne, and P. J. Dziedzic, *Anal. Chem.*, 1982, **54**, 2219.
[48] V. J. Caldecourt, D. Zakett, and J. C. Tou, *Int. J. Mass Spectrom. Ion Phys.*, 1983, **49**, 233.
[49] M. Tsuchiya, T. Taira, H. Kuwabara, and T. Nonaka, *Int. J. Mass Spectrom. Ion Phys.*, 1983, **46**, 355.
[50] M. Tsuchiya and H. Kuwabara, *Anal. Chem.*, 1984, **56**, 14.
[51] M. Mitsui, H. Kambara, M. Kojima, H. Tonuta, K. Katoh, and K. Satoh, *Anal. Chem.*, 1983, **55**, 477.

molecular weight can be identified by the difference in their ionization energies, and trace impurities can be determined quantitatively down to about 10 p.p.t.

Electrohydrodynamic ionization (EHI) is a similar high-pressure liquid-ionization technique that has also been examined.[52] In this work, the EHI inlet was provided with some visual control so that the processes occurring at the end of the inlet capillary could be photographed. Of particular interest was the effect on the sample meniscus inside the capillary of changing the voltage between the capillary and its counter electrode. An EHI source with a smaller than normal emitter aperture, an increased extraction electrode aperture, and a higher potential across the source enabled the extraction of ions from solution in volatile solvents.[53] Under more normal conditions and with a less volatile sample, PEG 4000, quasi-molecular ions with up to four positive charges have been observed, and triply and quadruply charged sodium adducts of the PEG 4000 have been sampled without fragmentation.[54] From the abundances of these ions, an average molecular weight was calculated for the sample, which was in good agreement with the ASTM standard method.

In an attempt to reduce the cost of high-pressure CI sources, a description has been given of a source that has the major parts manufactured quickly and cheaply in aluminium.[55] The internal surfaces of the aluminium source are then coated with a clean, non-reactive, and highly conducting surface comprising a layer of gold. The gold is vacuum-deposited to a thickness of 100 nm, although no attempt has been made to study the effects of changing the thickness of the gold layer. Although the gold adheres moderately well to the aluminium, the source cannot be cleaned by conventional methods; the sources have a lifetime of about four months. Operation of the CI source below 100 °C, which simplifies the analysis of some thermally labile molecules, can be achieved by using filaments of either lanthanum hexaboride or thin rhenium wire.[56] The lanthanum hexaboride filaments are unfortunately susceptible to poisoning by solvents, especially methanol.

The analysis of inorganic materials has for many years been carried out by spark-source mass spectrometry and, despite the emergence of new techniques that threaten to usurp spark-source mass spectrometry and despite a lack of enthusiasm for commercial support of the latter technique, it continues to be widely used. In a recent series of experiments that examined the effects of altering the pulse length, frequency, and voltage of the spark, it was found that the pulse length and frequency had no effect on the spectrum obtained; the relationship between breakdown voltage, spark gap, and sample matrix, however, showed that with increasing breakdown voltage the abundance of multiply charged ions decreased whereas that of cluster ions increased.[57] Additionally,

[52] M. L. Alexandrov, L. N. Gall, N. V. Krasnov, V. I. Nikolaev, V. A. Pavlenko, and V. A. Shkurov, *Int. J. Mass Spectrom. Ion Processes*, 1983, **54**, 231.

[53] S. L. Murawski and K. D. Cook, *Anal. Chem.*, 1984, **56**, 1015.

[54] K. W. S. Chan and K. D. Cook, *Org. Mass Spectrom.*, 1983, **18**, 423.

[55] K. Blom, J. J. Hilliard, H. S. Gold, and B. Munson, *Anal. Chem.*, 1982, **54**, 1898.

[56] L. Kelner, S. P. Markey, H. M. Fales, C. K. Crawford, and P. A. Cole, *Int. J. Mass Spectrom. Ion Phys.*, 1983, **46**, 3.

[57] L. Vos and R. van Grieken, *Int. J. Mass Spectrom. Ion Processes*, 1983/4, **55**, 233.

the absolute and relative responses for different elements were affected in different ways.

One of the recent pretenders to the spark-source throne in inorganic analysis is FAB. Inorganic materials can be analysed by FAB mass spectrometry, often using the intact specimen and without the need for a liquid matrix.[58] In one example of the direct FAB analysis of an inorganic material, the ratios of lead isotopes were measured to better than 1% precision. However, the technique that has emerged to challenge spark-source mass spectrometry most strongly uses a d.c. glow discharge, usually in argon.[59] This method provides a more stable ion beam than does the spark source, and it produces ions that are less multiply charged; it also permits a quicker sample turnround. In comparison with optical-emission spectroscopy, it produces simpler spectra with a uniform sensitivity for all elements. The standard d.c. glow-discharge source can be improved by the superimposition of a magnetic field either perpendicular to or parallel with the electric field.[60] This affords an enhancement of the signal arising from sputtered neutral species since the lifetimes of the electrons in the discharge are increased, thus increasing the probability of ionization of those neutral species. The superimposed magnetic field also allows the discharge to be operated at a lower source pressure. In an evaluation of a pulsed glow-discharge source, the higher-current and -voltage conditions used gave rise to larger atomic and ionic populations of the cathodic material and thus to more efficient utilization of the sample.[61] A glow-discharge source has also been designed with a special quartz window to allow the optical study of the atomic population within the source simultaneously with mass-spectrometric sampling.[62] Thus, this instrument is capable of simultaneous atomic-absorption spectrometry and mass spectrometry (AAS/MS). A Penning-discharge cold-cathode ion source has been modified for use in the analysis of inorganic materials.[63] The sample, supplied as a foil or as a small piece, is placed in an aperture in the cathode; an aperture in an opposite cathode provides a means for the exit of ionized sputtered sample atoms and molecules.

The combined development of inductively coupled plasma source mass spectrometry (ICP/MS) has been described in a review article.[64] Whereas previously published results have been obtained in the boundary-layer sampling mode, some preliminary results have been reported using continuum (bulk-plasma) sampling.[65] Boundary-layer sampling is reported to give better signal-to-noise ratios and better detection limits than continuum-mode sampling but to suffer some major disadvantages, such as aperture blockage at high salt concen-

[58] See refs. 18 and 19.

[59] J. E. Cantle, E. F. Hall, C. J. Shaw, and P. J. Turner, *Int. J. Mass Spectrom. Ion Phys.*, 1983, **46**, 11.

[60] B. L. Bentz and W. W. Harrison, *Anal. Chem.*, 1982, **54**, 1644.

[61] P. J. Savickas and W. W. Harrison, presented at the 32nd Annual Conference on Mass Spectrometry and Allied Topics, San Antonio, 27 May to 1 June, 1984.

[62] T. J. Loving and W. W. Harrison, *Anal. Chem.*, 1983, **55**, 1523; T. J. Loving and W. W. Harrison, *Anal. Chem.*, 1983, **55**, 1526.

[63] D. L. Swingler, *Int. J. Mass Spectrom. Ion Processes*, 1983, **54**, 341.

[64] A. L. Gray and A. R. Date, *Analyst (London)*, 1983, **108**, 1033.

[65] A. R. Date and A. L. Gray, *Analyst (London)*, 1983, **108**, 159.

trations, the formation of oxide and hydroxide complexes, and some serious ionization suppression, effects that are not encountered in continuum sampling.

3 Sample Introduction

A simple alternative to desorption techniques for the direct analysis of underivatized peptides, carbohydrates, and nucleotides comprises the introduction of solutions of these compounds from a pyrex capillary with a small outlet orifice.[66] The capillary can be mounted in the standard solids sample probe in a Finnigan quadrupole instrument, operating in the CI mode. An exhaustive study of the instrumental parameters necessary for the optimum use of commercial direct-inlet probes has been carried out.[67] In this study, probe tips manufactured from gold, stainless steel, platinum, neutral glass, copper, and silver were tested in conditions of both electrical isolation from and electrical connection with the ion-source potential. Additionally, the gold probes were manufactured with several differently shaped tips: straight, twisted wire, finned and flat, horizontal spiral. The conclusions were drawn that, using cysteine as the reference material, no significant differences in the spectra were observed with probes of different materials; differently shaped probe tips produced comparable signal levels but different evaporation profiles. A similar study of the effects of the shape of the probe tip on the resulting spectrum concluded that a long, thin gold wire gives a better result than a thicker, shorter wire of the same mass.[68] Other shapes examined included a loop (on to which solutions can be injected) and a flat spiral for solid samples.

A temperature-programmed fractionation inlet system that provides a version of pyrolysis/mass spectrometry useful in the thermal analysis of polymers has been described.[69] The inlet system is interfaced with the mass spectrometer's ion source through a jet separator, and the whole fractionation process, including desorption, evaporation, distillation, and pyrolysis, is carried out under a stream of helium.[70] Since the system can be dismantled for cleaning without disturbing the mass spectrometer, it is ideally suited for handling involatile, untreated, or messy samples. The prevention of serious source contamination, even under the arduous conditions of high sample throughput, is also a feature of a Curie-point pyrolysis/mass-spectrometry system.[71] A probe containing a platinum filament, which has been used for linear temperature-programmed thermal-degradation mass spectrometry, has been improved by the redesign of the probe and controller to allow an accurate measurement to be made of the decomposition temperature of the sample.[72] Polymers have also been character-

[66] R. C. Dougherty, J. de Kanel, and C. Grimm, presented at the 32nd Annual Conference on Mass Spectrometry and Allied Topics, San Antonio, 27 May to 1 June, 1984.
[67] P. Traldi, U. Vertori, and F. Dragoni, *Org. Mass Spectrom.*, 1982, **17**, 587.
[68] E. Constantin and R. Hueber, *Org. Mass Spectrom.*, 1982, **17**, 460.
[69] K.-h. C. Chan, R. S. Tse, and S. C. Wong, *Int. J. Mass Spectrom. Ion Phys.*, 1983, **46**, 119.
[70] K.-h. C. Chan, R. S. Tse, and S. C. Wong, *Anal. Chem.*, 1982, **54**, 1238.
[71] R. Tsao and K. J. Voorhees, *Anal. Chem.*, 1984, **56**, 368.
[72] T. H. Risby, J. A. Yergey, and J. J. Scocca, *Anal. Chem.*, 1982, **54**, 2228.

ized using thermogravimetry coupled through an all-glass transfer line with a ball-and-socket joint to a triple-quadrupole mass spectrometer having an API source.[73] The API source permits the routine use of an oxidative purge gas, while the ball-and-socket joint permits any one of several thermogravimetric analysers to be easily coupled and uncoupled.

An interesting method of elemental analysis has been developed by interposing a microwave plasma unit between a gas chromatograph and a high-resolution mass spectrometer.[74] In the microwave plasma, complex molecules are converted into a few simple neutral species, the quantity and composition of which are determined by the elements present in the original molecule.[75] Determination of the elemental composition of the original molecule is therefore possible by the separation, identification, and quantification of these simple neutral species. A similar system has also been used to simplify the detection and quantification of compounds containing radioactive carbon.[76] An inlet system has been developed for the measurement of carbon monoxide in combustion emissions, which comprises a liquid-nitrogen cold trap through which the low-pressure sample stream is passed.[77] The cold trap removes all organic species except CH_4 and CO; the resulting stream, containing only N_2, O_2, CH_4, and CO, is passed over a platinum catalyst. By precise control of the temperature of the catalyst at 560 °C, the CO is selectively oxidized to CO_2 (leaving the CH_4 unaffected), which can then be measured in the presence of the N_2.

Jet separators, whether used in a standard configuration as part of a GC/MS interface or in a non-standard manner as a special type of inlet system, have small orifices that are prone to clogging by minute particles, and various methods have been described for cleaning out these jet orifices. One method that has recently been described uses a high-voltage spark to vaporize the blockage.[78] A full description is given of how to set up the apparatus, including a wise warning that the voltages used are lethal. If this method (and all the other proprietary methods) fails, then the blockage will probably succumb to treatment with 10% HF solution, which, it is supposed, dissolves the glass in the area of the blockage so that the offending particle can be removed by application of a vacuum.[79] As well as pointing out that a separator with a permanently blocked jet is less useful than one with slightly enlarged orifices, the article contains graphic warnings about the health hazards of HF and about the use of too concentrated a solution of HF, which could do permanent damage to the jet separator.

[73] B. Shushan and R. B. Prime, presented at the 31st Annual Conference on Mass Spectrometry and Allied Topics, Boston, 8–13 May, 1983.
[74] R. A. Heppner and D. W. Proctor, presented at the 32nd Annual Conference on Mass Spectrometry and Allied Topics, San Antonio, 27 May to 1 June, 1984.
[75] R. A. Heppner, *Anal. Chem.*, 1983, **55**, 2170.
[76] S. P. Markey and F. P. Abramson, *Anal. Chem.*, 1982, **54**, 2375.
[77] L. P. Haack, J. W. Butler, and A. D. Colvin, *Anal. Chem.*, 1982, **54**, 2547.
[78] F. L. Cardinali and L. E. Lowe, *Anal. Chem.*, 1982, **54**, 1454.
[79] W. E. Wentworth, Y.-c. Chen, A. Zlatkis, and B. S. Middleditch, *Anal. Chem.*, 1982, **54**, 1895.

Porous polymeric membranes mounted on flexible probes can be used as inlet systems to mass spectrometers for the measurement of trace gases; these simple, inexpensive devices have been used for the monitoring of trace contaminants in both aqueous solutions and air.[80] An obvious application of this technique is in pollutant analysis;[81] a mobile mass spectrometer fitted with a membrane inlet system has been designed specifically for the simultaneous measurement of 22 different organic pollutants in air, or of up to 12 on a surface when a specially designed surface sampler is used. An API mass spectrometer has been adapted for a respiratory application using a special inlet system; the instrument enables the extraction of organic materials directly from the gaseous mixture.[82]

The successful marriage between the separative powers of gas chromatography and the analytical powers of mass spectrometry is well established. The marriage of high-performance liquid chromatography with mass spectrometry has been a much more difficult one that, after several false starts, has only recently begun to settle down. Developments, trends, and applications of GC/MS and LC/MS are discussed in Chapter 9 of this volume. The widely used technique of thin-layer chromatography (TLC) is less well suited to direct interfacing with the mass spectrometer than either of the other aforementioned chromatographic techniques, and most of the mass spectrometry of TLC extracts has been carried out on an off-line basis, generally using direct-probe techniques. One such method enables the TLC spot to be analysed directly by FAB mass spectrometry.[83] Once the sample has been located on the TLC plate by, say, a UV scanner, the FAB sample-introduction probe, with the target covered with a layer of double-sided sticky tape, is pressed on to the TLC spot. If the area of the plate containing the sample is lightly scored prior to the sampling process, then sample is removed without the removal of an excess of absorbent. A small amount (1–2 μl) of a suitable solvent (*e.g.* chloroform or methanol) and 2–5 μl of a chosen matrix material are then added to the sample adhering to the FAB targets, and FABMS is carried out in the normal way.

Attempts have occasionally been made to develop a simple on-line TLC/MS system, but so far none has been sufficiently attractive to warrant commercial development. A device has recently been described that scans a 1 cm $\times$ 10 cm TLC plate past a source of desorption energy, which can be either infrared/visible radiation from a tungsten-filament lamp or radiation from a pulsed CO_2 laser.[84] CI reactant gas is swept over the surface of the TLC plate and transports the desorbed sample through a heated transfer line inserted through the direct sample-inlet port into the ion source. The detection level of this device has been quoted at about 1 ng; quantitative determinations using an internal standard give around 20% reproducibility.

[80] G. J. Kallos and N. H. Mahle, *Anal. Chem.*, 1983, **55**, 813.

[81] J. Franzen, G. Heinen, G. Matz, and G. Weiss, presented at the 31st Annual Conference on Mass Spectrometry and Allied Topics, Boston, 8–13 May, 1983.

[82] F. M. Benoit, W. R. Davidson, A. M. Lovatt, S. Nacson, and A. Ngo, *Anal. Chem.*, 1983, **55**, 805.

[83] T. T. Chang, J. O. Lay, jun., and R. J. Francel, *Anal. Chem.*, 1984, **56**, 111.

[84] L. Ramaley, M. E. Nearing, M.-A. Vaughan, R. G. Ackman, and W. D. Jamieson, *Anal. Chem.*, 1983, **55**, 2285.

4 High-mass Analysis

The increased activity in the study of high-molecular-weight compounds occasioned by the almost universal acceptance of FABMS has continued unabated during the period under review. However, the unprecedented instrumental development described in the previous report of this series has not continued to such a great extent. Lessons learned by the manufacturers in developing medium-performance instruments for high-mass analysis have simply been transferred to the large instruments. Thus, for instance, the magnet design incorporating increased radius and reduced sector angle, which took VG Analytical's MM7070H from a mass range of 850 daltons to that of 2000 daltons for the 7070E, has been incorporated in the ZAB,[85] a mass range of 12 000 daltons being claimed for the ZAB-SE at full accelerating potential. Similarly, the segmented, inhomogeneous-field magnet technology used first on the Kratos MS80RF[86] has been adapted for use with the MS50, the new MS50RF having a mass range of 10 000 daltons at full accelerating potential.[87] This magnet technology has also been scaled down for incorporation into the smallest of the Kratos range of instruments, the MS25RF now being provided with a very useful 2000 dalton mass range under normal operating conditions.[88]

However, some serious questions have been raised about the measurement of molecular weights of large molecules. A recent review article on FABMS provides an excellent summary of the problems attendant on measuring the masses of large molecules, and it raises questions of whether unit mass resolution at high mass is really desirable.[89] A more detailed account of the problems involved is given in a discussion of the distributions of naturally occurring isotopes in large molecules.[90] In this discussion the suggestion is made that, because of the extremely low abundance of the monoisotopic molecular ion in high-molecular-weight materials compared with the abundance of ions containing more than one naturally occurring isotope of each element, and because of the multiplicity of isobaric ions caused by isotope overlap, the average mass of the isotope cluster, calculated to the best possible accuracy, and the width and distribution of the molecular-ion cluster are exploited to assign and confirm molecular weights, using the average chemical weights of the elements involved. Such a suggestion lends weight to the advocates of time-of-flight (TOF) instrumentation, which, with its almost unlimited mass range but very limited resolution, is capable of accurate measurement of the average mass of a molecu-

[85] R. H. Bateman, R. S. Bordoli, B. N. Green, M. Barber, A. N. Tyler, K. Faull, and J. D. Barchas, presented at the 32nd Annual Conference on Mass Spectrometry and Allied Topics, San Antonio, 27 May to 1 June, 1984, VG Analytical, Provisional Information VGA, TN.1/84.

[86] D. R. Denne, C. J. Wakefield, R. S. Stradling, D. Hazelby, and L. E. Moore, *Int. J. Mass Spectrom. Ion Phys.*, 1983, **46**, 47.

[87] J. S. Cottrell, L. C. E. Taylor, and D. H. Williams, presented at the 32nd Annual Conference on Mass Spectrometry and Allied Topics, San Antonio, 27 May to 1 June, 1984.

[88] Kratos Analytical Specification, A357-0684EM, 1984.

[89] C. Fenselau, J. Yergey, and D. Heller, *Int. J. Mass Spectrom. Ion Phys.*, 1983, **53**, 5.

[90] J. Yergey, D. Heller, G. Hansen, R. J. Cotter, and C. Fenselau, *Anal. Chem.*, 1983, **55**, 353.

lar-ion envelope. The recent commercial development of ^{252}Cf plasma desorption on a small TOF analyser exploits this concept.[91] High-mass analysis can also be achieved on a TOF instrument using a caesium primary-ion source with a pulsed primary beam, using pulses of a few ns duration spaced 200 ns apart.[92] However, despite the known problems of measuring the masses of larger molecules, which are not eased by increased resolution, several successful attempts have been made to increase the resolution of TOF instruments. For example, a simulation study of a new focusing principle for the TOF mass spectrometer, velocity compaction, predicts that unit mass resolution at 10 000 daltons should be achievable with a simple single-stage instrument.[93] An actual increase in performance has been achieved in a TOF mass spectrometer with a drift length of 1.6 m, to give a resolution of 3000 (fwhm definition).[94] This compares favourably with a resolution of 250 obtained with a standard instrument with a 1.5 m drift tube. The device that ensures this increase in performance is the 'reflectron', which compensates for the difference in times of flight of ions of differing energies by means of electrostatic fields and an ion reflector, and results in ions of greater velocity penetrating further into the reflectron. Thus, faster ions travel a greater distance and spend more time in the reflector system than slower ones; this compensates for the initial spread in ion velocities. A simpler solution has been adopted on a Bendix 12-101 TOF mass spectrometer by incorporating an energy filter (actually the electrostatic sector from a DuPont 21-491B) in front of the TOF drift tube.[95] This device also removes the dependence on the ion's starting position and reduces the energy spread, giving rise to higher resolution. It is suggested that this instrument may also have applications for MS/MS. A discussion of the effects of ion fragmentations in the drift tube of a TOF mass spectrometer points out that, in contrast with sector instruments and quadrupole mass filters, fragmenting particles that have no energy change are not lost to the spectrum in a TOF instrument.[96] By applying the appropriate potentials to deceleration grids incorporated in the drift tube, however, metastable ions can be rejected and only stable ions detected.

5 Reaction Studies

By combining the principles of operation of the TOF mass spectrometer with those of a double-focusing sector instrument, a novel type of multi-dimensional mass spectrometer can be achieved.[97] By pulsing the ion beam and using time-

[91] B. Sundqvist, I. Kamensky, P. Hakansson, J. Kjellberg, M. Salehpour, S. Widdiyasekera, J. Fohlman, P. A. Peterson, and P. Roepstorff, *Biomed. Mass Spectrom.*, 1984, **11**, 242; Bio-Ion Nordic AB Technical Literature, 1984.

[92] In ref. 22.

[93] M. L. Muga, presented at the 32nd Annual Conference on Mass Spectrometry and Allied Topics, San Antonio, 27 May to 1 June, 1984.

[94] D. M. Lubman, W. E. Bell, and M. N. Kronick, *Anal. Chem.*, 1983, **55**, 1437.

[95] J. D. Pinkston, J. Allison, and J. T. Watson, presented at the 32nd Annual Conference on Mass Spectrometry and Allied Topics, San Antonio, 27 May to 1 June, 1984.

[96] B. T. Chait, *Int. J. Mass Spectrom. Ion Phys.*, 1983, **53**, 227.

[97] C. G. Enke, J. T. Stults, J. F. Holland, J. D. Pinkston, J. Allison, and J. T. Watson, *Int. J. Mass Spectrom. Ion Phys.*, 1983, **46**, 229.

resolved detection techniques in a normal magnetic-sector instrument, simultaneous measurements of momentum and velocity can be made; this provides energy-dependent ion-mass assignments.[98] Thus, product ions formed from the same precursor ion appear at the same arrival time (*i.e.* they have the same time of flight through the instrument) but are dispersed according to momentum (they arrive at different magnetic fields). Stable ions have shorter flight times than metastable ions because of their greater initial velocity. Further developments and applications of this technique should be very interesting.

The study of reactions occurring in the first field-free region of a double-focusing mass spectrometer by scanning the two sectors in a constant ratio (the linked-scan technique) continues, with both hardware and software having been described to generate the accurate functions necessary for the successful transmission of the required ions. A digital real-time mass-scale linearizer and function generator incorporating a mass display and mass marker operates between m/z 0.0 and m/z 1638.3 without the need for a data system.[99] Twelve types of scan are provided for instruments of either conventional or reverse geometry. A low-cost digital electric-sector scanner designed to enable ion-kinetic-energy spectroscopy (IKES) and energy-release experiments to be carried out on older instruments has been described.[100] A new type of linked scan, available only with reverse-geometry instruments (in which ions pass through the electric sector after the magnetic sector), is available to study fragmentations in the field-free region between the two sectors of such instruments.[101] The performance of the scan, B^2/E = constant (B = magnetic field, E = electric-sector voltage), has been investigated by detailed experiment, and its major advantage has been shown to be the high resolving power available for the study of precursor ions of a chosen fragment ion. In the experiments described, two precursor ions of $C_7H_7^+$ differing in mass by only about 270 p.p.m. were successfully separated with a 20% valley; the possibilities of accurate mass determinations of such precursor ions are also discussed.

The linked scan that generates fragment-ion spectra from a given precursor ion uses a constant ratio of B/E. Under normal circumstances, in order to generate fragment-ion scans from different precursor ions, different B/E vectors must be calculated, one for each new precursor ion. However, the same B/E vector can be used for different precursor ions more easily, simply by altering the accelerating voltage to a new value so that the desired precursor ion is transmitted.[102] This elegant solution to a potentially time-consuming problem also has applications in selected metastable-peak monitoring (the monitoring of specific reactions using predetermined values of B and E), especially with the use of internal standards.

[98] J. T. Stults, C. G. Enke, and J. F. Holland, *Anal. Chem.*, 1983, **55**, 1323.
[99] F. Friedli, *Org. Mass Spectrom.*, 1984, **19**, 183.
[100] M. B. Alberti, P. Traldi, R. Cappadonna, and G. Moneti, *Int. J. Mass Spectrom. Ion Processes*, 1984, **57**, 141.
[101] R. K. Boyd, C. J. Porter, and J. H. Beynon, *Int. J. Mass Spectrom. Ion Phys.*, 1982, **44**, 199.
[102] N. W. Davies, J. C. Bignall, and R. W. Lincolne, *Org. Mass Spectrom.*, 1982, **17**, 451.

A method of ensuring that all fragment ions of all possible precursor ions are recorded in a single experiment is ion-decomposition mapping.[103] The ion-decomposition (or metastable) map is produced by a rapid, repetitive scan of the magnetic field in combination with a stepwise decrement of the electric-sector voltage under computer control at the end of each magnet scan.[104] A map of the complete B–E plane is recorded by the data system and can be reproduced as a three-dimensional perspective diagram.[105] Any vector of B and E can then be extracted from the stored information and reproduced as a spectrum; product-ion scans (B/E = constant), precursor-ion scans (B^2/E = constant), and scans of constant neutral loss can be reproduced in this way for any ion in the mass spectrum.

An alternative method of producing the metastable map comprises the superimposition of a fast, repetitive scan of the electric-sector voltage (2–5 s full scale) on a slow, continuous magnet scan (2000–5000 s decade^{-1}).[106] A Hall-effect probe senses the magnetic field, and the output signal is used both for mass assignment and for controlling the length of the electric-sector voltage scans. These are kept as short as possible whilst covering in full the region of interest; for instance, in the region of the molecular ion the electric-sector voltage scan can be kept relatively short.

An alternative method of presentation of the B–E surface employs a contour map.[107] In an article examining the graphical representation of metastable-ion data, perspective diagrams and contour maps are highlighted as being particularly useful in linked-scan studies because they enable artefact peaks due to the encroachment into the spectrum of, for example, neighbouring transitions to be identified.[108] The presence of interference peaks in linked-scan spectra and in spectra from mass-analysed ion-kinetic-energy spectrometry (MIKES) can otherwise often cause problems in interpretation.[109] The MIKE spectrometer is simply a reverse-geometry double-focusing instrument incorporating a collision region (usually a gas cell) between the magnetic and electric sectors configured so that the source and magnet comprise a single-focusing mass spectrometer, while the electric sector can be used to produce an energy spectrum of the ions arising from fragmentations taking place in the gas cell. The voltage scan of the electric sector usually becomes non-linear in the low-voltage range, which leads to mis-assignment of masses in this region.[110] This can be overcome by coupling the electric-sector voltage with the X-scale of an X–Y recorder so that any irregularities in the voltage scan are cancelled. By coupling the voltage through a differential amplifier supplying X-scale expansion and back-off, a zooming system is achieved that enables the fine detail of the MIKES peaks to be

[103] R. S. Mason, M. J. Farncombe, K. R. Jennings, and J. Scrivens, *Int. J. Mass Spectrom. Ion Phys.*, 1983, **48**, 415.
[104] M. J. Farncombe, R. S. Mason, K. R. Jennings, and J. Scrivens, *Int. J. Mass Spectrom. Ion Phys.*, 1982, **44**, 91.
[105] M. Cosgrove, D. Hazelby, G. A. Warburton, R. S. Stradling, and C. J. Chapman, *Int. J. Mass Spectrom. Ion Phys.*, 1983, **46**, 89.
[106] A. Fraefel and J. Seibl, *Org. Mass Spectrom.*, 1982, **17**, 448.
[107] A. Fraefel and J. Seibl, *Int. J. Mass Spectrom. Ion Phys.*, 1983, **46**, 87.
[108] C. G. Macdonald and M. J. Lacey, *Org. Mass Spectrom.*, 1984, **19**, 55.
[109] C. Guenat, F. Maquin, and D. Stahl, *Int. J. Mass Spectrom. Ion Processes*, 1984, **59**, 121.
[110] S. Daolio, C. Pagura, P. Traldi, and E. Vecchi, *Org. Mass Spectrom.*, 1982, **17**, 647.

examined.[111] Coupling this system with a commercial mass marker also permits accurate calibration of the mass scale simply and reliably. A similar use of an *X–Y* recorder in the MIKES experiment allows the influence of the collision-gas pressure on the product-ion peaks to be recorded with high precision by connecting the output of the ionization gauge, which is used to measure the pressure in the collision cell, to the *X*-terminal of the recorder, the output of the mass-spectrometer amplifier being connected to the other terminal.[112]

The MIKE spectrometer is a simple form of MS/MS instrument, where the magnet comprises the first analyser and the electric sector the second analyser. This configuration is only one of several that have been constructed to examine the analytical possibilities of MS/MS. One instrument, of EQ geometry (where the first analyser consists of an electric sector and the second consists of a quadrupole filter), has been built to study the dissociation of ions.[113] The instrument is capable of various scanning procedures to enable precursor-ion, product-ion, and constant neutral-loss scans to be recorded, although the constructors comment that the addition of a magnetic sector between the ion source and electric sector would increase the resolution with which the precursor ions can be selected. Such an instrument, a tandem mass spectrometer of BEQQ geometry, has been constructed.[114] The first analyser in this case comprises a high-resolution reverse-geometry double-focusing (BE) mass spectrometer, while the second is a quadrupole instrument; the use of a quadrupolar gas cell operating in the r.f.-only mode ensures a minimum of scattering in the collision cell. The important characteristic of this type of hybrid instrument, the high precursor-ion resolution, determined by the performance of the initial double-focusing system, is also featured in instruments of EBQQ geometry.[115] Such hybrid instruments are extremely versatile, providing MS/MS without compromising the high-resolution performance of the double-focusing system. The addition of a collision cell (including deceleration and focusing lenses) and a quadrupole analyser to an AEI MS9 double-focusing mass spectrometer enables a precursor-ion resolution of 5000 to be achieved whilst retaining the fast-scanning capabilities of the third (quadrupole) sector.[116] An AEI MS30 double-focusing mass spectrometer has been similarly modified, using a quadrupole ion trap (QUISTOR) as the collision cell and a small quadrupole analyser as the final mass analyser.[117]

[111] F. Dragoni, S. Daolio, P. Traldi, and E. Vecchi, *Org. Mass Spectrom.*, 1983, **18**, 86.

[112] M. Rabrenovic, G. W. Trott, M. S. Kim, and J. H. Beynon, *Org. Mass Spectrom.*, 1984, **19**, 203.

[113] F. M. Harris, G. A. Kennan, P. D. Bolton, S. B. Davies, S. Singh, and J. H. Beynon, *Int. J. Mass Spectrom. Ion Processes*, 1984, **58**, 273.

[114] J. D. Ciupek, J. R. O'Lear, R. G. Cooks, P. Dobberstein, and A. E. Schoen, presented at the 32nd Annual Conference on Mass Spectrometry and Allied Topics, San Antonio, 27 May to 1 June, 1984.

[115] L. C. E. Taylor and R. S. Stradling, presented at the 32nd Annual Conference on Mass Spectrometry and Allied Topics, San Antonio, 27 May to 1 June, 1984.

[116] F. L. DeRoos, D. L. Miller, and J. E. Tabor, presented at the 31st Annual Conference on Mass Spectrometry and Allied Topics, Boston, 8–13 May, 1983.

[117] M. Ho, R. J. Hughes, E. Kazdan, P. J. Matthews, A. B. Young, and R. E. March, presented at the 32nd Annual Conference on Mass Spectrometry and Allied Topics, San Antonio, 27 May to 1 June, 1984.

A commercial medium-performance double-focusing instrument of conventional geometry, JEOL DX300, has been converted to an MS/MS and MIKES system (EBE) by the addition of a small electric sector, of 90° angle and 12.75 cm radius, which aids high ion transmission in the MS/MS mode.[118] As with other instruments of this type, the addition of the third sector does not affect the performance of the original EB system. A new type of MS/MS system, which is claimed to offer some advantages over more conventional types (BEB, EBE, BEQ, or EBQ), uses crossed magnetic and electric fields as a component of the second mass analyser.[119] The configuration of this instrument, E–B–E–($E \times B$), provides a rapid-scanning second mass analyser with an unlimited mass range and a resolution of greater than 600 for product-ion detection. A commercial four-sector instrument of configuration BEEB comprises two identical double-focusing mass spectrometers in tandem, the first being of reverse geometry and the second of conventional geometry.[120] The central collision cell includes a retarding lens so that resolution can be maintained into the second mass spectrometer. Thus, mass selection of precursor ions at very high resolution is combined with high-resolution product-ion analysis.

The simplest MS/MS systems involve the choice of a quadrupole mass filter as the first analyser. Although this limits the mass range and resolution of precursor-ion selection, quadrupole instruments present few problems in computer control and automation. The addition of a quadrupole gas cell and quadrupole filter as second analyser produces the 'triple-quadrupole' MS/MS system. Although several of these are now available from commercial manufacturers, none has the degree of computer control of a triple-quadrupole assembly recently described.[121] Full computer control is provided in this system of all source and axial potentials; dual computers are used for data acquisition and data processing, and the system has a built-in capability for self-adaptive control of experiments. A triple-quadrupole system for combined tandem mass spectrometry and parallel mass spectrometry incorporates a collision cell that can be interchanged with an ion source.[122] With the collision cell in place, the standard triple-quadrupole system can be used for MS/MS studies; by turning a handle to remove the collision cell and replace it with a specially designed ion source, a parallel system is produced, providing simultaneous analysis in the two outer quadrupole analysers. The general concepts and uses of parallel mass spectrometry have been described with reference to a parallel system in which one unit is a quadrupole analyser and the other a magnetic-sector instrument,

118 F. Kunihiro, Y. Kammei, M. Naito, and Y. Itagaki, *Int. J. Mass Spectrom. Ion Phys.*, 1983, **46**, 151.
119 F. Kunihiro, Y. Naito, Y. Kammei, and Y. Itagaki, presented at the 32nd Annual Conference on Mass Spectrometry and Allied Topics, San Antonio, 27 May to 1 June, 1984.
120 J. R. Hass, B. N. Green, R. H. Bateman, and P. A. Bott, presented at the 32nd Annual Conference on Mass Spectrometry and Allied Topics, San Antonio, 27 May to 1 June, 1984.
121 C. M. Wong, R. W. Crawford, V. C. Barton, H. R. Brand, K. W. Neufeld, and J. E. Bowman, *Rev. Sci. Instrum.*, 1983, **54**, 996.
122 C. Chang and T. O. Tiernan, presented at the 32nd Annual Conference on Mass Spectrometry and Allied Topics, San Antonio, 27 May to 1 June, 1984.

both with a common ion source and each operated independently under computer control.[123] A novel MS/MS system has been produced by combining a quadrupole analyser with a TOF mass spectrometer and a thermal-ion source.[124] The instrument presents the possibility of product-ion MS/MS studies on transient species, for example from laser ionization methods, and the potential for the rapid acquisition of MS/MS spectra to permit on-line use with GC methods. Normal mass spectra are produced either by scanning the quadrupole instrument without gating the TOF analyser or by operating the quadrupole instrument in the r.f-only mode, so that all ions are transmitted, and carrying out analysis by time-of-flight.

6 Detection Systems

In a comprehensive discussion of the available types of mass spectrometer and their compatibility with capillary GC, the conclusion was drawn that ion statistical problems would eventually become apparent with the use of sequential detection, so that the future in this field is with array detectors, giving simultaneous detection of all ions; the time-array detector coupled with a TOF mass spectrometer was put forward as a possible useful combination.[125] The earliest and most widely used simultaneous detector is the photoplate, which has a unique combination of properties.[126] Not only does the photoplate detector make possible the recording of a complete high-resolution spectrum but it does so with good sensitivity, in part due to its integrating character of detection, so that during the elution of a GC peak, for example, integration can take place over the total elution area. However, the photoplate suffers the major inconvenience of the non-immediacy of the results – the photoplate must be removed and developed before the results can be assessed – and this has obvious repercussions for real-time control of experimental conditions. A technique has been developed more recently to allow integration of the ion signal over prolonged elution of the sample, but using sequential detection. This technique uses a multi-channel analyser to record the high-resolution mass spectrum over a limited mass range; by using a peak-matching unit to switch between two groups of ions, the sample and reference ions can be superimposed and accurate mass measurements calculated to an error of less than 5 p.p.m., even on compounds giving ion currents too small for normal peak-matching methods to be used.[127] A similar technique has been used to achieve not only accurate mass measurements of high-mass ions but also some quantitative measurements.[128] In a

[123] C. Chang, presented at the 32nd Annual Conference on Mass Spectrometry and Allied Topics, San Antonio, 27 May to 1 June, 1984.

[124] D. E. Goeringer and G. L. Glish, presented at the 32nd Annual Conference on Mass Spectrometry and Allied Topics, San Antonio, 27 May to 1 June, 1984.

[125] J. F. Holland, C. G. Enke, J. Allison, J. T. Stults, J. D. Pinkston, B. Newcom, and J. T. Watson, *Anal. Chem.*, 1983, **55**, 997A.

[126] J. Freudenthal and F. W. Jacobs, *Int. J. Mass Spectrom. Ion Phys.*, 1983, **46**, 115.

[127] K. L. Clay and R. C. Murphy, *Int. J. Mass Spectrom. Ion Phys.*, 1983, **53**, 327.

[128] L. Grotjahn and L. C. E. Taylor, presented at the 32nd Annual Conference on Mass Spectrometry and Allied Topics, San Antonio, 27 May to 1 June, 1984.

differential study of a multi-channel analyser used in the single-pass and multiple-pass modes, no substantial improvement was noted over a conventional detector in the single-pass mode whereas a significant reduction in baseline noise was observed in the multiple-pass mode.[129]

A true simultaneous detector comprising a microchannel-plate detector has been fitted to a TOF mass spectrometer.[130] The noted improvement in detector repeatability gives rise to an improvement in quantitative applications; in addition, the detector demonstrated good day-to-day stability, and its larger acceptance area provided enhanced stability. A similar microchannel-plate detector system has demonstrated improved efficiency for the detection of massive molecular ions.[131] The channeltron electron-multiplier array (CEMA) has also provided a very sensitive method of simultaneous ion detection.[132] In this electro-optical device the signal is converted into electrons and amplified; the electrons are then converted into photons in a phosphor layer and transmitted through the vacuum wall by a fibre-optic slab. The line spectrum so produced is imaged onto a photodiode array that provides the final readout of the signal.

Attempts to improve the sensitivity of normal electron-multiplier detection systems have centred around the incorporation of conversion dynodes, which are now fitted to most commercial mass spectrometers, either as a standard fitment or as an optional accessory. A modification to the detection system of MAT 311A and 731 mass spectrometers has been described in which a conversion dynode, manufactured from aluminium foil, has been fitted, with some alterations to the detector connections.[133] The system's capability includes a fast manual switchover between positive- and negative-ion detection. The major benefit is an increase in the sensitivity of negative-ion detection resulting in a nine-fold increase of the m/z 91 ion in the negative-ion FAB spectrum of glycerol, with no measurable decrease in ion current in the positive-ion mode. A detailed comparison of the performance of a conversion-dynode/electron-multiplier system with a Faraday-cup system showed that the performance of the electron-multiplier system is independent of the conversion-dynode voltage over the range -3 to -5 kV.[134] At lower voltages the conversion dynode loses efficiency because of ions failing to strike it, although movement of the conversion dynode closer to the ion beam lowers the voltage at which the sensitivity reduction occurs. Interestingly, the sensitivity of the system changes little with time; cleaning the conversion dynode after four months' operation had little effect on its performance. An improvement in the signal-to-noise level for weak signals has been observed on a Kratos MS50 high-resolution instrument by fitting

[129] D. L. Smith and J. A. McCloskey, *Int. J. Mass Spectrom. Ion Phys.*, 1983, **48**, 237.
[130] K. A. Lincoln, presented at the 32nd Annual Conference on Mass Spectrometry and Allied Topics, San Antonio, 27 May to 1 June, 1984.
[131] J. C. Hill, C. J. McNeal, and R. D. Macfarlane, presented at the 32nd Annual Conference on Mass Spectrometry and Allied Topics, San Antonio, 27 May to 1 June, 1984.
[132] C. J. Louter, A. N. Buijserd, and A. J. H. Boerboom, *Int. J. Mass Spectrom. Ion Phys.*, 1983, **46**, 131.
[133] R. M. Milberg, J. C. Cook, and L. Brayton, presented at the 31st Annual Conference on Mass Spectrometry and Allied Topics, Boston, 8–13 May, 1983.
[134] J. L. Homes and J. E. Szulejko, *Org. Mass Spectrom.*, 1983, **18**, 273.

an ion-counting system in addition to the normal detector system.[135] Although transparent to the normal operation of the instrument, the increase in capability has enabled trace analysis at picogram levels and isotope-ratio measurements at resolutions in excess of 30 000.

7 Other Techniques

Despite the ability of mass-spectrometry data systems to acquire and measure accurately the masses of all ions in a mass spectrum, even under the arduous conditions of rapid-scanning high-resolution capillary GC/MS, the technique of mass measurement by peak matching continues to be used. One of the problems of the manual peak-matching method is the subjectivity of the end-point, especially when matching peaks either of widely disparate sizes or of very small sizes, where ion statistics cause a varying signal. A method of overcoming this subjectivity is to use computer control to determine the end-point; such a program has been devised to carry out automatic peak matching in four passes. The first pass is used to set the gain factors to produce sample and reference peaks of similar heights, the second pass makes an initial mass-ratio adjustment to centre each peak, the third pass is used to make the final adjustment to match the peaks, and the final pass calculates the centroid of each peak and produces an accurate mass measurement.[136] The program also contains an off-line deconvolution routine for the separation and measurement of partially resolved multiplets.

Peak-matching units have also been used in single- or selected-ion monitoring, particularly at elevated resolutions. For the study of two peaks, the peak-matching unit can be modified to incorporate a signal-splitting device so that the two signals produced by the peak-matching unit appear on different outputs; a signal-and-hold amplifier is also incorporated to smooth the output.[137] Where the signals differ greatly in intensity, for instance in the simultaneous measurement of reactant ion and product ion in chemical ionization, a gain-change circuit, which allows the gain of the electron-multiplier detector to be changed by up to two orders of magnitude, may be incorporated into the amplifier system.[138]

Finally, the problems of mass scale linearization of magnetic-sector mass spectrometers without the need of a data system can be overcome by means of a commercially available digital system that includes not only the linearized but also a semi-automatic calibration process and subsequent mass marking.[139] The system uses calibration tables stored in non-volatile memory, each table containing up to 30 reference peaks from one of several common reference compounds. The calibration process comprises simple push-button registration of manually tuned ions.

[135] D. Giblin and R. Lapp, presented at the 31st Annual Conference on Mass Spectrometry and Allied Topics, Boston, 8–13 May, 1983.

[136] D. E. Giblin and R. L. Lapp, presented at the 32nd Annual Conference on Mass Spectrometry and Allied Topics, San Antonio, 27 May to 1 June, 1984.

[137] C. Pagura, S. Daolio, P. Traldi, and L. Doretti, *Org. Mass Spectrom.*, 1984, **19**, 204.

[138] C. J. Kallos, V. Caldecourt, and J. C. Tou, *Anal. Chem.*, 1982, **54**, 1313.

[139] F. Frieldi and E. Beck, *Org. Mass Spectrom.*, 1982, **17**, 646.

5

Applications of Computers and Microprocessors in Mass Spectrometry

BY J. R. CHAPMAN

1 Introduction

The use of a data system with a mass spectrometer (MS/DS) has brought obvious advantages in terms of data acquisition and subsequent data processing. In data acquisition the requirement for a data system arose from the ability of the mass spectrometer to produce enormous amounts of data, usually in a short time. For example, rapid repetitive scanning needed with high-resolution chromatographic inlets is practicable only with a data system. Again, complete high-resolution scans giving accurate mass data on every peak in a mass spectrum are possible only with a data system. A more complete integration of the mass spectrometer and data system, with the data system taking over many control facilities in addition to data acquisition and processing, brings further advantages. Some of these are discussed in Section 2 (Instrumentation) with special emphasis on the very high level of integration and feedback control now available with triple-quadrupole instruments.

The microcomputer or microprocessor now seems to be omnipresent because it is relatively inexpensive and can take on so many different roles in the laboratory. It may, for example, function as the basis of systems capable of data acquisition (at a reduced specification) and control or as a unit of a larger, distributed processing system. Both of these applications are included in Section 2. In addition, however, the microcomputer may be used to provide remote or directly linked work stations or to handle more general information.

Following the general treatment of instrumentation given in Section 2, hardware (and software, which is more specifically directed towards the process of data acquisition) is described in Section 3 (Data Acquisition). Subdivisions of this section are based on the relevant operating modes for the mass spectrometer. Some references to data-processing procedures may also be found in this section.

Major advances in data processing are discussed in Section 4. Particular emphasis in this review is given to reverse library search techniques and their extension to mixture analysis and quantitative analysis together with the processing of pyrolysis–mass-spectrometry data as an example of the application of pattern-recognition techniques.

This review covers the published literature from July 1982 to June 1984. It is hoped that the review will indicate to the reader those areas where the application of data systems and data-processing techniques has made progress during this period and where further reading can be found in readily available literature. The regular contributions to *Analytical Chemistry*, edited by Dessy, under the title 'A/C Interface' are recommended as sources of current information generally relevant to the use of computing facilities in the analytical-chemistry laboratory.

2 Instrumentation

Quadrupole and Sector Mass Spectrometers, Sample-inlet Systems. – Wong and co-workers[1] in an article describing a computer-controlled triple-quadrupole (TQ) instrument have listed four phases in the evolution of the use of computers to control mass spectrometers. The first three of these are (i) the 'programmed switch/recorder' phase in which the computer replaces some manual switches required for operation and acts as a data-logging device, (ii) the 'feedback switch/recorder' phase in which the computer can also make some programmed choices depending on an input parameter such as pressure or temperature, and (iii) the 'controller' phase in which the computer reads and/or sets all important operating parameters using D/A and A/D converters.

Most current data systems offer facilities that lie within phases (ii) and (iii) as described. Most also use some form of what is termed distributed processing with the objective of putting computing power where it is needed. For example, the TQ system described by Wong uses two LSI 11/23 computers. One of these is a single-user system dedicated to instrument control and data acquisition since it is the latter process, with the high data rates common in mass spectrometry, that imposes most load on processing facilities. The other computer runs under a time-sharing, multi-tasking system (*i.e.* many different programs can be run simultaneously) for data processing, manipulation, and storage. A particular feature of this system is a high-speed (0.5×10^6 words s^{-1}) parallel data link between the two processors.

Software for the TQ system carries out all major functions in line with the definition of phase (iii) above. These features are tuning *via* an operator-interactive process, calibration of the mass scale, recording and modification of all instrument parameters, and acquisition of spectral data. Flexibility is provided by a combination of manual and data-system control over many instrument parameters.

A more detailed description of the data-acquisition process using a triple-quadrupole instrument is given in another publication.[2] In this case, instrumental parameters relevant to data acquisition are organized into inlet descriptors, which refer to inlet (*e.g.* GC) conditions that change slowly during the entire process, and configuration descriptors. Configuration descriptors may be changed within milliseconds to allow different analysis modes, *e.g.* precursor-ion scan, product-ion scan, constant neutral-loss scan, or normal scan, as successive scans.

A particularly interesting application of this experimental flexibility is in sample screening using what is termed 'data-dependent configuration control'.

[1] C. M. Wong, R. W. Crawford, V. C. Barton, H. R. Brand, W. Neufeld, and J. E. Bowman, *Rev. Sci. Instrum.*, 1983, **54**, 996.

[2] H. Kirby, S. Sokolow, and U. Steiner in 'Tandem Mass Spectrometry', ed. F. W. McLafferty, Wiley, New York, 1983, p. 337.

In this mode the instrument is reconfigured dynamically to concentrate data acquisition on mass ranges that actually contain data. Every minute or so a normal scan over the whole mass range is taken, from which an algorithm selects the 20–30 most significant peaks and then directs the recording of product-ion spectra from each of these masses. This form of operation corresponds to phase (iv) described by Wong, *i.e.* the 'self-adaptive controller' in which the computer is taught to adjust operational parameters of the instrument to conform with an information feedback loop based upon the experimental data it is currently acquiring.

As previously mentioned, most data systems make use of distributed processing. A further example is provided by what is described as a distributed intelligent microprocessor configuration designed for use with a high-resolution double-focusing mass spectrometer.[3] This system is based on a host computer that controls two microprocessors. One microprocessor is a fast data preprocessor concerned with many aspects of the data-acquisition process; the second is dedicated to the control of various electrical and mechanical functions of the mass spectrometer, *e.g.* source operating mode, resolution control, and sample-inlet control. The data preprocessor, which is connected to the host computer *via* a high-speed parallel link, is designed to carry out data-acquisition processes to a very high specification. Off-loading these tasks allows more efficient use of central processor time in the host computer so that data acquisition and processing can occur together without compromising the mass-spectrometer operating speed. The second processor, which is connected *via* a serial link, permits an operator-interactive tuning process together with recording and control of instrument parameters including those relating to inlet systems such as a gas chromatograph. In addition to conventional scanning, various other operating modes such as peak matching, mass measurement, and metastable-ion analysis may be mediated by this microprocessor system.

A different manifestation of distributed processing is provided by the use of satellite computers that are subservient to a host computer and each of which is used to control individual instruments in a multiple-mass-spectrometer installation. A published description of a satellite system (MASDAT) for the control of low-resolution fast-scanning instruments[4] is again based on LSI 11 hardware and allows up to three satellite systems to be connected to the host computer. Satellite processing systems are also generally available commercially.

It is, however, often the case in laboratories with multiple mass spectrometers that data-processing facilities have not been designed as satellite systems that can be expanded as more instruments are purchased. A new MS/DS system may be installed in a laboratory where an older mass spectrometer with less powerful computing facilities already exists. An article by Jamieson and co-workers[5] demonstrates how an existing system may be integrated into and make use of up-to-date computing facilities particularly when off-line data processing is suitable. In this case a Varian 620/L-100 computer that had been used for the

[3] R. S. Stradling, P. A. Ryan, and J. D. Wood, *Comput. Enhanced Spectrosc.*, 1983, **1**, 25.
[4] E. Ziegler, D. Henneberg, H. Lenk, B. Weimann, and I. Wronka, *Anal. Chim. Acta*, 1982, **144**, 1.

automatic reading of high-resolution data previously recorded on photoplates was interfaced to provide hardwired data transfers to an INCOS system. Software was provided to reformat the 620/L data in INCOS format prior to transfer so that it could then be processed using INCOS program sets. The same article[5] describes several extensions and modifications that allow more effective use of INCOS software for the processing of scanning as well as of photoplate data.

In many instances, rather than trying to update older, less powerful computing facilities, it may be a case of providing some form of automation and/or data processing where none at all exists. Since microcomputers have become so much more accessible in the form of personal computers, there is a burgeoning interest in using even the most basic micro to provide some, albeit limited, computing facilities. A simple example is the interfacing of a small quadrupole mass spectrometer operating in the selected-ion-monitoring mode to a Commodore 32K Pet microcomputer.[6] Up to 13 masses may be monitored in a 1 s cycle together with an externally controlled experimental variable such as sample temperature. Two 12-bit D/A converters provide the voltages that control mass selection, whilst data signals are read following a voltage-to-frequency conversion.

A more ambitious application of a personal computer provides temperature control of a direct-insertion probe and data acquisition in the selected-ion-monitoring mode at high resolution using a 13-year-old double-focusing mass spectrometer.[7] A specially constructed interface between the personal computer (64K TRS-80-II) and the mass spectrometer uses an eight-channel, 12-bit A/D converter to read data signals and provides a number of digital outputs for control purposes. The personal computer performs all data acquisition in the selected-ion-monitoring mode, but full mass spectra are recorded using a conventional data system. Control of the direct-insertion probe provides either evaporation at a constant total ion current or a fixed temperature/time function for repeat quantitative determinations.

A less conventional inlet system for a mass spectrometer is an elemental GC.[8] A semi-automatic interface has been designed to link an elemental GC, which determines nitrogen on the Dumas principle, to an isotope-ratio mass spectrometer in order to permit the rapid determination of total nitrogen and nitrogen-isotope ratios. The interface can be made fully automatic and used for the determination of a wide range of light-gas stable isotopes with alternative instrumentation.

Other papers dealing with control of sample-inlet systems have described the integration of an autosampler–GC/MS instrument under data-system control to provide an automatic method for the routine analysis of organic compounds in complex environments,[9,10] a personal-computer-based data-acquisition and

[5] W. D. Jamieson, E. Lewis, and F. G. Mason, *Int. J. Mass Spectrom. Ion Phys.*, 1983, **46**, 71.
[6] D. Dahlgren, J. Arnold, and J. C. Hemminger, *J. Vac. Sci. Technol., Part A*, 1983, **1**, 81.
[7] K. Varmuza, H. Lohninger, and E. R. Schmid, *Int. J. Mass Spectrom. Ion Phys.*, 1983, **46**, 93.
[8] T. Preston and N. J. P. Owens, *Analyst (London)*, 1983, **108**, 971.
[9] P. A. Ryan, D. R. Denne, C. J. Wakefield, G. A. Warburton, and D. Hazelby, *Int. J. Mass Spectrom. Ion Phys.*, 1983, **48**, 283.
[10] D. T. Green, *Int. J. Mass Spectrom. Ion Phys.*, 1983, **48**, 43.

-processing system for a combined thermal analyser/mass spectrometer,[11] and improvements in software for a computer-controlled, temperature-programmable platinum-filament pyrolysis probe.[12]

Other Types of Mass Spectrometer. – Time-of-flight mass spectrometry has recently been brought back to public attention in two rather different fields. A classical time-of-flight (TOF) instrument operates in a discontinuous mode in which a discrete ion packet, initially comprising ions of a variety of m/z values, is pulsed into the field-free region of the flight tube. Each ion has a characteristic, mass-dependent velocity so that its time of flight to the detector can be correlated with its mass. Another ion packet is pulsed out of the source only when all of the ions from the preceding packet have reached the detector, perhaps every 100 microseconds.

TOF instruments are presently limited by their data-acquisition facilities so that only a single time window corresponding to one mass number is sampled for each pulse of the source. A complete spectrum is obtained by sampling time windows at different time delays after successive source pulses and combining the resulting data. In an interesting discussion on the limitations of current fast-scanning GC/MS instruments, Enke and co-workers[13] show that magnetic, quadrupole, and conventional TOF instruments are all limited to approximately 4–10 scan cycles per second. They put forward the concept of a faster-scanning instrument based on a proposal for an integrating transient recorder for time-array detection in TOF mass spectrometry. With this technique, which provides an initial sampling rate of 200 MHz, all of the ions extracted with each pulse may be usefully detected, and significantly higher scan-repetition rates then become available.

A potential application of the integrating transient recorder that may be of more immediate interest is time-resolved magnetic-dispersion mass spectrometry.[14] In this technique, ion-beam pulsing allows flight times to be measured in a magnetic-sector mass spectrometer, resulting in simultaneous velocity and momentum analysis of the ions. Time-array detection using the integrating transient recorder makes possible, in a single scan of the magnet, the detection of ions with any flight time at all possible values of the magnetic field. The result is a data matrix from which all metastable-ion data can be obtained as in metastable mapping with a conventional double-focusing sector instrument (*cf.* Section 3 – Metastable-ion Scanning) but which can be recorded over a much reduced time-scale.

Time-of-flight mass spectrometry is also the method of choice for analysing ions produced by ^{252}Cf plasma desorption.[15] This technique, which is still the most powerful method for the production of molecular ions of high-molecular-weight species, desorbs ions from thin sample layers excited by high-energy

[11] P. A. Barnes, S. B. Warrington, and S. W. Taylor, *Anal. Proc. (London)*, 1984, **21**, 98.
[12] T. H. Risby, J. A. Yergey, and J. J. Scocca, *Anal. Chem.*, 1982, **54**, 2228.
[13] J. F. Holland, C. G. Enke, J. Allison, J. T. Stults, J. D. Pinkston, B. Newcome, and J. T. Watson, *Anal. Chem.*, 1983, **55**, 997A.
[14] J. T. Stults, C. G. Enke, and J. F. Holland, *Anal. Chem.*, 1983, **55**, 1323.
[15] R. D. Macfarlane, *Anal. Chem.*, 1983, **55**, 1247A.

fission fragments derived from ^{252}Cf. Each fission event sends an energetic fragment towards the sample, but also another fragment in the opposite direction. Recording devices for ^{252}Cf TOF mass spectrometry use the detection of this backwards-moving fragment as the start signal for the timing process. The ionization process is, in effect, naturally pulsed. Macfarlane and co-workers[16] have built a multi-stop digitizer with a resolution of 78 picoseconds and 900 nanoseconds dead-time for ^{252}Cf experiments and, in a separate review,[15] have shown how this instrumentation with attendant data-processing facilities can be used both for the acquisition of spectra and for more detailed studies of the fission process itself. Another recent article describes a less sophisticated timer with a resolution of 6 nanoseconds designed for TOF–field-desorption experiments and controlled by a desk-top computer (HP85).[17]

Some software developments in Fourier-transform mass spectrometry (FTMS) have also been published during this period. In an initial investigation into the practicality of capillary GC/FTMS the equivalent of a total-ion-current trace has been used as a thresholding device to determine whether or not to store time-domain data, since storage available is below the level required for all the spectra from a complete GC analysis.[18] A zoom transform used to increase the number of data points across a mass peak in processing and therefore to improve mass-measurement accuracy has also been reported.[19]

Process Control. – Because of its exceptional long-term stability, a non-scanning magnetic-sector instrument with multiple collectors is favoured for process control where the compounds to be monitored are known beforehand.[20,21] Whistler and Schaefer[20] have described such an instrument where a microprocessor controls calibration, data acquisition, conversion and output of the data in terms of process-stream composition, and all diagnostic functions. Stability is also improved by the use of a new, thermoelectrically cooled, hermetically sealed pump.

Whistler and Schaefer give details of an ingenious automatic method of real-time compensation, without recalibration, for any changes in ion-source sensitivity. Based on the assumption that changes in source sensitivity affect each component equally, so that, whilst absolute peak sizes may change, the ratio of peak sizes for different components does not, the output from variable-gain amplifiers associated with each collector is scaled through summing resistors to create an analogue of the total sample. This sum is compared to a fixed reference (*e.g.* 10 volts), and any difference is fed *via* an error amplifier to the variable-gain amplifiers. These amplifiers make proportional changes in each output to drive the total error to zero. This technique of ratioing and summing to 100% has enabled calibration to remain valid from one to six months. The instrument also

[16] B. T. Turko, R. D. Macfarlane, and C. J. McNeal, *Int. J. Mass Spectrom. Ion Phys.*, 1983, **53**, 353.
[17] H. Reinmueller, *J. Phys. E.*, 1983, **16**, 1228.
[18] R. L. White and C. L. Wilkins, *Anal. Chem.*, 1982, **54**, 2443.
[19] T. J. Francl, R. L. Hunter, and R. T. McIver, jun., *Anal. Chem.*, 1983, **55**, 2094.
[20] W. J. Whistler and K. Schaefer, *Int. J. Mass Spectrom. Ion Phys.*, 1983, **46**, 159.
[21] O. R. Savin, V. F. Shkurdoda, and V. I. Simonovskii, *Int. J. Mass Spectrom. Ion Phys.*, 1983, **46**, 163.

incorporates automatic, real-time correction for interference from fragment ions. For example, correction is made for the contribution from the isobaric CO_2 fragment ion in the determination of nitrogen at m/z 28.

Another recent publication gives some details of a computer-controlled peak-switching quadrupole mass filter used to monitor liquid and gas phases directly during a fermentation process.[22] Two sampling interfaces based on a permeable membrane, one for the liquid phase and one for the gas, are separated from the mass-spectrometer vacuum system by computer-controlled valves. The computer also controls data acquisition and processing.

An illustration of feedback control is found in the real-time monitoring of a sputter deposition process.[23] The sputtering of tantalum is given as an example when a precise amount of a reactive gas, in this case oxygen, is required to achieve critical film properties. The measured ratio of Ta^+ to TaO^+ is used to provide feedback control for a pressure controller and servo inlet-valve system that regulates the supply of oxygen to the sputtering chamber. The ratio of ion abundances is preferred to absolute abundances for feedback control because of variations in the multiplier gain.

3 Data Acquisition

Metastable-ion Scanning. – The study of metastable ions, in particular ions that decompose in one of the field-free regions following the ion source, provides details of the processes involved in the formation and further reaction of fragment ions. As such, metastable ions constitute important aids in structural interpretation and in studies relating to the chemistry of gas-phase ions. The technique known as 'metastable mapping' attempts to provide in a single data-acquisition operation all metastable-ion data relevant to the compound being ionized.

In one implementation of metastable mapping on a double-focusing magnetic-sector instrument, rapid scans of the electrostatic analyser are superimposed on a very slow continuous scan of the magnetic field.[24,25] A Hall-effect probe senses the magnetic field, and the probe's output is then used by the data system for mass assignment and for control of the length of the electrostatic-analyser scans. In an alternative implementation, also on a double-focusing magnetic-sector instrument, fast repetitive magnet scans are effected whilst the data system reduces the electrostatic-analyser voltage only between each of these scans.[26–28] In either case the instrument scans include all combinations of the electrostatic- and magnetic-field strengths necessary for the recording of metastable ions of various masses and energies.

[22] E. Pungor, jun., E. J. Schaefer, C. L. Cooney, and J. C. Weaver, *Eur. J. Appl. Microbiol. Biotechnol.*, 1983, **18**, 135.
[23] B. F. T. Bolker, *Solid State Technol.*, 1983, **26**(12), 115.
[24] A. Fraefel and J. Seibl, *Org. Mass Spectrom.*, 1982, **17**, 448.
[25] A. Fraefel and J. Seibl, *Int. J. Mass Spectrom. Ion Phys.*, 1983, **51**, 245.
[26] M. J. Farncombe, K. R. Jennings, and R. S. Mason, *Org. Mass Spectrom.*, 1983, **18**, 612.
[27] M. J. Farncombe, R. S. Mason, K. R. Jennings, and J. Scrivens, *Int. J. Mass Spectrom. Ion Phys.*, 1982, **44**, 91.
[28] M. Cosgrove, D. Hazelby, G. A. Warburton, R. S. Stradling, and C. J. Chapman, *Int. J. Mass Spectrom. Ion Phys.*, 1983, **46**, 89.

Subsequent data processing allows the whole map to be presented in a three-dimensional representation where the co-ordinates are, for example, precursor and product mass of each fragmentation, and abundance. Alternatively, projections along any linked-scan line may be extracted and presented by the data system including reconstructions of conventional B/E, B^2/E, and constant neutral-loss scans. A commercially available data system provides extensive facilities for the acquisition and processing of metastable-map data.[28] The main disadvantage of metastable mapping is the length of time for which it is necessary to maintain a stable beam to complete data acquisition. Mapping of selected regions of the B–E plane and the use of time-resolved magnetic-dispersion mass spectrometry (Section 2 – Other Types of Mass Spectrometer) are possible solutions to this problem.

Linked scanning may be carried out with very high accuracy by means of a new purpose-built microprocessor that does not require the use of a data system. Working in conjunction with a Hall probe or field plate, the unit can deliver a digital value of a linearized mass scale every 60 microseconds and generate functions appropriate to various linked scans.[29,30] The output of the unit may be used to control the electrostatic analyser during the scan and to mark the mass scale on the output of a suitable recording device.

Selected-ion Monitoring. – With the realization that in many cases metastable-peak monitoring offers considerably greater selectivity than the monitoring of normal peaks at low resolution,[31] papers describing adaptations of existing hardware to carry out this technique using double-focusing magnetic-sector instruments have continued to appear. A particularly simple, although limited, method allows the monitoring of a number of metastable peaks with the same product ion without any requirements for magnet-switching facilities. In this case the accelerating voltage, uncoupled from the electrostatic-analyser voltage, is increased under data-system control by a factor (precursor-ion mass/product-ion mass) to focus the metastable ion. Standard selected-ion-monitoring software is used to process the data.[32]

Another publication introduces a more generally applicable method of selected-metastable-peak monitoring that permits the recording of one transition for the sample and another for the isotopically labelled internal standard, again using a double-focusing magnetic-sector instrument. The method is based on setting up a single ratio of magnetic-field strength to electrostatic-analyser voltage (B/E ratio) and then using magnet switching combined with accelerating voltage switching between the appropriate precursor ions.[33] A linked-scan unit is required to accomplish magnet switching, but, unlike an earlier method,[34] no instrumental modifications are necessary.

Selectivity in selected-ion monitoring can also be improved by operation at high resolution, particularly if facilities for exact mass determination of the peak

[29] F. Friedli, *Org. Mass Spectrom.*, 1984, **19**, 183.
[30] F. Friedli and E. Beck, *Org. Mass Spectrom.*, 1982, **17**, 646.
[31] J. R. Chapman, *Int. J. Mass Spectrom. Ion Phys.*, 1982, **45**, 207.
[32] G. A. Warburton and J. E. Zumberge, *Anal. Chem.*, 1983, **55**, 123.
[33] N. W. Davies, J. C. Bignall, and M. S. Roberts, *Biomed. Mass Spectrom.*, 1983, **10**, 646.
[34] D. A. Durden, *Anal. Chem.*, 1982, **54**, 666.

being monitored are available. Hardware described by Hass and co-workers[35,36] to carry out this type of analysis is based on a digital multiple-ion detector unit that enables automatic peak matching of up to seven ions with respect to a reference compound and that is coupled to a data system. The reference compound serves as a 'lock mass' because on each cycle its position in the peak-matching window is checked and a correction is applied if it is not centred. The window size is usually twice the peak width. Peak-profile data from the sample windows are recorded and stored for subsequent retrieval and exact mass determination.

Another high-resolution selected-ion-monitoring package has been introduced in which the data system is used to synchronize a calibrated sweep of the electrostatic analyser with the start of data acquisition at each mass.[31] The abundance data may then be displayed as a conventional mass chromatogram when plotted against time, or it may be plotted against mass to give a peak profile as in the previous example. Facilities for highly accurate measurement of the profile are not available in this case, but a new method of data processing that can be used to locate partially resolved components is proposed. Improved selected-ion-monitoring software for low-resolution operation with a Hewlett Packard 5992A GC/MS instrument has recently been described.[37]

Fast-atom Bombardment and Field Desorption. – A common problem in extending techniques such as fast-atom bombardment (FAB) or field desorption (FD) to the analysis of high-mass species or to high-resolution operation is the accurate measurement of small and often fluctuating ion currents. A versatile multi-channel analyser unit reported very recently by Rinehart and co-workers[38] is based on a PDP-8A minicomputer and an ion-counting interface with some subsequent data processing on a VAX 11/780. The PDP-8A memory is divided into two equal segments, one for sample data and one for reference data. Each segment is further divided into channels representing time intervals in the mass scan. As each scan is recorded, the new data are summed with the memory contents so that the spectra in memory are an average of data acquired. The software is independent of the mass spectrometer used, but the instrument must provide highly reproducible narrow scans. The scan mode may be accelerating voltage scanning or magnetic scanning on field-regulated instruments.

The facility for the averaging of multiple scans is particularly useful in field desorption where weak and transient ion currents are encountered, and it also allows the averaging of data from multiple loadings of a field-desorption emitter. Another application of this facility is to isotope-ratio measurement from samples introduced *via* a gas chromatograph. The acquisition and averaging of fast (0.3 s) multiple scans across the entire GC peak provide a means of nullifying the effect of changes in sample concentration on the measured isotope ratio and are

[35] Y. Tondeur, J. R. Hass, D. J. Harvan, and P. W. Albro, *Anal. Chem.*, 1984, **56**, 373.
[36] Y. Tondeur, J. R. Hass, D. J. Harvan, P. W. Albro, and J. D. McKinney, *J. Agric. Food Chem.*, 1984, **32**, 406.
[37] L. C. Dickson, F. W. Karasek, and R. E. Clement, *J. Chromatogr.*, 1983, **280**, 23.
[38] C. R. Snelling, jun., J. C. Cook, R. M. Milberg, M. E. Hemling, and K. L. Rinehart, jun., *Anal. Chem.*, 1984, **56**, 1474.

perfectly suited to the multi-channel-analyser approach described by these authors.[38]

Accurate mass measurement uses either a narrow scan in which the unknown peak is bracketed by two reference peaks or an additional dual-channel peak-matching unit when the mass range would otherwise be too large. Mass measurement has been carried out in a number of ionization modes using a reference spectrum produced in the electron-impact-ionization mode when a combined ion source is available. Mass-measurement accuracy in the FD and FAB modes has routinely been at the 5 p.p.m. level.

A further example of a multi-channel analyser used in conjunction with a peak-matching unit has been provided by Murphy and co-workers[39] for FAB mass measurement. In this case the hardware multi-channel analyser (Tracor Northern) is calibrated using reference compound peaks prior to sample-data acquisition. Mass measurement is a visual process at present but routinely provides better than 5 p.p.m. accuracy. Other papers describing FAB mass measurement by peak matching[40] and magnet scanning[41] have appeared during the period covered by this review.

Another publication[42] discusses the process of data-system calibration for high-mass, nominal-mass-resolution FAB operation in considerable detail. Procedures for calibration over the mass range 393–5941 u using caesium iodide/glycerol as a reference are given. The use of poly(ethylene glycol) has been recommended as a lower-mass (up to 1400 u) calibrant for FAB operation.[43]

Miscellaneous. – Two improvements in scanning-data acquisition have been suggested. In a very comprehensive investigation, Scheppele[44] has reported that the use of neutral rather than ionic masses (*i.e.* ignoring the mass of the electron) in scan-law calibration for accurate mass measurement at high resolution introduces insignificant mass-measurement errors provided that the scan law has the proper functional form. In this case, the form is an expression of the scan-law rate constant as an eighth-order polynomial whose parameters are evaluated by a least-squares analysis of reference peak data.

An adaptation of the top-hat filter as a peak-detection method in low-resolution GC/MS data acquisition has also been reported.[45] This report is limited to a comparison with a slope-detection algorithm that can fail with small peaks since these may be either detected as multiplets or rejected because they fall below a minimum-peak-width criterion. These difficulties are minimized with the top-hat filter, which is also relatively simple to implement. Further investigation of both of these improvements in data-acquisition procedures would be of interest.

[39] K. L. Clay and R. C. Murphy, *Int. J. Mass Spectrom. Ion Phys.*, 1983, **53**, 327.
[40] J. M. Gilliam, P. W. Landis, and J. L. Occolowitz, *Anal. Chem.*, 1983, **55**, 1531.
[41] R. P. Morgan and M. L. Reed, *Org. Mass Spectrom.*, 1982, **17**, 537.
[42] A. M. Buko, L. R. Phillips, and B. A. Fraser, *Biomed. Mass Spectrom.*, 1983, **10**, 324.
[43] L. J. Goad, M. C. Prescott, and M. E. Rose, *Org. Mass Spectrom.*, 1984, **19**, 101.
[44] S. E. Scheppele, Q. G. Grindstaff, R. D. Grigsby, S. R. McDonald, and C. S. Hwang, *Int. J. Mass Spectrom. Ion Phys.*, 1983, **49**, 179.
[45] A. D. Graddon, J. F. Smith, and J. D. Morrison, *Comput. Enhanced Spectrosc.*, 1983, **1**, 167.

4 Data Processing

Gas Chromatography/Mass Spectrometry. – A new data-reduction environment modelled on a stack-orientated programmable calculator has been proposed.[46] The environment uses a four-register last-in/first-out data stack, which may hold spectra or entire chromatographic data, together with data-processing, display, and stack-manipulative functions. The system is flexible since functions may be strung together and executed in any appropriate manner and new functions may easily be added for future expansion.

A recent publication has described extensions to software for the plotting of a new type of specific ion chromatogram, the isotope-cluster chromatogram, from gas-chromatographic/mass-spectrometric (GC/MS) data.[47] A user-entered isotope pattern, which may cover natural or artificial isotope mixtures, is searched for. Each of the recorded spectra is broken down into parts containing exactly the number of masses present in the required isotope pattern, and a reverse-library-search algorithm is used to determine any similarity between each part and the entered pattern. The total score, which is a sum of all similarity indices for each spectrum, is then plotted *versus* scan number to give a very specific isotope-cluster chromatogram. Applications to ^{13}C, ^{15}N, and deuterium labelling are reported.

Edwards[48] has described various types of two-dimensional digital filters including band-pass filters that should be suitable for noise filtering and peak location in a two-dimensional data set, *e.g.* GC/MS data with abundance as a function of m/z value and time. No applications to spectral data are included in the published report.

High-resolution Data. – A lengthy publication from Scheppele[49] describes an efficient program that assigns elemental formulae and Z numbers to accurate mass data and that has been designed to minimize the output of formulae of no chemical significance. Unlike existing methods that use combinatorial algorithms to generate appropriate formulae, this approach uses a search algorithm together with pre-generated data bases containing formulae and fractional masses calculated on the basis of homologous units such as $^{12}C^{1}H_2$. Applications of this program to experimental high-resolution data are presented in a subsequent paper.[44]

Van Katwijk[50] has described an improved method for the characterization of complex oil fractions from their high-resolution mass spectra. In this method, peak data are retained as digital samples rather than centroided to find peak positions. A least-squares procedure is then used to find the best fit between candidate ion species with their isotopes and the recorded data. Application to a complete scan is laborious but yields results at one fifth of the resolving power that would have been required using the centroiding technique.

[46] E. F. Reus, D. W. Peterson, and R. Ellis, *Int. J. Mass Spectrom. Ion Phys.*, 1983, **46**, 97.
[47] R. J. Anderegg, *J. Chromatogr.*, 1983, **275**, 154.
[48] T. R. Edwards, *Anal. Chem.*, 1982, **54**, 1519.
[49] S. E. Scheppele, K. C. Chung, and C. S. Hwang, *Int. J. Mass Spectrom. Ion Phys.*, 1983, **49**, 143.
[50] J. Van Katwijk, *Proc. Inst. Pet., London*, 1982 (2, Petroanal. '81), 171.

Metastable-ion Data. – See Section 3 – Metastable-ion Scanning.

Library Searching. – Most routine library-search systems are now based on reverse-search methods in which successive library spectra are used as the basis for comparison. The other extreme is the forward or interpretative library search in which the unknown is used as the basis for comparison. The reverse search uses each library spectrum in turn to answer the question 'Is the unknown this particular library compound?' and in doing so maximizes the likelihood of identification, particularly from impure spectra, so long as the unknown has a corresponding spectrum in the library. The forward search attempts to answer the question 'What is the unknown compound?' and is particularly useful when a spectrum corresponding to that of the unknown is absent from the library.[51]

McLafferty[52] has provided an update on the best-known reverse-search system, the so-called probability-based matching (PBM) system. A library data base containing 41 000 spectra has been ordered according to the mass of the most 'significant' peak in each spectrum. The most significant peaks in the unknown spectrum then determine which sections of the data base are searched. Significance is based on both mass and abundance values. With an ordered file an average search time of 10 seconds per unknown is possible, with 100% and 97.5% of the identifications being found in 7.5% and 1.6% of the file, respectively. For unknown spectra that are not in this data base, a forward-search system (STIRS) is used off-line. This forward search is able to offer a reliability of identification figure for 600 common substructures and an estimate of molecular weight.

Henneberg[53] has reported the extension of an interpretative library-search system (SISCOM) so that a reverse library search may be carried out on a relatively large list of similar spectra provided by the interpretative system. This approach, whilst providing a useful extension to an existing system, is probably less efficient than the usual procedure in which an interpretative search is restricted to those spectra not identified by the reverse search. An interesting feature in this case is provided by the use of pattern-recognition methods to estimate the weighting of variables in the reverse-search retrieval algorithm.

The use of relative-retention-time data in GC/MS operation is a powerful adjunct to any library-search procedure. The Mass Spectral Information System (MSIS) analyses GC/MS data by automatically locating GC peaks and performing background subtractions prior to identification using library-search and relative-retention-time data.[54] A novel feature of this system is that the spectra and retention times of unidentified compounds can be saved and used for future comparisons so that the user is made aware of unidentified compounds that have been seen before.

[51] J. R. Chapman, 'Computers in Mass Spectrometry', Academic Press, London, 1978.

[52] F. W. McLafferty, S. Cheng, K. M. Dully, C. J. Guo, I. K. Mun, D. W. Peterson, S. O. Russo, D. A. Salvucci, and J. W. Serum, *Int. J. Mass Spectrom. Ion Phys.*, 1983, 47, 317.

[53] L. Domokos, D. Henneberg, and B. Weimann, *Anal. Chim. Acta,* 1983, **150**, 37.

[54] D. P. Martinsen, F. L. Tobin, and B. H. Song, *Am. Lab. (Fairfield, Conn.)*, 1983, **15**, 124.

An article by Rasmussen[55] describes the use of library-search and retention data in a method in which an unknown mixture is analysed on two columns having widely different separating characteristics. A special algorithm that matches the spectra from the two analyses gives two mass spectra and two Kovats' retention indices for many components. These data are matched with a small (174 entries), specialized library of known spectra using the retention indices as 'windows' for the library spectra to be examined. The use of a small library leads to fast search times with only minimal computing facilities (HP Model 9830A calculator). The use of two columns increases the number of components that are clearly separated and helps in the assignment of isomeric structures as well as providing very reliable identification of components when used in conjunction with a specialized library.

Another report,[56] in an assessment of the analysis of 20 000 GC/MS data files over 2.5 years, stresses that it is principally the reliability of lower-quality matches, *i.e.* those using distorted or impure unknown spectra, that is enhanced by the use of retention data. The analysis scheme on which the assessment is based is an automatic method for the extraction and matching of spectra; this method uses the CLEANUP program of Dromey[57] to provide pure spectra from components separated by two or more scan cycles and then submits these data to a PBM method followed by a retention-data match.

One of the most difficult problems in analysis using a library-search procedure arises, even in GC/MS analysis, when a mass spectrum is due to a mixture of compounds. This problem has been reviewed recently,[31] and two approaches to the problem, *viz.* use of different GC stationary phases and use of the Dromey CLEANUP program, have already been alluded to in this section. An article specifically directed towards the identification, using a library-search system, of components in 'mixture spectra' has appeared recently.[58] The basis of the search is a slightly modified PBM system commencing with a comprehensive five-step pre-search that greatly reduces the number of spectra submitted to the main search. After identification of one component, a suitably scaled version of the corresponding library spectrum is subtracted from the unknown spectrum to leave a new sample spectrum, which is again subjected to a pre-search/main-search analysis. This spectrum-stripping procedure successfully identified 70% of the binary mixtures submitted with a complete analysis of ten mixed spectra in approximately two minutes. The first component was identified in all cases, but a search for the second component was limited by the difficulty in calculating a reliable subtracted spectrum.

The spectrum-stripping procedure can be applied when a mixture is completely unresolved in time, whereas a spectrum clean-up procedure, such as that published by Dromey, which is applied prior to the library search, does demand a minimum time resolution of the components.[31] A calculation of the contribution of the

[55] P. Rasmussen, *Anal. Chem.*, 1983, **55**, 1331.

[56] W. M. Shackleford, D. M. Cline, L. Faas, and G. Kurth, *Anal. Chim. Acta*, 1983, **146**, 15.

[57] R. G. Dromey, M. J. Stefik, T. C. Rindfleisch, and A. M. Duffield, *Anal. Chem.*, 1976, **48**, 821.

[58] A. Yashuhara, J. Shindo, H. Ito, and T. Mizoguchi, *Comput. Enhanced Spectrosc.*, 1983, **1**, 117.

identified library spectrum to the unknown spectrum can be used as a measure of the presence or absence of other components. The use of a faster-scanning mass spectrometer is, within the limits imposed by ion statistics, another approach to mixture resolution, and in this connection several authors have published statistical estimates of the number of multiplet peaks in multi-component chromatograms.[59,60]

The selectivity of the mass spectrometer allows a quantitative determination as well as a qualitative identification to be carried out on a compound, even though it may be unresolved from one or more other components of a complex mixture. An increasingly common technique that allows automated quantification in GC/MS analyses uses a reverse library search, together with retention data, to identify components and internal standards followed by a quantitative determination based on the abundances of major designated ions in the library spectrum.[9,10,61] Calculation of concentrations is based on a previously acquired calibration.

Whole chromatograms representing a GC/MS analysis may be subjected to a comparison procedure similar to that used for individual spectra in a standard library search. A recent application of this technique is the correlation of crude oils using the repetitive-scanning GC/MS analysis of each oil as a fingerprint.[62] The software developed for this method initially converts all the scan data to the form of mass chromatograms and then aligns chromatograms for comparison using the retention times of known compounds or standards. A correction is then made for mass discrimination based on the spectrum of a reference compound run together with the analysis. A search of the corrected data is used to eliminate regions that do not contain an abundance maximum for at least one significant m/z value. A correlation index based on the difference in spectra in non-eliminated areas may then be calculated for sample pairs, and the samples finally may be listed in order of decreasing similarity. Extensions to this software enable biomarkers responsible for any similarity, or lack of similarity, to be pinpointed. Similar, although less comprehensive, software has been described by other authors.[63]

In this discussion of library-search procedures considerable attention has been devoted to the quality of the unknown spectra and to the difficulties encountered with 'mixture spectra'. It is, of course, equally important that quality control should be applied to the reference spectra that form the library data base. In a recent and very laudable investigation, over 1400 electron-impact spectra of selected organic compounds were measured under carefully defined conditions.[64] An assessment of spectrum quality based on nine separate calculated factors showed that, whilst most factors exhibited low standard deviation and

[59] J. M. Davis and J. C. Giddings, *Anal. Chem.*, 1983, **55**, 418.
[60] D. P. Herman, M.-F. Gonnord, and G. Guiochon, *Anal. Chem.*, 1984, **56**, 995.
[61] J. J. Vrbanac, C. C. Sweeley, and J. D. Pinkston, *Biomed. Mass Spectrom.*, 1983, **10**, 155.
[62] D. A. Flory, H. A. Lichtenstein, K. Biemann, J. E. Biller, and C. Barker, *Oil Gas J.*, 1983, **81**(3), 91.
[63] A. K. Aldridge, *Proc. Inst. Pet., London*, 1982, (2, Petroanal. '81), 192.
[64] G. W. A. Milne, W. L. Budde, S. R. Heller, D. P. Martinsen, and R. G. Oldham, *Org. Mass Spectrom.*, 1982, **17**, 547.

approached the ideal value of 1.0, three factors relating to purity of reference sample did not do so. The conclusion drawn was that mechanical parameters such as instrument tuning and calibration can be reasonably controlled but chemical parameters such as sample purity present much more difficulty.

The most comprehensive mass-spectrometry data base is the collection of 79 525 different electron-impact spectra for 67 510 compounds available for probability-based matching.[46] Recent references to the use of other types of mass-spectral data or the association of other spectroscopic techniques with mass-spectral library search have covered collisionally activated decomposition spectra,[65] charge-exchange spectra,[66] a specialist library of drugs, poisons, and their metabolites,[67] and FTIR.[68–70]

Pattern Recognition. – Kowalski and Bender[71] have given the following description of the general problem addressed by pattern-recognition techniques: Can an obscure property of a collection of objects be detected and/or predicted using indirect measurements (in this case mass spectra), made on the objects, that are known to be related to the property *via* some unknown relationship?

An excellent example of the application of pattern-recognition techniques, and one that has received considerable attention, is in the use of pyrolysis–mass spectrometry for the classification of polymeric materials.[72] Recognition of the complexity as well as the selectivity and reproducibility of pyrolysis reactions determined the need for pattern-recognition software to allow the rapid classification of the pyrolysis spectra. The success of this method can be judged from the range of recent applications, which covers the characterization of coals,[73, 74] air particulates,[74] smoke aerosols,[75] whole bacteria,[76–78] and synthetic polymers[79] as well as forensic studies[72] amongst many others.

The data-processing procedure for pyrolysis–mass spectra, which is conveniently based on commercial software packages,[80,81] comprises the following

[65] M. T. Cheng, G. H. Kruppa, F. W. McLafferty, and D. A. Cooper, *Anal. Chem.*, 1982, **54**, 2204.

[66] D. F. Hunt and J. P. Gale, *Anal. Chem.*, 1984, **56**, 1111.

[67] H. Maurer and K. Pfleger, *Fresenius' Z. Anal. Chem.*, **1983**, **314**, 586.

[68] S. S. Williams, R. B. Lam, and T. L. Isenhour, *Anal. Chem.*, 1983, **55**, 1117.

[69] K. H. Shafer, T. L. Hayes, J. W. Brasch, and R. J. Jakobsen, *Anal. Chem.*, 1984, **56**, 237.

[70] D. F. Gurka, M. Hiatt, and R. Titus, *Anal. Chem.*, 1984, **56**, 1102.

[71] B. R. Kowalski and C. F. Bender, *J. Am. Chem. Soc.*, 1973, **95**, 686.

[72] J. Haverkamp and P. G. Kistemaker, *Int. J. Mass Spectrom. Ion Phys.*, 1982, **45**, 275.

[73] H. L. C. Meuzelaar, A. M. Harper, G. R. Hill, and P. H. Given, *Fuel*, 1984, **63**, 640.

[74] K. J. Voorhees and S. L. Durfee, *Colo. Sch. Mines Q.*, 1983, **78**, 23.

[75] R. Tsao and K. J. Voorhees, *Anal. Chem.*, 1984, **56**, 368.

[76] J. Haverkamp, G. Wieten, and D. G. Groothuis, *Int. J. Mass Spectrom. Ion Phys.*, 1983, **47**, 67.

[77] G. Wieten, H. L. C. Meuzelaar, and J. Haverkamp in 'Gas Chromatogr./Mass Spectrom. Appl. Microbiol.', ed. G. Odham and L. Larsson, Plenum, New York, 1984, p. 335.

[78] R. Hoogerbrugge, S. J. Willig, and P. G. Kistemaker, *Anal. Chem.*, 1983, **55**, 1710.

[79] R. P. Lattimer, J. B. Pausch, and H. L. C. Meuzelaar, *Macromolecules*, 1983, **16**, 1896.

[80] A. M. Harper, D. L. Duewer, B. R. Kowalski, and J. L. Fasching in 'Chemometrics: Theory and Applications', ed. B. R. Kowalski, Am. Chem. Soc. Symp. Ser., No. 52, American Chemical Society, Washington D.C., 1977.

[81] N. H. Nie, C. H. G. Hull, J. G. Jenkins, K. Steinbrenner, and W. H. Brent, 'Statistical Package for the Social Sciences', 2nd Ed., McGraw-Hill, New York, 1975.

steps.[66] Each spectrum is normalized to its total ion current. Next, each m/z value is autoscaled over all spectra so that the mean peak size at each m/z value over all the spectra is zero and the standard deviation is unity. Autoscaling helps to eliminate the dominating effect of the largest, but not necessarily most significant, peaks in the subsequent analysis. It is then necessary, using data from known samples, to select those m/z values that can most readily distinguish the various sample classes. A simple test in this instance is the Fisher ratio, *i.e.* the ratio of between-class and within-class variances. This ratio may be used to select m/z values or to weight the abundances. Using the reduced data containing N m/z values, a single point representing the spectrum is calculated in N-dimensional space. Finally, a non-linear mapping technique is used to give the best two-dimensional representation of the data from N-dimensional space. The map may be interpreted by comparing the distance between points: the closer the points on the map, the more similar the samples. Distinct classes should be well separated.

The autoscaled data may alternatively be analysed by factor-analysis procedures, which transform the original data into a new set of independent variables.[72,74] The first factor contains the largest amount of variance in the original data, whilst successive factors account for the greatest possible residual variance. Since peaks in a mass spectrum are typically highly correlated, factor analysis is an efficient alternative for the reduction of data dimensionality. The principal factors provided by this method no longer contain chemical information in an easily interpretable form. However, graphical-rotation techniques may then be applied to form new factors, which may be more readily interpreted to provide information concerning the physico-chemical basis of the differences between the spectra from different classes.[74,82] Both this application of factor analysis to the interpretation of pyrolysis–mass spectra and its use in the classification of unknown samples on the basis of pyrolysis–mass spectra may be made more efficient by the construction of linear-discriminant functions that optimally describe the differences between known classes with respect to within-class differences.[72,78,82]

Several other applications of pattern recognition have been described in addition to pyrolysis–mass-spectral studies. Van der Greef[83] has used non-linear mapping and factor analysis to classify wines and urine samples based on the field-desorption or fast-atom bombardment profiles of involatile constituents. The same author has also used a very simple extension of the Fisher-ratio method for the identification of the metabolites of xenobiotics in crude-urine extracts from a field-desorption profile analysis.[84] Crude oils may be classified on the basis of their geographical origin using factor or discriminant analysis of hydroxide-ion chemical-ionization spectra.[85] The accurate molar compositions of binary-

[82] W. Windig, J. Haverkamp, and P. G. Kistemaker, *Anal. Chem.*, 1983, **55**, 81.
[83] J. Van der Greef, A. C. Tas, J. Bouwman, and M. C. Ten Noever de Brauw, *Anal. Chim. Acta*, 1983, **150**, 45.
[84] J. Van der Greef and D. C. Leegwater, *Biomed. Mass Spectrom.*, 1983, **10**, 1.
[85] P. Burke, K. R. Jennings, R. P. Morgan, and C. A. Gilchrist, *Anal. Chem.*, 1982, **54**, 1304.

mixture mass spectra acquired during the elution of poorly separated GC peaks can be measured using a factor-analysis-based method.[86]

5 Miscellaneous Software

Isotopic-distribution Calculation. – Yergey[87] has described a general computer program designed to calculate and display the isotopic contributions for any molecular formula whilst retaining the isotopic distribution, exact mass, and absolute abundance for all individual peaks at each mass. This program is especially applicable to higher-mass (>1000 u) calculations where the molecular-ion cluster becomes a complex distribution in which each peak at a given nominal mass contains numerous isotopic contributions. The program is made more practical by directly calculating only the unique permutations for each element and by operating within a relative-abundance threshold. The molecular distribution of bovine insulin is presented as an example of the program output.

A simpler program that is restricted to a display of the relative abundances at each nominal mass and that will consider up to three isotopically enriched elements has also been reported.[88] In another recent publication, Tou[89] has considered the calculation of isotope patterns seen in scans that record fragment-ion spectra when multi-isotopic elements such as chlorine and bromine are present.

Molecular-weight Calculation. – Fales and co-authors[90] have also considered the difficulties in molecular-weight determination caused by the complex distribution of isotope peaks at higher masses. Their viewpoint is that of the user of an instrument of comparatively low resolution. To this end they have provided a Basic program that lists nominal, precise (single isotopic), and average molecular weights for the molecular ion and all fragment ions. The average molecular weight is equivalent to the centroid of an unresolved isotope distribution.

Quantitative Analysis. – Numerous computational procedures for the quantitative analysis of unseparated mixtures from scan data continue to be reported. Applications include gas analysis,[91–95] hydrocarbon-type analysis,[93] and inorganic analysis using secondary-ion mass spectrometry.[94] A Bayesian statistical method that offers better precision than more conventional least-squares procedures has been described.[95]

[86] M. A. Sharaf and B. R. Kowalski, *Anal. Chem.*, 1982, **54**, 1291.
[87] J. A. Yergey, *Int. J. Mass Spectrom. Ion Phys.*, 1983, **52**, 337.
[88] M. L. Brownawell, *J. Chem. Educ.*, 1982, **59**, 663.
[89] J. C. Tou, *Anal. Chem.*, 1983, **55**, 367.
[90] R. Hill, H. Fales, C. McNeil, and R. Macfarlane, *Biomed. Mass Spectrom.*, 1983, **10**, 505.
[91] J. Voogd, E. Huiting, G. J. van Rossum, J. M. Petri, and L. Lelion, *Int. J. Mass Spectrom. Ion Phys.*, 1983, **48**, 7.
[92] W. K. Schorr, H. Duschner, and K. Starke, *Int. J. Mass Spectrom. Ion Phys.*, 1983, **48**, 11.
[93] M. Story, B. Laser, and K. Habfast, *Int. J. Mass Spectrom. Ion Phys.*, 1983, **48**, 47.
[94] W. Steiger, F. G. Ruedenauer, J. Antal, and S. Kugler, *Vacuum*, 1983, **33**, 321.
[95] L. M. Karrer, H. L. Gordon, S. M. Rothstein, J. M. Miller, and T. R. B. Jones, *Anal. Chem.*, 1983, **55**, 1745.

A brief publication by de Leenheer and co-workers[96] describes a general computational approach to the problem of calibration-curve non-linearity in quantitative determination by isotope-dilution GC/MS. The use of a polynomial model provides an improved accuracy of analysis with no requirement for an initial estimate of labelling purity for the internal standard or of the influence of naturally occurring isotopes.

Resolution Enhancement. – A function-domain iterative technique and a Fourier-transform method, both including data smoothing, have been evaluated as a basis for mathematical enhancement of mass-spectrometer resolution.[97]

Structural Interpretation. – Extensions to a program for the determination of amino acid sequences in peptides have been reported.[98] Data from field-desorption and field-desorption–collisional-activation experiments can be accommodated. A structure-cutting program accepts a chemical structure entered *via* a graphics tablet and creates a fragment pool by cleaving selected bonds one, two, or three at a time.[99] The program can then be used in an interactive manner for assessment of the proposed structure by input of observed fragment-ion masses when fragment structures with the same mass will be displayed.

Education. – A computer-assisted learning package provides an interactive teaching aid that reviews the principles associated with the interpretation of mass spectra such as molecular ions, fragmentation, isotope patterns, elemental compositions, and metastable ions.[100] The package, which is written in Basic for the Apple II microcomputer, is also able to display a range of simulated mass-spectral data for examination and interpretation.

A compute-based system designed to provide users of analytical instrumentation with current information has GC/MS as one of its modules.[101] The function of the modules lies in an area between computer-assisted learning and large library-search systems. Each module is subdivided into five major classifications, *viz.* 'How it Works', 'What it Does', 'Nuts and Bolts' (*e.g.* bibliography, short courses, required training, service and maintenance, manufacturers, and prices), 'Applications', and 'What's New'. The system is presently accessed through network terminals, but programs for use on personal computers are being developed.

Software Availability. – Many authors of papers included in this review indicate their willingness to provide interested users with information such as program listings. Readers are advised to check the original publication in any particular case. Further information regarding software sharing may be found in reference 102.

[96] J. A. Jonckheere and A. P. De Leenheer, *Anal. Chem.*, **1983**, **55**, 153.
[97] J. W. Ioup, G. E. Ioup, G. H. Rayborn, jun., G. M. Wood, jun., and B. T. Upchurch, *Int. J. Mass Spectrom. Ion Processes*, 1983, **55**, 93.
[98] T. Matsuo, T. Sakurai, H. Matsuda, H. Wollnik, and I. Katakuse, *Biomed. Mass Spectrom.*, 1983, **10**, 57.
[99] J. Figueras, *Anal. Chim. Acta*, 1983, **146**, 29.
[100] B. Semmens, *Educ. Chem.*, 1982, **19**, 178.
[101] F. A. Settle and M. Pleva, *TrAC, Trends Anal. Chem. (Pers. Ed.)*, 1982, **1**, 242.
[102] 32nd Annual Conference on Mass Spectrometry and Allied Topics, San Antonio, Texas, 1984, p. 904.

6

Fourier-transform Ion Cyclotron Resonance

BY N. M. M. NIBBERING

1 Introduction

Fourier-transform ion cyclotron resonance (FT-ICR) is one area in mass spectrometry that has grown rapidly in the past few years. This method, first developed by Comisarow and Marshall in 1974,[1] is now applied increasingly to studies of analytical importance and gas-phase ion chemistry. This assertion is born out by the appearance of the proceedings of a symposium devoted to this field[2] and to reviews[3,4] or chapters[5–7] in books on FT-ICR and its applications. Furthermore, two commercial firms have put FT-ICR mass spectrometers on the market.[8,9] The FT-ICR method is free of the limitations of the conventional ICR method, which are low mass resolution, slow scanning speeds, and limited mass range. This has led to the new name Fourier-transform mass spectrometry (FTMS),[3] although the term FT-ICR indicates the physical principles of the method more precisely.

Described first are the basic principles of the method coupled to recent developments of its performance. Next to be surveyed is the emerging area of analytical applications. The section thereafter covers studies of ion/molecule reactions with the emphasis on the chemistry involved. Some concluding remarks are given at the end of the chapter.

[1] (*a*) M. B. Comisarow and A. G. Marshall, *Chem. Phys. Lett.*, 1974, **25**, 282; (*b*) M. B. Comisarow and A. G. Marshall, *Can. J. Chem.*, 1974, **52**, 1997.
[2] 'Lecture Notes in Chemistry 31, Ion Cyclotron Resonance Spectrometry II', ed. H. Hartmann and K.-P. Wanczek, Springer-Verlag, Berlin, 1982.
[3] C. L. Wilkins and M. L. Gross, *Anal. Chem.*, 1981, **53**, 1661A.
[4] C. L. Johlman, R. L. White, and C. L. Wilkins, *Mass Spectrom. Rev.*, 1983, **2**, 389.
[5] K.-P. Wanczek in 'Dynamic Mass Spectrometry', ed. D. Price and J. F. J. Todd, Heyden, London, 1981, Vol. 6, p. 14.
[6] R. T. McIver, jun. and W. D. Bowers in 'Tandem Mass Spectrometry', ed. F. W. McLafferty, Wiley, New York, 1983, p. 287.
[7] S. Ingemann, J. C. Kleingeld, and N. M. M. Nibbering in 'Ionic Processes in the Gas Phase', ed. M. A. Almoster Ferreira, Reidel, Dordrecht, 1984, p. 87.
[8] Nicolet Analytical Instruments, Madison, Wisconsin, U.S.A.: (*a*) 'FT-MS 1000 Fourier Transform Mass Spectrometer', 1982, and (*b*) 'FT-MS 2000 Fourier Transform Mass Spectrometer', 1984.
[9] Spectrospin AG, Zürich, Switzerland: 'CMS 47 Cyclotron Resonance Mass Spectrometer', 1982.

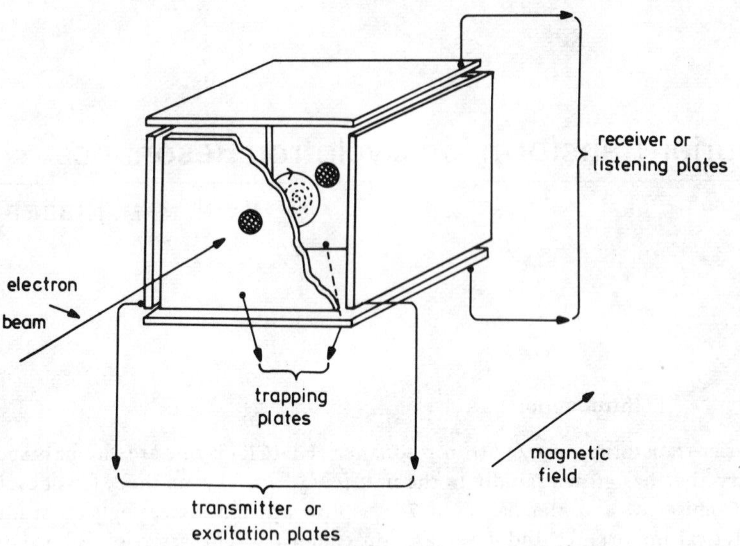

Figure 1 *Schematic drawing of a cubic FT-ICR trapped-ion cell*

2 Various Aspects of FT-ICR

Basic Principles. – These can easily be explained on the basis of Figure 1. Ions are generated in a cell, which is commonly cubic in geometry[10] ($2.54 \times 2.54 \times 2.54$ cm^3) and which is located in a high-vacuum chamber between the poles of an electromagnet or in a superconducting magnet. The ions travel in circular paths perpendicular to the direction of the magnet field, B, and are trapped in the cell by a potential of about −1 V, in the case of anions, being applied to its trapping plates. The angular or cyclotron frequency ω_c of the ions – having low translational energies and random phases in their cyclotron motion – is given by equation 1,

$$\omega_c = v/r = qB/m \tag{1}$$

where q is the charge, v the velocity, m the mass of the ion, and r the radius of its circular path. For example, at a magnetic-field strength of 1.2 T, the mass range m/z 10–1000 corresponds to a frequency band from about 1.8 MHz (m/z 10) to 18 kHz (m/z 1000), and the radius of the circular path of a thermal $Ar^{+\cdot}$ ion is 0.01–0.02 cm. After a certain trapping time of the ions, which for example may vary from 0.05 to 5 s following ion generation, the ions are accelerated coherently in phase to larger orbits by applying a radiofrequency (RF) excitation pulse to the transmitter plates of the cell and sweeping the frequency of the excitation voltage across the frequency range of the ions. A typical sweep rate to cover the above-mentioned mass range would be 1–2 MHz ms^{-1}, known as a frequency

[10] M. B. Comisarow, *Int. J. Mass Spectrom. Ion Phys.*, 1981, **37**, 251.

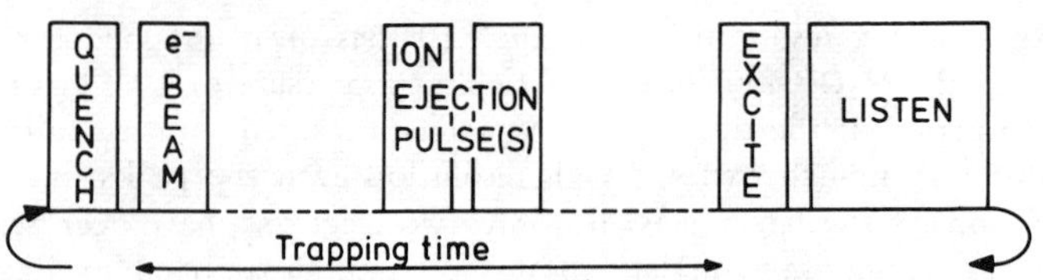

Figure 2 *Typical sequence of pulses in an FT-ICR trapped-ion cell. During the trapping-time period, mass-selected ions or electrons may be ejected from the cell*

'chirp'.[11] The coherently in-phase accelerated ions move up and down between the top and bottom or receiver plates of the cell, generating image currents in the circuit connecting these plates. These image currents will decay with time as the coherent motion of the ions is destroyed by ion/molecule collisions. The analogue signal induced in the receiver plates is thus a transient that is converted, after amplification, into a digital signal and stored in the memory of a digital computer. Ions are then removed from the cell with a quench pulse by inverting the polarity of the potential applied to one of the trapping plates. The pulse sequence, schematically shown in Figure 2, is then repeated for a chosen number of times to improve the signal-to-noise (S/N) ratio.[12] Finally, the summed transient and digitized time-domain signal is subjected to Fourier transformation to generate an ICR frequency-domain spectrum, which subsequently can be converted into a FT-ICR mass spectrum.

Recently it has been shown that broad-band frequency excitation can be achieved by pseudo-random-noise[13] and random-noise[14] excitation. In pseudo-random-noise excitation, a phase coherence from scan to scan is retained so that rapid averaging of the time-domain signal is possible as in the case of chirp excitation. In random-noise excitation, this phase coherence no longer exists, so, to improve the S/N ratio, accumulation of separate scans must be carried out in the frequency domain. Random-noise excitation will therefore be more time-consuming, but it would also lead to more accurate measurements of ion abundance than pseudo-random-noise excitation.

Mass Resolution and Accurate Mass Measurements. — The operating pressures in FT-ICR are low and preferably of the order of 10^{-6} Pa. In these circumstances the motion of translationally excited ions can persist for a relatively long time, giving relatively long-lived transients. This permits very precise measurements of effective cyclotron-resonance frequencies of the ions and hence very high to ultra-high resolution. In the zero-pressure limit the resolution $m/\Delta m$ is proportional to the magnetic-field strength and the acquisition period of the time-domain signal and inversely proportional to ion mass. In a so-called narrow-band

[11] A. G. Marshall and D. C. Roe, *J. Chem. Phys.*, 1980, **73**, 1581.
[12] A. G. Marshall, *Anal. Chem.*, 1979, **51**, 1710.
[13] A. G. Marshall, T.-C. L. Wang, and T. L. Ricca, *Chem. Phys. Lett.*, 1984, **108**, 63.
[14] C. F. Ijames and C. L. Wilkins, *Chem. Phys. Lett.*, 1984, **108**, 58.

mode experiment a resolution of $m/\Delta m > 10^8$ has been obtained for the m/z 18 signal of the $H_2O^{+\cdot}$ ion in a 4.7 T field at a pressure of 10^{-8} Pa and a 51 s acquisition period of the time-domain signal as measured from its full width at half-height.[15] With such extreme high resolution even the peaks due to $^{35}Cl^+$ and $^{35}Cl^-$ ions, differing in mass by only two electrons, have been resolved.[16] However, to acquire a complete mass spectrum, broad-band excitation and detection must be applied. In that case the mass resolution becomes relatively low because of the limited number of acquired data points for the time-domain signal, which in turn is limited by practical parameters such as the ADC rate, the size of the computer memory, and the time required to perform the Fourier transformation on a given set of time-domain data. High resolution is, of course, essential to separate interfering peaks from each other when elemental compositions are to be assigned. However, in the absence of such interferences, accurate mass determinations can also be made under the relatively low-resolution conditions of broad-band FT-ICR spectra. The first requirement then is to be able to measure the peak centroids as accurately as possible. This can be achieved if there are enough data points across the peaks; however, this is usually not the case. This problem can be overcome by application of mathematical procedures, such as interpolation,[17] zoom transform,[18] or zero-filling.[19] It seems that the procedure of zoom transform could be the method of choice for determining the peak centroids in broad-band FT-ICR spectra.[18] These peak centroids correspond to effective ICR frequencies, which differ from those predicted by equation (1) because of various effects. These effects have been the subject of a number of recent studies, for example space charge,[20–22] trapping potential and magnetic-field stabilities,[23] and magnetic-field inhomogeneity.[24] Other studies have dealt with the dynamic behaviour of ions as a function of time following their formation[25,26] and are important contributions to a better understanding of the basis of ICR frequency shifts. It has become clear from all these studies that the ion densities in the FT-ICR cell should be low, *i.e.* about 10^5–10^6, although other cell designs could change these numbers. For example, it has been shown that a

[15] M. Allemann, H. Kellerhals, and K.-P. Wanczek, *Int. J. Mass Spectrom. Ion Phys.*, 1983, **46**, 139.
[16] M. Allemann, H. Kellerhals, and K.-P. Wanczek in ref. 2, p. 380.
[17] C. Giancaspro and M. B. Comisarow, *Appl. Spectrosc.*, 1983, **37**, 153.
[18] T. J. Francl, R. L. Hunter, and R. T. McIver, jun., *Anal. Chem.*, 1983, **55**, 2094.
[19] M. B. Comisarow and J. D. Melka, *Anal. Chem.*, 1979, **51**, 2198.
[20] J. B. Jeffries, S. E. Barlow, and G. H. Dunn, *Int. J. Mass Spectrom. Ion Processes*, 1983, **54**, 169.
[21] T. J. Francl, M. G. Sherman, R. L. Hunter, M. J. Locke, W. D. Bowers, and R. T. McIver, jun., *Int. J. Mass Spectrom. Ion Processes*, 1983, **54**, 189.
[22] E. B. Ledford, jun., D. L. Rempel, and M. L. Gross, *Int. J. Mass Spectrom. Ion Processes*, 1983/1984, **55**, 143.
[23] R. L. White, E. C. Onyiriuka, and C. L. Wilkins, *Anal. Chem.*, 1983, **55**, 339.
[24] D. Schuch, K.-M. Chung, and H. Hartmann, *Int. J. Mass Spectrom. Ion Processes*, 1984, **56**, 109.
[25] T. J. Francl, E. K. Fukuda, and R. T. McIver, jun., *Int. J. Mass Spectrom. Ion Phys.*, 1983, **50**, 151.
[26] R. C. Dunbar, J. H. Chen, and J. D. Hays, *Int. J. Mass Spectrom. Ion Processes*, 1984, **57**, 39.

trapped-ion cell elongated in the dimension parallel to the magnetic field can store about 50 times more ions before space-charge effects are observed,[27] thus increasing the dynamic range of FT-ICR (10^2–10^3), which is still small compared to that of conventional mass spectrometers (10^5–10^6). Moreover, the trapping potentials applied to the plates of such an elongated cell appear to shift the ICR frequencies only slightly.[27] This would make this cell well suited for accurate mass determinations, allowing the elemental compositions of ions to be assigned with reasonable confidence. Nevertheless, it has been shown possible to perform accurate mass measurements with the cubic cell over a mass range of m/z 15–160 with an average error of 6 p.p.m. after mass calibration, but in the absence of calibrant[23] and over a mass range of m/z 18–170 with an average error of 1.5 p.p.m. by use of low-intensity (~1%) sidebands for mass calibration.[16]

Ion Abundances. – Clearly, accurate measurement of ion abundances is important for both analytical applications of and ion/molecule chemistry studies by FT-ICR. Assuming that the operating pressures are sufficiently low to allow relatively long-lived transients, as discussed in the preceding section, peak-height errors can be reduced to less than 2% if correction is made for uneven excitation of ions with different masses and if care is taken that there are enough data points across the peaks. Correction for the uneven excitation, due to the finite length of the excitation pulse, can be achieved by a deconvolution procedure.[11] Enough data points across the peaks can be obtained by using the mathematical procedures of interpolation,[17] zoom transform,[18] or zero-filling,[19] mentioned in the preceding section. Another method, which requires less computer time than zero-filling, involves the multiplication of the time-domain signal by a sum of cosines to give a weighted transient[28] prior to Fourier transformation. Also in this case peak-height errors have been found to be less than 2% on the basis of isotopic-abundance measurements.

3 Analytical Applications

Ionization Methods. – In the past 15 years many ionization methods have been developed in mass spectrometry to analyse not only volatile but also non-volatile compounds, which frequently have relatively high molecular weights.[29] Unfortunately, several of them are not compatible with FT-ICR because they operate either at high potential or at relatively high pressure, which is one of the major factors that adversely affect both the resolution and the sensitivity of the method.[30] There are, however, instrumental developments, to be mentioned at the end of the section, that promise to make some of the ionization methods compatible with FT-ICR.

One of the ionization methods that has successfully afforded ions in the ICR cell is multi-photon ionization of naphthalene, fluoranthene, and triphenylene

[27] R. L. Hunter, M. G. Sherman, and R. T. McIver, jun., *Int. J. Mass Spectrom. Ion Phys.*, 1983, **50**, 259.
[28] A. J. Noest and C. W. F. Kort, *Comput. Chem.*, 1982, **6**, 115.
[29] See for a review N. M. M. Nibbering, *J. Chromatogr.*, 1982, **251**, 93.
[30] M. B. Comisarow in ref. 2, p. 484.

by an excimer laser beam.[31] In a similar experiment it has been shown that perdeuteriobiphenyl could be ionized selectively in the presence of an excess of di-isopropyl sulphide, argon, or toluene by multi-photon ionization.[32] Pulsed laser beams have also been used to generate metal ions from metal surfaces to study their reactions with organic compounds, as will be discussed in the section on ion/molecule reactions. Interesting results have been obtained in laser-desorption experiments on some underivatized dipeptides and dicarboxylic acids.[33,34] For example, a resolution of 10 000 has been obtained for the peak of the succinate anion at m/z 117 as measured from its full width at half-height and at a pressure of 10^{-5} Pa.[33]

Another ionization method that has been shown to be compatible with FT-ICR is desorption ionization by bombarding non-volatile molecules, such as Vitamin B_{12}, with a 2–3 keV Cs^+ beam.[35] The results obtained from these experiments[35] support the assertion that FT-ICR might become the method of choice for the analysis of compounds of high molecular weight.

Finally, it should be noted that at short trapping times (see Figure 2) a FT-ICR spectrum will resemble an electron-impact mass spectrum. At longer trapping times ion/molecule reaction products will be formed. This constitutes the basis for studies of ion/molecule reactions in low-pressure conditions (*vide infra*), but it has also been exploited in analytical chemistry for preparing in the cell the well known chemical ionization (CI) reactant ions H_3O^+, CH_5^+, and NH_4^+, which were then allowed to react with a number of substrates to yield prominent $(M + H)^+$ peaks.[36,37] Methyl nitrite has also been put forward as a low-pressure CI reagent, affording $(M + NO)^+$ ions (often as the sole ions at long trapping times) from a variety of substrates.[38]

Collision-induced Dissociation. – Fragmentations of mass-selected ions by collisions with inert gases in sector-type or quadrupole mass spectrometers have found extensive applications in both analytical and gas-phase ion-chemistry studies.[39] The method, known as MS/MS, has recently been reviewed with regard to instrumentation,[40] mass resolution,[41] and its application to the chemistry of gas-phase ions.[42]

In FT-ICR, mass-selected ions can be isolated by ejecting all other ions from the cell with so-called ion-ejection pulses. This can be achieved by application of

[31] M. P. Irion, W. D. Bowers, R. L. Hunter, F. S. Rowland, and R. T. McIver, jun., *Chem. Phys. Lett.*, 1982, **93**, 375.
[32] T. J. Carlin and B. S. Freiser, *Anal. Chem.*, 1983, **55**, 955.
[33] D. A. McCrery, E. B. Ledford, jun., and M. L. Gross, *Anal. Chem.*, 1982, **54**, 1435.
[34] E. Onyiriuka, R. L. White, D. A. McCrery, M. L. Gross, and C. L. Wilkins, *Int. J. Mass Spectrom. Ion Phys.*, 1983, **46**, 135.
[35] M. E. Castro and D. H. Russell, *Anal. Chem.*, 1984, **56**, 578.
[36] S. Ghaderi, P. S. Kulkarni, E. B. Ledford, jun., C. L. Wilkins, and M. L. Gross, *Anal. Chem.*, 1981, **53**, 428.
[37] M. L. Gross and C. L. Wilkins in ref. 2, p. 392.
[38] W. D. Reents, jun., R. C. Burnier, R. B. Cody, and B. S. Freiser, *Anal. Chem.*, 1982, **54**, 1245.
[39] 'Tandem Mass Spectrometry', ed. F. W. McLafferty, Wiley-Interscience, New York, 1983.
[40] R. A. Yost and D. D. Fetterolf, *Mass Spectrom. Rev.*, 1983, **2**, 1.
[41] F. W. Crow, K. B. Tomer, and M. L. Gross, *Mass Spectrom. Rev.*, 1983, **2**, 47.
[42] K. Levsen and H. Schwarz, *Mass Spectrom. Rev.*, 1983, **2**, 77.

appropriate RF pulses to the transmitter plates, whereby the ions to be ejected are accelerated to such large orbits that they strike the plates of the cell (see Figure 2). However, the resolution available to isolate ions of a particular m/z value in the cell for collision-induced dissociation (CID) or ion/molecule reaction studies is much lower than that for a triple sector or tandem double-focusing mass spectrometer,[43] being a few hundreds *versus* 15 000–50 000, respectively. This limited so-called 'front-end' resolution of FT-ICR is due to the frequency bandwidth associated with the length of the RF pulse, the narrowest bandwidths being attained with long pulses.[44] Yet, unit 'front-end' resolution is achievable with careful adjustment of the RF levels and excitation times (in the ms range) of the ion-ejection pulses.[44,45] Another method to eject all ions from the cell with the exception of ions with a particular m/z value is based upon a 180° phase shift of the ion-ejection pulse at the cyclotron frequency of the ion to be isolated in the cell.[46,47] This ion-ejection method has been applied successfully in ion/molecule reaction studies[48] (see also the section on ion/molecule reactions), and methods have been proposed to increase its mass selectivity.[47]

The ions with a particular m/z value isolated in the cell can then be accelerated by use of an appropriate RF pulse. The maximum translational energy acquired by the ions (in excess of their thermal energy) is given by equation 2.[49]

$$E_{tr}(\max) = q^2 r^2 B^2 / 2m \tag{2}$$

This equation shows that the maximum attainable translational energy is proportional to both the square of the cell radius and the square of the magnetic field and inversely proportional to the ionic mass. For example, an ion of m/z 100 can be accelerated to a maximum kinetic energy of 152 eV in a cubic inch cell ($r = 1.27$ cm) when placed in a magnetic field of 1.4 T, whereas the maximum kinetic energy becomes 7.5 keV with $r = 2.5$ cm and a magnetic field of 5 T. Equation 2 also shows that the maximum kinetic energy is independent of the amplitude of the RF pulse used, although it should be chosen to be sufficiently large to avoid collisions of the ions during the acceleration period. After acceleration the ions can collide with either their parent molecules or target gases such as He, Ar, or N_2, resulting in collision-induced dissociation as demonstrated in a number of papers.[49–50] As noted above, the kinetic energies of ions accelerated in a cell with $r = 1.27$ cm at 1.4 T will be in the order of a few hundred eV, so that low-energy CID processes are expected to occur. This has been verified to be the case on the basis of a comparative study of energy-resolved tandem and FT-ICR mass spectrometry concerning the collision-energy dependence of

[43] (*a*) M. L. Gross and D. H. Russell in ref. 39, p. 255; (*b*) P. J. Todd, D. C. McGilvery, M. A. Baldwin, and F. W. McLafferty in ref. 39, p. 271.
[44] R. B. Cody, R. C. Burnier, C. J. Cassady, and B. S. Freiser, *Anal. Chem.*, 1982, **54**, 2225.
[45] J. C. Kleingeld and N. M. M. Nibbering, *Org. Mass Spectrom.*, 1982, **17**, 136.
[46] A. J. Noest and C. W. F. Kort, *Comput. Chem.*, 1983, **7**, 81.
[47] A. G. Marshall, T. C. Lin Wang, and T. Lebatuan Ricca, *Chem. Phys. Lett.*, 1984, **105**, 233.
[48] J. C. Kleingeld and N. M. M. Nibbering, *Tetrahedron*, 1983, **39**, 4193.
[49] R. B. Cody, R. C. Burnier, and B. S. Freiser, *Anal. Chem.*, 1982, **54**, 96.
[50] R. B. Cody and B. S. Freiser, *Int. J. Mass Spectrom. Ion Phys.*, 1982, **41**, 199.

MS/MS spectra.[51] High-energy (keV) CID processes are attainable in larger ICR cells and at higher magnetic-field strengths, as demonstrated by the formation of $C_6H_5^{2+}$ *via* a successive charge stripping and fragmentation of $C_6H_6^{+\bullet}$.[52]

It is also possible to observe consecutive CID processes in FT-ICR. The primary fragment ions generated by CID from the mass-selected parent ions are then accelerated with another appropriate RF pulse to generate secondary fragment ions by CID, *etc.*[44] In this way it has been possible to fragment the molecular ion of acetophenone in four consecutive CID reactions (MS/MS/MS/MS) to the fragment ion *m/z* 50 as summarized in equation 3.[53]

$$\underset{m/z\ 120}{PhCOMe^{+\bullet}} \xrightarrow{-Me^{\bullet}} \underset{m/z\ 105}{PhCO^{+}} \xrightarrow{-CO} \underset{m/z\ 77}{Ph^{+}} \xrightarrow{-C_2H_2} \underset{m/z\ 51}{C_4H_3^{+}} \xrightarrow{-H^{\bullet}} \underset{m/z\ 50}{C_4H_2^{+\bullet}} \quad (3)$$

In such experiments the pressure is higher than normal because of the required collisions. This reduces the mass resolution of the daughter ions, although early CID studies provided mass resolutions of about 3000 for the *m/z* 43 CID daughter ions formed from acetone and 2-chloropropane at 0.9 T[54] and of 9200 for the *m/z* 866 CID daughter ion formed from the *m/z* 1166 ion of tris(perfluoroheptyl)-*sym*-triazine at 1.9 T,[55] as measured from the widths of the corresponding peaks at half-height. Another result of the increased pressure in CID experiments is that a so-called 'uncontrolled' CID may occur if the bandwidth of the RF pulse to eject the daughter ions overlaps the resonance frequency of the parent ion.[44]

Both reduced mass resolution and uncontrolled CID can be overcome by the method of pulsed-valve addition of the collision gas.[56] In this method ions with a particular *m/z* value, which have been isolated under low-pressure conditions as described above, are accelerated during the period of the instantaneous high pressure in the cell, whereas detection of the formed fragment ions occurs at such a delayed time that the pressure is sufficiently low to obtain high-resolution conditions. In this way a mass resolution of greater than 20 000 has been obtained at 0.9 T for *m/z* 105 CID daughter ions.[56] The method of pulsed-valve addition of gases has further advantages, such as higher yields of daughter ions upon CID *via* multiple collisions,[56,57] no ionization of the target gas during generation of the parent ions, formation of cluster ions not attainable in low-pressure conditions, the feasibility of studying ion/molecule reactions at high pressure as with conventional CI mass spectrometry, *etc.*[56] Moreover, the pulsed-valve method has been exploited successfully as an interface for GC/FT-ICR,[58] which will be discussed below.

In conclusion, the mass resolution of CID fragment ions in FT-ICR has been shown to be much higher than that of a triple sector or tandem double-focusing

[51] S. A. McLucky, L. Sallans, R. B. Cody, R. C. Burnier, S. Verma, B. S. Freiser, and R. G. Cooks, *Int. J. Mass Spectrom. Ion Phys.*, 1982, **44**, 215.
[52] D. L. Bricker, T. A. Adams, jun., and D. H. Russell, *Anal. Chem.*, 1983, **55**, 2417.
[53] J. C. Kleingeld and N. M. M. Nibbering, unpublished results.
[54] R. B. Cody and B. S. Freiser, *Anal. Chem.*, 1982, **54**, 1431.
[55] R. L. White and C. L. Wilkins, *Anal. Chem.*, 1982, **54**, 2211.
[56] T. J. Carlin and B. S. Freiser, *Anal. Chem.*, 1983, **55**, 571.
[57] R. C. Burnier, R. B. Cody, and B. S. Freiser, *J. Am. Chem. Soc.*, 1982, **104**, 7436.
[58] T. M. Sack and M. L. Gross, *Anal. Chem.*, 1983, **55**, 2419.

mass spectrometer,[43] where the last-mentioned instrument would have a so-called 'back-end' resolution of over 10 000.

GC/FT-ICR. – In principle FT-ICR is an ideal candidate for coupling with GC because of its fast scanning capabilities and high-to-ultra-high mass-resolving power. Nevertheless, few papers on GC/FT-ICR have appeared since the first pioneering experiments were performed.[59,37] Of course, the main problem is to maintain the pressure in the cell at a low enough level to preserve the high performance of FT-ICR. In line with this, it has been found that support-coated open tubular (SCOT) capillary columns should be coupled to FT-ICR *via* a jet separator rather than directly.[59,37] With that set-up, low-resolution spectra (*ca.* 1000 full width at half-height for m/z 78) covering a mass range from m/z 38 to m/z 200 and consisting of 100 signal-averaged scans could be collected at a rate of one per second.[60] Furthermore, it has been shown possible to monitor the molecular ions of benzene and bromobenzene during a GC separation using peak switching from m/z 78 to m/z 156 and back every 285 ms and to obtain a resolution of 40 000 for m/z 78 and 20 000 for m/z 156 measured as full widths of the peaks at half-height.[60] As mentioned earlier, the dynamic range of FT-ICR is limited (10^2–10^3) so that trace analysis of complex mixtures by FT-ICR is best performed with GC/FT-ICR rather than the CID/FT-ICR technique discussed in the previous section. Such an analysis has been performed on a commercial lacquer thinner, where the complete analysing system consisted of a linked GC/FTIR/FT-ICR arrangement,[61] which has recently been developed further.[62] Quite promising for the development of GC/FT-ICR is the use of a pulsed-valve interface to introduce the GC effluent into the ICR cell.[58] The detection limit of naphthalene, at a mass resolution of 32 000 full width at half-height for its molecular ion, was established to be about 1 ng. Compared to previously published GC/FT-ICR results, this constitutes a factor of 2.5 improvement in resolution and of 6 in sensitivity for the pulsed-valve interface.[58]

Instrumental Developments. – It will be clear from the material presented in the previous sections that the best performance of FT-ICR is obtained at pressures lower than 10^{-6} Pa. These are not compatible with most of the ionization methods[29] developed in the past 15 years that have increased greatly the analytical applicability of mass spectrometry. Although it is attractive to generate and to detect ions along with a variety of experiments in the same physical space, *i.e.* the cell, the analytical applicability of FT-ICR would increase substantially if the regions of ion generation and ion detection were separated physically from each other. In other words, the FT-ICR cell can best be used as a detector. Two instrumental developments along the above-mentioned lines have been reported

59 E. B. Ledford, jun., R. L. White, S. Ghaderi, C. L. Wilkins, and M. L. Gross, *Anal. Chem.*, 1980, **52**, 2450.
60 R. L. White and C. L. Wilkins, *Anal. Chem.*, 1982, **54**, 2443.
61 C. L. Wilkins, G. N. Giss, R. L. White, G. M. Brissey, and E. C. Onyiriuka, *Anal. Chem.*, 1982, **54**, 2260.
62 D. A. Laude, jun., G. M. Brissey, C. F. Ijames, R. S. Brown, and C. L. Wilkins, *Anal. Chem.*, 1984, **56**, 1163.

recently. One of them involves the coupling of a quadrupole mass spectrometer to an FT-ICR instrument;[63] the other makes use of a split cell.[8b] One half of this cell operates at relatively high pressure and is used as the ion-generation compartment from where the ions migrate to and are trapped in the other half of the cell, which operates at low pressure and serves as the detection region. These developments are undoubtedly very important for making the FT-ICR method applicable to a wide variety of analytical problems.

4 Ion/Molecule Reactions

General Picture. – Prior to discussion of recent studies it is appropriate to give in a qualitative way a general picture of the course of ion/molecule reactions under low-pressure conditions.[64] First, a loose ion/molecule complex is formed that has gained energy provided by the long-range ion/induced-dipole and ion/dipole interactions between the approaching reactants, *i.e.* solvation energy. This energy is not dissipated to other molecules because of the absence of third-body collisions at the low pressures used in FT-ICR (unless the previously mentioned pulsed-valve addition of reagent or chemically inert gases[56] is applied). In the absence of third-body collisions, the solvation energy can then be used by the initially formed loose ion/molecule complex to undergo reactions and to overcome the associated barriers on the potential-energy surface, sometimes leading to products that at first sight are unexpected. The heights of the barriers will depend on the type of reaction and on the nature of the reactants, which explains why rate constants of ion/molecule reactions can span the range from being almost collision controlled to being too slow to be observed.[65] These barriers, however, should not be higher than the energy of the initial reactants at infinite separation. Furthermore, the overall ion/molecule reactions should be either exothermic or thermoneutral in order to lead to products.

Organic Ion/Molecule Reactions. – FT-ICR has proven in the past few years to have extraordinary potential for studying the chemistry involved in gas-phase ion/molecule reactions. Snapshots of the ion distribution present in the cell can be taken at different trapping times (see Figure 2), so that a time-resolved view of the progress and associated chemistry of the ion/molecule reactions can be obtained. The ion-ejection methods described in the previous section permit removal of unwanted ions from the cell and study of the chemistry of mass-selected ions in bimolecular reactions. Moreover, the structures of the mass-selected ions can be probed by either suitable ion/molecule reactions or collision-induced dissociation. In this way a variety of ion/molecule reactions has been studied recently, involving both positive ions (to be discussed first) and negative ions.

[63] W. D. Bowers, R. L. Hunter, and R. T. McIver, jun., *Ind. Res. Dev.*, 1983, November Issue, p. 124.
[64] N. M. M. Nibbering, *Kemia Kemi*, 1984, **11**, 11.
[65] M. J. Pellerite and J. I. Brauman, *J. Am. Chem. Soc.*, 1980, **102**, 5993 and references cited therein.

Protonated methanol has been shown to react with neutral methanol *via* an S_N2 reaction to give protonated dimethyl ether.[45] In such a reaction the protonated dimethyl ether would contain the oxygen atom of the neutral methanol, whereas the oxygen atom in the displaced molecule of water would originate from the protonated methanol species when it is assumed that no rapid proton transfer occurs between the reactants in the collision complex. The reactants therefore should be labelled differently with regard to oxygen in order to prove the S_N2 mechanism. This has been achieved in the following way. First, ions were generated from methanol containing the oxygen isotopes at the natural-abundance level by a relatively long electron-beam pulse. This results in the formation of abundant protonated methanol species, including $Me^{18}OH_2^+$ ions, after an appropriate trapping time. The ^{18}O-containing ions were then isolated selectively in the cell by application of a series of ion-ejection pulses to remove all other ions from the cell, so that the $Me^{18}OH_2^+$ ions were in a bath gas consisting of 99.8% [^{16}O]methanol. They were then allowed to react with the bath-gas molecules for about 300 ms, under continuous ejection of regenerated protonated [^{16}O]methanol species, to give $C_2H_7{}^{16}O^+$ and $C_2H_7{}^{18}O^+$ product ions in a ratio of approximately 3.5 : 1, respectively. Thus, most of the $Me^{18}OH_2^+$ ions are attacked by the neutral methanol molecules *via* the S_N2 mechanism when forming the protonated dimethyl ether species, *i.e.* $C_2H_7{}^{16}O^+$. The formation of the other product ions, $C_2H_7{}^{18}O^+$, is presumably due to a relatively slow proton-transfer reaction in the collision complex prior to the S_N2 reaction.[45]

The phenyl-substituted 2-aza-allenium ions $PhCH{=}\overset{+}{N}{=}CH_2$, $PhCH{=}\overset{+}{N}{=}CHPh$, and $Ph_2C{=}\overset{+}{N}{=}CH_2$ generated from some benzylidene amines by electron impact and α-cleavage of an alkyl radical could only be deprotonated by the use of the superbase $(Me_2N)_2C{=}NCMe_3$. This reaction in combination with deuterium labelling has been used to probe the structures of the phenyl-substituted 2-aza-allenium ions.[66]

$C_3H_5^+$ ions generated from cyclopropyl, allyl, and 1- and 2-propenyl bromide molecular ions by loss of the bromine atom all react with benzene or substituted benzenes by a formal CH transfer.[67] Other experiments such as the measurement of rate constants for these reactions by pulsed ICR and CID and charge stripping of the corresponding adduct ions, prepared and pressure stabilized in a chemical ionization source, have shown that two $C_3H_5^+$ ion structures are stable, *i.e.* the allyl and the 2-propenyl cations.[67]

The molecular ions of various C_5H_{10} isomers have been studied with regard to their reactivity towards neutral ammonia.[68] Some of them could be distinguished easily by use of this chemical probe. For example, ionized ethylcyclopropane and 1,1-dimethylcyclopropane afforded product ions of m/z 18, 30, and 58, *trans*-1,2-dimethylcyclopropane the product ions of m/z 18, 30, and 44, and 1-pentene the product ions of m/z 18 and 44, whereas *cis*-2-pentene, 2-methyl-2-butene, 2-methyl-1-butene, 3-methyl-1-butene, cyclopentane, and methylcyclobutane only gave the product ion of m/z 18.[68]

[66] J. C. Kleingeld, N. M. M. Nibbering, H. Halim, H. Schwarz, and E.-U. Würthwein, *Chem. Ber.*, 1983, **116**, 3877.
[67] J. O. Lay, jun. and M. L. Gross, *J. Am. Chem. Soc.*, 1983, **105**, 3445.
[68] D. L. Miller and M. L. Gross, *Org. Mass Spectrom.*, 1983, **18**, 239.

Five unique $C_7H_7O^+$ ion structures, including the hydroxybenzyl ion, the hydroxytropylium ion, the protonated benzaldehyde ion, the methylphenyloxy ion, and the *O*-phenylated formaldehyde ion, could be identified on the basis of ion/molecule reactions in combination with photodissociation and low-energy CID experiments.[69]

As part of a nuclear-decay and radiolytic study on the formation and behaviour of dialkylphenyloxonium ions in the gas phase, the reactions between the $MeFMe^+$ ion and propyl phenyl and isopropyl phenyl ethers have been studied by FT-ICR.[70] Attack of $MeFMe^+$ on both ethers quantitatively and irreversibly yields a product ion of m/z 109, corresponding possibly to protonated anisole or isomeric cresols. A bimolecular rate constant of $(2.3 \pm 0.2) \times 10^{-9}$ cm^3 molecule^{-1} s^{-1} has been measured for both reactions.[70]

In the area of negative ions, some studies on hydrogen–deuterium exchange have been reported. The $(M-H)^-$ ions of methyl phenyl ether, generated by abstraction of one of the ring hydrogen atoms with hydroxide ions, undergo such an exchange in the presence of D_2O.[71] This exchange is possible because of the solvation energy available to the $(M-H)^- \cdot D_2O$ complex, which can be as large as 80 kJ mol^{-1} and which allows deuteron transfer from the less acidic D_2O molecule to the $(M-H)^-$ ion in the complex. A subsequent proton transfer to the OD^- ion in the complex yields a H/D-exchanged $(M-H)^-$ ion, which can either leave the complex or undergo a second H/D exchange with the HDO molecule generated in the complex. The H/D-exchanged $(M-H)^-$ ion can then encounter another D_2O molecule for further H/D exchange, eventually resulting in the exchange of all ring hydrogen atoms. Interestingly, the methyl hydrogen atoms were shown to be exchanged as well. This does not involve a direct proton abstraction from the methyl group by OH^- because even the stronger base NH_2^- is not capable of inducing this reaction, as shown by D-labelling. However, the $(M-H)^-$ ions generated from $PhOCD_3$ with NH_2^- eliminate CD_2O and CHDO in a 1:1 ratio. This can easily be explained by an equilibrated exchange between the methyl deuterium atoms and one hydrogen atom, most probably from the *ortho* position, in the $(M-H)^-$ ions prior to the loss of formaldehyde. This interpretation implies the rare occurrence of a gas-phase primary carbanion,[72] which must have a lifetime sufficiently long to account for the equilibrated methyl/*ortho* hydrogen-atom exchange. Thus, the ring hydrogen atoms of the $(M-H)^-$ ions of methyl phenyl ether exchange in an intermolecular process with D_2O, whereas the methyl hydrogen atoms do so in an intramolecular process.[71] Similar observations have been made for the $(M-H)^-$ ions of 2-fluorophenyl and 4-fluorophenyl methyl ether in the presence of D_2O.[73]

Studies on H/D exchange have been performed also on the $(M-H)^-$ ions derived from cycloheptatriene and toluene.[74] Both ions behave rather similarly

[69] C. J. Cassady, B. S. Freiser, and D. H. Russell, *Org. Mass Spectrom.*, 1983, **18**, 378.
[70] S. Fornarini and M. Speranza, *J. Chem. Soc., Perkin Trans. 2*, 1984, 171.
[71] J. C. Kleingeld and N. M. M. Nibbering, *Tetrahedron*, 1983, **39**, 4193.
[72] A. J. Noest and N. M. M. Nibbering, *J. Am. Chem. Soc.*, 1980, **102**, 6427.
[73] S. Ingemann and N. M. M. Nibbering, *J. Org. Chem.*, 1983, **48**, 183.
[74] R. L. White, C. L. Wilkins, J. J. Heitkamp, and S. W. Staley, *J. Am. Chem. Soc.*, 1983, **105**, 4868.

in the presence of either ND_3 or D_2O. The level of H/D exchange is relatively low with ND_3 because of the large acidity difference between ammonia and toluene or cycloheptatriene in the gas phase. However, with D_2O the $(M-H)^-$ ions rapidly exchange two hydrogen atoms for deuterium, pointing to a benzyl anion structure. In the presence of CD_3OD the $(M-H)^-$ ions behave differently: those from toluene still exchange two hydrogen atoms rapidly, but those from cycloheptatriene undergo a low level of exchange of one hydrogen atom and a negligible amount of exchange of two and three hydrogen atoms. It seems therefore that the $(M-H)^-$ ion from cycloheptatriene rearranges to a benzyl anion structure if hydroxide ion is used as base. From studies of $[7,7\text{-}^2H_2]$- and $[1,2,3,4,5,6\text{-}^2H_6]$-cycloheptatriene, it has been concluded that the cycloheptatrienyl anion is an intermediate, which first rearranges to the benzyl anion before the H/D exchange occurs. Although details of this rearrangement are not known, it has been speculated that all the events of proton abstraction, rearrangement, and exchange may well occur within the OH^-/cycloheptatriene complex once it has been formed.[74]

Nucleophilic aromatic substitution has been shown to occur in reactions between various nucleophiles and some alkyl pentafluorophenyl ethers.[75,76] For example, OH^- reacts with pentafluorophenyl methyl ether mainly by attack upon the fluorine-substituted carbon atoms (*ca.* 75%) and by S_N2 (20%) and *ipso* (5%) substitution (Scheme 1*a*, *b*, and *c*, respectively). Also, for other nucleophiles such as alkoxide, thiolate, enolate, and (un)substituted allyl anions, the attack upon the fluorine-substituted carbon atoms (Scheme 1*a*) is in most cases the dominant reaction channel. In this attack F^- ion is displaced *via* a σ-anion complex to form a F^- ion/molecule complex, which is not observed to dissociate into F^- as ionic product. Instead, the displaced F^- ion attacks the newly formed molecule to lead eventually to products (Scheme 1*a*). Depending upon the nature of the original nucleophile the attack by the displaced F^- ion can be proton transfer, S_N2 substitution, or $E2$ elimination. The lifetimes of the F^- ion/molecule complexes formed are not known, but they must be sufficiently long to allow attack by the F^- ion. In this respect it is noteworthy that, in the reaction of CD_3O^- with C_6F_5OMe, eventually the ions $MeOC_6F_4O^-$ and $CD_3OC_6F_4O^-$ are generated in a 1 : 1 ratio, showing that there is sufficient time for the displaced F^- ion to attack both the MeO and the CD_3O group of the newly formed $MeOC_6F_4OCD_3$ molecule with equal probability.[75] However, it should also be noted here that it is not known whether the F^- ion is displaced from one position only or from all the positions of the pentafluorophenyl ring.[75,76]

Hydroxide ions have been found to react with phenyl trifluoroacetate to give trifluoroacetate and phenoxy anions in addition to $(M-H)^-$ and CF_3^- ions.[77] The formation of the trifluoroacetate anions can be explained by a so-called $B_{Ac}2$ mechanism, where the OH^- ion attacks the carbonyl carbon atom. When the tetrahedral intermediate thus formed dissociates into trifluoroacetic acid and

[75] S. Ingemann, N. M. M. Nibbering, S. A. Sullivan, and C. H. DePuy, *J. Am. Chem. Soc.*, 1982, **104**, 6520.

[76] S. Ingemann and N. M. M. Nibbering, *Nouv. J. Chim.*, 1984, **8**, 299.

[77] J. C. Kleingeld and N. M. M. Nibbering in ref. 2, p. 209.

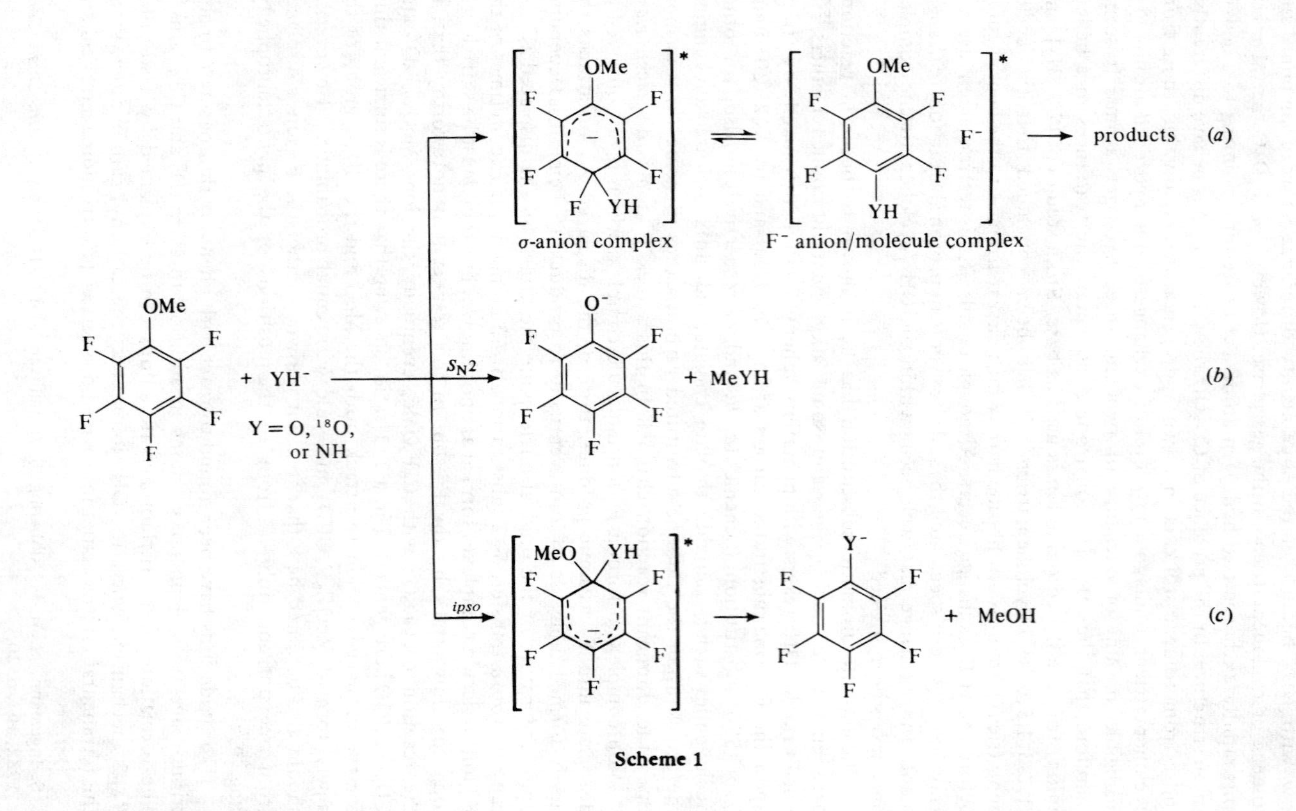
OMe
F
F
F
F
F
+ YH⁻
Y = O, ¹⁸O,
or NH
S_N2
ipso
σ-anion complex
YH
F⁻ anion/molecule complex
products
(a)
O⁻
+ MeYH
(b)
MeO
YH
Y⁻
+ MeOH
(c)

Scheme 1

the phenoxy anion, a proton transfer will occur between these two species, as trifluoroacetic acid is more acidic than phenol in the gas phase.[78] The phenoxy anion should therefore be formed by another mechanism, for which the S_N2 mechanism has been proposed.[77] With this mechanism, the OH^- ion would attack the carbon atom of the CF_3 group to form CF_3OH and carbon monoxide and the phenoxy anion, the two last species *via* the transient moiety of $PhO\bar{C}{=}O$. Ejection experiments on the collision complex of hydroxide ion/phenyl trifluoroacetate indeed support the view that the trifluoroacetate and phenoxy anions are formed by different routes.[77] However, the phenoxy anion might also be formed by dissociation of the tetrahedral intermediate, generated by attack of OH^- upon the carbonyl carbon atom, into phenyl bicarbonate and CF_3^- with concomitant proton transfer between these two species and decomposition of the generated phenyl bicarbonate anion into CO_2 and PhO^-.

In the reaction of the superoxide ion $O_2^{-\cdot}$ with phenyl acetate five product ions are formed: the acetate, peroxyacetate, and phenoxy anions and ions of *m/z* 126 and 108.[79] The last two ions correspond to loss of ketene and acetic acid from the collision complex, respectively. Upon reaction with $^{18}O_2^{-\cdot}$ a single ^{18}O is incorporated in the acetate and *m/z* 108 ions whereas two ^{18}O atoms are incorporated in the peroxyacetate and *m/z* 126 ions.[79] The formation of the peroxyacetate ion shows that the $O_2^{-\cdot}$ ion attacks the ester function, where the resulting tetrahedral intermediate upon decomposition can yield either the peroxyacetate or the phenoxy anion. The generation of the other ions requires bond formation between one of the oxygen atoms of the $O_2^{-\cdot}$ ion and the aromatic ring at some stage of the reaction. For example, the $O_2^{-\cdot}$ ion could attack the aromatic ring directly, where *ortho* attack would account most easily for the product ions formed. However, it is also possible that first a nucleophilic addition of $O_2^{-\cdot}$ occurs to the ester carbonyl group, generating a tetrahedral intermediate, and that the terminal oxygen atom of the added $O_2^{-\cdot}$ ion subsequently attacks the *ortho* position of the phenyl ring to form a bicyclic ion. The latter could then decompose *via* several steps to the observed product ions. Analogous product ions and incorporations of ^{18}O atoms have been found in the reactions of $O_2^{-\cdot}$ and $^{18}O_2^{-\cdot}$, respectively, with phenyl benzoate with the exception of the ketene loss from the collision complex, which in this case can no longer take place.[79]

The $(M-H)^-$ ion $Et_2B\bar{C}HMe$, generated from triethylborane by electron impact, has been found to react slowly with its precursor by hydride transfer to produce Et_3BH^-, but rapidly with water or with molecular oxygen to yield diethylborinate anions, Et_2BO^-, which are observed at *m/z* 84 and 85.[80] The latter ions are observed also when the $(M-H)^-$ ions of triethylborane react with carbon dioxide. In that case, however, they could be either $H_2BCH(Me)CO_2^-$ or Et_2BO^-, depending on whether two ethylene molecules or methylketene are eliminated from the activated $Et_2BCH(Me)CO_2^-$ complex, respectively. High-resolution mass measurements have shown that the ions formed have the elemen-

[78] J. E. Bartmess and R. T. McIver, jun. in 'Gas Phase Ion Chemistry', ed. M. T. Bowers, Academic Press, New York, 1979, Vol. 2, Ch. 11.

[79] C. L. Johlman, R. L. White, D. T. Sawyer, and C. L. Wilkins, *J. Am. Chem. Soc.*, 1983, **105**, 2091.

tal composition corresponding to that of Et_2BO^-, so that the $(M - H)^-$ ions of triethylborane apparently react with carbon dioxide *via* a Wittig-type reaction.[80]

Reactions have been carried out with the 2-formyl-[1,1-$^2H_2$]allyl anion, generated from 2-trideuteriomethylpropenal by deuteron abstraction, and some unsaturated substrates to test whether cycloaddition product ions are generated.[81] With the substrates hexafluorobenzene and methyl pentafluorophenyl ether the 2-formyl-[1,1-$^2H_2$]allyl anion appears to react by expulsion of (HF + DF) and (HF + DF + Me$^•$), respectively. No product ions are observed that would correspond to loss of either 2HF, *c.q.* (2HF + Me$^•$), or 2DF, *c.q.* (2DF + Me$^•$), from the collision complexes. These observations strongly indicate that indeed a formal [4 + 2] cycloaddition reaction between the 2-formyl-[1,1-$^2H_2$]allyl anion and the mentioned aromatic substrates has occurred, although it is not known whether this has taken place in a concerted or stepwise fashion.[81]

In solution, carbonyl compounds can be reduced with alkoxide ions by a reaction known as the Meerwein–Ponndorf–Verley reduction. Similarly, it has been found possible to reduce formaldehyde, benzaldehyde, 2,2-dimethylpropanal, and 1-adamantyl aldehyde with methoxide ions in the gas phase *via* a hydride-transfer reaction.[82] An interesting hydride-transfer step has been observed in the reaction of hydroxide ion with formaldehyde.[83] Once the corresponding collision complex is formed, the solvation energy allows a proton transfer from the formaldehyde molecule to OH^-. However, the water-solvated formyl anion generated in this way cannot separate because water is more acidic than formaldehyde in the gas phase. Instead, a hydride transfer from the formyl anion to the water molecule in the complex takes place, leading to H_3O^- and carbon monoxide as the products. It has been shown by ^{18}O-labelling experiments that the H_3O^- ion exclusively contains the oxygen atom of the hydroxide ion, whereas D-labelling experiments have indicated clearly that the hydrogen atoms in H_3O^- are not equivalent. For example, the H_2DO^- ion generated by reaction of OD^- with formaldehyde transfers a hydride, but not a deuteride, to formaldehyde. These observations are consistent with the view that the H_3O^- ion can best be described as a hydride ion solvated by a water molecule.[83]

It has also been possible to form NH_4^- ions in the gas phase by reaction of the amide ion with formaldehyde.[84] In this case the proton abstraction from formaldehyde by NH_2^-, which is a stronger base than OH^-, is exothermic and results in the formation of HCO^-. This ion then transfers a hydride to ammonia in a subsequent ion/molecule reaction to give NH_4^- and carbon monoxide. D-Labelling experiments have proven that the hydride ion transferred to ammonia retains its identity so that the NH_4^- ion can best be described as a hydride ion solvated by an ammonia molecule[84] (*cf.* the H_3O^- ion above).

[80] C. L. Johlman, C. F. Ijames, C. L. Wilkins, and T. H. Morton, *J. Org. Chem.*, 1983, **48** 2628.

[81] J. C. Kleingeld and N. M. M. Nibbering, *Recl. Trav. Chim. Pays-Bas*, 1984, **103**, 87.

[82] S. Ingemann, J. C. Kleingeld, and N. M. M. Nibbering, *J. Chem. Soc., Chem. Commun.*, 1982, 1009.

[83] J. C. Kleingeld and N. M. M. Nibbering, *Int. J. Mass Spectrom. Ion Phys.*, 1983, **49**, 311.

[84] J. C. Kleingeld, S. Ingemann, J. E. Jalonen, and N. M. M. Nibbering, *J. Am. Chem. Soc.*, 1983, **105**, 2474.

Organometallic Ion/Molecule Reactions. – The FT-ICR method is particularly compatible with laser ionization in that a complete mass spectrum may be obtained from one laser pulse. Hence, this combination has facilitated the study of gas-phase ion/molecule reactions between metal ions and organic substrates. The metal ions are generated by focusing the laser beam on to a high-purity foil of the appropriate metal attached to one of the cell plates. These ions can then be trapped for times ranging from tenths of seconds to several seconds in the presence of sample gas molecules, thus leading to organometallic ion/molecule reaction products.

Ti^+ metal ions have been found to react with ferrocene by a metal-exchange reaction to produce titanocene cations, whereas Rh^+ metal ions generate rhodocene cations in reactions with nickelocene.[85]

The chemistry involved in reactions of Fe^+, Ti^+, and Rh^+ with various alkanes has been studied.[86,87] The Ti^+ and Rh^+ ions behave rather similarly in their reactions with alkanes in that dehydrogenation is by far the dominant reaction. Such a reaction has been proposed to proceed *via* an initial insertion, *i.e.* oxidative addition, of the metal ion into a C–H bond followed by a β-hydride shift on to the metal with reductive elimination of a hydrogen molecule. This corresponds to an overall 1,2-elimination process and would lead to an olefin–metal-ion complex. Support for such an ion structure has been obtained from CID studies on $MC_nH_{2n}^+$ ions generated by reactions of M^+ ions (M = Fe, Co, or Ni) with a series of linear alkanes.[88,89] In the reductive elimination of a hydrogen molecule from propane activated by Rh^+ ions, the corresponding propene–Rh^+ complex seems to interconvert with a π-allyl–Rh^+–hydride complex.[87] This ion-structure proposal was based upon the exchange of five hydrogen atoms for deuterium in the $RhC_3H_6^+$ ions in the presence of D_2.[87] Such a H/D exchange, however, has not been observed to occur for the $MC_3H_6^+$ ions when M = Fe, Co, or Ni.[89]

The reductive elimination of a hydrogen molecule can also take place by an oxidative addition of a metal ion across a C–C bond as the initial step. After a β-hydride shift on to the metal following this step, an alkane can be lost to give an olefin–metal-ion complex as observed in reactions of Fe^+, Co^+, and Ni^+ with a series of linear alkanes.[88,89] However, after the first β-hydride shift a second β-hydride shift on to the metal can also take place, eventually leading to loss of a hydrogen molecule and formation of a bis(olefin)–metal-ion complex.[88,89] Evidence for the latter structure has been obtained from CID experiments and reactions with molecules such as hydrogen cyanide and acetonitrile, which have been observed to exchange for one or two olefin molecules of the bis(olefin)–metal-ion complex.[88,89]

As a general trend in the addition of the above metal ions across either C–H or C–C bonds of various alkanes, it can be noted that Ti^+[86] and Rh^+[87] insert predominantly into C–H bonds, Fe^+[86,89] into both C–H and C–C bonds with

[85] D. B. Jacobson, G. D. Byrd, and B. S. Freiser, *J. Am. Chem. Soc.*, 1982, **104**, 2320.
[86] G. D. Byrd, R. C. Burnier, and B. S. Freiser, *J. Am. Chem. Soc.*, 1982, **104**, 3565.
[87] G. D. Byrd and B. S. Freiser, *J. Am. Chem. Soc.*, 1982, **104**, 5944.
[88] D. B. Jacobson and B. S. Freiser, *J. Am. Chem. Soc.*, 1983, **105**, 736.
[89] D. B. Jacobson and B. S. Freiser, *J. Am. Chem. Soc.*, 1983, **105**, 5197.

not much selectivity, and Ni^+ [88,89] and Co^+ [89] into C–C bonds, both latter metal ions being highly selective against insertion into terminal C–C bonds.

The reactions of Fe^+, Co^+, and Ni^+ with cyclic hydrocarbons including cyclobutane, cyclopentane, and cyclohexane have also been studied.[90] The structures of the product ions have been probed by CID and H/D-exchange reactions. All three metal ions react with cyclobutane to generate $MC_2H_4^+$ and $MC_4H_6^+$ ions by elimination of a molecule of ethylene and hydrogen from the collision complexes, respectively. Both reactions have been proposed to proceed *via* initial insertion of the metal ion into a C–C bond, generating a metallacyclopentane intermediate. This then decomposes either by symmetric ring cleavage, resulting in loss of ethene, or by dehydrogenation *via* a 1-butene–metal-ion complex to produce eventually a butadiene–metal-ion complex.[90] Support for this route of dehydrogenation has been obtained from studies of the dehydrogenation of metallacyclopentanes synthesized by decarbonylation of specifically deuterium-labelled cyclopentanone in the gas phase by Fe^+, Co^+, and Ni^+.[90] These metal ions react with cyclopentane and cyclohexane predominantly by dehydrogenation. CID studies support the formation of cyclopentene–metal-ion and cyclohexene–metal-ion complexes, respectively.[90] Also in these cases an initial insertion of the metal ions into a C–C bond has been proposed to generate metallacyclohexane and metallacycloheptane, respectively, rather than insertion into a C–H bond, which is of minor importance or does not occur at all in reactions of the metal ions with linear alkanes.[89] These metallacyclic ions would then lose a hydrogen molecule by a successive α-hydride transfer onto the metal with a concurrent ring contraction and a β-hydride transfer on to the metal, eventually leading to the cycloalkene–metal-ion complexes.[90] In addition to dehydrogenation, the Fe^+, Co^+, and Ni^+ ions react with cyclopentane and cyclohexane to a minor extent by olefin elimination. This reaction has also been proposed to proceed *via* the intermediacy of the corresponding metallacyclic ions.[90] Corroboration for this proposal has been obtained from a study concerning the generation and decomposition of metal (Fe, Co, and Ni) ion metallacycles in the gas phase.[91] Metallacyclopentane ions rearrange by an initial 1,3-H shift to a 1-butene–metal-ion complex, whereas metallacyclohexane and metallacycloheptane ions rearrange by initial 1,4-H and 1,5-H shifts to give 1-pentene– and 1-hexene–metal-ion complexes, respectively, prior to decomposition by, amongst others, olefin elimination *via* bis(olefin)–metal-ion complexes.[91]

Various olefins ranging from C_2H_4 to C_6H_{12} isomers have been allowed to react with Fe^+ ions.[92] The chemistry involved has been found to be complex, but the olefin elimination in reactions of Fe^+ with 1-pentene and 1-hexene is in line with the proposals of such intermediate complexes for the olefin elimination in reactions of Fe^+, Co^+, and Ni^+ with cyclopentane and cyclohexane mentioned above.

The reactions of Co^+ and Rh^+ ions with toluene and its isomers cycloheptatriene and norbornadiene have been described.[93] They are very different for

[90] D. B. Jacobson and B. S. Freiser, *J. Am. Chem. Soc.*, 1983, **105**, 7492.
[91] D. B. Jacobson and B. S. Freiser, *Organometallics*, 1984, **3**, 513.
[92] D. B. Jacobson and B. S. Freiser, *J. Am. Chem. Soc.*, 1983, **105**, 7484.
[93] D. B. Jacobson, G. D. Byrd, and B. S. Freiser, *Inorg. Chem.*, 1984, **23**, 553.

toluene in that Co^+ exclusively gives the condensation product $CoC_7H_8^+$ whereas Rh^+ yields only the product ion $RhC_7H_6^+$ by dehydrogenation. The latter ion decomposes predominantly to RhC^+ and for the remaining part to Rh^+ upon CID, suggesting that at some stage a carbide–benzene $C\text{-}Rh^+\text{-}C_6H_6$ complex is generated. Rh^+ reacts with cycloheptatriene to give predominantly $RhC_7H_6^+$ and RhC^+ ions, the latter being generated even more by reaction of Rh^+ with norbornadiene in addition to C–C bond-cleavage products. The observation that RhC^+ ions are not formed by reaction of Rh^+ with toluene, but are increasingly generated in reactions of Rh^+ with cycloheptatriene and norbornadiene, respectively, can be accounted for on energetic considerations.[93] Co^+ reacts with cycloheptatriene and norbornadiene, predominantly producing C–C bond-cleavage products.[93]

It has also been found that the $RhC_7H_6^+$ ions, which are generated by Rh^+-induced dehydrogenation of toluene and which undergo H/D exchanges in the presence of deuterium, can be hydrogenated with excess hydrogen or with a variety of simple hydrocarbons, such as ethane.[94] The resulting $RhC_7H_8^+$ ions have been shown to be $Rh(MePh)^+$ by CID experiments. Arguments have been put forward on the basis of ^{13}C-labelling combined with CID that the non-decomposing $RhC_7H_6^+$ ions, generated by Rh^+-induced dehydrogenation of toluene, may be a mixture of $PhCHRh^+$ and $c\text{-}C_7H_6\text{—}Rh^+$ ions.[94]

Reactions of FeD^+, CoD^+, and NiD^+ ions, produced by reactions of the bare-metal ions with trideuteriomethylnitrite by loss of NO and CD_2O in two successive steps, with various hydrocarbons have been investigated.[95] In sharp contrast to the bare-metal ions Fe^+, Co^+, and Ni^+, which add nearly exclusively across C–C bonds of linear alkanes,[89] the corresponding metal hydrides have been found to favour initial oxidative addition across C–H bonds.[95] Furthermore, the observed apparent order of reactivity $NiD^+ > CoD^+ > FeD^+$ [95] is the opposite of what would be expected on the basis of the bare-metal reactivity.[89]

In a similar way it has been found that FeO^+ ions, generated by reaction of Fe^+ with N_2O, are more reactive towards linear, cyclic, and branched alkanes than Fe^+.[96] The majority of the product ions observed could be explained by initial C–H insertion and loss of H_2O to generate an activated Fe^+–olefin complex that subsequently decomposes.[96]

Negative metal ions such as Cr^-, Fe^-, Co^-, Mo^-, and W^- have also been prepared in a FT-ICR cell.[97] For example, Cr^- has been made by electron bombardment of $Cr(CO)_6$ to give $Cr(CO)_5^-$, which upon CID loses CO molecules sequentially to give eventually Cr^-. The proton affinity of Cr^- has been determined to be 1423 ± 13 kJ mol^{-1}, which, combined with the measured electron affinity of Cr and the ionization energy of hydrogen, gives an energy of 175 ± 13 kJ mol^{-1} for the Cr–H bond.[97]

[94] D. B. Jacobson and B. S. Freiser, *J. Am. Chem. Soc.*, 1984, **106**, 1159.
[95] T. J. Carlin, L. Sallans, C. J. Cassady, D. B. Jacobson, and B. S. Freiser, *J. Am. Chem. Soc.*, 1983, **105**, 6320.
[96] T. C. Jackson, D. B. Jacobson, and B. S. Freiser, *J. Am. Chem. Soc.*, 1984, **106**, 1252.
[97] L. Sallans, K. Lane, R. R. Squires, and B. S. Freiser, *J. Am. Chem. Soc.*, 1983, **105**, 6352.

5 Concluding Remarks

Fourier-transform ICR has undergone rapid developments in the past few years. The analytical applicability of this method, however, is not yet as broad as sector mass spectrometers, but it is expected to grow in the forthcoming years. For example, the FT-ICR method seems to be promising for the analysis of compounds of relatively high molecular weight under high-resolution conditions.

There is no doubt that the FT-ICR method has extraordinary potential for detailed study of ion/molecule reactions under low-pressure conditions, which will be exploited further in the future. However, at this point it should also be realized that in the past 15 years many studies have been performed successfully with the traditional ICR method, including the determination of proton affinities,[98] electron affinities,[99] and acidities[78] of molecules in the absence of solvent molecules, and of gas-phase ion/molecule reaction mechanisms.[100–102]

[98] D. H. Aue and M. T. Bowers in 'Gas Phase Ion Chemistry', ed. M. T. Bowers, Academic Press, New York, 1979, Vol. 2, Ch. 9.
[99] B. K. Janousek and J. I. Brauman in 'Gas Phase Ion Chemistry', ed. M. T. Bowers, Academic Press, New York, 1979, Vol. 2, Ch. 10.
[100] N. M. M. Nibbering, *Recl. Trav. Chim. Pays-Bas*, 1981, **100**, 297.
[101] J. H. Bowie, *Acc. Chem. Res.*, 1980, **13**, 76.
[102] J. H. Bowie, *Mass Spectrom. Rev.*, 1984, **3**, 1.

7

Reactions of Organic Negative Ions in the Gas Phase

BY J. H. BOWIE

1 Introduction

The material contained in this review has been selected from work listed in the *Mass Spectrometry Bulletin*[1] in the period April 1982 to June 1984. The policy used in the last review[2] has been largely retained – *selective* references are listed in each section, but only those of *particular* interest in the context of this chapter are considered in any detail. It is becoming increasingly common for certain journals and publishers to publish conference papers. To date material of this type, not having been refereed, has not been included in this series of chapters. In this review interesting material pertaining to new analytical techniques has been included where appropriate.

During the review period the analytical methods of negative-ion chemical ionization (NICI) and negative-ion fast-atom bombardment (FAB) have been widely applied to structural problems. In addition, a considerable volume of work pertaining to both theoretical and applied aspects of ion/molecule reactions has been published.

Several books[3–5] and reviews[6–19] relating to certain aspects of negative-ion chemistry have appeared during the review period. Reviews on negative-ion

1 *Mass Spectrometry Bulletin*, Mass Spectrometry Data Centre, The Royal Society of Chemistry, The University of Nottingham, Nottingham.

2 J. H. Bowie in 'Mass Spectrometry', ed. R. A. W. Johnstone (Specialist Periodical Reports), The Royal Society of Chemistry, London, 1984, Vol. 7, p. 151.

3 I. Howe, D. H. Williams, and R. D. Bowen, 'Mass Spectrometry. Principles and Applications', 2nd Ed., McGraw-Hill, New York, St.Louis, San Francisco, 1981.

4 F. W. McLafferty, 'Tandem Mass Spectrometry', J. Wiley, New York, 1983.

5 A. G. Harrison, 'Chemical Ionization Mass Spectrometry', CRC Press Inc., Florida, 1983.

6 A. L. Burlingame, A. Dell, and D. H. Russell, *Anal. Chem.*, 1982, **54**, 363R.

7 M. E. Rose and R. A. W. Johnstone in 'Mass Spectrometry for Chemists and Biochemists', Cambridge University Press, Cambridge, 1982, p. 78.

8 J. A. Page, *Anal. Proc.*, 1982, **19**, 307.

9 N. M. M. Nibbering, *J. Chromatogr.*, 1982, **251**, 93.

10 M. McKeown and R. T. Brackmann in 'Mass Spectrometry. Part B', Marcel Dekker Inc., New York, 1980, Ch. 4.

11 D. L. McCorkle, L. G. Christophorou, and S. R. Hunter in 'Electron and Ion Swarms', Proc. 2nd Int. Swarm Seminar, Oak Ridge, Tennessee, 22–23 July, 1981, Pergamon Press, Oxford, 1981, p. 21.

12 M. Jarman, *Annu. Rep. Prog. Chem., Sect. B*, 1981, **78**, 3.

13 D. F. Hunt, A. M. Buko, J. M. Ballard, J. Shabinowitz, and A. B. Giordani, 'Sequence Analysis of Polypeptides in Soft Ionization Biological Mass Spectrometry', Proc. Chem.

molecule reactions[20] and negative-ion mass spectrometry of organometallic[21] and organic compounds[22] have appeared recently. Other reviews on specific topics are listed separately in particular sections of this chapter.

2 Negative Ions Formed by Electron Capture (or Dissociative Electron Capture): Fragmentation Mechanisms

Aspects of the spectra of $C_3H_3^-$,[23] perfluoroalkanes,[24] fluoro- and chloromethanes,[25,26] pentafluorophenyl compounds,[27] fluoroethers and fluorosulphides,[28] isomeric nonanols,[29] α-dicarbonyl parent negative ions,[30,31] aromatic acids and esters,[32,33] terephthalate polyesters,[34] vinylene carbonate,[35] dioximes of α-diketones,[36] ketosulphides,[37] plumieride and plumieridine,[38–40] triterpenes,[41]

Soc. Symp. Adv. Mass Spectrom. Soft Ionization Methods, London, July, 1980, Heyden and Sons, London, 1981, p. 85.

14 H. G. Grohmann, M. Scheutwinkel-Reich, A. M. Preiss, and H.-J. Stan, 'Recent Developments in Mass Spectrometry in Biochemistry, Medicine and Environmental Research', Anal. Chem. Symp. Ser. No. 12, Proc. 8th Int. Symp. Mass Spectrom. Biochem., Med. Environ. Res., Venice, 18–19 June, 1981, Elsevier, Amsterdam, 1983, p. 117.

15 V. N. Reinhold and S. A. Carr, *Mass Spectrom. Rev.*, 1983, **2**, 153.

16 J. R. Chapman in 'Organophosphorus Chemistry', ed. D. W. Hutchinson and J. A. Miller (Specialist Periodical Reports), The Royal Society of Chemistry, London, 1983, Vol. 14, p. 278.

17 C. Brunneé, *Spectra*, 1983, **9**, 5.

18 C. Brunneé, *Spectra*, 1983, **9**, 10.

19 H. Budzikiewicz, *Spectra*, 1983, **9**, 44.

20 J. H. Bowie, *Mass Spectrom. Rev.*, 1984, **3**, 1.

21 I. K. Gregor and M. Guilhaus, *Mass Spectrom. Rev.*, 1984, **3**, 39.

22 J. H. Bowie, *Mass Spectrom. Rev.*, 1984, **3**, 161.

23 J. M. Oakes and G. B. Ellison, *J. Am. Chem. Soc.*, 1983, **105**, 2969.

24 S. M. Spyrou, J. Sauers, and L. G. Christophorou, *J. Chem. Phys.*, 1983, **78**, 7200.

25 E. Illenberger, *Ber. Bunsenges. Phys. Chem.*, 1982, **86**, 252.

26 H.-U. Sheunemann, M. Heni, E. Illenberger, and H. Baumgaertel, *Ber. Bunsenges. Phys. Chem.*, 1982, **86**, 321.

27 I. K. Gregor and M. Guilhaus, *J. Fluorine Chem.*, 1983, **23**, 549.

28 S. M. Spyrou and L. G. Christophorou, *Bull. Am. Phys. Soc.*, 1983, **28**, 185.

29 J. M. Knox and A. B. Denison, *Int. J. Mass Spectrom. Ion Phys.*, 1983, **49**, 123.

30 R. N. Compton, L. G. Christophorou, G. S. Hurst, and P. W. Reinhardt, *J. Chem. Phys.*, 1966, **45**, 4634; R. N. Compton and I. Bouby, *C. R. Hebd. Seances Acad. Sci., Ser. C*, 1967, **264**, 1153; A. Hadjiantoniou, L. G. Christophorou, and J. G. Carter, *J. Chem. Soc., Faraday Trans. 2*, 1973, **69**, 1704, 1713; J. G. Dillard, J. H. O'Toole, and M. A. Ogliaruso, *Org. Mass Spectrom.*, 1975 **10**, 728; J. H. Bowie and S. Janposri, *Aust. J. Chem.*, 1975, **28**, 2169.

31 J. H. Bowie, T. Blumenthal, M. H. Laffer, S. Janposri, and G. E. Gream, *Aust. J. Chem.*, 1984, **37**, 1447.

32 H. Budzikiewicz and D. Stöckl, *Org. Mass Spectrom.*, 1982, **17**, 470.

33 H. Budzikiewicz, A. Poppe, and D. Stöckl, *Int. J. Mass Spectrom. Ion Phys.*, 1983, **47**, 217; H. Budzikiewicz and A. Poppe, *Monatsh. Chem.*, 1983, **114**, 89.

34 R. E. Adams, *J. Polym. Sci., Chem. Ed.*, 1982, **20**, 119.

35 R. N. Compton, P. W. Reinhardt, and H. C. Schweinler, *Int. J. Mass Spectrom. Ion Phys.*, 1983, **49**, 113.

36 J. Charalambous, G. Soobramanien, A. D. Stylianoir, G. Manini, L. Operti, and G. A. Vaglio, *Org. Mass Spectrom.*, 1983, **18**, 406.

37 I. I. Furlei, V. I. Dronov, Yu. E. Nikitin, Yu. E. Murinov, and V. I. Khrostenko, *Bull. Acad. Sci. U.S.S.R., Div. Chem. Sci. (Engl. Transl.)*, 1983, **32**, 599.

steroids,[42] gibberellins,[43] 5-halogenated pyrimidine glucuronides,[44] triazenes,[45] purines labelled with ^{3}H,[46] nucleosides,[47] anabasine and lupinine alkaloids,[48] mercapto acids and esters,[49,50] and various RS^- and RSO^- (R = aryl) species[51,52] have been reported. Because of the difficulties in defining NICI (see next section), some negative ions produced by electron capture are considered in the NICI section.

Photoelectron spectroscopy suggests that the ion $C_3H_3^-$ has an allenyl ($CH_2{=}C{=}CH^-$) rather than a propargyl ($HC{\equiv}C{-}CH_2^-$) structure.[23] Perfluorodimethyl ether fragments to yield CF_3O^- and F^- at ionization energies between 0 and 10 eV.[28] Isomeric nonanols yield $(M-H)^-$ ions at both 70 and 6–15 eV, but the major fragment ions are $O^{-\bullet}$ (base peak) and HO^-.[29] Parent negative ions of α-dicarbonyl compounds generally do not undergo unimolecular fragmentation,[30] but the major collision-induced decomposition is β-cleavage [see structure (1)].[31] The dodecane-6,7-dione parent negative ion undergoes collision-induced cleavage of all carbon–carbon bonds shown in structure (2); it is suggested that fragmentations A $\rightarrow$ C involve cyclization reactions of the radical centre in the charged dicarbonyl unit.[31] The spectra of structures (3) (R = H or Me) show pronounced $(M-H)^-$ ions, with additional fragmentation as illustrated.[32,33] The decomposing

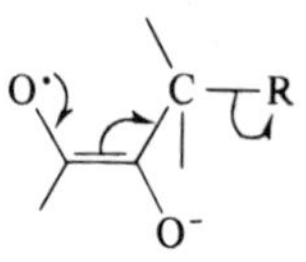

(1)

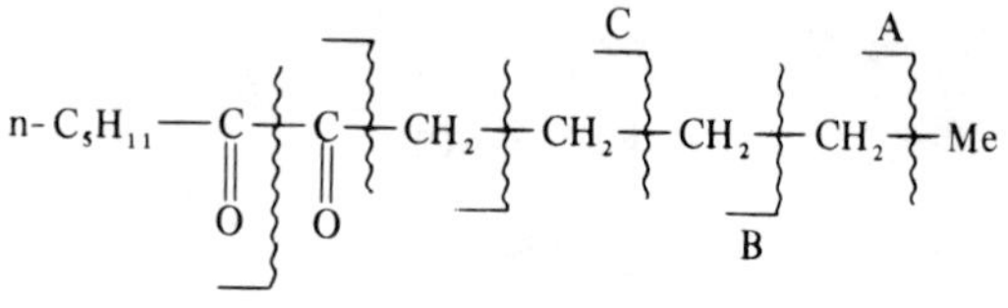

(2)

[38] M. von Ardenne, *Z. Angew. Phys.*, 1959, **11**, 121; M. von Ardenne, K. Steinfelder, and R. Tümmler, 'Electronenanlagerungs-Massenspektrographie Organischer Substanzen', Springer Verlag, Berlin, Heidelberg, New York, 1971; see also J. H. Bowie and B. D. Williams in 'Int. Rev. Sci., Mass Spectrom. Phys. Chem.', Ser. 2, ed. A. Maccoll, Butterworths, London, 1975, Vol. 5, p. 96, 115.

[39] D. Voigt, G. Adam, and W. Shade, *Org. Mass Spectrom.*, 1981, **16**, 85.

[40] W. Schliemann and G. Adam, *Phytochemistry*, 1982, **21**, 1438.

[41] K. P. Madhusudanan and C. Singh, *Indian J. Chem., Sect. B*, 1982, **21**, 675.

[42] I. I. Furlei, L. A. Baltina, and G. A. Tolstikov, *Chem. Nat. Compd. (U.S.S.R.)*, 1982, **18**, 435.

[43] M. Lischewski, *Z. Chem.*, 1984, **24**, 64.

[44] R. Mahrwald and W. Schwarz, *J. Prakt. Chem.*, 1982, **324**, 177.

[45] V. G. Voronin, A. P. Pleshkova, A. I. Ermakov, A. A. Sorokin, I. D. Muravskaya, N. F. Karaseva, and S. Sh. Trokhova, *J. Org. Chem. U.S.S.R. (Engl. Transl.)*, 1982, **18**, 1437; *Zh. Org. Khim.*, 1982, **18**, 1644.

[46] M. Kiessling, W. Schwarz, and R. Tümmler, *Radiochem. Radioanal. Lett.*, 1982, **52**, 65.

[47] D. L. Smith, K. H. Schram, and J. A. McCloskey, *Biomed. Mass Spectrom.*, 1983, **10**, 269.

[48] V. A. Mazuñov, V. I. Khvostenko, E. Kh. Timbekov, Kh. A. Aslanov, and G. A. Tolstikov, *Bull. Acad. Sci. U.S.S.R., Div. Chem. Sci. (Engl. Transl.)*, 1982, **30**, 1845.

[49] J. H. Bowie, F. Duus, S.-Ø. Lawesson, F. C. V. Larsson, and J. Ø. Madsen, *Aust. J. Chem.*, 1969, **22**, 153.

[50] J. H. Bowie, M. B. Stringer, F. Duus, S.-Ø. Lawesson, F. C. V. Larsson, and J. Ø. Madsen, *Aust. J. Chem.*, 1984, **37**, 1619.

[51] J. A. Benbow, J. H. Bowie, and G. Klass, *Org. Mass Spectrom.*, 1977, **12**, 432.

[52] J. H. Bowie and M. B. Stringer, *Org. Mass Spectrom.*, 1985, **20**, 138.

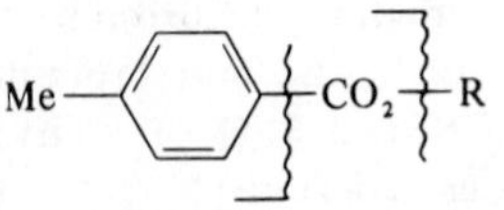

(3)

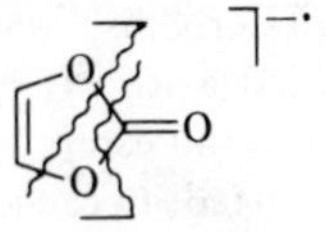

(4)

parent ion of vinylene carbonate gives (at 10 eV) the two stable ions shown in structure (4).[35]

The major fragmentation of the molecular anions of α-diketone dioximes (5) (R = Me or Ph) is loss of H_2O. In addition, it is suggested that loss of H_2 forms structure (6), which may then eliminate O, RCNO, and RCN competitively.[36] The negative-ion spectra (von Ardenne source[38]) of plumieride (7; R = glucosyl)[39] and plumieridine (7; R = H)[40] both exhibit molecular ions and complex fragmentation including $(M - H_2O)^{-\cdot}$ and $(M - CO_2)^{-\cdot}$. The glucoside fragments additionally as shown in structure (7). Certain gibberellins (von Ardenne source[38]) yield both $M^{-\cdot}$ and $(M - MeCO_2H)^{-\cdot}$ ions.[43] The secondary-electron-capture spectra (70 eV) of nucleosides [*e.g.* uridine (8)] yield $(M - H)^-$ ions, with the base peak being produced as shown in structure (8).[47]

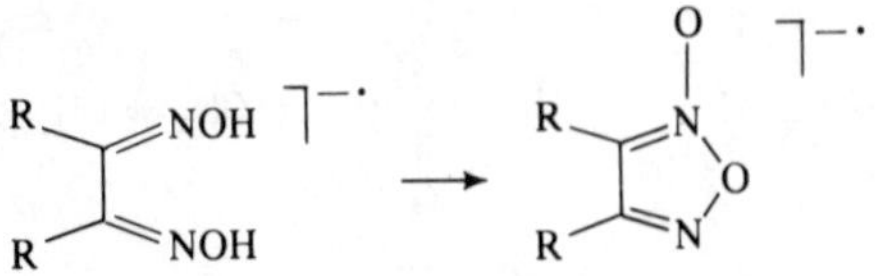

(5)

(6)

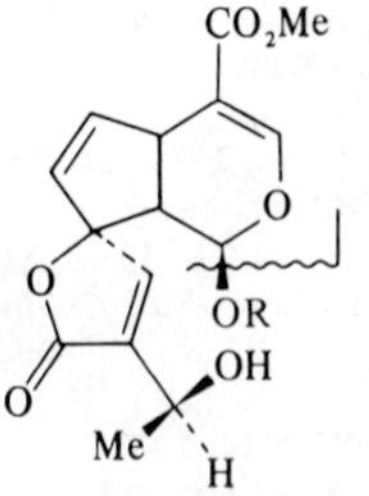

(7)

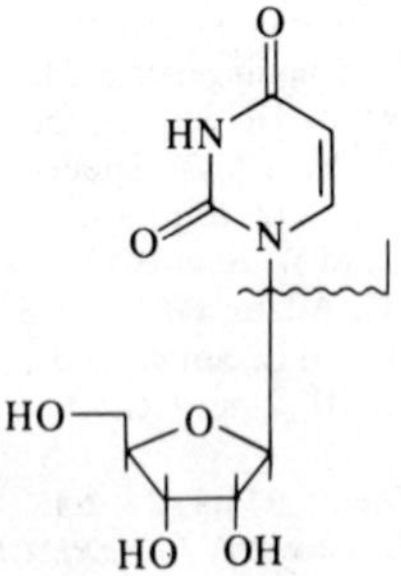

(8)

Perfluorothiol ethers (*e.g.* CF_3SMe) fragment to give CF_3S^- and F^-.[28] The earlier report[49] that mercapto acids form parent negative ions by secondary electron capture has been corrected.[50] Such compounds do not yield $M^{-\cdot}$ ions but form $(M - H)^-$ ions in high yield. The basic unimolecular cleavage of thioglycollates is shown in structures (9) and (10). $(M - H)^-$ ions derived from structure (10) undergo the characteristic collision-induced rearrangement shown in structure (11) (R = alkyl or aryl).[50] Aryl sulphide ions are stable and undergo no unimolecular decomposition,[51] but complex collision-induced dissociations

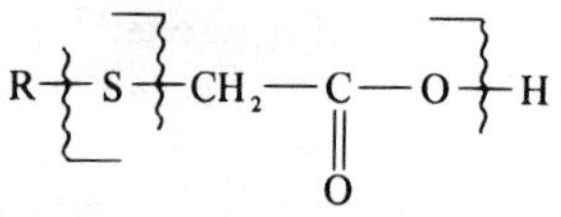

R–S–CH₂–C(=O)–O–H

(9)

H–S–CH₂–C(=O)–O–R

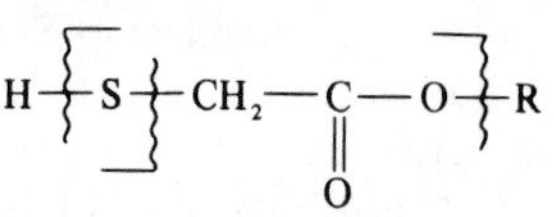

(10)

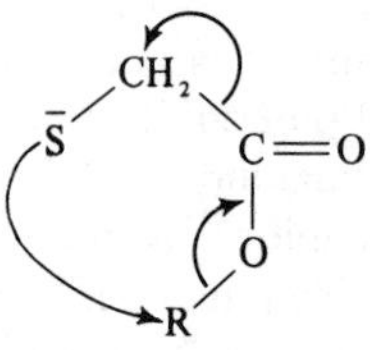

(11)

do occur.[52] For example, PhS^- yields HS^- and C_2HS^- without prior hydrogen scrambling of the aromatic ring. In contrast, $PhCH_2S^-$ undergoes benzylic/phenyl *o*-hydrogen interchange prior to elimination of CH_2S. Similarly, $PhCH_2SO^-$ also undergoes benzylic/phenyl *o*-hydrogen interchange prior to collision-induced losses of H_2O and SO.[52]

3 Negative-ion Chemical Ionization

There is still debate concerning the definition of negative-ion chemical-ionization mass spectrometry. The Reporter believes that NICI should be defined as involving a reaction between some negative ion and the neutral molecule under study to produce a negatively charged species derived from the neutral molecule, *e.g.* the HO^-/NICI and X^- (X = halogen)/NICI techniques. The term NICI should not be used to include electron-capture or dissociative-electron-capture spectra obtained when the sample captures an electron produced by ionization of some neutral species (*e.g.* methane) added for this purpose. But methane/ or isobutane/'NICI' spectra are often difficult to classify. Many such spectra are produced by electron capture, others are produced by impurities in the gas (*e.g.* $O^{-\bullet}$ from O_2 or HO^- from H_2O),[53,54] halogenated compounds may yield X^-, which may in turn produce $(M + X)^-$ species, and, finally, radical reactions may produce negative ions under methane/NICI conditions.[55,56] Consequently, all ions formed by any one of these processes are considered in this section.

Methane/Isobutane Negative-ion Chemical Ionization. – The following compounds have been studied by methane or isobutane NICI mass spectrometry (in certain instances GC or LC has been used to separate components before intro-

[53] K. R. Jennings in 'Mass Spectrometry', ed. R. A. W. Johnstone (Specialist Periodical Reports), The Chemical Society, London, 1977, Vol. 4, p. 203.
[54] D. Stöckl and H. Budzikiewicz, *Org. Mass Spectrom.*, 1982, **17**, 470.
[55] C. N. McEwen and M. A. Rudat, *J. Am. Chem. Soc.*, 1981, **103**, 4343.
[56] D. Stöckl and H. Budzikiewicz, *Org. Mass Spectrom.*, 1982, **17**, 376.

duction into the chemical-ionization source): benzo[a]pyrene,[57] polycyclic aromatic hydrocarbons,[58] chlorinated paraffins,[59,60] chlorinated compounds in Arctic air,[61] chlordane in Hawaiian fish,[62] chlorophenols,[63] halogenated pesticides,[64,65] polychlorinated biphenyls in marine sediments,[66] polybrominated biphenyls in serum,[67] bile acids,[68] terephthalate polymers,[69] pentafluorobenzyl esters,[1] derivatives of prostanoids,[70] 6-oxoprostaglandins,[71] tetrahydrocannabinol derivatives,[72] silylated catecholamines,[73] anthracyclin antiobiotics,[74] histamine,[75] fluorinated derivatives of dopamine,[76] amino acids,[77] underivatized peptides,[78] 1-(2-chloroethyl)-3-(2,6-dioxo-3-piperidyl)-1-nitrosourea,[79] 2-*o*-chlorobenzoyl-4-chloro-*N*-methyl-*N*′-glycylglycinanilide,[80] diazepam,[81] medazolam[82] and imidazobenzodiazepine-3-carboxamide[83] in human plasma, 2-(4′,6′-bistrifluoromethyl-2′-pyrimidinyl)tetrahydroisoquinoline,[84] triazine herbicides,[85] tetrahydroprotoberberine alkaloids,[86] nucleotides,[87] nitroglycerin in human plasma,[88,89]

[57] S. Daishima, Y. Iida, A. Shibata, and K. Furuya, *Bunseki Kagaku*, 1983, **32**, 761.
[58] Y. Iida and S. Daishima, *Chem. Lett.*, 1983, 273.
[59] M. D. Mueller and P. P. Schmid, *J. High Resolut. Chromatogr. Chromatogr. Commun.*, 1984, **7**, 33.
[60] J. Gjos and K. O. Gustavsen, *Anal. Chem.*, 1982, **54**, 1316.
[61] M. Oehme, *Trends Anal. Chem.*, 1982, **1**, 321.
[62] M. A. Ribick and J. Zajicek, *Chemosphere*, 1983, **12**, 1229.
[63] S. Shang-Zhi and A. M. Duffield, *J. Chromatogr.*, 1984, **284**, 157.
[64] M. Guilhaus and I. K. Gregor, *Talanta*, 1984, **31**, 55.
[65] L. D. Betowski, H. M. Webb, and A. D. Sauter, *Biomed. Mass Spectrom.*, 1983, **10**, 369.
[66] E. Lewis and W. D. Jamieson, *Int. J. Mass Spectrom. Ion Phys.*, 1983, **48**, 303.
[67] J. Roboz, J. Greaves, J. F. Holland, and J. G. Bekesi, *Anal. Chem.*, 1982, **54**, 1104.
[68] N. Yamaga, H. Yamasaki, T. Hara, K. Alachi, and K Shimizu, *Koenshu-Iyo Masu Kentyukai*, 1981, **6**, 93 (*Chem. Abstr.*, 1982, **96**, 965 139).
[69] R. E. Adams, *Anal. Chem.*, 1983, **55**, 414.
[70] K. A. Waddell, I. A. Blair, and J. Wellby, *Biomed. Mass Spectrom.*, 1983, **10**, 83.
[71] S. E. Barrow, K. A. Waddell, M. Ennis, C. T. Dollery, and I. A. Blair, *J. Chromatogr.*, 1982, **239**, 71.
[72] R. L. Foltz, K. M. McGinnis, and D. M. Chinn, *Biomed. Mass Spectrom.*, 1983, **10**, 316.
[73] J. T. Martin, J. D. Barchas, and K. F. Faull, *Anal. Chem.*, 1982, **54**, 1806.
[74] R. G. Smith, *Anal. Chem.*, 1982, **54**, 2006.
[75] L. J. Roberts and J. A. Oates, *Anal. Biochem.*, 1984, **136**, 258.
[76] P. L. Wood, *Biomed. Mass Spectrom.*, 1982, **9**, 302.
[77] T. Hayaski, H. Naruse, Y. Iida, and S. Daishima, *Shitsuryo Bunseki*, 1983, **31**, 205.
[78] C. N. Kenyon, *Biomed. Mass Spectrom.*, 1983, **10**, 535.
[79] R. G. Smith, L. K. Cheung, L. G. Feun, and T. L. Loo, *Biomed. Mass Spectrom.*, 1983, **10**, 404.
[80] S. Hashimoto, E. Sakurai, M. Mizobuchi, S. Takahashi, K. Yamomoto, and T. Momose, *Biomed. Mass Spectrom.*, 1982, **9**, 546.
[81] W. A. Garland and B. J. Miwa, *Biomed. Mass Spectrom.*, 1983, **10**, 126.
[82] F. Rubio, B. J. Miwa, and W. A. Garland, *J. Chromatogr.*, 1982, **233**, 157.
[83] F. Rubio, B. J. Miwa, and W. A. Garland, *J. Chromatogr.*, 1982, **233**, 167.
[84] S. Murray and K. A. Waddell, *Biomed. Mass Spectrom.*, 1982, **9**, 466.
[85] C. E. Parker, C. A. Haney, D. J. Harvan, and J. R. Hass, *J. Chromatogr.*, 1982, **242**, 77.
[86] K. P. Madhusudan, S. Gupta, and D. S. Bhakani, *Indian J. Chem., Sect. B*, 1983, **22**, 128.
[87] H. Seto, T. Okuda, T. Takesui, and T. Ikemura, *Shitsuryo Bunseki*, 1983, **31**, 197.
[88] G. Idzu, M. Ishibashi, H. Miyazaki, and K. Yamomoto, *J. Chromatogr.*, 1982, **229**, 327.
[89] J. A. Settlage, W. Gielsdorf, and H. Jaeger, *J. High Resolut. Chromatogr. Chromatogr. Commun.*, 1983, **6**, 68.

explosives,[90–92] nitro-aromatic compounds,[93–95] aromatic thioethers,[96] and organophosphorus pesticides.[97,98]

The majority of studies listed above are concerned with the detection of $M^{-\bullet}$, $(M-H)^-$, or quasi-molecular ions; few are concerned with fragmentation mechanisms. A selection of examples of CH_4/NICI spectra follows. Perfluoroacetates of tetrahydrocannabinol derivatives yield pronounced molecular ions [*e.g.* structure (12)] together with $CF_3CO_2^-$.[72] The anthracyclin antibiotic (13) yields the molecular negative ion as the base peak of the spectrum with major fragmentation occurring through the glycosidic linkage [structure (13), fragmentation A] and with subsequent loss of $2H_2O$ from the aglycone ion.[74] The pyrimidinyltetrahydroisoquinoline (14) gives only a parent negative ion,[84] whereas the trinitro compound (15) yields NO_2^-, which reacts with the neutral species to form an $(M + NO_2)^-$ ion.[92] Perfluoroaromatic thioethers form molecular anions that fragment as shown in structure (16),[96] while organophosphorus pesticides fragment as in structure (17),[97,98] as reported earlier.[99]

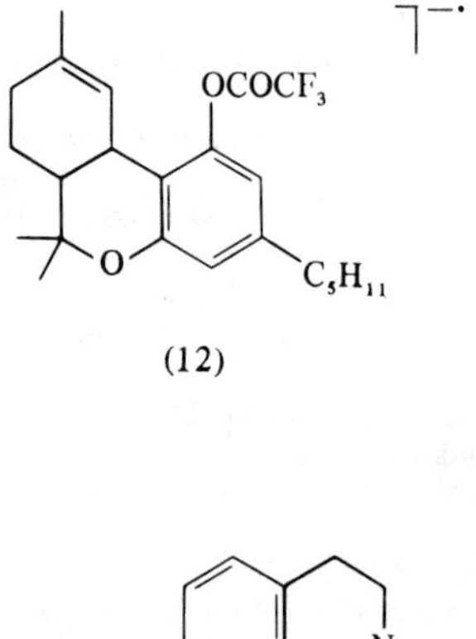

OCOCF$_3$ C$_5$H$_{11}$

(12)

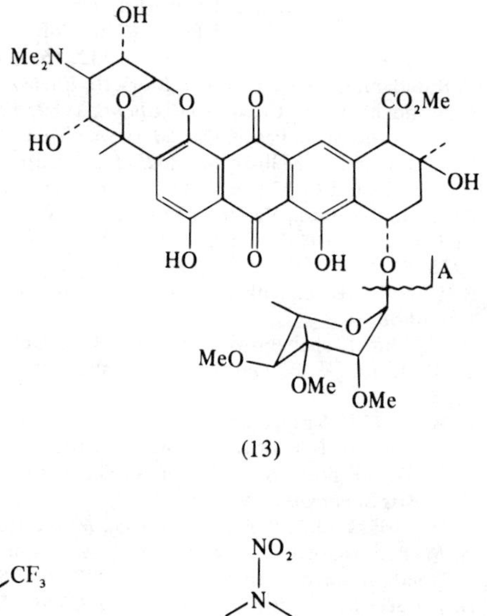

OH Me$_2$N O O O CO$_2$Me HO OH HO O OH O A MeO OMe OMe

(13)

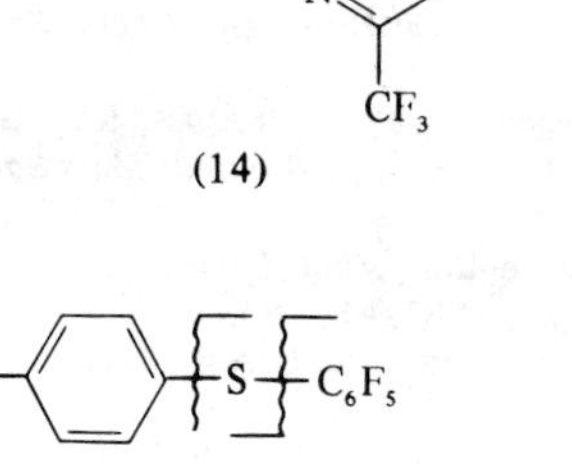

N N N CF$_3$ CF$_3$

(14)

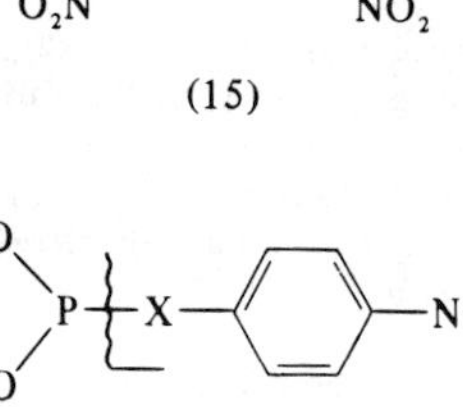

NO$_2$ N N N O$_2$N NO$_2$

(15)

Cl S C$_6$F$_5$

(16)

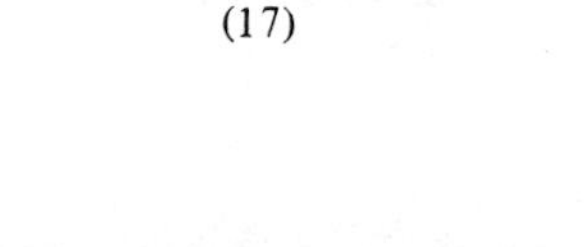

EtO EtO P X NO$_2$

(17)

Hyroxide-ion Negative-ion Chemical Ionization. – The following systems have been studied using HO^- negative-ion chemical ionization: $C_7H_7^-$ ions from cycloheptatriene,[100] ketones,[101,102] carboxylic acids,[101,102] hydroxy acids,[103] carboxylic acids in urine,[104] fatty-acid esters,[105,106] bornyl C_4–C_5 esters,[107] diesters,[108] cyclohexane-1,4-diol,[109] androstane diols,[110] trichothecenes,[111] erythromycin A,[112] cardenolides,[113] pyrrolizidine alkaloids,[114] barbituric acid derivatives,[115] and nitrotriphenylcyclohexene.[116]

Reaction of cycloheptatriene with HO^- (or NH_2^-) yields the tropylium anion (18), which is suggested to rearrange by a bimolecular mechanism [perhaps through the collision complex (19)] to the benzyl anion (20).[100] Further reactions of $(M - H)^-$ species derived from ketones and carboxylic acids have been reported[102] (*cf.* Vol. 7 of this title, p. 157, and ref. 101). The cyclohexanone

[90] J. Yinon and S. Zitrin, 'The Analysis of Explosives', Pergamon Press, Elmsford, New York, and Oxford, 1981.

[91] J. Yinon, *Mass Spectrom. Rev.*, 1982, **1**, 257.

[92] J. Yinon, D. J. Harvan, and J. R. Hass, *Org. Mass Spectrom.*, 1982, **17**, 321.

[93] D. L. Newton, M. D. Erickson, K. B. Tomer, E. D. Pellizzari, P. Gentry, and R. B. Zweidinger, *Environ. Sci. Technol.*, 1982, **16**, 206.

[94] T. Ramdahl, G. Becher, and A. Bjorseth, *Environ. Sci. Technol.*, 1982, **16**, 861.

[95] T. Ramdahl and K. Urdal, *Anal. Chem.*, 1982, **54**, 2292.

[96] E. Lewis and M. E. Peach, *Org. Mass Spectrom.*, 1982, **17**, 597.

[97] H.-J. Stan and G. Kellner, *Biomed. Mass Spectrom.*, 1982, **9**, 483.

[98] H.-J. Stan and G. Kellner, 'Recent Developments in Food Analysis. Organophosphorus Pesticide Residues', Verlag Chemie GmbH, Weinheim, New York, 1982, p. 183.

[99] H. J. Meyer, F. C. V. Larsson, S.-Ø. Lawesson, and J. H. Bowie, *Bull. Soc. Chim. Belg.*, 1978, **87**, 517.

[100] R. L. White, C. L. Wilkins, J. J. Heitkamp, and S. W. Staley, *J. Am. Chem. Soc.*, 1983, **105**, 4868.

[101] D. F. Hunt, J. Shabanowitz, and A. B. Giordani, *Anal. Chem.*, 1980, **52**, 386.

[102] D. F. Hunt, A. B. Giordani, J. Shabanowitz, and G. Rhodes, *J. Org. Chem.*, 1982, **47**, 738.

[103] Z. M. Li, T. L. Kruger, and R. G. Cooks, *Org. Mass Spectrom.*, 1982, **17**, 519.

[104] D. F. Hunt, A. B. Giordani, G. Rhodes, and D. A. Herold, *Clin. Chem.*, 1982, **28**, 2387.

[105] A. M. Bambagiotti, S. A. Coran, V. Giannellini, F. F. Vincieri, S. Daolio, and P. Traldi, *Org. Mass Spectrom.*, 1983, **18**, 133.

[106] H. Hendriks and A. P. Bruins, *Biomed. Mass Spectrom.*, 1983, **10**, 377.

[107] A. M. Bambagiotti, V. Giannellini, S. A. Coran, F. F. Vincieri, G. Moneti, A. Selva, and P. Traedi, *Biomed. Mass Spectrom.*, 1982, **9**, 495.

[108] D. J. Burinsky and R. G. Cooks, *J. Org. Chem.*, 1982, **47**, 4864.

[109] T. Gäumann, D. Stahl, and J. C. Tabet, *Org. Mass Spectrom.*, 1983, **18**, 263.

[110] J. C. Belocil, M. Butranne, D. Stahl, and J. C. Tabet, *J. Am. Chem. Soc.*, 1983, **105**, 1355.

[111] W. C. Brumley, D. Andrzejewski, E. W. Trucksess, P. A. Dreifuss, J. A. G. Roach, R. M. Eppley, F. S. Thomas, C. W. Thorpe, and J. A. Sphon, *Biomed. Mass Spectrom.*, 1982, **9**, 451.

[112] M. Dedieu, C. Juin, P. J. Arpino, and G. Guinochon, *Anal. Chem.*, 1982, **54**, 2372.

[113] A. P. Bruins, *Int. J. Mass Spectrom. Ion Phys.*, 1983, **48**, 185.

[114] P. A. Dreifuss, W. C. Brumley, J. A. Sphon, and E. A. Caress, *Anal. Chem.*, 1983, **55**, 1036.

[115] D. J. Burinsky and R. G. Cooks, *Org. Mass Spectrom.*, 1983, **18**, 410.

[116] K. G. Das, M. Mallaiah, K. P. Madhusudhanan, and A. P. Bruins, *Tetrahedron*, 1982, **38**, 2285.

$(M-H)^-$ ion eliminates both H_2 and C_2H_4 under collisional activation. The former process is complex since structure (21) eliminates both H_2 and HD. The latter elimination may be represented by the retro-Diels–Alder process (22) → (23).[102]

$(M-H)^-$ ions from hydroxy acids may polymerize on a hot probe tip; for example, the highest mass ion produced from lactic acid is $[6M - 4H_2O - H]^-$.[103] A triple-quadrupole instrument may be used to analyse carboxylic acids in crude urine residues: $(M-H)^-$ ions are formed by HO^-/NICI, collision activation of these ions occurs in quadrupole two, and the fragmentations $[-CO_2, -(CO_2 + H_2O)]$ are monitored by quadrupole three.[104] The collision-induced MIKE spectrum of the $(M-H)^-$ ion from methyl homo-γ-linoleate (24) shows cleavage of the majority of C–C bonds in the molecule. Simple cleavage is indicated by solid lines; cleavages represented by dotted lines occur with additional hydrogen loss.[105] The only other example of such extensive side-chain cleavage is that for dodecan-6,7-dione [*cf.* (2)].[31] The gas-phase Dieckmann ester-type condensation (25) → (27) has been reported.[108] Only the $(M-H)^-$ ions derived from the *trans* diols of androstanes (28) eliminate RH (R = H, D, C_2H, C_2H_3, or Et).[110]

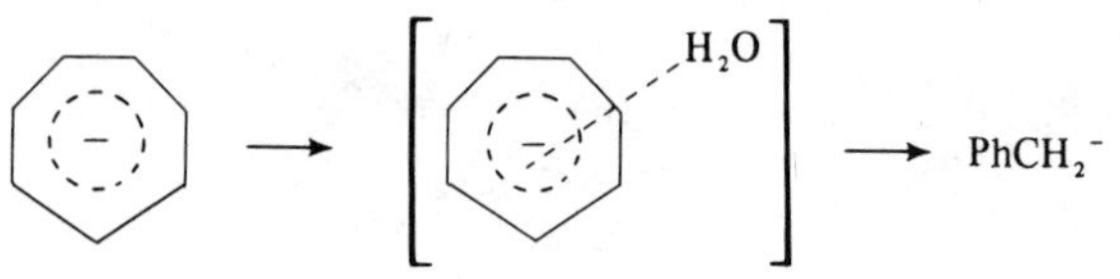

(18) → (19) → $PhCH_2^-$ (20)

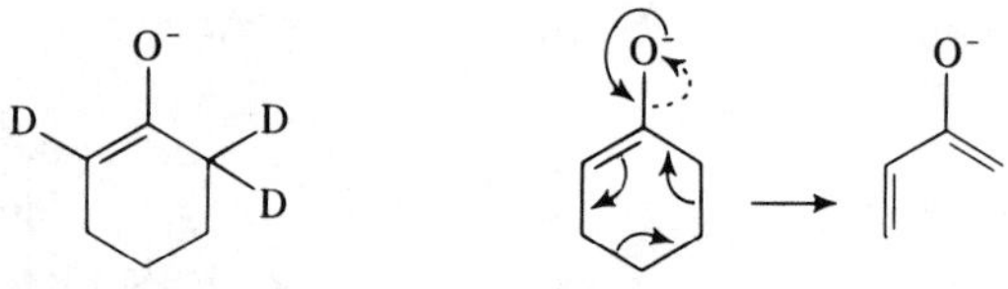

(21) (22) → (23)

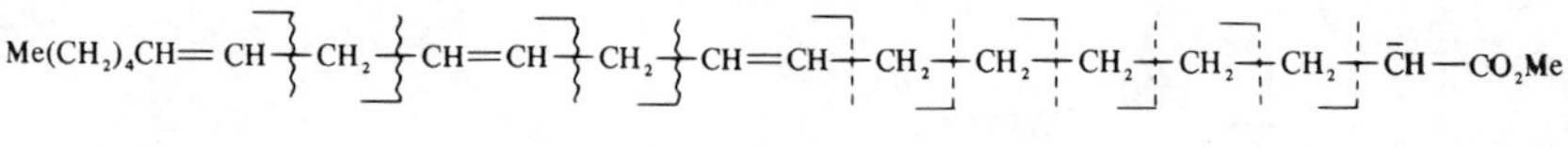

$Me(CH_2)_4CH{=}CH-CH_2-CH{=}CH-CH_2-CH{=}CH-CH_2-CH_2-CH_2-CH_2-CH_2-\bar{C}H-CO_2Me$

(24)

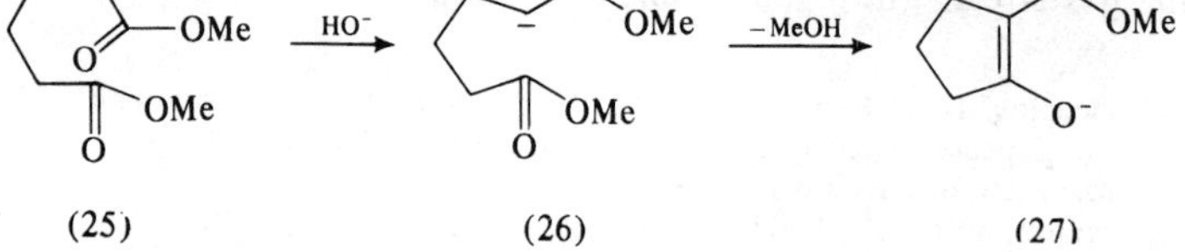

(25) $\xrightarrow{HO^-}$ (26) $\xrightarrow{-MeOH}$ (27)

It is suggested that formation of a ketolate by the mechanism in structure (29) rationalizes this reaction.

HO^-/NICI mass spectrometry is particularly useful for structure determination of naturally occurring compounds containing (at least) one 'acidic' hydrogen. For example, verrucarol (30) shows $(M-H)^-$, $[(M-H)-H_2O]^-$, and $[(M-H)-CH_2O]^-$ ions together with the fragmentation shown in structure (30).[111] Senecionine (31) exhibits an $(M-H)^-$ ion and the fragmentations shown.[114] Cardenolides [*e.g.* structure (32)] also yield $(M-H)^-$ ions together with the characteristic fragmentations shown in structure (32).

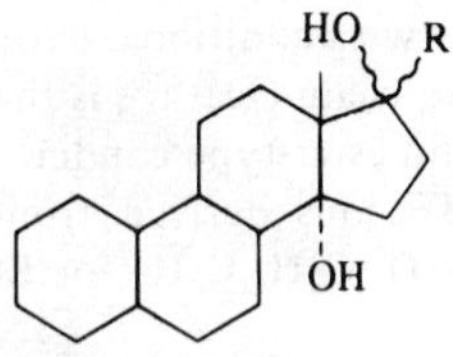

(28)

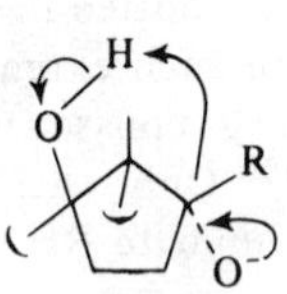

(29)

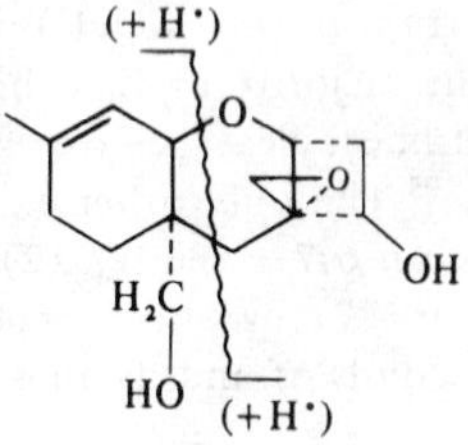

(30)

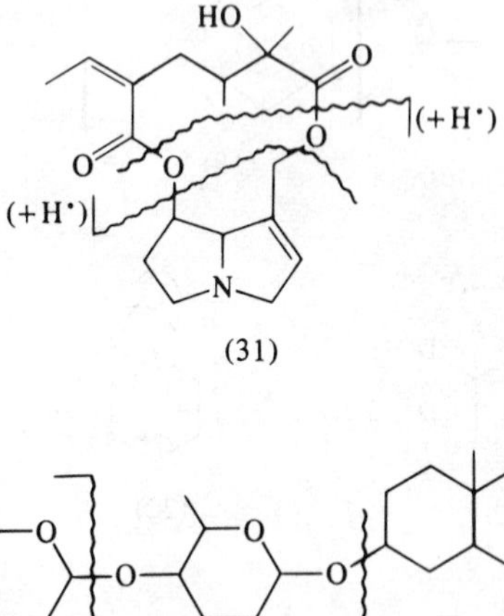

(31)

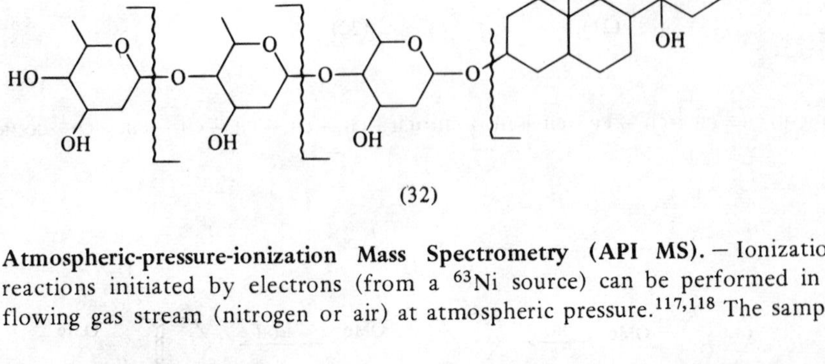

(32)

Atmospheric-pressure-ionization Mass Spectrometry (API MS). – Ionization reactions initiated by electrons (from a ^{63}Ni source) can be performed in a flowing gas stream (nitrogen or air) at atmospheric pressure.[117,118] The sample

[117] E. C. Horning, D. I. Carroll, I. Dzidic, R. N. Stillwell, A. Zlatkis, and C. F. Poole in 'Electron Capture. Theory and Practice in Chromatography', Elsevier Sci. Publ. Co., Amsterdam, Oxford, New York, 1981, p. 359.

[118] C. J. Proctor and J. F. J. Todd, *Org. Mass Spectrom.*, 1983, **18**, 509.

is directed into the source in a solvent[119,120] or *via* a gas chromatogram.[121,122] The species $O_2^{-\bullet}$, $O^{-\bullet}$, and HO^- are thought to yield the $M^{-\bullet}$ and $(M-H)^-$ ions observed using this analytical technique.[123] The technique has been used for a number of systems, including carboxylic acids,[123,124] chloro- and nitrobenzenes,[123,125,126] tetrachlorodibenzofurans,[127] dibenzo-*p*-dioxins,[121,122,128,129] barbiturates,[119,120] and explosives.[123] Negative-ion API MS has also been used for breath analysis[130] and for the formation of $(M-H)^-$ ions from nucleotides, ions that can be further analysed by their collisional-activation MIKE spectra.[131]

Miscellaneous Negative-ion Chemical-ionization Methods. – Alkyl nitrites, which are widely used precursors of alkoxide negative ions (*cf.* reference 20), can be used as chemical-ionization agents to produce RO^-/NICI spectra.[132] The F^-/NICI technique[133] has been extended to a recent study of esters.[134]

4 Negative-ion Fast-atom-bombardment Mass Spectrometry

Fast-atom bombardment is becoming the soft ionization technique of choice for mass-spectrometric structure determination of labile compounds, involatile compounds including salts, and high-molecular-weight molecules, particularly those of biological importance (see Vol. 7 of this title, p. 153, also ref. 22). Several recent reviews[135–137] have highlighted work in this area, including the following chapter of this volume.

119 E. C. Horning, M. C. Horning, D. I. Carroll, I. Dzidic, and R. N. Stillwell, *Anal. Chem.*, 1973, **45**, 936.

120 D. I. Carroll, I. Dzidic, R. N. Stillwell, M. G. Horning, and E. C. Horning, *Anal. Chem.*, 1974, **46**, 706.

121 R. K. Mitchum, G. F. Moler, and W. A. Korfmacher, *Anal. Chem.*, 1980, **52**, 2278.

122 R. K. Mitchum, W. A. Korfmacher, G. F. Moler, and D. L. Stalling, *Anal. Chem.*, 1982, **54**, 719.

123 B. A. Thomson, W. R. Davidson, and A. M. Lovett, *Environ. Health Perspect.*, 1980, **36**, 77.

124 S. H. Kim and F. W. Karasek, *J. Chromatogr.*, 1982, **234**, 13.

125 I. Dzidic, D. I. Carroll, R. N. Stillwell, and E. C. Horning, *J. Am. Chem. Soc.*, 1974, **96**, 5258.

126 K. Tanaka, G. I. Mackay, J. D. Payzant, and D. K. Bohme, *Can. J. Chem.*, 1976, **54**, 1643.

127 W. A. Korfmacher, R. K. Mitchum, F. D. Hillman, and T. Mazer, *Chemosphere*, 1983, **12**, 1243.

128 R. K. Mitchum, W. A. Korfmacher, G. F. Moler, and D. L. Stalling, *Anal. Chem.*, 1982, **54**, 719.

129 R. K. Mitchum, W. A. Korfmacher, and G. F. Moler, *Int. J. Mass Spectrom. Ion Phys.*, 1983, **48**, 307.

130 F. M. Benoit, W. R. Davidson, A. M. Lovett, S. Nacson, and A. Ngo, *Anal. Chem.*, 1983, **55**, 805.

131 B. A. Thomson, J. V. Irebarne, and P. J. Dziedzic, *Anal. Chem.*, 1982, **54**, 2219.

132 G. Caldwell and J. E. Bartmess, *Org. Mass Spectrom.*, 1982, **17**, 456.

133 T. O. Tiernan, C. Chang, and C. C. Cheng, *Environ. Health Perspect.*, 1980, **36**, 47.

134 H. F. Grützmacher and B. Grotemayer, *Org. Mass Spectrom.*, 1984, **19**, 136.

135 M. Barber, R. S. Bordoli, and R. D. Sedgewick in 'Soft Ionization Biological Mass Spectrometry', Proc. Chem. Soc. Symp. Adv. Mass Spectrom. Soft Ionization Methods, London, July, 1980, Heyden and Son, London, 1981, p. 137.

136 J. M. Miller, *J. Organomet. Chem.*, 1983, **249**, 299.

137 K. L. Rinehart, *Science*, 1982, **218**, 254.

The ionization process for negative ions is not yet fully understood. Usually $(M + H)^+$ and $(M - H)^-$ ions are produced from samples dissolved in glycerol when the argon beam impinges on the sample. Gycerol may protonate the sample to produce $(M + H)^+$ ions, while the residual alkoxide ion RO^- may deprotonate sample molecules to produce $(M - H)^-$ ions. This cannot be the full story since negative-ion FAB spectra may be obtained from certain molecules not acidic enough to be deprotonated by RO^-. Also some fragmentation processes must involve high-energy processes.

Negative-ion FAB mass spectra have been determined for carboxylic acids,[138,139] steroid sulphates,[140,141] phosphate esters of ecdysone,[142] clerodin,[143] diospyrol glycosides,[144] vitamin D glucuronides,[145] polysaccharides,[146,147] glucans,[148] carbohydrates derived from glycoproteins,[149] penicillins,[150] fluphenazine oxide derivatives,[151] arthreonam,[152] phosphonated and sulphonated azo dyestuffs,[153,154] naphthalene sulphonic acids,[155] glucosinolates.[156] vitamin B_{12} derivatives,[157] corrins,[158] chlorophylls,[159] anionic surfactants,[160] various peptides,[161–171] hydro-

[138] K. B. Tomer, F. W. Crow, and M. L. Gross, *J. Am. Chem. Soc.*, 1983, **105**, 5487.
[139] J. W. Dallinga, N. M. M. Nibbering, J. Van Der Greef, and M. C. Ten Noever De Braun, *Org. Mass Spectrom.*, 1984, **19**, 10.
[140] S. J. Gaskell, B. G. Brownsley, P. W. Brooks, and B. N. Green, *Int. J. Mass Spectrom. Ion Phys.*, 1983, **46**, 435.
[141] J. Gaskell, B. G. Brownsey, P. W. Brooks, and B. N. Green, *Biomed. Mass Spectrom.*, 1983, **10**, 215.
[142] R. E. Isaac, M. E. Rose, H. H. Rees, and T. W. Goodwin, *Biochem. J.*, 1983, **213**, 533.
[143] N. Rastrup Andersen and P. R. Rasmussen, *Tetrahedron Lett.*, 1984, **25**, 465.
[144] S. Paphassarang, M. Becchi, and J. Raynaud, *Tetrahedron Lett.*, 1984, **25**, 523.
[145] I. Jardine, G. F. Scanlan, V. R. Mattox, and R. Kumar, *Biomed. Mass Spectrom.*, 1984, **11**, 4.
[146] A. Dell and C. E. Ballou, *Carbohydr. Res.*, 1983, **120**, 95.
[147] A. Dell and C. E. Ballou, *Biomed. Mass Spectrom.*, 1983, **10**, 50.
[148] A. Dell, W. S. York, M. McNeil, A. G. Darvill, and P. Albersheim, *Carbohydr. Res.*, 1983, **117**, 185.
[149] J. P. Kamerling, W. Heerma, J. F. G. Vliegenthart, B. N. Green, I. A. S. Lewis, G. Strecker, and G. Spik, *Biomed. Mass Spectrom.*, 1983, **10**, 420.
[150] J. L. Gower, *Int. J. Mass Spectrom. Ion Phys.*, 1983, **46**, 431.
[151] P. Brooks and W. F. Heyes, *Biomed. Mass Spectrom.*, 1982, **9**, 522.
[152] A. I. Cohen, P. T. Funke, and B. N. Green, *J. Pharm. Sci.*, 1982, **71**, 1065.
[153] J. J. Monaghan, M. Barber, R. S. Bordoli, R. D. Sedgewick, and A. N. Tyler, *Org. Mass Spectrom.*, 1983, **18**, 75.
[154] J. J. Monaghan, M. Barber, R. S. Bordoli, R. D. Sedgewick, and A. N. Tyler, *Org. Mass Spectrom.*, 1982, **17**, 569.
[155] J. J. Monaghan, M. Barber, R. S. Bordoli, R. D. Sedgewick, and A. N. Tyler, *Org. Mass Spectrom.*, 1982, **17**, 529.
[156] G. R. Fenwick, J. Eagles, and R. Self, *Org. Mass Spectrom.*, 1982, **17**, 544.
[157] B. Kraeutler, R. Stepanck, and G. Holze, *Helv. Chim. Acta*, 1983, **66**, 44.
[158] H. Schwarz, K. Eckart, and L. C. E. Taylor, *Org. Mass Spectrom.*, 1982, **17**, 458.
[159] R. G. Brereton, M. B. Bazzaz, S. Santikarn, and D. H. Williams, *Tetrahedron Lett.*, 1983, 5775.
[160] P. A. Lyon, W. L. Stibbings, F. W. Crow, K. B. Tomer, D. L. Lippstreu, and M. L. Gross, *Anal. Chem.*, 1984, **56**, 8.
[161] G. V. Garner, D. B. Gordon, L. W. Tetler, and R. D. Sedgewick, *Org. Mass Spectrom.*, 1983, **18**, 486.
[162] K. B. Tomer, F. W. Crow, H. W. Knoche, and M. L. Gross, *Anal. Chem.*, 1983, **55**, 1033.
[163] D. H. Williams, S. Santikarn, F. De Angelis, R. J. Smith, D. G. Reid, P. B. Oelrichs, and J. K. MacLeod, *J. Chem. Soc.*, *Perkin Trans. l*, 1983, 1869.

xamate-containing siderophores,[172] glycoproteins,[173] and nucleotides/nucleosides.[174–179]

Most negative-ion FAB spectra contain $(M-H)^-$ ions that are used principally for molecular-weight determination. However, simple cleavage is noted in certain systems, for example the facile α-cleavage to oxygen that occurs for compounds containing X–O (X = C, S, P, *etc.*) bonds. In some instances the collision-induced MIKE spectra of $(M-H)^-$ ions yield valuable structural information. A selection of examples follows. The collision-induced MIKE spectra of the $(M-H)^-$ ion (33) shows β-cleavage to the double bond,[138] in marked contrast to the complex fragmentation of structure (24).[105] $(M-H)^-$ ions formed from maleic acid are more stable than those formed from fumaric acid because of intramolecular hydrogen bonding [see structure (34)].[139] The negative-ion FAB spectrum of diospyrol glycoside (35) shows an $(M-H)^-$ ion together with fragment ions defined by the sequence $[(M-H)^- -$ apiose $-$

n-C_7H_5 … $C_5H_{10}CO_2^-$

(33)

(34)

[164] W. A. Konig, M. Aydin, U. Schulze, U. Rapp, M. Höhn, P. Pesch, and V. N. Kalikhevitch, *Int. J. Mass Spectrom. Ion Phys.*, 1983, **46**, 403.

[165] M. Barber, R. S. Bordoli, R. D. Sedgewick, and A. N. Tyler, *Biomed. Mass Spectrom.*, 1982, **9**, 208.

[166] A. M. Buko, L. R. Phillips, and B. A. Fraser, *Biomed. Mass Spectrom.*, 1983, **10**, 387.

[167] K. L. Rinehart, J. C. Cook, R. C. Pauley, L. A. Gaudioso, H. Meng, M. L. Moore, J. B. Gloer, G. R. Wilson, R. E. Gutowsky, P. D. Zierath, L. S. Shield, L. H. Li, H. E. Renis, J. P. McGovren, and P. G. Canonico, *Pure Appl. Chem.*, 1982, **54**, 2409.

[168] M. Barber, R. S. Bordoli, R. D. Sedgewick, and A. N. Tyler, 'Recent Developments in Mass Spectrometry in Biochemistry, Medicine and Environmental Research', Anal. Chem. Symp. Ser. No. 12, Proc. 8th Int. Symp. Mass Spectrom. Biochem., Med. Environ. Res., Venice, Jan., 1981, Elsevier, Amsterdam, 1983, p. 233.

[169] H. Brueckner, G. Jung, and M. Przybylski, *Chromatographia*, 1983, **17**, 679.

[170] M. E. Rose, M. C. Prescott, A. H. Wilby, and I. J. Galpin, *Biomed. Mass Spectrom.*, 1984, **11**, 10.

[171] J. M. Wasylyk, J. E. Biskupiak, C. E. Costello, and C. M. Ireland, *J. Org. Chem.*, 1983, **48**, 4445.

[172] A. Dell, R. C. Hider, M. Barber, R. S. Bordoli, R. D. Sedgewick, A. N. Tyler, and J. B. Neilands, *Biomed. Mass Spectrom.*, 1982, **9**, 158.

[173] J. P. Kamerling, W. Heerma, J. F. G. Vliegenthart, B. N. Green, I. A. S. Lewis, G. Strecker, and G. Spik, *Biomed. Mass Spectrom.*, 1983, **10**, 420.

[174] C. Fenselau, V. T. Vu, R. J. Cotter, G. Hausen, D. Heller, T. Chen, and O. M. Colvin, *Spectrosc. Int. J.*, 1982, **1**, 132.

[175] G. Sindona, N. Uccella, and K. Weclawek, *J. Chem. Res. (S)*, 1982, 184.

[176] L. Grotjahn, R. Frank, and H. Blöcker, *Int. J. Mass Spectrom. Ion Phys.*, 1983, **46**, 439.

[177] M. Panico, G. Sindona, and N. Uccella, *J. Am. Chem. Soc.*, 1983, **105**, 5607.

[178] N. Neri, G. Sindona, and N. Uccella, *Gazz. Chim. Ital.*, 1983, **113**, 197.

[179] J. Eagles, C. Javanaud, and R. Self, *Biomed. Mass Spectrom.*, 1984, **11**, 41.

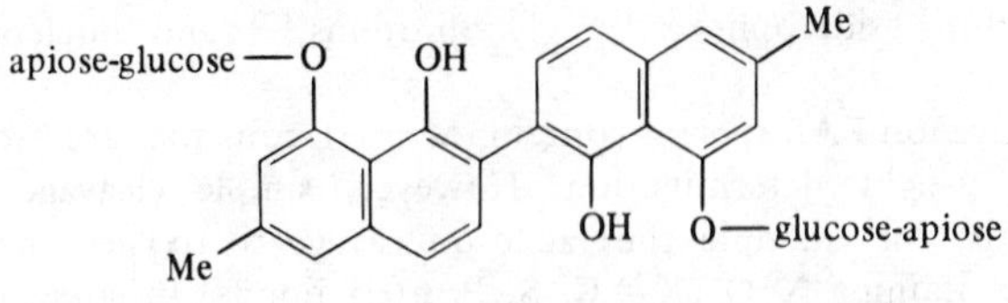

apiose-glucose—O OH Me OH O—glucose-apiose Me

(35)

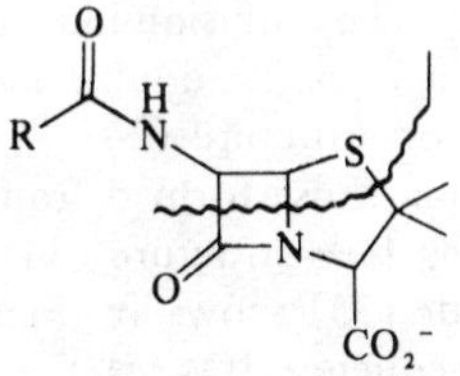

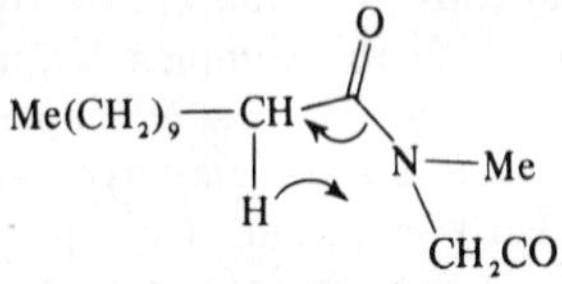

(36)

$Me(CH_2)_9$—CH N—Me H $CH_2CO_2^-$

(37)

glucose − apiose − (glucose − H^+)].[144] Both penicillins and cephalosporins fragment by complex ring cleavage [*e.g.* structure (36)].[150] Vitamin B_{12},[157] corrins,[158] and chlorophylls[159] yield $(M-H)^-$ ions.

The major collision-induced dissociation of the $(M-Na)^-$ ion of sodium laurylsarcosinate occurs by the rearrangement shown in structure (37).[160] The major fragment ions produced from butyloxycarbonyl amino acids are produced as shown in structure (38).[161] Collision-induced MIKE spectra of peptide $(M-H)^-$ ions, for example ornithine-containing lipids[162] and angiotensin 1 [see structure (39)],[165] give useful sequencing information. Sequencing information

(38)

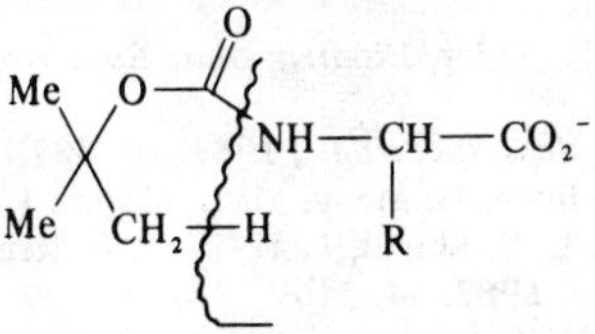

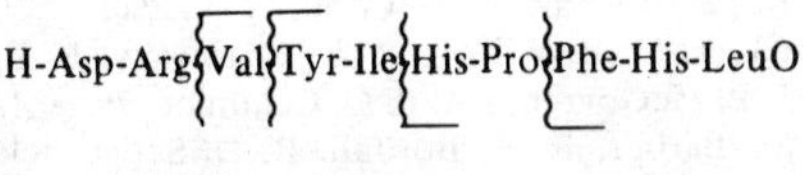

H-Asp-Arg-Val-Tyr-Ile-His-Pro-Phe-His-LeuO⁻

(39)

of nucleotides may be obtained from the normal negative-ion FAB spectrum because of facile cleavage of C–O and P–O bonds.[174,177] An illustration is shown in structure (40) for the simple case of NAD.[174]

5 Other Ionization Techniques

Laser-induced mass spectrometry produces $(M-H)^-$ ions of carboxylic acids,[180] amines,[181] peptides,[182] and cobalamines.[183] Laser mass spectrometry has also been

180 D. A. McCrery, E. B. Ledford, and M. L. Gross, *Anal. Chem.*, 1982, **54**, 1435; R. J. Day, A. L. Forbes, and D. M. Hercules, *Spectrosc. Lett.*, 1981, **14**, 703.
181 J. K. De Waele, E. F. Vansant, P. Van Espen, and F. C. Adams, *Anal. Chem.*, 1983, **55**, 671.
182 E. Onyiriuka, R. L. White, D. A. McCrery, M. L. Gross, and C. L. Wilkins, *Int. J. Mass Spectrom. Ion Phys.*, 1983, **46**, 135.
183 S. W. Graham, P. Dowd, and D. M. Hercules, *Anal. Chem.*, 1982, **54**, 649.

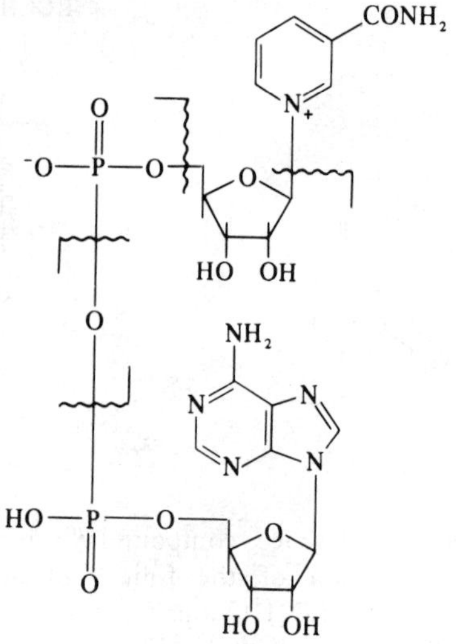

CONH2 N+ O ⁻O—P—O O HO OH O NH2 N N N N HO—P—O O O HO OH

(40)

used to analyse nitro-aromatics,[184] coal-shale samples,[185] quaternary ammonium salts,[186] and polymers.[187] The $MeCOCH_2^-$ ion undergoes laser-induced decomposition to C_2HO^- and methane,[188] whereas $CF_3COCH_2^-$ eliminates both CF_3H and CH_2CO.[189] A combination of laser ICR and deuterium-isotope effects has led to the suggestion that the loss of methane from the t-butoxide negative ion occurs by a stepwise mechanism.[190]

Several reviews of negative-ion field desorption (see Vol. 7 of this title, p. 152) have appeared.[191,192] Studies of corrins,[193] anionic surfactants,[194] and nucleosides[195] have been reported.

Several reviews on secondary-ion mass spectrometry (SIMS – see Vol. 7 of this title, p. 152) have been published.[196–199] Steroids,[200] bacterial products from

[184] F. P. Novak, K. Balasanmugam, V. Viswanadham, C. Parker, Z. A. Wilk, D. Mattern, and D. M. Hercules, *Int. J. Mass Spectrom. Ion Phys.*, 1983, **53**, 135.
[185] N. E. Vanderborgh and C. E. Roland Jones, *Anal. Chem.*, 1983, **55**, 527.
[186] K. Balasanmugam and D. M. Hercules, *Spectrosc. Lett.*, 1983, **16**, 1.
[187] S. W. Graham and D. M. Hercules, *Spectrosc. Lett.*, 1982, **15**, 1.
[188] R. F. Foster, W. Tumas, and J. I. Brauman, *J. Chem. Phys.*, 1983, **79**, 4644.
[189] C. R. Moylan, J. M. Jasinski, and J. I. Brauman, *Chem. Phys. Lett.*, 1983, **98**, 1.
[190] W. Tumas, R. F. Foster, M. J. Pellerite, and J. I. Brauman, *J. Am. Chem. Soc.*, 1983, **105**, 7464.
[191] F. W. Röllgen, *Trends Anal. Chem.*, 1982, **1**, 304.
[192] F. W. Röllgen in 'Ion Formation from Organic Solids', Springer Ser. Chem. Phys., Springer Verlag, 1983, Vol. 25, p. 2.
[193] H. M. Schiebel and H.-R. Shulten, *Biomed. Mass Spectrom.*, 1982, **9**, 354.
[194] P. Dähling, F. W. Röllgen, J. J. Zwinselman, R. H. Fokkens, and N. M. M. Nibbering, *Fresenius' Z. Anal. Chem.*, 1982, **312**, 319.
[195] P. Dähling, K. H. Ott, F. W. Röllgen, J. J. Zwinselman, R. H. Fokkens, and N. M. M. Nibbering, *Int. J. Mass Spectrom. Ion Phys.*, 1983, **46**, 301.
[196] 'SIMS III', Proc. 3rd Int. Conf. SIMS, Budapest, Springer Ser. Chem. Phys., Springer Verlag, 1982, Vol. 19.

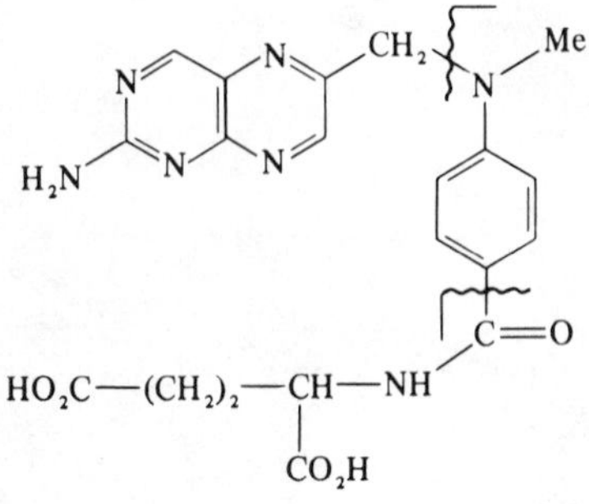

(41)

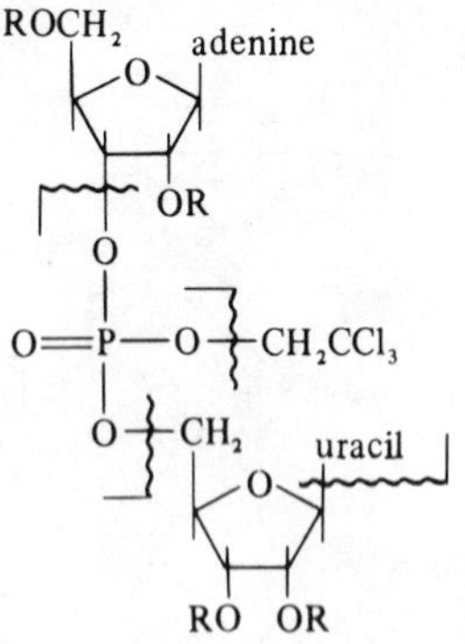

(42)

granaticin,[201] and biological reference compounds[202] have been studied by negative-ion SIMS. The spectrum of the folic acid derivative (41) shows $(M-H)^-$ and $[(M-H)-CO_2-HCO_2H]^-$ ions together with the fragmentations shown.[203] Negative-ion SIMS spectra of nucleosides are similar to those produced by negative-ion FAB. For example, structure (42) ($R = SiMe_2Bu^t$) shows an $(M-H)^-$ ion and the fragmentations indicated[204] [*cf.* the negative-ion FAB spectrum of structure (40)[174]]. Bombardment of molecular solids by Ar^+ ions produces interesting cluster ions; for example, water yields $HO^-(H_2O)_n$ (n = 0–5)[205] whereas carbon dioxide gives $CO_2^-(CO_2)_n$ (n = 1–9).[206] Caesium cations may also be used as a SIMS reagent, and the resulting negative-ion SIMS spectra of many high-molecular-weight compounds are similar to negative-ion FAB spectra.[207–209]

^{252}Californium plasma spectrometry (Vol. 7 of this title, p. 152) has been reviewed.[210,211] This method has been used to determine molecular weight

[197] A. Benninghoven, *Trends Anal. Chem.*, 1982, **1**, 311.
[198] A. Benninghoven, *Int. J. Mass Spectrom. Ion Phys.*, 1983, **46**, 459; 1983, **53**, 85.
[199] S. M. Scheifers, R. C. Hollar, K. L. Busch, and R. G. Cooks, *Int. Lab.*, 1982, **14**, 12.
[200] W. Sichtermann, A. Eicke, M. Junack, and A. Benninghoven, *Fresenius' Z. Anal. Chem.*, 1982, **311**, 410.
[201] M. Junack, E. Korman, A. Eicke, W. Sichtermann, A. Benninghoven, and H. Pape, *Fresenius' Z. Anal. Chem.*, 1982, **311**, 411.
[202] G. O. Ramseyer and G. H. Morrison, *Anal. Chem.*, 1983, **55**, 1963.
[203] A. Elcke, V. Anders, M. Junack, W. Sichtermann, and A. Benninghoven, *Anal. Chem.*, 1983, **55**, 178.
[204] W. Ens, K. G. Standing, J. B. Westmore, K. K. Ogilvie, and M. J. Nemer, *Anal. Chem.*, 1982, **54**, 960.
[205] J. Marien and E. De Pauw, *J. Chem. Soc., Chem Commun.*, 1982, 949.
[206] R. G. Orth, H. J. Jonkman, and J. Michl, *Int. J. Mass Spectrom. Ion Phys.*, 1982, **43**, 41.
[207] W. Aberth, K. M. Straub, and A. L. Burlingame, *Anal. Chem.*, 1982, **54**, 2029.
[208] C. N. McEwen, *Anal. Chem.*, 1983, **55**, 967.
[209] R. Beavis, W. Ens, K. G. Standing, and J. B. Westmore, *Int. J. Mass Spectrom. Ion Phys.*, 1983, **46**, 47.
[210] R. D. Macfarlane in 'Soft Ionization Biological Mass Spectrometry', Proc. Chem. Soc. Symp. Adv. Mass Spectrom. Soft Ionization Methods, London, July, 1980, Heyden and Son, London, 1981, p. 110.
[211] H. M. Fales, *Int. J. Mass Spectrom. Ion. Phys.*, 1983, **53**, 159.

(43)

(44)

as high as 5000–8000 daltons. $(M-H)^-$ ions are produced, and fragmentation is not generally as pronounced as it is in the corresponding positive-ion spectra (in common with other 'soft ionization' spectra described above). $(M-H)^-$ ions are reported for structure (43) (fragmentations indicated),[212] grisorixine,[213] chlorophyll *a*[214] (also $M^{-\bullet}$ [215]), various peptides[216,217] (also ^{127}I plasma spectrum of insulin[218]), and nucleotides.[219,220] Fragmentation of nucleotide $(M-H)^-$ ions is very similar for ^{252}Cf spectra [see structure (44)],[220] negative-ion SIMS of structure (42),[204] and negative-ion FAB mass spectra of structure (40).[174]

The collision-induced charge-inversion technique (Vol. 7 of this title, p. 157)[221] continues to be used in structural studies of both negative and positive ions. Examples include the allyl anion and cation,[222] enolate ions,[223] diesters (the negative-ion Dieckmann condensation),[108] and barbituric acid derivatives.[115] The $C_2H_2N^+$ species produced by charge inversion from $^-CH_2CN$ and by elimination of $H^\bullet$ from $MeCN^{+\bullet}$ have different structures: the former eliminates CH_2 and the latter C.[224] In the reaction $PhS^- \rightarrow PhS^+ \rightarrow$ products, aromatic hydrogen

[212] P. Viguy, M. Spiro, F. Gaboriau, Y. Le Beyec, S. D. Negra, J. Cadet, and L. Voiturier, *Int. J. Mass Spectrom. Ion Phys.*, 1983, **53**, 69.

[213] L. David, S. D. Negra, D. Fraisse, G. Heminet, I. Lorthiois, Y. Le Beyec, and J. C. Tabet, *Int. J. Mass Spectrom. Ion Phys.*, 1983, **46**, 391.

[214] B. J. Chait and F. H. Field, *J. Am. Chem. Soc.*, 1982, **104**, 5519.

[215] J. E. Hunt, R. D. Macfarlane, J. J. Katz, and R. C. Dougherty, *J. Am. Chem. Soc.*, 1981, **103**, 6775.

[216] B. T. Chait, B. F. Gisen, and F. H. Field, *J. Am. Chem. Soc.*, 1982, **104**, 5157.

[217] I. Kamensky, J. Fohlman, P. Hakansson, J. Kjellberg, P. Peterson, and B. Sundqvist, *Int. J. Mass Spectrom. Ion Phys.*, 1983, **46**, 467.

[218] P. Hakansson, I. Kamensky, B. Sundqvist, J. Fohlman, P. Peterson, C. J. McNeal, and R. D. Macfarlane, *J. Am. Chem. Soc.*, 1982, **104**, 2948.

[219] C. J. McNeal, K. K. Ogilvie, N. Y. Theriault, and M. J. Nemer, *J. Am. Chem. Soc.*, 1982, **104**, 972.

[220] C. J. McNeal, K. K. Ogilvie, N. Y. Theriault, and M. J. Nemer, *J. Am. Chem. Soc.*, 1982, **104**, 976.

[221] J. H. Bowie and T. Blumenthal, *J. Am. Chem. Soc.*, 1975, **97**, 2959.

[222] T. A. Lehman, M. M. Bursey, D. J. Harvan, J. R. Hass, D. Liotter, and L. Waykole, *Org. Mass Spectrom.*, 1982, **17**, 607.

[223] T. A. Lehman, M. M. Bursey, and J. R. Hass, *Org. Mass Spectrom.*, 1983, **18**, 373.

[224] M. M. Bursey, D. J. Harvan, C. E. Parker, and J. R. Hass, *J. Am. Chem. Soc.*, 1983, **105**, 6801.

scrambling does not occur for the anion but accompanies fragmentation of the cation.[52] Charge-inversion reactions can be studied using triple-quadrupole instruments.[225]

6 Ion/Molecule Reactions

This section reviews negative-ion/molecule reactions studied principally by ion-cyclotron-resonance (ICR) and flowing-afterglow (FA) spectrometry. Studies using high-pressure sources (including some using NICI conditions) are also reported here. Many interesting studies have been published during the review period, some of which have been highlighted in recent reviews.[20, 226–229]

Acidity Measurements. – A number of studies have been concerned with various aspects of acidity. The abundances of products observed in the MIKE spectra of alkanol alkoxides for the two reactions $[R^1O\cdots H\cdots OR^2]^- \rightarrow R^1O^- + R^2OH$ and $R^2O^- + R^1OH$ can be directly related to the two rates, which in turn indicate the relative acidity of alcohols.[230, 231] A similar study has been reported for ions $[R^1CO_2\cdots H\cdots O_2CR^2]^-$, which demonstrates, for example, that the propionate anion has a lower proton affinity than the acetate anion.[232] The gas-phase acidities of glycine derivatives have been reported.[233] H/D exchange of carbanions indicates a relative acidity sequence aldehydes > amides > esters.[234] H/D exchange reactions have been observed for other negative ions including allyl,[235] arylallyl,[236] and anions derived from alkyl phenyl ethers.[237] Phenylallyl and phenylcyclopropyl anions may be differentiated by the respective H/D exchange sequences of structure (45) → (46) and structure (47) → (48).[236] H/D exchange with weakly acidic neutral molecules is defined by the reaction sequence $DO^- + MH \rightarrow [DO^-\cdot MH] \rightarrow [DOH\cdot M^-] \rightarrow [HO^-\cdot MD] \rightarrow HO^- + MD$.[238] Exchange occurs more

225 D. J. Douglas and B. Shushan, *Org. Mass Spectrom.*, 1982, **17**, 198.

226 S. Ingemann, J. C. Kleingeld, and N. M. M. Nibbering in 'Ionic Processes in the Gas Phase', Proc. NATO Adv. Study Inst. Chem. Ions Gas Phase, Vimiero, Portugal, 6–17 Sept., 1982, Reidel Publ. Co., Dordrecht, Boston, Lancaster, 1984, p. 87.

227 D. K. Bohme in 'Ionic Processes in the Gas Phase', Proc. NATO Adv. Study Inst. Chem. Ions Gas Phase, Vimiero, Portugal, 6–17 Sept., 1982, Reidel Publ. Co., Dordrecht, Boston, Lancaster, 1984, p. 111.

228 C. H. DePuy in 'Ionic Processes in the Gas Phase', Proc. NATO Adv. Study Inst. Chem. Ions Gas Phase, Vimiero, Portugal, 6–17 Sept., 1982, Reidel Publ. Co., Dordrecht, Boston, Lancaster, 1984, p. 227.

229 G. Boand, R. Houriet, and T. Gäumann, 'I.C.R. Spectrometry. 11. Reactions of Alkoxide Anions', Lecture Notes Chem., Springer Verlag, Berlin, 1982, Vol. 31, 195.

230 G. Boand, R. Houriet, and T. Gäumann, *J. Am. Chem. Soc.*, 1983, **105**, 2203.

231 S. A. McLuckey, D. Cameron, and R. G. Cooks, *J. Am. Chem. Soc.*, 1981, **103**, 1313.

232 L. G. Wright, S. A. McLuckey, R. G. Cooks, and K. V. Wood, *Int. J. Mass Spectrom. Ion Phys.*, 1982, **42**, 115.

233 M. J. Locke and R. T. McIver, *J. Am. Chem. Soc.*, 1983, **105**, 4226.

234 J. E. Bartmess, G. Caldwell, and M. D. Rozeboom, *J. Am. Chem. Soc.*, 1983, **105**, 340.

235 R. R. Squires, V. M. Bierbaum, J. J. Grabowski, and C. H. DePuy, *J. Am. Chem. Soc.*, 1983, **105**, 5185.

236 A. H. Andrist, C. H. DePuy, and R. R. Squires, *J. Am. Chem. Soc.*, 1984, **106**, 845.

237 J. C. Kleingeld and N. M. M. Nibbering, *Tetrahedron*, 1983, **39**, 4193.

238 J. J. Grabowski, C. H. DePuy, and V. M. Bierbaum, *J. Am. Chem. Soc.*, 1983, **105**, 2565.

$$\mathrm{Ph}\text{-allyl}^{-}\ (\mathbf{45}) \xrightarrow{D_2O} (\mathbf{46})$$

$$\mathrm{Ph}\text{-cyclopropyl}^{-}\ (\mathbf{47}) \xrightarrow{D_2O} (\mathbf{48})$$

readily for molecules with either large dipole moments or high polarizability. The rate of exchange is dependent upon the relative basicities of DO^- and M^- and upon the relative energies of the two complexes $[DO^- \cdot MH]$ and $[DOH \cdot M^-]$.[238] Reactions of F^-, Cl^-, and I^- with Broensted acids have been reported.[239–242]

Reactions of Negative Ions with Organic Molecules. – The following studies have been reported: the formation of NH_4^- in the system $NH_2^-/NH_3/CH_2O$,[243] pyrolysis of alkoxide ions,[244] reactions of CF_3^- with carbonyl compounds,[245,246] reactions of alkoxide ions with acrylates,[247] reactions of $PhN^{-\bullet}$ and $S^{-\bullet}$ with sp^3 and sp^2 carbon,[248] gas-phase hydrolysis of esters,[249] phenyl acetate and phenyl benzoate reactions,[250,251] nucleophilic addition to fluoroanisoles,[252,253] reactions of HO^- and NH_2^- with cyclic ethers,[254] the absence of an α-effect in gas-phase reactions of HO_2^-,[255] the gas-phase anionic oxy-Cope rearrangement,[256] nitrile

[239] J. W. Larson and T. B. McMahon, *J. Am. Chem. Soc.*, 1982, **104**, 5848.
[240] J. W. Larson and T. B. McMahon, *J. Am. Chem. Soc.*, 1983, **105**, 2944.
[241] J. W. Larson and T. B. McMahon, *J. Am. Chem. Soc.*, 1984, **106**, 517.
[242] G. Caldwell and P. Kebarle, *J. Am. Chem. Soc.*, 1984, **106**, 967.
[243] J. C. Kleingeld, S. Ingemann, J. E. Jalonen, and N. M. M. Nibbering, *J. Am. Chem. Soc.*, 1983, **105**, 2474; J. C. Kleingeld and N. M. M. Nibbering, *Int. J. Mass Spectrom. Ion Phys.*, 1983, **49**, 311.
[244] G. Boand, R. Houreit, and T. Gäumann, *Int. J. Mass Spectrom. Ion Phys.*, 1983, **52**, 95.
[245] R. N. McDonald and A. K. Chowdhury, *J. Am. Chem. Soc.*, 1983, **105**, 7267.
[246] R. N. McDonald and A. K. Chowdhury, *J. Am. Chem. Soc.*, 1983, **105**, 2194.
[247] G. Klass, J. C. Sheldon, and J. H. Bowie, *J. Chem. Soc., Perkin Trans. 2*, 1983, 1337.
[248] R. N. McDonald and A. K. Chowdhury, *J. Am. Chem. Soc.*, 1983, **105**, 198.
[249] K. Takashima, S. M. Jose, A. T. Do Amarel, and J. M. Riveros, *J. Chem. Soc., Chem. Commun.*, 1983, 1255.
[250] J. C. Kleingeld, N. M. M. Nibbering, J. J. Grabowski, C. H. DePuy, E. K. Fukada, and R. T. McIver, *Tetrahedron Lett.*, 1982, **23**, 4755.
[251] C. L. Johlman, R. L. White, D. J. Sawyer, and C. L. Wilkins, *J. Am. Chem. Soc.*, 1983, **105**, 2091.
[252] S. Ingemann and N. M. M. Nibbering, *J. Org. Chem.*, 1983, **48**, 183.
[253] S. Ingemann, N. M. M. Nibbering, S. A. Sullivan, and C. H. DePuy, *J. Am. Chem. Soc.*, 1982, **104**, 6520.
[254] C. H. DePuy, E. C. Beidle, and V. M. Bierbaum, *J. Am. Chem. Soc.*, 1982, **104**, 6483.
[255] C. H. DePuy, E. W. Della, J. Filley, J. J. Grabowski, and V. M. Bierbaum, *J. Am. Chem. Soc.*, 1983, **105**, 2481.
[256] M. D. Rozeboom, J. P. Keplinger, and J. L. Bartmess, *J. Am. Chem. Soc.*, 1984, **106**, 1025.

reactions,[257,258] reduction reactions of RO^- (R = alkyl or CF_3),[259–262] a gas-phase Wittig reaction,[263] silicon negative-ion chemistry in the atmosphere,[264] organo-silicon reactions,[265] and reactions of H_2P^- with organic substrates.[266,267]

The reaction of CF_3^- with acetone produces a collision-stabilized adduct that is proposed to have the tetrahedral structure $Me_2(CF_3)CO^-$ rather than the structure $[CF_3H \cdots {}^-CH_2COMe]$.[245] The ion CF_3^- initiates oligomerization of methylacrylate to form structure (49), which eliminates methanol by a Dieckmann cyclization [*cf.* structure (25) → (27)[108]].[246] A decomposing tetrahedral species is formed in the MeO^-/methylacrylate system since CD_3O^- reacts with $CH_2{=}CDCO_2Me$ to yield a decomposing intermediate (50), which eliminates MeOD and CD_3OD to yield structures (51) and (52).[247] Superoxide ion reacts

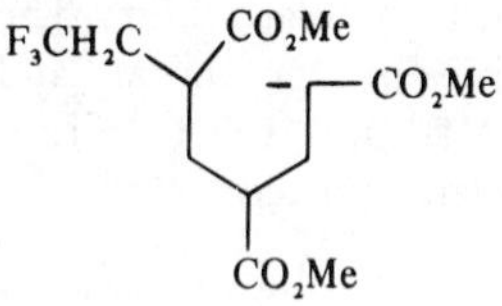

(49) 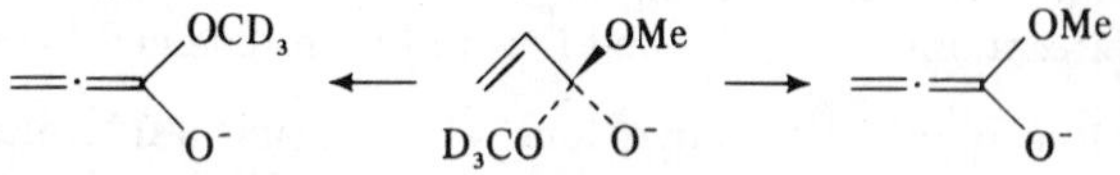 (51) (50) (52)

with phenylacetate to give $MeCO_2^-$, $MeCO_3^-$, PhO^-, and the two ions $C_6H_6O_3^{-\bullet}$ and $C_6H_4O_2^{-\bullet}$. It is suggested that the latter ions correspond to structures (54) and (55) and that they are formed from the initial intermediate (53).[251] The $H^{18}O^-$ ion reacts with C_6F_5OMe to yield $C_6F_4(OMe)^{18}O^-$ (S_NAr), $C_6F_5{}^{18}O^-$ (*ipso*), and $C_6F_5O^-$ (S_N2).[253] Reaction of structure (56) with MeO^- yields structure (57), which in the presence of methanol rearranges to structure (58) by an oxy-Cope reaction. It is suggested that the formation of structure (58) from (57) proceeds through a number of methanol-solvated reactive intermediates.[256] Anionic-catalysed polymerization reactions are observed for acrylonitrile [*cf.* structure (49)[246]].[258]

[257] J. F. Paulson and F. Dale, *Bull. Am. Phys. Soc.*, 1982, **27**, 108.
[258] R. N. McDonald and A. K. Chowdhury, *J. Am. Chem. Soc.*, 1982, **104**, 2675.
[259] S. Ingemann, J. C. Kleingeld, and N. M. M. Nibbering, *J. Chem. Soc., Chem. Commun.*, 1982, 1009.
[260] C. H. DePuy, V. M. Bierbaum, R. J. Schmitt, and R. H. Shapiro, *J. Am. Chem. Soc.*, 1978, **100**, 2920.
[261] J. C. Sheldon, J. H. Bowie, and R. N. Hayes, *Nouveau J. Chim.*, 1984, **8**, 79.
[262] J. C. Sheldon, G. J. Currie, J. Lahnstein, R. N. Hayes, and J. H. Bowie, *Nouv. J. Chim.*, in press.
[263] C. L. Johlman, C. F. Ijames, C. L. Wilkins, and T. H. Morton, *J. Org. Chem.*, 1983, **48**, 2628.
[264] A. A. Viggiano, F. Arnold, D. W. Fahey, F. C. Fehsefeld, and E. E. Ferguson, *Atomindex*, 1982, **13**, 693 328.
[265] R. R. Squires and C. H. DePuy, *Org. Mass Spectrom.*, 1982, **17**, 187.
[266] D. R. Anderson, V. M. Bierbaum, and C. H. DePuy, *J. Am. Chem. Soc.*, 1983, **105**, 4244.
[267] D. R. Anderson, V. M. Bierbaum, and C. H. DePuy, *J. Am. Chem. Soc.*, 1983, **105**, 5185.

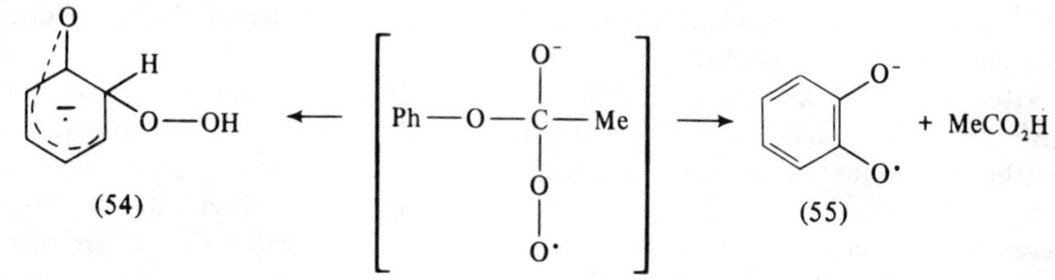

(54) ← [Ph—O—C(O⁻)(Me)—O—O·] (53) → (55) + $MeCO_2H$

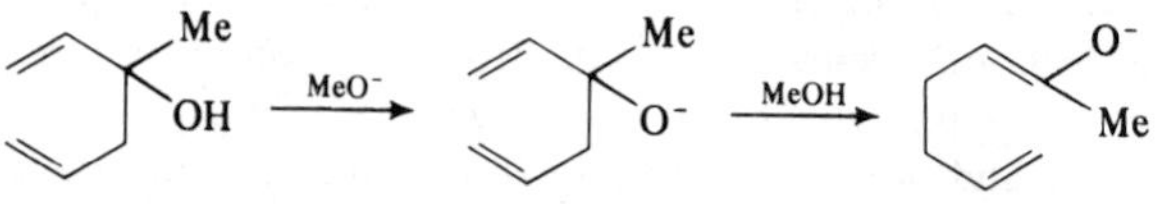

(56) $\xrightarrow{MeO^-}$ (57) $\xrightarrow{MeOH}$ (58)

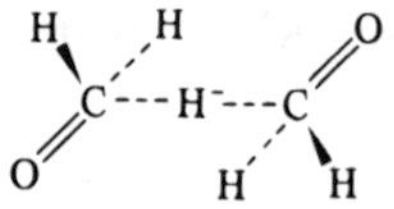

(59)  (60)

Although alkoxide negative ions generally react as bases or nucleophiles (through oxygen) with organic substrates, reduction by hydride-ion transfer has also been reported. The hydride-ion transfer is often pronounced when the substrate has no 'acidic' hydrogens. For example, CD_3O^- reacts with formaldehyde by the reaction $CD_3O^- + CH_2O \rightarrow CH_2DO^- + CD_2O$[259] (*cf.* reaction of $C_6H_7^-$ with formaldehyde[260]); *ab initio* calculations (4-31G) indicate that the transition state for this reaction has structure (59).[261] The CF_3O^- species transfers F^- to a variety of carbonyl compounds in a similar manner.[261] The methoxide anion also transfers H^- to CO_2, CS_2, and SO_2.[262] In these cases *ab initio* calculations suggest that structure (60) (A = C or S, X = O or S) is the key reactive intermediate.[262] The ion H_2P^- is a weaker nucleophile than H_2N^- and undergoes nucleophilic substitution with organosilanes and sp^3 and sp^2 carbon.[266,267]

The Reactions of Solvated Nucleophiles. – A number of groups is studying the difference in reactivity between nucleophiles, Nu^-, and their solvated counterparts, $[Nu^-(HNu)_n]$, in order to bridge the gap between gas-phase and solution chemistry. Work prior to 1982 in this area has been reviewed.[20] Bohme and colleagues have led the way in this area and have demonstrated the changes in reaction rates and reactivity with increasing solvation.[268–272] A recent example concerns the relative basicities of $[HO^- \cdots HOH]$ and $[MeO^- \cdots HOMe]$: the

[268] K. Tanaka, G. I. Mackay, J. D. Payzant, and D. K. Bohme, *Can. J. Chem.*, 1976, **54**, 1643.
[269] D. K. Bohme and G. I. Mackay, *J. Am. Chem. Soc.*, 1981, **103**, 978.
[270] D. K. Bohme, A. B. Rakshit, and G. I. Mackay, *J. Am. Chem. Soc.*, 1982, **104**, 1100.
[271] G. I. Mackay, A. B. Rakshit, and D. K. Bohme, *Can. J. Chem.*, 1982, **60**, 2594.
[272] G. Klass, J. C. Sheldon, and J. H. Bowie, *Aust. J. Chem.*, 1982, **35**, 2471.

former reacts with acetylene to yield $[C_2H^- \cdots HOH]$, whereas the latter does not yield $[C_2H^- \cdots HOMe]$.[269]

Reaction between $[MeO^- \cdots DOMe]$ and carbonyl compounds >CH–CO– gives both $[M + MeO^-]$ and $[(M - H + D) + MeO^-]$ adducts.[272] The observation of the two adducts indicates a reaction sequence in which H and D have equilibrated. Theoretical calculations suggest initial formation of structure (61), which rearranges to give a symmetrical disolvated ion (62), which finally produces structure (63) and methanol.[272] This type of reaction occurs for a variety of solvated nucleophiles and a variety of >CH–CO– and >CH–CN substrates.[273,274] In contrast, the reaction of $[HO^- \cdots HOH]$ with acetonitrile appears to be quite different. The reaction occurs without hydrogen equilibration to form structure (64); the mechanistic pathway is not known with certainty.[275]

Alkoxide ions and solvated alkoxide ions react very differently with silyl ethers. The reactions with alkoxides have been described (Vol. 7 of this title, p. 161; see also refs. 20 and 276). Alkanol-alkoxides $[R^1O^- \cdots HOR^2]$ $(R^2 > R^1)$ undergo the specific four-centre reaction (65) → (66).[277] Carbon ethers undergo the same reaction to a lesser extent; in this case $[R^1O^- \cdots HOR^2]$ can add in either direction.[278]

(61) → (62) →→ (63) HOMe + MeOH

$$MeCN + [HO^- \text{---} HOH] \longrightarrow [MeOH \text{---} {}^-CN] + H_2O$$

(64)

$$[R^1O^- \text{---} HOR^2] + Me_3Si\text{—}OR^3 \; (65) \longrightarrow Me_3SiOR^1 + [R^2O^- \text{---} HOR^3] \; (66)$$

The Rates of Nucleophilic Reactions in the Gas Phase. – Although the reactions occurring between nucleophiles and either trigonal or tetrahedral carbon in the gas phase are normally exothermic or thermoneutral, the rates of such reactions

273 J. H. Bowie and J. C. Sheldon, *Aust. J. Chem.*, 1983, **36**, 289.
274 J. C. Sheldon, J. H. Bowie, and M. B. Stringer, *Spectrosc. Int. J.*, 1983, **2**, 43.
275 G. Caldwell, M. D. Rozeboom, J. P. Kiplinger, and J. E. Bartmess, *J. Am. Chem. Soc.*, 1984, **106**, 809.
276 C. H. DePuy and V. M. Bierbaum, *Acc. Chem. Res.*, 1981, **14**, 146.
277 R. N. Hayes, J. H. Bowie, and G. Klass, *J. Chem. Soc., Perkin Trans. 2*, 1984, 1167.
278 R. N. Hayes and J. H. Bowie, *J. Chem. Soc., Perkin Trans. 2*, 1985, in press.

range from almost collision controlled to too slow to be observed.[268, 279–283] Brauman explains this variation in observed rates (with particular reference to tetrahedral carbon) by (*a*) a central barrier whose crest may be even lower in energy than reactants and (*b*) the internal-energy levels of the transition state being more widely spaced than those of the reactants so that the back reaction is favoured by entropy.[281–283] A recent study of S_N2 reactions has addressed the problem of the temperature dependence of the reaction as a function of the energy difference between reactants and transition state.[284]

In contrast, nucleophilic substitution and displacement at silicon appear to proceed at or near the collision rate. For example, nucleophiles react with trimethylchlorosilane in the gas phase at the collision rate for exothermic reactions or not at all for endothermic reactions.[285] *Ab initio* calculations indicate that HO^- and MeO^- should initially attack alkylsilanes at hydrogen and that there is then a small barrier to be surmounted before the silicon is approached. Even so, these reactions are collision controlled,[286] presumably because of the large negative energy difference between reactants and transition states.

[279] D. K. Bohme and L. B. Young, *J. Am. Chem. Soc.*, 1970, **92**, 7354; D. K. Bohme, G. I. Mackay, and J. D. Payzant, *J. Am. Chem. Soc.*, 1974, **96**, 4027.
[280] J. I. Brauman, W. N. Olmstead, and C. A. Lieder, *J. Am. Chem. Soc.*, 1974, **96**, 4030; W. N. Olmstead and J. I. Brauman, *J. Am. Chem. Soc.*, 1977, **99**, 4219.
[281] M. J. Pellerite and J. I. Brauman, *J. Am. Chem. Soc.*, 1980, **102**, 5993.
[282] M. J. Pellerite and J. I. Brauman, *J. Am. Chem. Soc.*, 1983, **105**, 2672.
[283] O. I. Asubiojo and J. I. Brauman, *J. Am. Chem. Soc.*, 1979, **101**, 3715.
[284] G. Caldwell, T. F. Magnera, and P. Kebarle, *J. Am. Chem. Soc.*, 1984, **106**, 959.
[285] R. Damrauer, C. H. DePuy, and V. M. Bierbaum, *Organometallics*, 1982, **1**, 1553.
[286] J. C. Sheldon, R. N. Hayes, and J. H. Bowie, *J. Am. Chem. Soc.*, 1985, **106**, 7711.

8
Fast-atom-bombardment Mass Spectrometry: Applications to Solution Chemistry

BY R. M. CAPRIOLI

1 Introduction

Fast-atom-bombardment mass spectrometry (FABMS) has had an enormous impact on many fields of chemistry, perhaps most profoundly in its biological applications, in only the few years since its inception. It has provided a method for the analysis of compounds that are highly polar and ionic, classes of molecules from which it is otherwise difficult or impossible to obtain meaningful spectra using well known ionization methods even with specific derivatization reactions. Of course, FAB is one amongst many new so-called soft-ionization techniques that have been used to analyse such compounds. Although some of the examples and concepts discussed in this chapter might also be appropriate to several of the other ionization methods, the scope of this chapter will be limited to a discussion of applications of FABMS.

One of the key features of the FAB method, in addition to the use of a neutral-atom beam, which has come to distinguish it from other ionization methods, is the use of a fluid sample-support matrix, most commonly glycerol or glycerol and water mixtures. It is often correctly stated that a major advantage of the liquid sample is its ability to renew constantly the surface with molecules from within the droplet as other material is sputtered from the surface during the atom-bombardment process. This indeed leads to ion currents from the sample droplet that can persist for 20–30 minutes or more. However, it is becoming apparent that the liquid sample support confers a second major advantage, and that is its ability to allow chemical reactions to occur as the bombardment and subsequent analysis are actually taking place. Even in the off-line mode, where samples are removed with time from a batch reaction taking place on the laboratory bench, the FAB technique facilitates the direct sampling of a reaction occurring in an aqueous solution without prior purification, concentration, or derivatization procedures.

This chapter will review the application of FABMS to the quantitative analysis of substances in aqueous solutions that have been generated from chemical reactions. Perhaps in its most widely applied role to date, FAB has been employed in the static analysis of components that are simply dissolved in aqueous glycerol prior to analysis, *e.g.* many polypeptides, oligonucleotides, polysaccharides, and other compounds that are insoluble in solvents of low dielectric constant. Since this qualitative description of FAB spectra is often discussed in articles and reviews, it will not be discussed in detail here. The greater part of this chapter will be concerned with reviewing those applications where FAB has been employed as the analytical tool for following dynamic aspects of chemical reactions, both batch reactions and those occurring on the probe inside the mass spectrometer.

2 Quantification

The quantitative determination of molecules in solution from within a droplet during FAB is a vital part of many applications of this technique. Although most of the concepts normally considered with other ionization methods still apply, some new aspects also come into play when FAB is used. These mainly result from the fact that the sputtered molecules originate from the surface of the droplet, but the composition of the molecules in the surface layers may differ from that of the interior of the droplet in a solution containing a mixture of compounds. Of course, the closer two molecules are to each other in their physical and chemical properties, the less one expects fractionation of the two at the surface. Thus, for FAB, as with other ionization modes, the best internal standards are stable isotopically labelled analogues of a compound in which only a few atoms have been replaced. For the types of compounds generally considered in biological applications, the substitution of a few deuterium atoms or ^{13}C atoms for the corresponding protons or ^{12}C atoms, respectively, should have little or no effect on the composition of the surface mixture. Some discussions of the theoretical aspects of the formation of surfaces have been published.[1]

One of the basic assumptions in quantitative mass spectrometry is that the ion current obtained from one compound is proportional to its concentration and, whether linear or not, that this is unaffected by the presence of moderate amounts of other compounds that might be present in the mixture. This would be especially true if some physical separation technique, such as gas chromatography, was used prior to ionization. Further, if an internal standard is employed that is chemically similar to the compound of interest, the ionization efficiencies of both compounds are assumed to be the same. However, these assumptions are not necessarily true with FAB; interactions amongst the various analytes, internal standard, and matrix components as well as the temperature and the nature of the probe can affect the composition of the surface being bombarded and therefore the ratio of the ions being sputtered. A few examples of the quantitative use of FABMS are briefly described below to illustrate some of these considerations.

In a study of the quantitative aspects of FAB, Beckner and Caprioli[2] measured the abundance of the $(M + H)^+$ ions of 4-aminobutyric acid (GABA) and its dideuterio analogue, 4-amino-[2,2-$^2H_2$]butyric acid or [$^2H_2$]GABA. Measurements of the isotope ratios were made for a range of GABA concentrations from 0.5 to 200 nmol μl^{-1}, with the [$^2H_2$]GABA concentration constant at one of

[1] J. W. Gibbs, 'The Collected Works of J. W. Gibbs', Longmans Green, New York, 1931, Vol. 1; J. J. Betts and B. A. Pethica, *Proc. Second Int. Congr. Surf. Activity*, 1957, **1**, 152; B. A. Pethica, *Trans. Faraday Soc.*, 1954, **50**, 413.

[2] C. F. Beckner and R. M. Caprioli, *Biomed. Mass Spectrom.*, 1984, **11**, 60.

eight different values ranging from 1.8 to 21.2 nmol μl^{-1}. For one such series of measurements, shown in Figure 1, ion-ratio measurements were made using 10.6 nmol μl^{-1} of $[^2H_2]$GABA as internal standard throughout the range of concentrations of GABA. It can be seen that at low concentrations of GABA, up to approximately 20 nmol μl^{-1}, the ion abundance of $(M + H)^+$ for GABA was linear with amount. At concentrations of GABA above this, however, the increase in ion current was not proportional to concentration, and above approximately 150 nmol μl^{-1} the ion abundance became nearly constant. Similarly, the $(M + H)^+$ ion abundance of $[^2H_2]$GABA remained constant only below a concentration of total analyte of about 20 nmol μl^{-1} and then decreased thereafter. This effect can be explained as a function of the packing of molecules at the surface of the droplet. In the range below 20 nmol μl^{-1} a tightly packed monolayer has not yet formed and the ion abundances of the two compounds behave independently of each other. However, as the analyte concentration increases beyond this level, two processes occur. As the monolayer begins to form, the analyte molecules replace that of the internal standard at the surface as more analyte is added; that is, the absolute number of molecules of internal standard begins to decrease while that of the analyte increases. At the same time, the

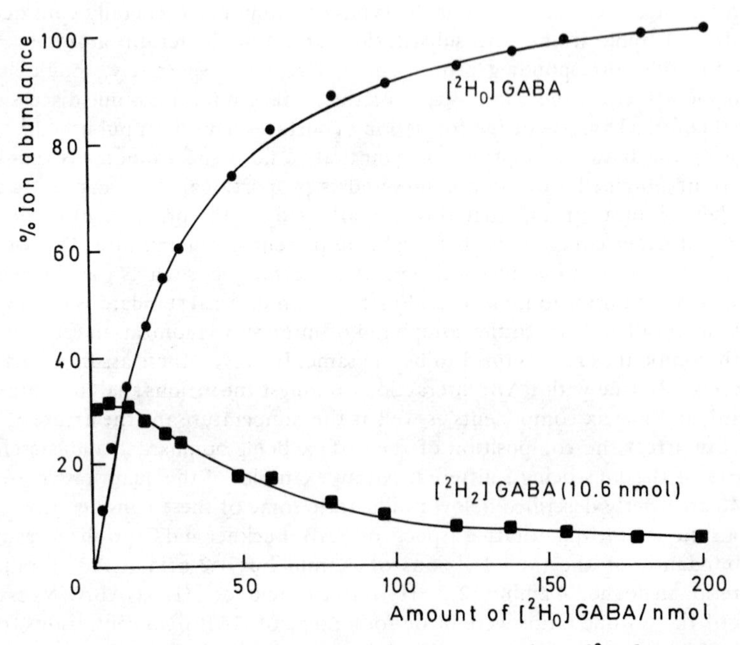

Figure 1 *Effect of increasing the amount of analyte* $[^2H_0]$*GABA on the* $(M + H)^+$ *ion abundances of both the analyte,* $[^2H_0]$*GABA, and the internal standard,* $[^2H_2]$*GABA. The amount of internal standard remained constant at* 10.6 nmol *as the concentration of analyte increased* (Reproduced with permission from *Biomed. Mass Spectrom.*, 1984, **11**, 60)

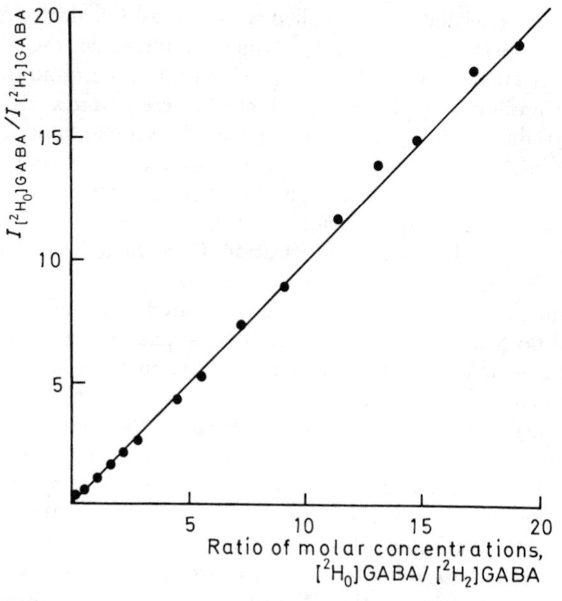

Figure 2 *Calibration curve of the ratios of ion abundances* (I) *of* $(M+H)^+$ *ions* vs. *the ratio of concentrations, analyte,* $[^2H_0]$*GABA, to internal standard,* $[^2H_2]$*GABA. Plotted in this manner, the theoretical slope has a value of* 1.0. *The experimentally determined curve shown above has a slope of* 1.09 *with a correlation coefficient of* 0.9996
(Reproduced with permission from *Biomed. Mass Spectrom.*, 1984, **11**, 60)

surface molecules redistribute themselves into a more tightly packed form, and thus the total number of molecules at the surface increases. This cannot continue indefinitely; finally no further packing of molecules can occur at the surface, and the measured ion abundances thus become constant. Most importantly from the point of view of quantification, however, the relative ion-abundance ratios correspond to the relative concentrations in the bulk solution, as seen in Figure 2. In other words, since the physical properties of the two compounds are for all practical purposes identical, neither one is preferentially concentrated at the surface and the calibration curve is linear. However, since the response of the internal standard is related to a given analyte concentration, it is important to produce a calibration curve for each level of internal standard used.

The neuropeptide leucine-enkephalin labelled with ^{18}O has been used as an internal standard to measure quantitatively levels of enkephalin peptides in biological tissues.[3] The isotope was incorporated into the carboxyl oxygen atoms

[3] D. M. Desiderio, M. Kai, F. S. Tanzer, J. Trimble, and C. Wakelyn, *J. Chromatogr.*, 1984, **342**, 245.

of the peptide by exchange with labelled water. The biological sample was first purified by high-performance liquid chromatography, and the fractions were analysed by FABMS. Because of the large chemical background of the sample, including that produced by glycerol, the ions of interest were analysed by means of MS/MS methods. Thus, on bombardment of the sample, m/z 556 [$(M + H)^+$ for leucine-enkephalin] was selected by the first analyser and subjected to collision-activated dissociation through which fragmentation of the peptide occurred. A C-terminal fragment ion, m/z 336, formed from cleavage of a peptide bond, was used for the quantification. Using linked-field (B/E) scanning methods, the ion abundances of this ion and of its ^{18}O-labelled counterpart produced from the internal standard were measured. The calibration curves so produced for both leucine- and methionine-enkephalin were linear through a range from 0.03 to 2 μg. The sensitivity of the method was approximately 30 ng g^{-1} of tissue.

The N-terminal analysis of proteins using FAB for the quantitative determination of dansylamino acids has been described.[4] The internal standard employed was 4-dansylaminobutyric acid. The mass spectra were characterized by abundant molecular-ion species, namely $(M + H)^+$, $(M[\text{Na salt}] + H)^+$, $(M[Na_2 \text{ salt}] + H)^+$, and $(M[Na_2 \text{ salt}] + Na)^+$ ions. The distribution of abundances amongst these specific ions was dramatically affected by the presence of salt, the pH, and other components in the matrix solution. For example, Figure 3 shows the effect of

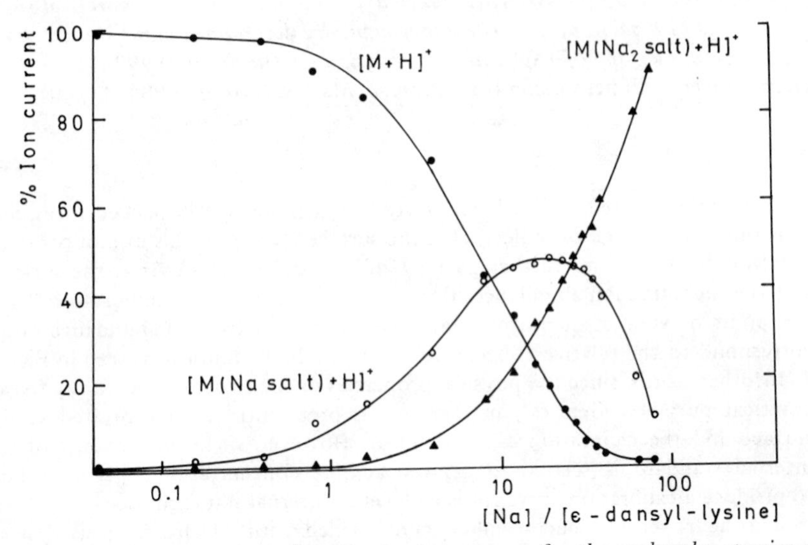

Figure 3 *Percentage of total molecular-ion current for the molecular species of dansyl-lysine as a function of sodium concentration. The dansyl-lysine concentration was* 25mM *in* 50% *glycerol,* pH 1.2
(Reproduced with permission from *Anal. Biochem.*, 1983, **130**, 328)

[4] C. F. Beckner and R. M. Caprioli, *Anal. Biochem.*, 1983, **130**, 328.

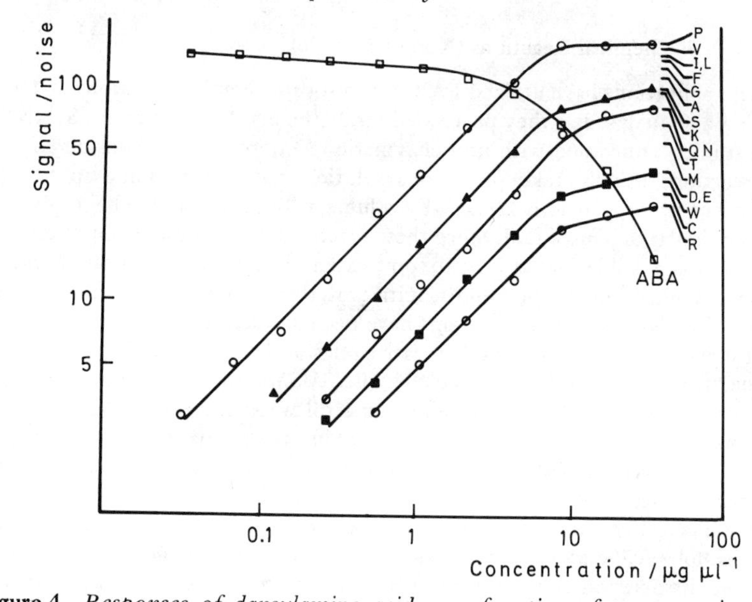

Figure 4 *Responses of dansylamino acids as a function of concentration. The internal standard, dansylaminobutyric acid (ABA), was present at a constant level of* 5.6mM. *For simplicity, only a few of the curves are plotted above, although all had the same shape. Their relative sensitivities are indicated to the right, using the one-letter code*
(Reproduced with permission from *Anal. Biochem.*, 1983, **130**, 328)

increasing the NaCl concentration on the distribution of the molecular species of dansyl-lysine. The dansyl-lysine concentration was kept constant at 25mM in 50% glycerol at pH 1.2. One can thus observe wide variations in the abundances of the $(M + H)^+$, $(M[\text{Na salt}] + H)^+$, and $(M[Na_2\ \text{salt}] + H)^+$ ions. Depending on the application, it may be advisable to sum the ion current for all the molecular species to quantify such compounds accurately. Figure 4 shows the standard curve for the various dansylamino acids with the internal standard, dansylaminobutyric acid, present at a constant level of 5.6mM. For simplicity, only a few of the standard curves have been plotted in the figure; all the curves are nearly identical in shape and vary only in their relative sensitivities as indicated in the right margin of the figure. It can be seen that the ion abundances of the various dansylamino acids are linear with concentration up to about 10 $\mu g\ \mu l^{-1}$ but level off thereafter and that there is a concomitant decrease in the ion abundance of the internal standard. However, as shown earlier, the ratio of the two increases linearly. This methodology was used to determine the N-terminal residue of the P-22 viral tail protein, which has a molecular weight of approximately 75 000 daltons. The results showed that threonine is the N-terminal residue and that, quantitatively, it was present at a level of 0.95 mol mol^{-1} of protein.

3 Chemical Reactions (Non-enzymatic)

Several applications have utilized FAB to investigate chemical reactions that were sampled and analysed as they proceeded inside the mass spectrometer. Several of these studies, concerned with non-enzymatic reactions, will be considered first. The reactions usually take place in a solution matrix containing appreciable amounts of water, generally 20–50% by volume, with the remainder being glycerol or some similar organic liquid. Since these reactions contain a substantial aqueous fraction, it is possible that at least to some extent the ionic distributions measured in the gas phase might be representative of those distributions in the liquid phase. Although only a few systems have been studied, it has been found that under certain conditions these quantitative relationships do apply.

One of the simpler reactions studied by FABMS was the dissociation of protons from a number of weak acids in mixed glycerol/water solutions.[5] Measurements were made of the ion abundances of acid/conjugate base pairs and apparent pK_a values determined. In order to relate these to published values in pure aqueous systems, the measured apparent pK_a values were corrected for the presence of the lower dielectric-constant component (*e.g.* glycerol) and also for ionic strength. For a simple acid-dissociation reaction, the relationship between the pK_a' (the apparent dissociation constant), the pH, and the ratio of acid to base forms of the weak acid at equilibrium can be expressed by the Henderson–Hasselbalch equation. If one considers dilute solutions (<0.2M) of these acids whose pK_a values lie between 3 and 10, then, with respect to ion abundances measured by FAB, the equation can be expressed:

$$(\mathrm{p}K_a')_{g50} = \mathrm{pH} + \log \frac{I_{(HA+H)^+}}{I_{(ANa+H)^+} + I_{(ANa+Na)^+}} \quad (1)$$

where $(\mathrm{p}K_a')_{g50}$ is the apparent dissociation constant of the acid in 50% glycerol (v/v), $I_{(HA+H)^+}$ is the ion abundance of the protonated weak acid, $I_{(ANa+H)^+}$ that of the protonated sodium salt of the conjugate base, and $I_{(ANa+Na)^+}$ that of the sodium addition ion of the sodium salt of the conjugate base. The effect of changing the pH on ion abundances is shown for tris(hydroxymethyl)methyl-aminopropanesulphonate (TAPS) in Figure 5. As expected from considerations of solution chemistry, when the pH is above the pK_a' (pK_a' = 8.4 for this compound), ions of the conjugate base dominate. Below the pK_a' value ions from the protonated acid predominate, and at the pK_a' value both acid and conjugate-base species are approximately equal. Further, when ion abundances are substituted into equation (1) and the appropriate correction is made for the presence of 50% glycerol, the measured value of pK_a' for this acid was 8.37 ± 0.04. Table 1 summarizes the data for measured pK_a (or pK_a') values for a number of different types of weak acids. The average deviation of the measured values from those reported in the literature is less than ±0.05 pK_a units. It is important to note that the method depends upon charge pairing in order to measure quantitatively all the ion species present. When there is an insufficient concentration of sodium

[5] R. M. Caprioli, *Anal. Chem.*, 1983, 55, 2387.

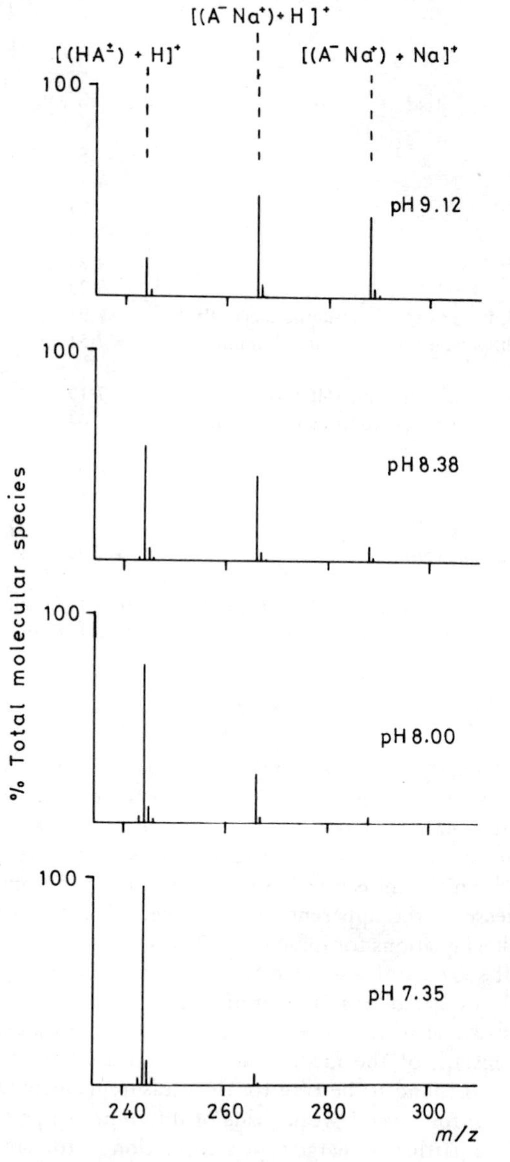

Figure 5 *Effect of* pH *on the molecular species of solutions of tris(hydroxymethyl)methylaminopropanesulphonic acid (TAPS) in* 50% *glycerol/water.* $HA^{\pm}$ *represents the protonated amine, a zwitterion of molecular weight* 243, *and* A^{-} *the conjugate base resulting from the dissociation of the proton*

(Reproduced with permission from *Anal. Chem.*, 1983, 55, 2387)

Table 1 pK_a *of acids determined by FABMS*

Acid	pK_a *Measured*	pK_a *Literature*[a]
Acetic	4.82	4.76
Hexanoic	4.87	4.88
Cacodylic	6.24	6.21
Hippuric	3.72	3.64
β-Phenylpropionic	4.24	4.37
Sulphanilic	3.23	3.17
N-2-Hydroxyethylpiperazinepropanesulphonic (EPPS)[b]	7.91	8.00
N-2-Hydroxyethylpiperazine-*N*'-2-ethanesulphonic (HEPES)[b]	7.51	7.55
3-(*N*-Morpholino)propanesulphonic (MOPS)[b]	7.19	7.15
Tris(hydroxymethyl)methylaminopropanesulphonic (TAPS)[b]	8.44	8.40
Fructose 6-phosphate (pK_2)	6.05	6.11
Glycylglycine (pK_1)	3.12	3.15
Penicillin G	2.67	2.64
Phosphorylethanolamine (pK_2)[b]	7.08	7.00

[a] From D. D. Perrin and B. Dempsy, 'Buffers for pH and Metal Ion Control', Chapman and Hall, London, 1974. [b] Measured and literature values are apparent constants (pK_a')

cations present to charge-pair adequately to the base species, then incorrect ion ratios will be recorded. This is shown in Figure 6 for the pK_1 for glycylglycine, where the apparent pK_a is plotted against the ionic strength through addition of NaCl. Only when enough Na^+ has been added, approximately at an ionic strength of 0.2 mol dm $^{-3}$, do the measured ion ratios reflect the appropriate pK_a' values. The linear decrease in the apparent pK_a is expected and can be described by the DeBye–Hückel equations for ionic interactions.

The effect of solvents of low dielectric constant on the pK_a values of weak acids has been investigated by a number of workers, using mixtures of water and organic solvents such as methanol, ethanol, and dioxane. In these cases, lowering the dielectric constant of the medium leads to an increase in the pK_a' of the acid.[6] This was also found to be true for the measurements obtained by FAB as shown in Figure 7 for several weak acids of different charge types. From this type of curve for a particular charge type a correction factor can be obtained for any desired concentration of glycerol in order to relate the measurement to that in pure water. Qualitatively, this effect is reasonable since the dissociation reaction involves formation of a cation and an anion through removal of a proton from the acid, giving a net increase in the free energy. This free energy should increase

[6] J. T. Edsall and J. Wyman, 'Biophysical Chemistry', Academic Press, New York, 1958, Vol. 1, p. 471.

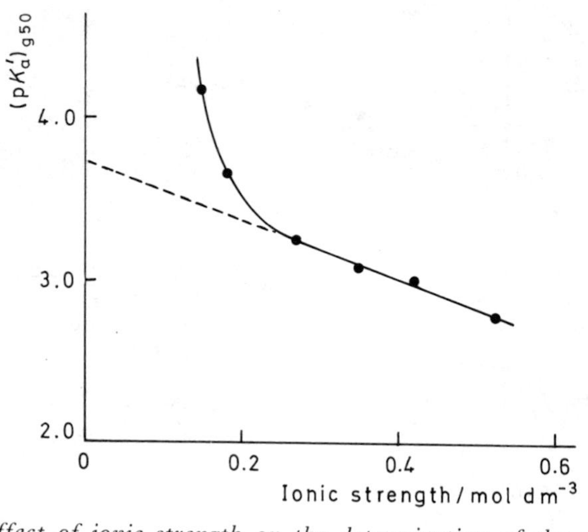

Figure 6 *Effect of ionic strength on the determination of the* pK_a' *of* 0.1M *glycylglycine in* 50% *aqueous glycerol. Addition of aliquots of* 2M NaCl *in 50% glycerol was used to increase ionic strength*
(Reproduced with permission from *Anal. Chem.*, 1983, 55, 2387)

as the dielectric constant of the medium decreases and would therefore result in an increase in the pK_a' of the acid. Quantitatively, this effect is complex. The Born equation[7] has been used in the case of simple systems to relate the difference of the pK_a in two different solvents with the dielectric constants of those solvents. Equation (2) is a modified form[8] of this equation:

$$[(pK_a)_m - (pK_a)_w] = \frac{122n}{r}\left(\frac{1}{D_m} - \frac{1}{D_w}\right) \tag{2}$$

where D is the dielectric constant, n the charge number, and r the average ionic radius, and m and w refer to the medium and water, respectively. This equation can only be considered an approximation since it ignores the free energy associated with ion–solvent interactions and the differences in the measurement of the activity of water in the partially aqueous solutions. Nevertheless, if one plots the apparent average ionic radius measured for a variety of acids *versus* their molecular weight, one obtains a reasonably good correlation as shown in Figure 8. Individual variations from the average curve shown in the figure as a dotted line are perhaps real and reflect the individual chemical properties of the compounds. It is stressed that, at best, values for r calculated from this equation are good only as they relate to one another and not as an absolute measurement of ionic radii.

[7] M. Born, *Z. Phys.*, 1920, **1**, 45.
[8] R. G. Bates, 'Determination of pH, Theory and Practice', Wiley, New York, 1973, p. 211.

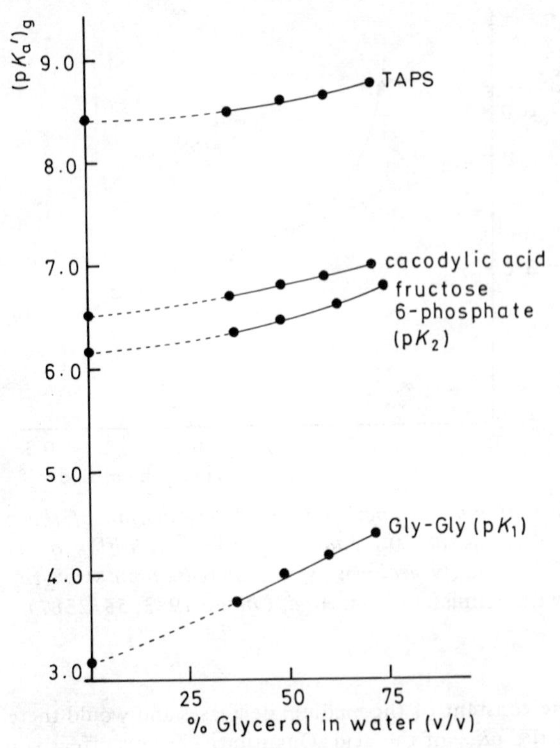

Figure 7 *Effect of glycerol on the* pK_a' *of weak acids determined by FABMS. The ordinate* $(pK_a')_g$ *is the measured (apparent) value determined for solutions containing glycerol at different concentrations. The dotted lines are the extrapolation of the experimentally determined curves to the published* pK_a' *values in water*
(Reproduced with permission from *Anal. Chem.*, 1983, **55**, 2387)

In a study of the FAB analysis of crown-ether/cation complexes, Johnstone, Rose, and co-workers[9,10] considered the mechanism of energy transfer leading to the detection of ionized species during the bombardment process. They considered the partition of kinetic energy transferred from a fast atom to a solution of the ligand and metal, in glycerol/water (1:2), and then compared the theoretical mathematical calculations with ion-ratio measurements made using FAB. They concluded from their mathematical considerations that, regardless of whether the transfer of kinetic energy to a desorbed ion is an equilibrium or a non-equilibrium process, the ion abundances measured by FAB will be proportional to the concentration of those species in solution at the bulk

[9] R. A. W. Johnstone, I. A. S. Lewis, and M. E. Rose, *Tetrahedron*, 1983, **39**, 1597.
[10] M. E. Rose, R. A. W. Johnstone, and C. Longstaff, *Proc. Thirteenth Meet. Br. Mass Spectrom. Soc.*, 1983, 28.

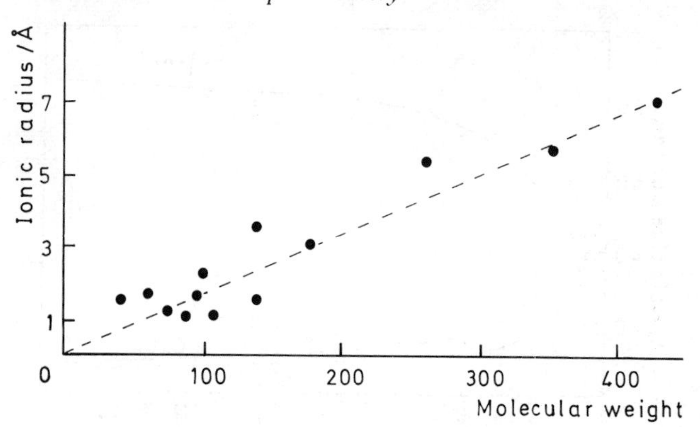

Figure 8 *Plot of the average ionic radii for weak acids determined from the modified Born equation [equation* (2) *in the text]* vs. *the molecular weight of the acids*

temperature of the solution. In one experiment where a solution of 18-crown-6 was titrated with KCl, they found that the measured values and theoretical curve were remarkably similar and were characterized by an initial linear portion that curved sharply and became nearly constant near and after a molar ratio of ligand : metal of 1.0. Moreover, the crossover point of the curves for the increase of the ligand–metal complex and the decrease of the unbound ligand occurred at 50% relative peak height and a molar ratio of 0.5, as expected from a consideration of the stability constants and experimental conditions used.

These investigators also considered the effect of mixtures of metal ions on the binding constants of some crown-ether ligands. If one considers the case for two metal ions, M_1^+ and M_2^+, having the same counter-ion and competing for the same ligand, then the stability constants may be expressed by equations (3):

$$K_1 = \frac{r}{(c-r-s)(a-r)} \quad \text{and} \quad K_2 = \frac{s}{(c-r-s)(b-s)} \tag{3}$$

where r is the concentration of the ligand–M_1 complex, s is that of the ligand-M_2 complex, and c, a, and b are the initial concentrations of crown ether, M_1^+, and M_2^+, respectively. Since c, a, b, K_1, and K_2 can be known for a given study, then it is possible to calculate the theoretical concentrations of complexes, *i.e.* r and s. The authors applied this approach to the study of the formation of complexes of crown ethers with K^+, Na^+, and Cs^+. As an example, a solution of dicyclohexyl-18-crown-6 in glycerol was treated with aliquots of an aqueous glycerol solution containing varying amounts of KCl, NaCl, and CsCl. Using FABMS, the ion abundances at m/z 395, 411, and 505, corresponding to ligand plus Na^+, K^+, and Cs^+, respectively, were measured. Figure 9b shows these experimental data, uncorrected for possible differences in sensitivities of the complexes and the effect of glycerol in the solution, and Figure 9a shows the calculated relative

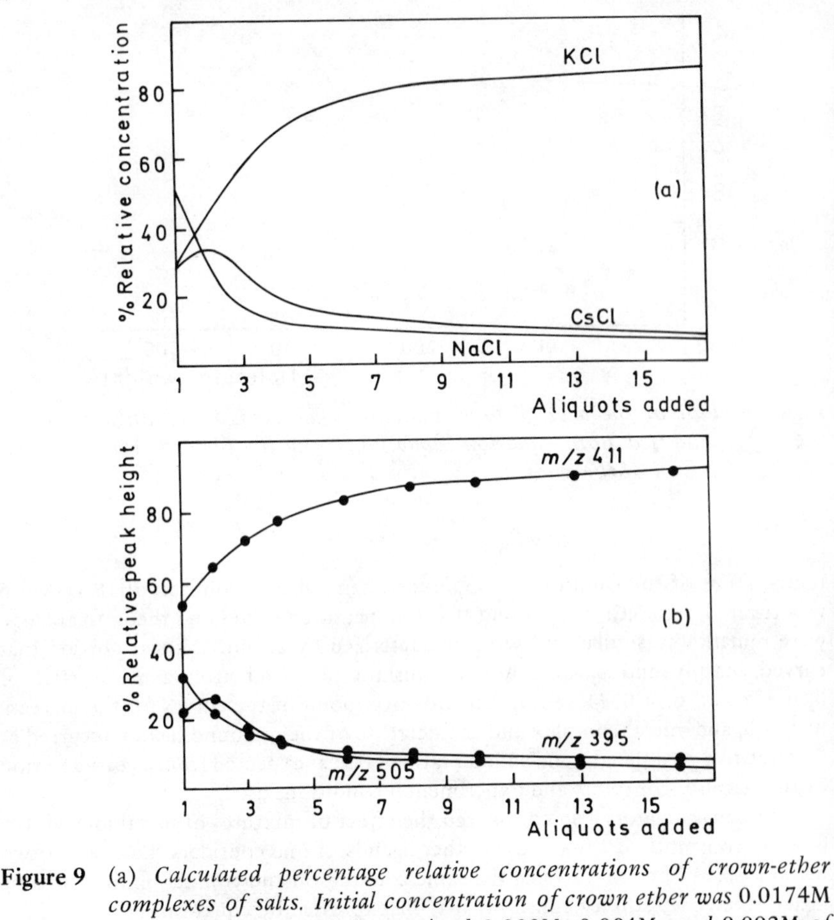

Figure 9 (a) *Calculated percentage relative concentrations of crown-ether complexes of salts. Initial concentration of crown ether was* 0.0174M *and each aliquot added contained* 0.002M, 0.004M, *and* 0.002M *of* KCl, NaCl, *and* CsCl, *respectively.* (b) *Observed percentage of above complexes with* K^+ (m/z 411), Na^+ (m/z 395), *and* Cs^+ (m/z 505)
(Reproduced with permission from *Tetrahedron*, 1983, **39**, 1597)

percentage concentrations of these complexes. Comparison of the two curves shows that these data are in excellent agreement.

In further work, Johnstone and Rose[11] used FAB to study the selectivity of formation of complexes between macrocyclic ligands and metal cations in aqueous glycerol solutions. Glycerol/water solutions (2 : 1 v/v) were prepared containing the iodides of several cations, each at 5.0mM concentration. Tetra-n-butylammonium chloride was used as an internal standard at 0.06mM concentration, and the appropriate ion abundances were measured. The data for the

[11] R. A. W. Johnstone and M. E. Rose, *J. Chem. Soc., Chem. Commun.*, 1983, 1268.

Table 2 *Relative abundance of [ligand/metal]$^+$ complexes (from ref.* 11*)*

Macrocyclic ligand[a]	K^+		Rb^+		Cs^+	
	Expt.	*Calc.*	*Expt.*	*Calc.*	*Expt.*	*Calc.*
15C5	59	34	31	27	10	39
18C6	57	65	36	27	7	8
DCH18C6	72	67	24	25	4	8
DB24C8	37	30	37	37	26	33
C222	75	77	25	23	0	0.1

[a] See original paper for complete chemical names of crown ethers

experimentally measured and calculation concentrations are given in Table 2. Values given in the table have not been corrected for differences in sensitivities for the various complexes or for the effect of glycerol in the solutions. Nevertheless, it can be seen that the experimental data closely parallel abundances calculated from published stability constants. The authors also measured the cation selectivity of the macrotetralide antibiotic nonactin for Na^+ and K^+. The measured complex ratio of [ligand–Na]/[ligand–K] was 1:17. Published ratios of affinity constants are 1:19 (30 °C in methanol) and 1:23 (30 °C in ethanol).

Kalinoski[12] investigated the use of FAB as a probe of condensed-phase reaction chemistry in the palladium(II)-mediated reaction of aryl and heteroaryl mercuric acetates with chiral glycals. The courses of these reactions are complex, with several postulated mechanisms involving undetected metal-containing complexes. Various metals were added to the reaction mixture and analysed directly by FABMS both to identify the nature of the complexes formed and also to follow the extent of the reaction. The results suggest the presence of σ-bonded palladium adducts containing a *cis*-β-hydrogen atom and the intermediacy of arylpalladium compounds formed by transmetallation in these reactions. Other complexes, not necessarily leading to the formation of products, were also identified by FAB.[12]

The oxidation–reduction processes that occur on bombardment of glycerol solutions with both atoms and ions have been investigated.[13] Although this paper primarily deals with secondary-ion mass spectrometry (SIMS) using a caesium-ion gun, preliminary results with FAB are also discussed. The investigators used various compounds, including inorganic salts, organometallic compounds, and cationic dyes, and have demonstrated that a reduction process occurs for these compounds resulting in a hydrogen-atom attachment to the ion molecular complex. For the inorganic salts, a lowering of the oxidation state of the cation takes place simultaneously with the protonation of an ion, whereas the reduction of a quinoid to a semiquinoid species is observed for the dyes. It is noted that the reduction process that occurs is inherent in the fact that the substance is

[12] H. T. Kalinoski, Ph.D. dissertation, Lehigh Univ., 1984.
[13] G. Pelzer, E. De Pauw, D. V. Dung, and J. Marien, *J. Phys. Chem.*, 1984, **88**, 5065.

dissolved in glycerol and is completely different from what is observed when the sample is in the solid state. It is likely that the reduction process occurring in FAB and SIMS of glycerol solutions could be governed by a simple redox equilibrium between hydrogen atoms produced by bombardment and the oxidized species present in solution. The experimental results presented can be interpreted qualitatively on the basis of the usual standard redox potential scale. A more quantitative treatment awaits the establishment of such a redox scale in glycerol.

Negative-ion FAB was used by Rose and co-workers[14] to investigate the reaction of boronic acids with triols, sugars, and nucleosides on the probe inside the mass spectrometer. Boronic acids react with trifunctional compounds with the appropriate configuration to give 5- to 8-membered-ring boronate cage compounds that are negatively charged (Scheme 1). The reaction matrix consisted of a

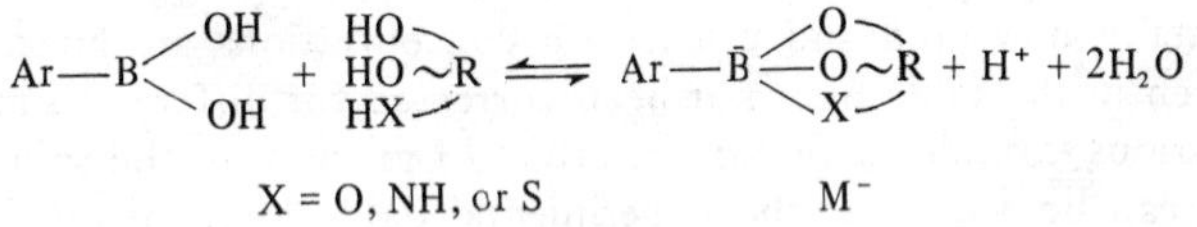

$$\mathrm{Ar{-}B(OH)_2} + \mathrm{HO}\sim\mathrm{R}(\mathrm{HO})(\mathrm{HX}) \rightleftharpoons \mathrm{Ar{-}\bar{B}(O)(O)(X)}\sim\mathrm{R} + \mathrm{H^+} + 2\mathrm{H_2O}$$

X = O, NH, or S $\qquad$ M^-

Scheme 1

boronic acid and, in the case of liquid substrates like glycerol, the reactant, which acted also as the solvent, or, in the case of solid substrates like carbohydrates, the reactant and poly(ethylene glycol) 200 as solvent. For example, when benzeneboronic acid was dissolved in glycerol and analysed by FAB, the negatively charged caged complex of the two was formed. Similar complexes were observed when the solvent was changed to thioglycerol, 3-aminopropane-1,2-diol, and triethanolamine. The authors employed this complex formation using 4-tolueneboronic acid to characterize glycosidic natural products; the extent of the reaction and the stability of the M^- ion of the complex are dependent on the orientation of the reacting hydroxy groups. In the case of the aldopentoses, formation of the complex produced M^- species of varying stability. The complexes of D(−)-ribose and D(−)-lyxose were much more stable, *i.e.* gave a much greater M^- ion abundance than that for D(−)-arabinose and D(+)-xylose. This is explained on the basis that the first two can, through their 6-membered-ring forms, react through stable chair conformations to give strain-free boronate cage compounds, whereas the complexes of the latter two pentoses would be required to adopt unfavourable boat conformations.

The same approach[14] was also applied to complex formation of the nucleosides adenosine and 2′-deoxyadenosine. It was concluded that the method has use in configurational analysis of the appropriate polyfunctional natural products, in obtaining useful FAB spectra of neutral compounds of this nature, and in increasing the sensitivity with which such compounds can be detected. It was further suggested that, since the data from complex formation of the aldopentoses appear to correlate well with known affinity-chromatographic behaviour of these

[14] M. E. Rose, C. Longstaff, and P. D. G. Dean, *Biomed. Mass Spectrom.*, 1983, **10**, 512.

complexes, it is possible to measure affinity constants under certain conditions using FABMS. Such a study would involve, for example, the competition between a given aldopentose, a fixed amount of glycerol, and 4-tolueneboronic acid. The relative affinity constants of the sugars for the boronic acid can be determined by measuring the appropriate ion abundances of each M^- of the sugar complexes relative to that of the glycerol complex. For the aldopentoses, the data show a ratio of affinities for D-ribose, D-lyxose, D-xylose, and D-arabinose of 100:1.8:1.1:1.9. The only published comparison available is for complexes of these sugars with boric acid, giving the respective ratios 100:0.4:0.1:0.01.

Perhaps the most important conclusions to be drawn from the work cited thus far have to do with the FAB process itself. The data clearly demonstrate the quantitative relationship between the ionic equilibrium in the liquid solution and the ionic distribution in the gas phase following atom bombardment under certain conditions. Regardless of the actual detailed mechanism leading to the ejection of ions into the gas phase, it is apparent that the process can be sensitive to subtle ionic changes in the medium. This suggests that the water/glycerol support solution usually employed in FAB can be used for the study of a variety of reactions normally carried out in aqueous solutions.

4 Enzyme Reactions

The reactions of enzymes with substrates have been followed using FABMS. The general methods used fall into two general categories, the first where samples are removed from batch reactions at timed intervals and these aliquots are subsequently analysed, and the second where the enzyme reaction actually takes place on the probe within the mass spectrometer and the products are analysed in real-time. Each approach has advantages in certain applications. For example, if the point of interest in a study is to monitor the products of the enzymatic reaction for the determination of structural aspects of the substrate, then real-time analysis is especially useful since it utilizes very small amounts of substrate, is extremely rapid, and can provide many sampling times. In these applications it is generally not of concern whether the kinetics of the reaction are altered as a result of the bombardment process or the glycerol that is present as long as the reaction proceeds over the time period for which the sample is in the spectrometer. However, if the interest in the study is to gather kinetic data from the reaction, especially when it is to be compared with data obtained from other techniques using purely aqueous solutions, then the batch reaction with sampling at intervals would be preferred since it could be performed under completely aqueous conditions and also allows for the addition of activators or inhibitors as the reaction is proceeding.

Hydrolysis of Substrates in Real-time. – Smith and Caprioli[15] demonstrated that enzymes could remain active in glycerol/water mixtures on the probe tip within the mass spectrometer. In an initial experiment, the enzyme trypsin was mixed with 0.05M Tris buffer (pH 8, containing 70% glycerol), coated onto the probe tip, and inserted into the mass spectrometer. The solution on the probe tip was

[15] L. A. Smith and R. M. Caprioli, *Biomed. Mass Spectrom.*, 1983, **10**, 98.

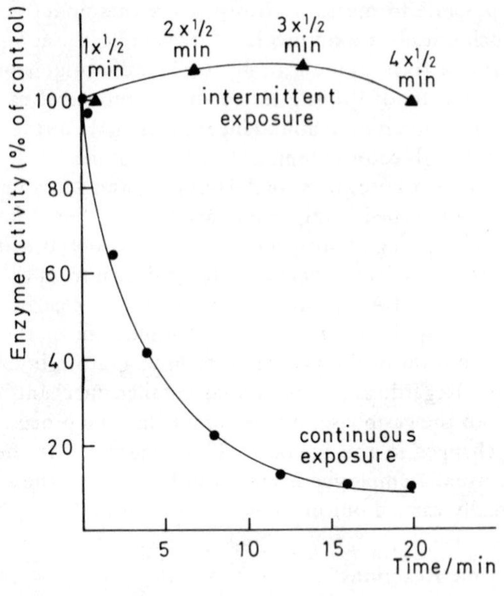

Figure 10 *Percent activity of control when samples of trypsin are exposed under low vacuum to xenon FAB for varied lengths of time. The circles denote samples exposed to a continuous beam, and the triangular points those exposed for* 0.5 min *periods between which the probe was maintained inside the mass spectrometer*
(Reproduced with permission from *Biomed. Mass Spectrom*., 1983, **10**, 98)

exposed to the neutral-atom beam for varying periods of time then washed into a larger volume of buffer, and the trypsin activity was assayed using standard biochemical methods. Exposure times differed from various periods of continuous exposure to multiple 0.5 minute exposures as the probe remained under vacuum. The results are given in Figure 10. Continuous exposure led to the rapid inactivation of the enzyme, with over 50% loss of activity occurring in the first 3–4 minutes. Intermittent exposure of short duration, however, gave no inactivation of the enzymatic activity even after 20 minutes. The slight increase in activity shown in the figure is probably real and is due to a warming of the solution during these short periods of bombardment. After 20 minutes the activity could be expected to fall since the glycerol content would be expected to be very high (>80%), thereby inhibiting the enzyme. Figure 11 shows the action of trypsin on one of its normal assay substrates, *p*-toluenesulphonyl-L-arginine methyl ester (TAME), at m/z 343 to give the free acid at m/z 329. The course of this reaction was followed over 24 minutes (this rate can be easily changed by altering the amount of enzyme added). This experiment was repeated for a number of enzymes including dipeptidyl aminopeptidase I, chymotrypsin,

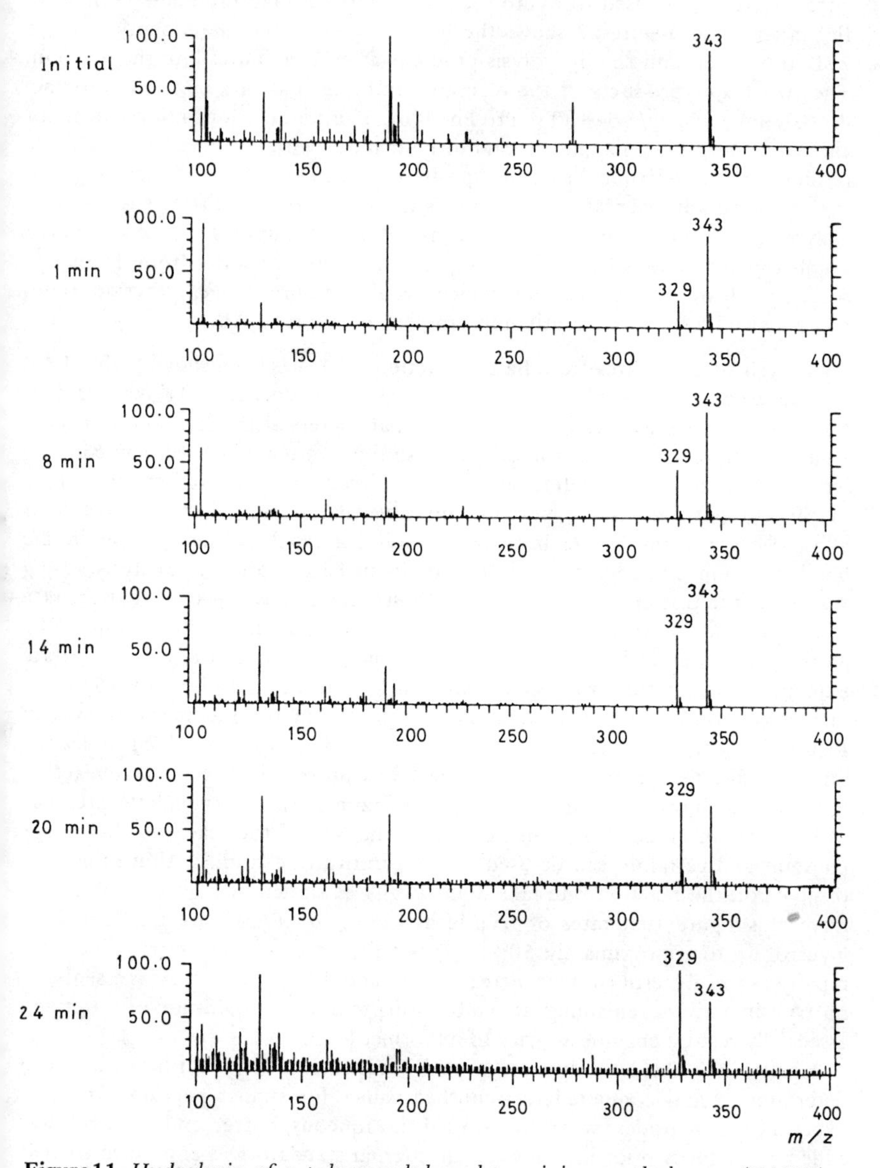

Figure 11 *Hydrolysis of* p-*toluenesulphonyl*-L-*arginine methyl ester (TAME) at* m/z 343 *to the product* p-*toluenesulphonyl*-L-*arginine at* m/z 329 *by trypsin as monitored by FABMS*

(Reproduced with permission from *Biomed. Mass Spectrom.*, 1983, **10**, 98)

carboxypeptidase Y, and proline-specific endopeptidase.[16] For example, the latter protease was used to hydrolyse β-casomorphin (Tyr-Pro-Phe-Pro-Gly-Pro-Ile) in real-time. Figure 12 shows the FAB analysis of the intact peptide at m/z 791 (top panel) and the hydrolysis products after 17 minutes into the reaction. The two major products of the reaction can be seen at m/z 678 (Tyr-Pro-Phe-Pro-Gly-Pro) and m/z 524 (Tyr-Pro-Phe-Pro). The proline in position two is not cleaved because the enzyme appears to be inactive with peptides having a single residue on the N-terminal side of a prolyl residue. One of the disadvantages of the real-time approach at the moment is the fact that it is difficult to keep an enzyme active on the probe for more than about 20 minutes even with limited bombardment periods because of the preferential loss of water from the matrix solution. However, applications which would require longer reaction times could be well served by a continuous sampling device for FAB.

Hydrolysis of Substrates from Batch Reactions. – It has been shown[17] that FAB can be used to measure accurately the rates of enzymatic reactions and that these rates can be used to calculate kinetic parameters of the reaction. For these studies, removal of aliquots for analysis with time from a batch reaction containing the assay solution is preferable because it circumvents the problem of changes in the composition of the reaction mixture that occur when the sample is allowed to stay inside the instrument. Glycerol can be either present in the batch reaction or added to each aliquot immediately prior to analysis. Using trypsin as the test enzyme and the TAME substrate, it was shown that the rate of enzyme catalysis measured by FAB using a solution already containing 67% glycerol was identical to that determined spectrophotometrically in a separate experiment using the same solution, giving a rate of hydrolysis of 164 μmol min^{-1} mg^{-1} of enzyme. The rate plot so generated by FABMS is shown in Figure 13. This is contrasted with the rate of hydrolysis determined under the same conditions, except that no glycerol was present in the original reaction mixture, which was 221 μmol min^{-1} mg^{-1} of enzyme. Thus, although the presence of this level of glycerol has some effect on the rate of the reaction, that rate is constant and certainly can be used in comparative rate studies. Above this level of glycerol, the reaction rate falls dramatically as shown in Figure 14. It is seen from this figure that rates of tryptic hydrolysis of TAME are unaffected by glycerol up to approximately 50% (v/v) but then the enzyme activity decreases rapidly as the glycerol content increases until about 90% where there is essentially no trypsin activity remaining. It is interesting to note that addition of water will essentially restore enzyme activity to its former level.

Kinetics for the hydrolysis of peptide substrates by trypsin have also been determined[18] and compared to published values for similar peptides. In these experiments, peptides were hydrolysed in aqueous buffer and glycerol was added to aliquots prior to analysis. An internal standard was employed so that the various ion abundances measured from each aliquot could be related to each

[16] R. M. Caprioli, L. A. Smith, and C. F. Beckner, *Int. J. Mass Spectrom. Ion Phys.*, 1983, **46**, 419.
[17] L. A. Smith and R. M. Caprioli, *Biomed. Mass Spectrom.*, 1984, **11**, 392.
[18] R. M. Caprioli and L. A. Smith, unpublished data.

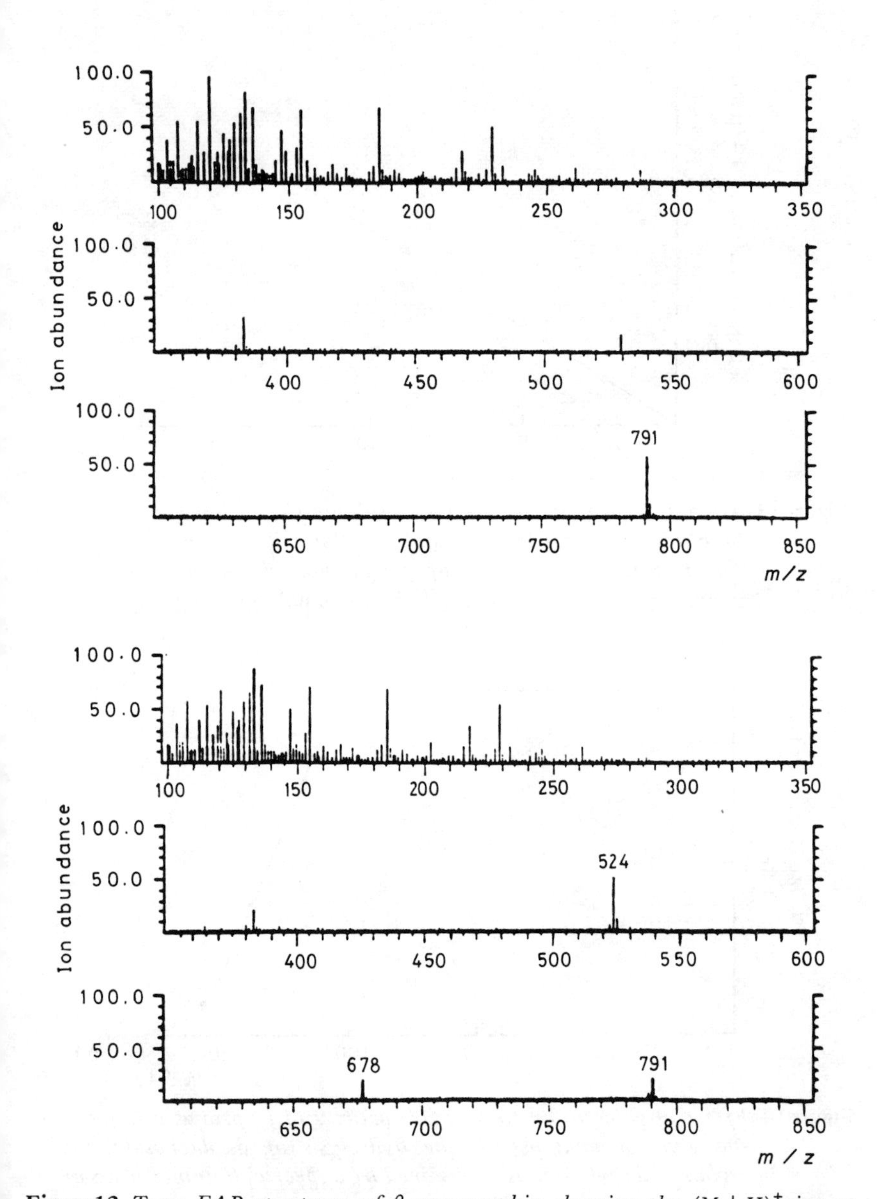

Figure 12 Top: *FAB spectrum of β-casomorphin showing the* $(M+H)^+$ *ion at* m/z 791. Bottom: *spectrum taken* 17 min *into the real-time hydrolysis of the peptide by proline-specific endopeptidase on the probe inside the mass spectrometer. Ions at* m/z 678 *and* 524 *correspond to peptides produced during the hydrolysis as described in the text*

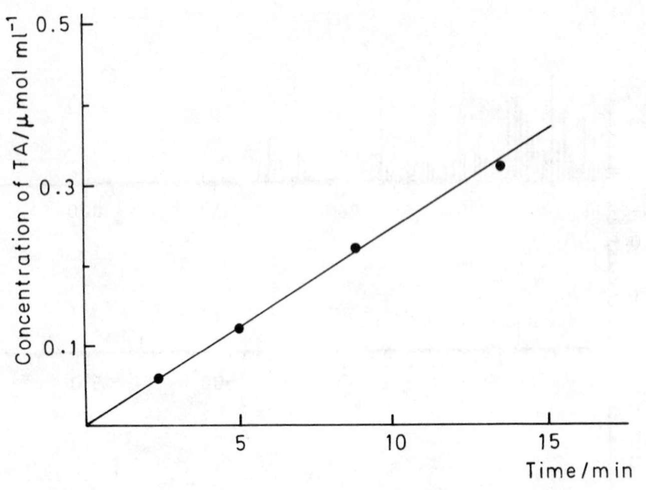

Figure 13 *Tryptic hydrolysis of* p-*toluenesulphonyl*-L-*arginine methyl ester monitored by FAB from a batch reaction with removal of aliquots at timed intervals. The rate of production of* p-*toluenesulphonyl*-L-*arginine (TA) was determined to be* 164 μmol min^{-1} mg^{-1} *of enzyme*

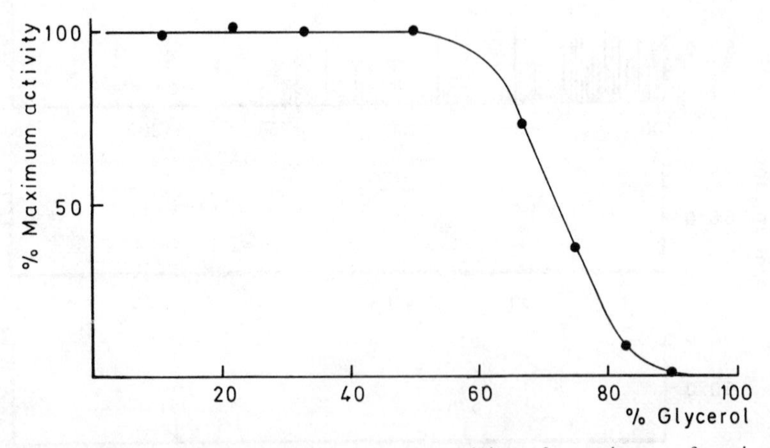

Figure 14 *Effect of glycerol on the enzyme activity of trypsin as a function of the glycerol content for the hydrolysis of* p-*toluenesulphonyl*-L-*arginine methyl ester as determined by a spectrophotometric assay*

other. Figure 15 shows the Lineweaver–Burke plot of the rate data determined over a range of substrate concentrations for the peptide Phe-Arg-Ser-Val. The X-intercept of the curve is equal to $-1/K_m$ and the Y-intercept to $1/V_{max}$. From these data, k_{cat} can be calculated and a measure of the activity of the enzyme for this substrate expressed as the ratio k_{cat}/K_m. Table 3 gives a comparison of

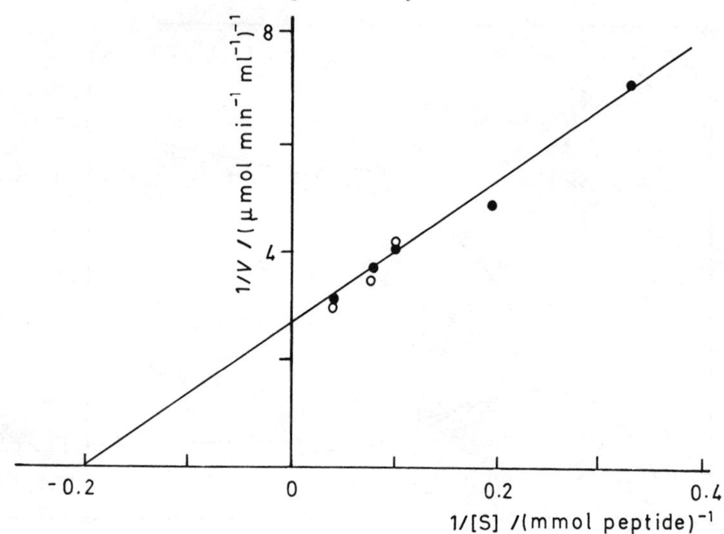

Figure 15 *Lineweaver–Burke (double reciprocal) plot of the rates of trypsin hydrolysis of Phe-Arg-Ser-Val* vs. *the initial substrate concentrations. The rates determined by FABMS are noted as closed circles, and those by quantitative GC/MS by open circles*

Table 3 *Kinetic constants for tryptic hydrolysis of peptides (from ref.* 18*)*

Peptide	*Method*[a]	K_m/mM	k_{cat}/s^{-1}	$k_{cat}/K_m/M^{-1} s^{-1}$
Met-Arg-Phe-Ala	FABMS	1.9	7.1	3622
Phe-Met-Arg-Phe(NH_2)	FABMS	5.9	69.0	11 600
Phe-Arg-Ser-Val	FABMS	5.0	99.2	19 840
Gly-Gly-Arg-Gly(NH_2)[b]	SPEC	–	–	3500
Ac-Gly-Arg-Val(OMe)[c]	SPEC	3.6	20	5600
Ala-Val-Arg-Gly(NH_2)[b]	SPEC	–	–	18 000
Leu-Val-Arg-Gly(NH_2)[b]	SPEC	–	–	21 600

[a]Abbreviations: FABMS fast-atom-bombardment mass spectrometry, SPEC spectrophotometry using ninhydrin. [b]From R. K. H. Liem and H. A. Scheraga, *Arch. Biochem. Biophys.*, 1974, **160**, 333. [c]From A. P. Lobo, J. D. Wus, S. M. Yu, and W. B. Lawson, *Arch. Biochem. Biophys.*, 1976, **177**, 235

kinetic data for several peptides determined by FAB in this manner together with data for other peptides determined by classical biochemical methods.

Conditions used in FABMS are ideal for sampling of enzymatic reactions in order to determine reaction products. This has been shown by several workers with respect to peptide sequencing. Self and Parente[19] used both leucine amino-

[19] R. Self and A. Parente, *Biomed. Mass Spectrom.*, 1983, **10**, 78.

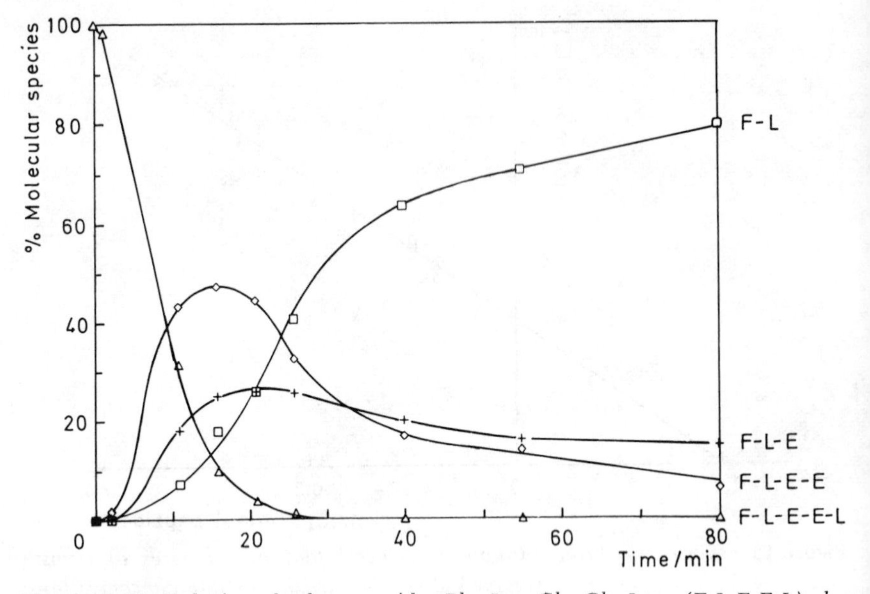

Figure 16 *Hydrolysis of the peptide Phe-Leu-Glu-Glu-Leu (F-L-E-E-L) by carboxypeptidase Y as monitored by FAB. The ions recorded were the* $(M+H)^+$ *ions for each of the newly formed peptides as the enzyme sequentially removed residues from the C-terminus*

peptidase M and carboxypeptidase Y to verify the sequences of the terminal portions of tryptic peptides of the coat protein of Papaya Mosiac Virus. The digestion was carried out in a volatile buffer with approximately 50 nmol of substrate and limited amounts of enzyme. Aliquots were removed with time, taken to dryness, mixed with glycerol, and analysed by FAB. Their data show that sequence information can be readily obtained in this manner and, further, that it is possible to use this method of analysis to study the relationship between the susceptibility of a particular peptide bond to enzymic cleavage and the effect of the proximity of other residues to that bond.

In a similar approach, Caprioli[20] showed that proteases could be used to determine polypeptide sequences by analysing directly aliquots of the digests by FABMS. For example, the hydrolysis of a pentapeptide with carboxypeptidase Y was allowed to take place in 0.05 mM Tris buffer, pH 5.5, containing 0.022 μmol μl^{-1} of peptide. The $(M + H)^+$ ions of the substrate and of the newly formed products were monitored with time. Figure 16 shows the hydrolysis of Phe-Leu-Glu-Glu-Leu (F-L-E-E-L). The various newly generated molecular species are readily observed by FABMS as a series of new ions appearing with time, each ion decreased in mass by an amount corresponding to the molecular weight of the appropriate amino acid minus 18 mass units. As expected,

[20] R. M. Caprioli, unpublished data.

the original protonated molecules of F-L-E-E-L are seen to decrease rapidly; those of F-L-E-E maximize first then decrease, those of F-L-E are next to maximize, and finally those of F-L. The sequence of this peptide is thus easily obtained. The use of the exopeptidases is particularly interesting because they provide the ability to sequence unambiguously a peptide without time-consuming and peptide-consuming derivatization reactions and without having to use mixed-isotope acetylating agents to tag the N-terminus in order to identify that series of ions in the spectrum. It remains to be determined what the practical limit in size of a peptidic substrate would be for an exopeptidase such as this before incomplete digestion or other factors might terminate the sequence procedure. However, a particular advantage of this method is that, in following the appearance of newly formed molecular species, one observes a series of ions of steadily decreasing mass. One is not concerned with the problems encountered with other methods that follow released residues where greatly differing rates of bond cleavage for specific residues and repeating residues lead to ambiguities in the reconstruction of the sequence.

Enzyme Equilibrium Measurements. – One of the enzyme reactions that has been well studied over the years using common biochemical methods is that of enolase, especially in terms of the effects of pH and metal ions on the equilibrium constant.[21] Enolase catalyses the reversible dehydration of 2-phosphoglyceric acid (2PG) to form phosphoenolpyruvate (PEP), as shown in Scheme 2. The

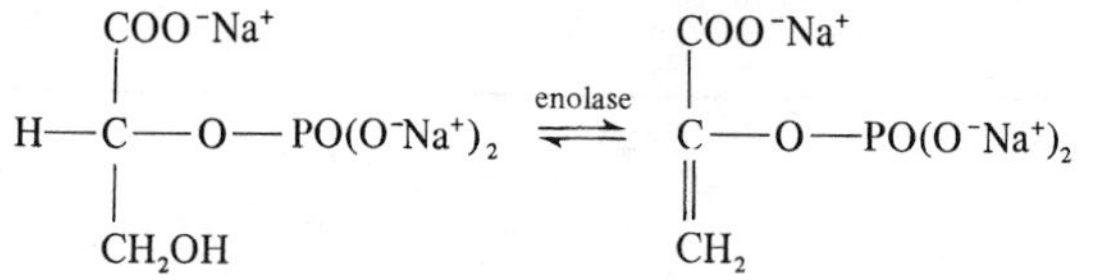

$$\mathrm{H{-}C(COO^-Na^+)(CH_2OH){-}O{-}PO(O^-Na^+)_2 \;\underset{}{\overset{\text{enolase}}{\rightleftharpoons}}\; C(COO^-Na^+)(=CH_2){-}O{-}PO(O^-Na^+)_2}$$

Sodium 2-phosphoglycerate (2PG) — Sodium phosphoenolpyruvate (PEP)

Scheme 2

enzyme is active at physiological pH only with the completely ionized forms of the substrates as shown in the scheme. Equation (4) is the pH-independent equilibrium expression for this reaction:

$$K_{eq} = \frac{[PEP^{3-}]}{[2PG^{3-}]} \tag{4}$$

However, classical biochemical methods cannot distinguish between the various ionized forms of these compounds and so measure the total concentration, that is the sum of all the ionized forms. Thus, the equilibrium constant measured in this way is an apparent constant [equation (5)]:

$$K_{eq(app)} = \frac{[PEP]_{total}}{[2PG]_{total}} = \frac{[PEP^{3-}] + [PEP^{2-}] + [PEP^{-}] + [PEP]}{[2PG^{3-}] + [2PG^{2-}] + [2PG^{-}] + [2PG]} \tag{5}$$

[21] F. Wold and C. E. Ballou, *J. Biol. Chem.*, 1975, **227**, 301.

At physiological pH this simplifies to equation (6):

$$K_{eq(app)} = K_{eq} \frac{(1 + [H^+])/K_{3(PEP)}}{(1 + [H^+])/K_{3(2PG)}} \quad (6)$$

Therefore, to calculate the pH-independent constant K_{eq} one must measure the apparent equilibrium constant from $[PEP]_{total}$ and $[2PG]_{total}$, the hydrogen-ion concentration, and the dissociation constants, K_3, for both compounds under the experimental conditions employed.

In contrast, the molecular specificity of FAB allows the appropriate ions in the equilibrium expression to be determined directly.[20] For example, data for the enolase reaction are shown in Figure 17. The reaction was initiated by adding 13.6 units of enzyme to a 0.15M PEP solution at pH 8.05 containing 4mM $MgSO_4$, 25mM imidazole, and 50% glycerol. The molecular species of the PEP trianion, $[M(Na_3\,salt) + H]^+$ and $[M(Na_3\,salt) + Na]^+$, and those for the 2PG trianion were monitored throughout the course of the reaction. A standard curve relating the ion abundance ratios to the actual millimolar concentration ratios was used for quantification. Therefore, the pH-independent equilibrium constant was obtained directly from equation (4) using FABMS and gave $K_{eq} = 5.0$. This is in very close agreement with that reported in the literature, 5.4, using similar conditions,[21] although it is difficult to compare these since the presence of metal

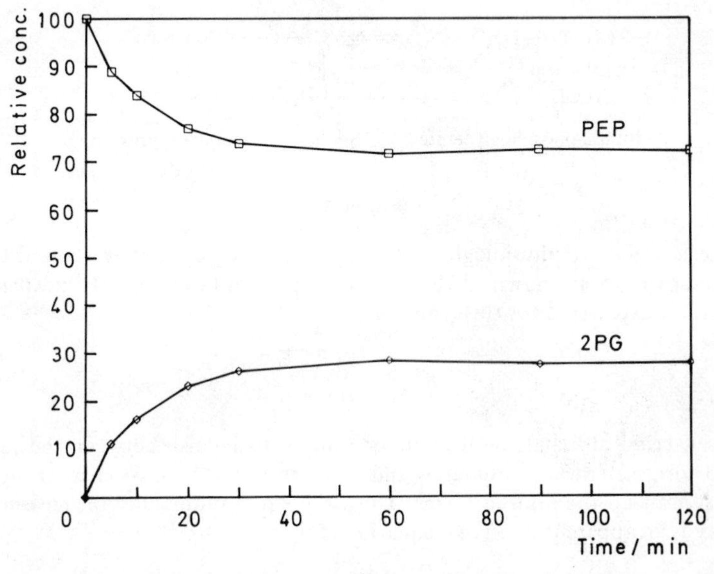

Figure 17 *The formation of the equilibrium condition for the reaction of enolase with phosphoenolpyruvate (PEP) to form 2-phosphoglyceric acid (2PG). The $[M(Na_3\,salt) + H]^+$ and $[M(Na_3\,salt) + Na]^+$ ions were monitored with time for both compounds*

ions, ionic strength, and temperature may have a significant effect on the measured value of this constant. Further, the above constants are metal-dependent constants; the metal- and pH-independent equilibrium constant can be calculated if the binding constant for the metal is known. Nevertheless, this study underscores the analytical power and specificity of FABMS in being able to quantify selected ionic species in a mixture of several ionic forms of a molecule produced by the dissociation of protons from groups within that molecule.

5 Conclusion

The impressive increase in the use of FABMS in many fields of analysis in just the few years since its introduction can be attributed to its ability to produce mass spectra from compounds that are ionic, polar, or otherwise termed intractable with regard to more established ionization methods. The specific advantages of the method are several; certainly one is that it is relatively inexpensive and extremely easy to install and use in existing instruments. Of great practical importance is the fact that FAB lends these attributes without the use of time-consuming and sample-costly derivatization processes that invariably produce side-products and complicate the mixture further. The spectra are characterized by abundant molecular species often with sufficient fragmentation to provide structural information. Further, the fact that a liquid-sample solution is employed (glycerol or other suitable solvent) provides a medium that allows the surface of the sample droplet to be constantly renewed with analyte as the surface layers are sputtered by the bombardment process. This leads to the production of spectra over a relatively long period of time, 30 minutes or more in certain cases, using only a few micrograms of sample.

One additional important advantage of the liquid-sample medium used by FAB is that it allows solution chemistry to take place as an integral part of the analytical process. The work described in this chapter utilizes this new ionization technique as a direct probe into the molecular dynamics occurring in the condensed phase. Several studies cited above have established that the ionic distribution formed in the gas phase through the atom-bombardment process can indeed be an accurate reflection of the ionic distribution in the liquid phase. Although there are many studies that must yet be done to understand fully the dynamics of the process and the conditions that must be maintained to ensure the integrity of the measurements, the analytical implications for many branches of chemistry, especially biochemical and medical applications, are enormous. There is little doubt that FABMS is making a major contribution to the biotechnological approach that is employed to characterize better our environment at the molecular level.

9

Gas Chromatography/Mass Spectrometry and High-performance Liquid Chromatography/Mass Spectrometry

BY M. E. ROSE

1 Introduction

For the first time since this title of periodicals began, the number of papers published in the field of gas chromatography/mass spectrometry (GC/MS) has fallen in the review period (June 1982–June 1984). An indication of the downward trend is shown by the number of entries each year in 'Gas Chromatography–Mass Spectrometry Abstracts',[1] which approaches comprehensive abstracting in GC/MS: 1020 (1980), 1049 (1981), 1156 (1982), and 1118 (1983). The decreasing publication rate is undoubtedly a good thing for a technique with the maturity of GC/MS. Even with the reduced amount of literature – there are still over 2000 references in the review period – the sheer amount of derivative material makes the search for the truly innovative physically demanding. The strategy adopted in this chapter in the previous Report[2] is retained here: the full range of applications of GC/MS is presented, but the citations within each category are restricted to those that advance methodology, represent novel applications, and/or appear interesting for the readership in the opinion of the Reporter. Mundane applications are omitted.

The same comments do not apply to high-performance liquid chromatography/mass spectrometry (HPLC/MS or generally LC/MS) since it cannot be described as a mature method. In this review period much research effort has been expended on improving the performance and extending applications of existing instrument designs. The literature shows little evidence of totally new approaches. However, the comment made in Volume 7 of this title[2] that no one interface is clearly superior for all LC/MS applications is still generally valid.

Introduced into the last Report[2] was a section on a specialized topic (metastable-ion techniques in conjunction with GC/MS and LC/MS) involving more detailed and critical coverage than is possible in the main body of the chapter. The practice is continued here, with application of combined chromatography/mass spectrometry to inorganic and organometallic compounds being the selected topic.

[1] 'Gas Chromatography–Mass Spectrometry Abstracts', P.R.M. Science and Technology Agency Ltd., London, 1982–1984, Vol. 13–15.

[2] M. E. Rose in 'Mass Spectrometry', ed. R. A. W. Johnstone (Specialist Periodical Reports), The Royal Society of Chemistry, London, 1984, Vol. 7, p. 196.

Specialized books and reviews of GC/MS and LC/MS are referred to in the appropriate sections, but some general texts are listed here. A book describing practical and technological aspects of GC/MS has appeared.[3] The proceedings of the Ninth International Mass Spectrometry Conference were published commendably soon after the meeting.[4] Other published proceedings occur in the fields of biochemistry, medicine and environmental research,[5,6] and nutrition science and food safety.[7] General mass spectrometry[8] and mass spectrometry in combination with separative methods[9,10] have been reviewed. In capillary-column gas chromatography[11] sample introduction is frequently the quality-determining factor, so careful consideration should be given to sampling methods. Whilst being outside the immediate scope of this Report, an article containing much sensible information on the various GC inlet methods is available.[12] A book dedicated to plasma chromatography[13] has been published, and mass spectrometers have been evaluated along with Fourier-transform infrared spectrometers as detectors for gas chromatographs.[14,15]

Being pertinent to chromatography/mass spectrometry but not in context elsewhere in this review, three reports of paper and thin-layer chromatography (TLC) in combination with mass spectrometry are noted here. It has been shown that, following development of papers or TLC plates, located 'spots' can be subjected to fast-atom-bombardment (FAB) mass spectrometry.[16,17] Portions of the paper were cut out and placed in turn on the FAB probe tip. After adding a glycerol/methanol matrix, FAB spectra of adsorbed penicillins were observed.[16] In the case of TLC the probe tip was covered with double-sided masking tape then pressed against the 'spot' of interest. After adding a liquid matrix to the particles adhering to the tape, FAB spectra were recorded apparently without background ions from the masking tape or the TLC adsorbent.[17] The method may be of interest particularly for pharmaceutical analyses. In a very different study[18] components on a TLC plate were desorbed by a

[3] G. M. Message, 'Practical Aspects of Gas Chromatography/Mass Spectrometry', Wiley, Chichester, 1984.

[4] E. R. Schmid, K. Varmuza, and I. Fogy, *Mass Spectrom. Adv.*, 1982, 9. (Reprinted from *Int. J. Mass Spectrom. Ion Phys.*, 1982, **45–48.**)

[5] 'Recent Developments in Mass Spectrometry in Biochemistry, Medicine and Environmental Research', ed. A. Frigerio, Elsevier, Amsterdam, 1983, Vol. 8.

[6] 'Chromatography and Mass Spectrometry in Biomedical Sciences, 2', ed. A. Frigerio, Elsevier, Amsterdam, 1983.

[7] 'Chromatography and Mass Spectrometry in Nutrition Science and Food Safety', ed. A. Frigerio and H. Milon, Elsevier, Amsterdam, 1984.

[8] A. L. Burlingame, J. O. Whitney, and D. H. Russell, *Anal. Chem.*, 1984, **56**, 417R.

[9] P. J. Arpino, *Int. J. Mass Spectrom. Ion Phys.*, 1982, **45**, 161.

[10] F. Heresch, *Oesterr. Chem.-Ztg.*, 1982, **83**, 189.

[11] M. L. Lee, F. J. Yang, and K. D. Bartle, 'Open Tubular Column Gas Chromatography. Theory and Practice', Wiley, Chichester, 1984.

[12] G. Schomburg, *J. Chromatogr. Sci.*, 1983, **21**, 97.

[13] T. W. Carr, 'Plasma Chromatography', Plenum Press, New York, 1984.

[14] S. L. Smith, *J. Chromatogr. Sci.*, 1984, **22**, 143.

[15] C. L. Wilkins, *Science*, 1983, **222**, 291.

[16] J. L. Gower, *Int. J. Mass Spectrom. Ion Phys.*, 1983, **46**, 431.

[17] T. T. Chang, J. O. Lay, jun., and R. J. Francel, *Anal. Chem.*, 1984, **56**, 109.

[18] L. Ramaley, M. E. Nearing, M. A. Vaughan, R. G. Ackman, and W. D. Jamieson, *Anal. Chem.*, 1983, **55**, 2285.

device that scans the developed plate past an incandescent lamp or laser beam and swept into a quadrupole mass spectrometer by chemical-ionization reactant gas (methane). This on-line TLC/MS system does not require previous knowledge of spot positions, does not destroy chromatograms, and provides a detection limit of about 10 ng for full spectra of aromatic hydrocarbons. The device has yet to prove itself for quantification of analytes and application beyond aromatic hydrocarbons.

2 Methodology: Gas Chromatography/Mass Spectrometry

Instrumentation. – The promise of commercial availability of ion traps as detectors for gas chromatography is significant when cost, complexity, and size of GC/MS equipment must be minimized.[19,20] The ion trap serves as a mass spectrometer because ions are held in a stable trajectory until they are ejected and detected in turn by adjustment of the voltages applied to the trap. The mass range, 10–650 u, can be scanned with unit resolution in 0.5 s, detection limits being in the low nanogramme range for many eluents. Operation appears to be very simple, controlled by a 16-bit personal computer. Standard data-processing facilities (mass chromatograms, library searching, calculation of peak areas, background subtraction) and selected-ion monitoring are available. Another interesting miniaturization study concerns the gas chromatograph rather than the mass spectrometer.[21] A miniature gas chromatograph has been incorporated within a direct-insertion probe for GC/MS, the flow rate of the carrier gas being sufficiently low to obviate the need for a splitter or separator. Applications are awaited.

At the other end of the financial scale, developments continue in the coupling of both a mass spectrometer and a Fourier-transform infrared spectrometer to a gas chromatograph. The advantages of the GC/FTIR/MS system have been expounded clearly, albeit with simple examples.[22] The two conceivable geometries for GC/FTIR/MS have been considered.[23] In the first, an adjustable microvalve was used to split the capillary-column GC effluent, 1% being introduced to a magnetic-sector mass spectrometer and 99% simultaneously to an FTIR spectrometer. A fairly complex mixture, peppermint oil, was examined. In the second configuration, lacquer-thinner components were led directly into the IR spectrometer and then through glass-lined tubing to a Fourier-transform mass spectrometer operating in either electron-impact (EI) or chemical-ionization (CI) mode. Both geometries offer advantages for analysis of complex mixtures,[23] but the chief problem (apart from expense) is that greater sensitivity is required for characterization of minor components. More recently, alternate recording of EI and CI spectra during a single GC analysis has been described, together with

[19] S. Evans and P. J. F. Watkins, *Lab. Pract.*, 1983, **32** (June), 41.
[20] G. C. Stafford, jun., P. E. Kelly, and D. C. Bradford, *Int. Lab.*, 1983, **13**(7), 84.
[21] J. E. Campana, *U.S. Pat. Appl. U.S.* 490 996, 1983.
[22] T. Hirschfeld, *Eur. Spectrosc. News*, 1984, **55**, 15.
[23] C. L. Wilkins, G. N. Giss, R. L. White, G. M. Brissey, and E. Onyiriuka, *Anal. Chem.*, 1982, **54**, 2260.

improved software for data processing.[24] Clearly, if capillary-column GC/MS generates large amounts of data, making data evaluation the most time-consuming part of an analysis, then the problem is exacerbated in GC/FTIR/MS experiments. It is thus heartening that approaches for the reduction of qualitative and quantitative data are in progress.[24,25] It is perhaps data processing that presently limits the great potential of the GC/FTIR/MS instrument, so considerable research effort is required in this area. In another combined instrument based on GC/MS an automated system for simultaneous thermal analysis and mass spectrometry has been described.[26]

Capillary-column supercritical-fluid chromatography/mass spectrometry (SFC/MS), sometimes called dense-gas chromatography/mass spectrometry, has potential advantages both over GC/MS, in that thermally labile and non-volatile compounds of high molecular weight are amenable to analysis, and over HPLC/MS in terms of simpler interfacing and higher chromatographic efficiency. A direct-fluid-injection interface, not unlike the direct-liquid-introduction system used for LC/MS, has been reported that transfers the total effluent from a wall-coated open tubular (WCOT) capillary SFC column into a CI tandem quadrupole mass spectrometer.[27,28] For polycyclic aromatic hydrocarbons, detection limits of 100 pg (full scanning) and 1 pg (selected-ion monitoring) were reported.[27] Styrene oligomers have also been studied, using n-pentane as mobile phase.[28] The flow rate of mobile phases (often more volatile than ones employed for HPLC) through capillary SFC columns is compatible with CI mass spectrometry. Application of SFC/MS to complex mixtures of 'difficult' samples has yet to be demonstrated; such work is encouraged. Current limitations stem basically from the state of knowledge of SFC technology, which is poor relative to that of GC and HPLC. In this respect, reviews of the theory, methods, and principles of SFC are recommended reading.[29] The SFC/MS combination can also be utilized for direct mass-spectrometric analysis of the products of supercritical-fluid extraction of bituminous coal, for example.[30]

The principal advantages of coupling Fourier-transform mass spectrometers and gas chromatographs (GC/FTMS) were outlined in the previous Report.[2] Maintenance of high vacuum is necessary for high-resolution Fourier-transform mass spectrometry, so large pumping speeds or specialized interface designs are required if the potential advantages of the method are not to be compromised. By using a jet separator in conjunction with a capillary GC column, high-resolution switching between peaks widely separated in mass was demonstrated by

[24] D. A. Laude, jun., G. M. Brissey, C. F. Ijames, R. S. Brown, and C. L. Wilkins, *Anal. Chem.*, 1984, **56**, 1163.
[25] S. S. Williams, R. B. Lam, D. T. Sparks, T. L. Isenhour, and J. R. Hass, *Anal. Chim. Acta*, 1982, **138**, 1.
[26] H. K. Yuen and G. W. Mappes, *Thermochim. Acta*, 1983, **70**, 269.
[27] R. D. Smith, W. D. Felix, J. C. Fjeldsted, and M. L. Lee, *Anal. Chem.*, 1982, **54**, 1883.
[28] R. D. Smith, J. C. Fjeldsted, and M. L. Lee, *J. Chromatogr.*, 1982, **247**, 231; *Int. J. Mass Spectrom. Ion Phys.*, 1983, **46**, 217.
[29] P. A. Peaden and M. L. Lee, *J. Chromatogr.*, 1983, **259**, 1; *J. Liq. Chromatogr.*, 1982, **5** (Suppl. 2), 179.
[30] R. D. Smith and H. R. Udseth, *Sep. Sci. Technol.*, 1983, **18**, 245; *Fuel*, 1983, **62**, 466.

GC/FTMS.[31] An alternative and viable approach for GC/FTMS relies on a small solenoid valve as an interface that pulses the GC effluent into the ion source.[32] The resulting reduction in pressure within the source preserves mass resolution and sensitivity.

Atmospheric-pressure ionization (API) mass spectrometry has been reviewed with regard to principles and applications.[33,34] The ionization conditions of API mass spectrometers can be made very similar to those obtaining in electron-capture detectors for GC. Hence GC/APIMS can be used to investigate the interactions that cause responses in electron-capture detectors.[35] In the absence of oxygen ionization occurs by simple electron capture, whereas charge transfer occurs in the presence of oxygen ($O_2^- + M \rightarrow M^- + O_2$). All isomers of the trifluoroacetyl derivatives of polycyclic aromatic amines underwent the same reactions, but the relative rates of the two processes are highly dependent on substitution location, providing a basis for distinguishing individual members. Newer methods of ionization that could be utilized in GC/MS are fast-atom bombardment (FAB) and laser multi-photon ionization. Volatized species can be ionized in the gas phase by fast-atom bombardment.[36] Positive-ion FAB of dihydroxybenzenes and nitrophenols in the gas phase resulted in molecular ions (comprising the base peaks) and fragmentation similar to that of EI. In the negative-ion mode the base peaks usually corresponded to molecular anions and there was far less fragmentation. Incorporation of this methodology in GC/MS would be interesting. Another bombardment method, laser multi-photon ionization,[37] has been examined for its potential in improving selectivity and sensitivity of capillary-column GC/MS.[38] For example, selective ionization, based on small differences in ionization energies, was demonstrated for chrysene and triphenylene co-eluting from a GC column.[38] The interfacing between the gas chromatograph and multi-photon-ionization time-of-flight mass spectrometer was described in detail. Detection limits as low as 200 fg and a good range of linearity were observed.[38] Wide application of the technique remains to be demonstrated, but development of the method may be fruitful. Conventional GC/CIMS has been utilized with some unconventional reactant gases. Three reports[39–41] have described methanol CI for GC/MS of neutral aromatic polar fractions of petroleum substitutes[39,40] and urinary dipeptides.[41] Use of deuteriated methanol leads to determination of the number of active hydrogens in phenols and differentiation of isomeric phenols and ethers.[39] For derivatized dipeptides, methanol CI is said to provide clear information on both molecular weight and

[31] R. L. White and C. L. Wilkins, *Anal. Chem.*, 1982, **54**, 2443.
[32] T. M. Sack and M. L. Gross, *Anal. Chem.*, 1983, **55**, 2419.
[33] A. E. Williams, *Org. Mass Spectrom.*, 1983, **18**, 509.
[34] R. K. Mitchum and W. A. Korfmacher, *Anal. Chem.*, 1983, **55**, 1485A.
[35] J. A. Campbell and E. P. Grimsrud, *J. Chromatogr.*, 1984, **284**, 27.
[36] B. Kralj, V. Kramer, and V. Vrscaj, *Int. J. Mass Spectrom. Ion Phys.*, 1983, **46**, 399.
[37] J. E. Wessel, *Gov. Rep. Announce. Index (U.S.)*, 1982, **82**, 1327.
[38] G. Rhodes, R. B. Opsal, J. T. Meek, and J. P. Reilly, *Anal. Chem.*, 1983, **55**, 280.
[39] M. V. Buchanan, *Anal. Chem.*, 1984, **56**, 546.
[40] M. V. Buchanan, G. L. Kao, B. D. Barkenbus, C. Ho, and M. R. Guerin, *Fuel*, 1983, **62**, 1177.
[41] C. Charpentier, R. A. W. Johnstone, A. Lemonnier, I. Myara, M. E. Rose, and D. Tuli, *Clin. Chim. Acta*, 1984, **138**, 299.

sequence of residues.[41] An extensive survey[42] of GC/CIMS with dimethyl ether as reactant gas indicates that the method may be useful for recognition of functional groups through the different reactions that they undergo with the reactant ions.

A useful discussion of the attainable scan speeds of magnetic-sector, quadrupole, time-of-flight, and Fourier-transform mass spectrometers and their suitability for acquiring mass spectra of components eluting as increasingly narrow peaks from high-resolution GC columns has been published.[43] Recent advances in micro-electronics have enabled an integrating transient recorder to be designed. The impressive time resolution provided virtually eliminates the scan-speed limitation in time-of-flight instruments. When time-array detection is applied to magnetic-sector mass spectrometers, ion flight times can be determined along with mass resolution, and this leads to assignment of precursor-/product-ion relationships.[43] In effect, the analysis is a so-called GC/MS/MS experiment and as such is covered in the section on metastable-ion techniques.

Incorporation of a low-pressure microwave-induced plasma into a GC/MS interface converts all organic compounds into a few simple neutral molecules like CO and CO_2.[44,45] The mass spectrometer then acts as an element-specific detector for GC, the elemental composition of eluents being computed once the identity and quantity of the products have been determined by conventional mass spectrometry. The use of the technique for determining elemental carbon/oxygen ratios in alcohols and ketones for example[44] and quantifying ^{14}C in methyl 1-[^{14}C]palmitate[45] has been described.

Two fused-silica capillary columns with different stationary phases have been connected in parallel to the same injector. The outlets of both columns were inserted directly into an ion source for methane chemical-ionization mass spectrometry of eluents.[46] The dual-column system allows retention-time data for two different columns to be gathered in a single analysis. However, the samples require extensive clean-up, chromatograms are complex, and the authors report that 'the analyst must know what particular compounds are to be identified'.[46] The system may have advantages for identifying isomers affording identical mass spectra. The observation of photochemical processes has been described by use of a visible-light source placed inside the GC oven. Photochemical reactions occurring within the column affected the chromatographic behaviour of eluents, and the products were identified by on-line mass spectrometry.[47] The observed reactions depend on the chemical properties of the eluent and stationary phase. The capabilities of a commercial high-resolution gas chromatograph coupled to a high-resolution mass spectrometer have been demonstrated with reference to a sample of lime oil.[48]

[42] T. Keough, *Anal. Chem.*, 1982, **54**, 2540.
[43] J. F. Holland, C. G. Enke, J. Allison, J. T. Stults, J. D. Pinkston, B. Newcome, and J. T. Watson, *Anal. Chem.*, 1983, **55**, 997A.
[44] R. A. Heppner, *Anal. Chem.*, 1983, **55**, 2170.
[45] S. P. Markey and F. P. Abramson, *Anal. Chem.*, 1982, **54**, 2375.
[46] L. R. Hogge and D. J. H. Olson, *J. Chromatogr. Sci.*, 1983, **21**, 524.
[47] W. G. Laster, J. B. Pawliszyn, and J. B. Phillips, *J. Chromatogr. Sci.*, 1982, **20**, 278.
[48] K. E. Hall and P. W. Brooks, *Anal. Proc.*, 1982, **19**, 246.

Instrumentation for mass spectrometry generally is described in Chapter 4, and instruments combining gas or liquid chromatographs with mass spectrometers for analysis of inorganic and organic compounds have been reviewed.[49]

Interfaces. – The requirements and performance of various currently available GC/MS interfaces have been discussed.[50] Improvements available through use of fused-silica columns were emphasized. As stated in the previous Report,[2] direct insertion of the end of a fused-silica capillary column into the ion source now seems to provide the optimum interface method.[51,52] Alternatively, deactivated fused-silica tubing can be positioned permanently between the GC oven and the ion source.[53] A zero dead-volume connector within the GC oven serves to couple capillary columns to the interface. Such a fused-silica transfer line can be incorporated into an open-split interface, allowing direct coupling or open-split modes of operation.[54] Testing and performance of the device have also been reported.[55] Cold spots and active sites were eliminated, allowing underivatized vitamin D_3 to be eluted with excellent peak shape. A simple interface comprising a drilled-out aluminium block mounted on the drilled oven door of a gas chromatograph has been used with fused-silica and glass capillary-column GC/MS.[56]

Despite being outmoded, the membrane separator has come under scrutiny. Because solubility of organic molecules in a silicone membrane dominates diffusivity in determining their total transmission, efficiency decreases with increasing temperature. There is no maximum transfer efficiency at one specific temperature. These conclusions have important ramifications for temperature programming and high-temperature operation of such membranes for GC/MS studies.[57] To minimize adsorption losses of organosulphur compounds in a membrane interface for GC/MS, an all-polymer system was preferred over the same device with a membrane housing constructed from glass.[58]

The importance of avoiding prolonged contact of analytes with metal surfaces in the source housing and ensuring inertness of glass restrictor interfaces has been stressed.[59] Without such precautions, methoxime trimethylsilyl derivatives of some steroids eluting from capillary columns were shown to undergo thermal degradation.[59] When using hydrogen as carrier and scavenger gas in an open-split interface for capillary-column GC/MS, partial catalytic hydrogenation of alkenes was noted to occur in the interface.[60] The problem was eliminated by substituting helium for hydrogen.

49 A. S. Kuzema and A. Pilipenko, *Khim. Tekhnol. Vody*, 1983, **5**, 79.
50 J. R. Chapman, *Anal. Proc.*, 1982, **19**, 235.
51 K. Rose, *J. Chromatogr.*, 1983, **259**, 445.
52 T. E. Jensen, R. Kaminsky, B. D. McVeety, T. J. Wozniak, and R. A. Hites, *Anal. Chem.*, 1982, **54**, 2388.
53 B. Y. Giang, *J. High Resolut. Chromatogr. Chromatogr. Commun.*, 1984, **7**, 137.
54 E. Wetzel and T. Kuster, *J. Chromatogr.*, 1983, **268**, 177.
55 T. Kuster and E. Wetzel, *Int. J. Mass Spectrom. Ion Phys.*, 1983, **46**, 173.
56 W. D. Koller, *J. High Resolut. Chromatogr. Chromatogr. Commun.*, 1984, **7**, 43.
57 C. C. Greenwalt, K. J. Voorhees, and J. H. Futrell, *Anal. Chem.*, 1983, **55**, 468.
58 M. Thompson and M. Stanisavljevic, *Talanta*, 1982, **29**, 867.
59 J. W. Honour, C. J. W. Brooks, and C. H. L. Shackleton, *Biomed. Mass Spectrom.*, 1982, **9**, 505.
60 G. C. Jamieson, *J. High Resolut. Chromatogr. Chromatogr. Commun.*, 1982, **5**, 632.

For packed-column GC/MS, a maximum transmission factor of 65% was achieved for a supersonic seeded beam separator based on a nozzle-skimmer assembly in a vacuum chamber.[61] Two different methods for clearing clogged glass-jet separators have been proposed.[62,63]

There have been several reports of gas-phase reactions on eluents brought about purposely within the GC/MS interface region (*i.e.* post-column derivatization or 'reaction GC/MS') as an aid to identification of analytes. Post-column hydrogenation of double bonds has proved valuable in two studies.[64,65] In one of these[64] alkenes were identified in three steps: (i) conventional GC/MS for determination of relative molar mass and degree of unsaturation, (ii) identification of the carbon skeleton by catalytic hydrogenation with H_2, and (iii) identification of double-bond position by hydrogenation with $^2H_2$. Similarly, monosubstituted cyclopropanes were distinguished from alkenes.[66] On-line dehydrogenation over a copper catalyst converts primary and secondary alcohols to aldehydes and ketones, respectively, while tertiary alcohols are unaffected or dehydrated.[67,68] The reaction can be brought about in the GC/MS interface or between the injection port and column and serves to distinguish the various alcohols, to locate the hydroxyl group in secondary alcohols, and to determine the relative molar masses of aliphatic alcohols that do not themselves afford stable molecular ions.

The Role of Data Systems. – All aspects of data systems in mass spectrometry are reviewed in Chapter 5. A high degree of automation is necessary for GC/MS,[69,70] so some reports specific to the combined instrument are covered briefly here.

New software for processing GC/MS data has been proposed.[71] It consists of a register stack of the 'last-in first-out' type and a set of functions to operate on the data stored in the stack (entire mass spectra and/or chromatographic data) in a very flexible fashion and then to display those data. Probability-based matching has been implemented for identification of GC eluents during analysis.[72] A search

[61] M. R. Cimberle, U. Garibaldi, P. Bottino, and F. Valerio in ref. 6, p. 215.
[62] F. L. Cardinali and L. E. Lowe, *Anal. Chem.*, 1982, **54**, 1454.
[63] W. E. Wentworth, Ya Chi Chen, A. Zlatkis, and B. S. Middleditch, *Anal. Chem.*, 1982, **54**, 1895.
[64] A. I. Mikaya, V. I. Smetanin, and V. G. Zaikin, *Izv. Akad. Nauk SSSR, Ser. Khim.*, 1982, 2270.
[65] P. M. Woollard and A. I. Mallet, *J. Chromatogr.*, 1984, **306**, 1.
[66] A. I. Mikaya, L. P. Medvedkova, V. G. Zaikin, V. M. Vdovin, and A. A. Kamyshova, *Izv. Akad. Nauk SSSR, Ser. Khim.*, 1983, 1181.
[67] A. I. Mikaya, A. V. Antonova, and V. G. Zaikin, *Izv. Akad. Nauk SSSR, Ser. Khim.*, 1982, 1436.
[68] A. I. Mikaya, V. I. Smetanin, V. G. Zaikin, A. V. Antonova, and N. S. Prostakov, *Org. Mass Spectrom.*, 1983, **18**, 99.
[69] J. R. Chapman, *Int. J. Mass Spectrom. Ion Phys.*, 1982, **45**, 207.
[70] D. T. Green, *Int. J. Mass Spectrom. Ion Phys.*, 1983, **48**, 43.
[71] E. F. Reus, D. W. Peterson, and R. Ellis, *Int. J. Mass Spectrom. Ion Phys.*, 1983, **46**, 97.
[72] F. W. McLafferty, S. Cheng, K. M. Dully, C.-J. Guo, I. K. Mun, D. W. Peterson, S. O. Russo, D. A. Salvucci, J. W. Serum, W. Staedeli, and D. B. Stauffer, *Int. J. Mass Spectrom. Ion Phys.*, 1983, **47**, 317.

of 41 000 reference spectra during GC/MS was possible in 10 seconds by careful ordering of the reference file. For compounds not represented in the library, the self-training interpretative and retrieval system was utilized in the off-line mode.[72] An automatic method that searches GC/MS data for specified isotope patterns has been expanded to allow the user to specify any pattern of ion abundances.[73] The procedure promises to be of value in life sciences where incubations with mixtures of isotopically labelled exogenous or endogenous compounds are often employed in tracer studies. Location of metabolite spectra in GC/MS data would be facilitated by searching for the characteristic isotope cluster.

A method has been described for automatic computation of methylene-unit retention-time indices in real time with a computerized GC/MS system.[74] The method eliminates the need to include a set of methylene-unit reference standards in each analysis. Quantitative resolution of unseparated GC peaks has been achieved without requiring assumptions of peak shapes or needing unique masses for the pure components that require deconvolution.[75] The procedure relies on curve resolution and the generalized inverse method. The summing of several selected ions for display as a 'multiple mass chromatogram' is often more characteristic of a compound class than is a single-ion profile. A programme has been reported[76] that allows up to ten masses to be summed for display or removed from the total ion-current chromatogram. The latter facility aids background correction by removal of, for example, air peaks.[76] Selected-ion monitoring at high mass-spectral resolution has been shown to be an efficient method of measuring the accurate masses of capillary-column eluents when using a peak-matching system under computer control.[77] Improved software that writes data gained during selected-ion monitoring in a compact form has increased significantly the capabilities of a low-cost GC/MS calculator system.[78]

Instrument control and data acquisition have not been ignored. Control of a double-focusing magnetic-sector mass spectrometer has been implemented effectively with distributed microprocessors under the overall control of a mini-computer.[79] The advantages of the so-called top-hat filter over slope-detection methods for acquisition of GC/MS data have been stated.[80] The method can now be applied to both quadrupole and magnetic-sector instruments.

Quantification. – Good reviews of quantitative mass spectrometry in biochemistry and medicine have appeared.[81,82] With reference particularly to steroid conjugates

[73] R. J. Anderegg, *J. Chromatogr.*, 1983, **275**, 154.
[74] I. L. Bromberg, J. A. Heininger, and A. G. Cherian, *Clin. Chem.*, 1982, **28**, 349.
[75] M. A. Sharaf and B. R. Kowalski, *Anal. Chem.*, 1982, **54**, 1291.
[76] M. G. Rawdon, *J. Chromatogr. Sci.*, 1984, **22**, 125.
[77] Y. Tondeur, J. R. Hass, D. J. Harvan, and P. W. Albro, *Anal. Chem.*, 1984, **56**, 373.
[78] L. C. Dickson, F. W. Karasek, and R. E. Clement, *J. Chromatogr.*, 1983, **280**, 23.
[79] R. S. Stradling, P. A. Ryan, and J. D. Wood, *Comput. Enhanced Spectrosc.*, 1983, **1**, 25.
[80] A. D. Graddon, J. F. Smith, and J. D. Morrison, *Comput. Enhanced Spectrosc.*, 1983, **1**, 167.
[81] A. P. De Leenheer, J. A. Jonckheere, and C. F. Gelijkens, *Int. J. Mass Spectrom. Ion Phys.*, 1982, **45**, 231.
[82] W. J. A. VandenHeuval, J. R. Carlin, and R. W. Walker, *J. Chromatogr. Sci.*, 1983, **21**, 119.

in saliva, selected-ion monitoring at high resolution and the still under-utilized method of metastable-peak monitoring have been reviewed.[83]

A useful, in-depth discussion of methods for calculating confidence limits in quantitative GC/MS is available.[84] Sources of variance are identified, their relative importance is considered, and methods to reduce such variance are examined. Instead of transforming non-linear data artificially to a linear model, it is possible to describe the relationship between isotope ratios and molar ratios in isotope-dilution GC/MS by means of a polynomial regression.[85] Computational facilities for higher-order polynomials are generally available for microcomputers, so accurate description of curvature in calibration graphs should now be widely facilitated. For interpolating analyte concentrations between calibration standards, linear and non-linear models have been compared.[86] Bias in calibration can be a major source of error, at least in one assay of priority pollutants in waste water.[87] Major factors affecting accuracy of such quantitative analyses have been evaluated.[87]

Matrix effects in quantitative mass spectrometry have attracted deserved attention.[88,89] It has been amply demonstrated over the years that detection limits are frequently determined more by selectivity than by sensitivity. However, great selectivity brings its own problems because peaks that are out of sight are often out of mind (see previous Report[2]). For example, an unidentified co-eluent interfered with highly selective assays of 2,3,7,8-tetrachlorodibenzodioxin (2,3,7,8-TCDD) even though the impurity did not yield ions in the mass region monitored.[88] Several possible mechanisms for the matrix effect were proposed, and it was recommended that quantitative work at trace levels should not be performed on crude extracts.[88]

Long-term precision studies on both retention times and quantitative response during capillary-column GC/MS indicated, not surprisingly, that retention times are best computed relative to the most closely eluting reference compound and that quantification is best based on response relative to an internal standard that is chemically similar to the analyte.[90] Other authors[91] agree that precision in measurements of retention time and peak area necessitates the use of internal standards and, further, advocate the use of multiple internal standards eluting within 10 minutes of the analyte retention time. It is said[91] that, whilst most clinical chemists use internal standards, environmental scientists on the whole do not. Affirming current thinking, an on-column injector was preferred for

[83] S. J. Gaskell, *Anal. Proc.*, 1982, **19**, 264.
[84] G. F. Moler, R. R. Delongchamp, W. A. Korfmacher, B. A. Pearce, and R. K. Mitchum, *Anal. Chem.*, 1983, **55**, 835.
[85] J. A. Jonckheere, A. P. De Leenheer, and H. L. Stevaert, *Anal. Chem.*, 1983, **55**, 153.
[86] W. T. Yap, R. Schaffer, H. S. Hertz, E. White, and M. J. Welch, *Biomed. Mass Spectrom.*, 1983, **10**, 262.
[87] C. J. Kirchmer, M. C. Winter, and B. A. Kelly, *Environ. Sci. Technol.*, 1983, **17**, 396.
[88] Y. Tondeur, P. W. Albro, J. R. Hass, D. J. Harvan, and J. L. Schroeder, *Anal. Chem.*, 1984, **56**, 1344.
[89] G. J. Kallos, V. Caldecourt, and J. C. Tou, *Anal. Chem.*, 1982, **54**, 1313.
[90] B. N. Colby, P. W. Ryan, and J. E. Wilkinson, *J. High Resolut. Chromatogr. Chromatogr. Commun.*, 1983, **6**, 72.
[91] D. E. Wells and A. A. Cowan, *J. Chromatogr.*, 1983, **279**, 209.

precise assays.[91] In a drug study[92] precision was again dependent on injection mode and internal standards, splitless injection with an isotopically labelled standard being more reliable than split injection with an homologous standard. An on-column injector was not utilized in this study.

Sampling Techniques. – Only papers of general interest are discussed in this section, the remainder being considered in the appropriate categories below. Collection, isolation, and concentration of samples for capillary-column GC have been discussed.[93] Another paper that does not cover mass spectrometry is nevertheless highly recommended and pertinent to both GC/MS and LC/MS.[94] The analyst is reminded that inappropriate choice of sampling and work-up procedure can invalidate results, even when the instrumental measurement itself is excellent. Some factors affecting sample preparation are quite subtle and possibly unexpected, so great care is necessary to obtain representative analytical samples, particularly from heterogeneous matrices like blood.[94]

A novel and quantitative extraction procedure that separates samples into acidic, basic, and neutral (non-extractable) fractions uses gel electrophoresis.[95] In another method borrowed from biological sciences, an immunoadsorption column was used as a rapid and selective procedure for extracting steroids from physiological fluids.[96,97] The method involves cellulose-coupled antisera raised against individual steroids and achieves extraction efficiencies of about 80% for oestradiol and cortisol.[96,97] Whilst HPLC is now a very popular means of sample isolation and purification prior to GC/MS, simpler methods are also of considerable value. For example, the use of Sep-Pak cartridges continues as with polar cortisol metabolites[98] and eicosanoids.[99] Adsorption on graphite and subsequent solvent elution has been utilized for enrichment of herbicides in water,[100] and an improved thermal-desorption device has been described for injecting a concentrated sample of chloromethane into a GC/MS system.[101] After optimization of temperature and purge time, a closed-loop stripping system achieved average recoveries of 97% for chloroalkanes in water.[102]

The efficiency of capillary cold traps has been called into question.[103] Compounds with boiling points under 70 °C and carried in a stream of carrier gas cannot be condensed quantitatively at the inlet of a GC column by simple cold trapping even at liquid-nitrogen temperature. Enhanced contact between the

[92] M. Desage, J. L. Brazier, F. Comet, and R. Guilluy, *Analusis*, 1983, **11**, 119.
[93] C. F. Poole and S. A. Schuette, *J. High Resolut. Chromatogr. Chromatogr. Commun.*, 1983, **6**, 526.
[94] R. Gill and A. C. Moffatt, *Anal. Proc.*, 1982, **19**, 170.
[95] C.-G. V. Hammar, *Anal. Biochem.*, 1983, **130**, 226.
[96] S. J. Gaskell, B. G. Brownsey, C. J. Collins, H. M. Leith, and G. C. Thorne, *Int. J. Mass Spectrom. Ion Phys.*, 1983, **48**, 245.
[97] S. J. Gaskell and B. G. Browney, *Clin. Chem.*, 1983, **29**, 677.
[98] J. W. Honour, J. Kent, and C. H. Shackleton, *Clin. Chim. Acta*, 1983, **129**, 229.
[99] D. Alber and J. C. Henry, *Int. J. Mass Spectrom. Ion Phys.*, 1983, **48**, 257.
[100] F. Mangani and F. Bruner, *Chromatographia*, 1983, **17**, 377.
[101] B. E. Wilkes, L. J. Priestley, jun., and L. K. Scholl, *Microchem. J.*, 1982, **27**, 420.
[102] W. E. Coleman, J. W. Munch, R. W. Slater, R. G. Melton, and F. C. Kopfler, *Environ. Sci. Technol.*, 1983, **17**, 571.
[103] J. W. Graydon and K. Grob, *J. Chromatogr.*, 1983, **254**, 265.

solid surface and the vapour would be necessary.[103] On the other hand, 'whole-column cryotrapping' at −80 °C has been claimed[104] to provide a quantitative trap for eluents ranging from 1,1-dichloroethene (boiling point 32 °C) to chlorobenzene (132 °C).

Chromatographic Aspects. – The advantages of capillary columns over packed columns for GC/MS are thoroughly appreciated but have been stressed once more.[105] It has been shown that use of fused silica of very high purity for capillary columns ensures improved thermal stability of the stationary phase and great inertness.[106] A practical alternative to fused silica is soft glass for manufacture of flexible capillary columns.[107] The chief advantage of the latter appears to be cost. Laboratory preparation of inert glass capillary columns has been described authoritatively.[108] Capillary columns with bores of just 0.1 mm afford better resolutions in shorter times than the commonly used columns with 0.25 mm internal diameter. Such a narrow-bore column was adapted for on-column injection and interfaced to a mass spectrometer.[109]

On-column injectors have a broader range of applicability and afford better precision and accuracy than splitless injectors.[110] The former inlets are the only injectors that cope satisfactorily with low levels of mixtures of wide boiling range. Jennings[111] has offered some useful advice on trouble-shooting in capillary-column GC, particularly on column bleed and deactivation. Deactivation of columns with octaphenylcyclotetrasiloxane leaves a thermostable layer that accepts thin coatings of liquid phases containing high percentages of phenyl groups.[112] Good performance of such columns operated up to 320 °C was observed for separation of polycyclic aromatic hydrocarbons.[112]

Selectivity and resolution of GC separations are improved by coupling a high-polarity packed column to a low-polarity capillary column and then to a mass spectrometer (GC/GC/MS).[113] This potentially useful system was described in the previous Report[2] and is now shown to permit full chromatographic resolution of 2,3,7,8-, 2,3,4,7,8-, and 1,2,3,7,8,9-polychlorodibenzofurans.[113] It is unfortunate that the authors persist in calling their method 'two-dimensional gas chromatography'. Another chromatographic coupling, that of on-line HPLC/GC,[114] also has considerable potential for time-saving in analysing complex mixtures without prior clean-up. Interfacing to mass spectrometry would be worth pursuing.

A theoretical treatment of the various chromatographic variables, particularly flow rates in packed-bed and open-tubular columns in relation to the allowable

[104] J. F. Pankow, *J. High Resolut. Chromatogr. Chromatogr. Commun.*, 1983, **6**, 292.
[105] J. Settlage and H. Jaeger, *J. Chromatogr. Sci.*, 1984, **22**, 192.
[106] H. Saito, *J. Chromatogr.*, 1982, **243**, 189.
[107] K. L. Ogan, C. Reese, and R. P. W. Scott, *J. Chromatogr. Sci.*, 1982, **20**, 425.
[108] K. Grob, G. Grob, W. Blum, and W. Walther, *J. Chromatogr.*, 1982, **244**, 197.
[109] F. I. Onuska, *J. Chromatogr.*, 1984, **289**, 207.
[110] F. I. Onuska, R. J. Kominar, and K. Terry, *J. Chromatogr. Sci.*, 1983, **21**, 512.
[111] W. Jennings, *J. Chromatogr. Sci.*, 1983, **21**, 337.
[112] R. Burrows, M. Cooke, and D. G. Gillespie, *J. Chromatogr.*, 1983, **260**, 168.
[113] W. V. Ligon, jun. and R. J. May, *J. Chromatogr.*, 1984, **294**, 77, 87.
[114] K. Grob, jun., D. Fröhlich, B. Schilling, H. P. Neukom, and P. Nägeli, *J. Chromatogr.*, 1984, **295**, 55.

inlet flow to a mass spectrometer, has led to optimization of GC/MS with respect to chromatographic resolution and mass-spectral sensitivity.[115] A mass spectrometer coupled to a packed-bed inlet has been used to investigate solute focusing due to the solvent effect.[116] A useful review has appeared of the various solvent effects operating during injection onto a capillary column.[117]

Capillary-column chromatography with bonded (non-extractable) stationary phases is to be recommended in GC/MS for extended upper-temperature limits, reduced column bleed, ability to cope with large volumes of solvent, and stability of the coating, which allows thick films and polar phases to be coated evenly. *In situ* generation of free-radical cross-polymerization can be used to immobilize silicone stationary phases to glass or fused-silica column walls.[118] Re-silylation of immobilized apolar coatings can result in much reduced bleed rates and improved overall column quality.[119] Capillary columns with thick films of stationary phases (1–8 μm) have the advantage of large loading capacity and, perhaps surprisingly, exhibit very good separation efficiencies.[120]

Liquid crystals form interesting stationary phases for GC. Owing to their rod-like shape and regular arrangement of their molecules, solutes are separated mainly on the basis of their length-to-width ratios and molecular planarity. A timely review of liquid-crystalline stationary phases has appeared.[121] A mesomeric phase has been found superior to conventional silicone phases for GC and GC/MS of acyl derivatives of various hydroxynaphthalenes.[122] Inclusion of β-cyclodextrin, well known for forming stereoselective molecular-inclusion complexes, in formamide stationary phases affects markedly elution patterns.[123] On such a GC column, xylene isomers and ethylbenzene were separated completely.[123] These new approaches to chromatographic selectivity have much to offer in GC/MS also.

The trimethylsilyl (TMS) ethers of *erythro* and *threo* forms of alkane-2,3-diols have been separated by capillary-column GC and GC/MS.[124] Conformational immobility around the bond between chiral carbon atoms, because of steric hindrance to rotation by the large TMS groups, was thought to be the basis of the chromatographic resolution. As was recommended in the previous Report,[2] application of chiral stationary phases for direct resolution of enantiomers[125] by GC/MS has continued and expanded. (The role of chiral derivatization is

[115] J. F. K. Huber, E. Matisova, and E. Kenndler, *Anal. Chem.*, 1982, **54**, 1297.
[116] V. Pretorius, E. R. Rohwer, K. H. Lawson, and P. J. Apps, *J. High Resolut. Chromatogr. Chromatogr. Commun.*, 1984, **7**, 142.
[117] K. Grob, jun., *J. Chromatogr.*, 1982, **251**, 235.
[118] S. R. Springston, K. Melda, and M. V. Novotny, *J. Chromatogr.*, 1983, **267**, 395.
[119] K. Grob and G. Grob, *J. High Resolut. Chromatogr. Chromatogr. Commun.*, 1982, **5**, 349.
[120] K. Grob and G. Grob, *J. High Resolut. Chromatogr. Chromatogr. Commun.*, 1983, **6**, 133.
[121] Z. Witkiewicz, *J. Chromatogr.*, 1982, **251**, 311.
[122] G. Chiavari and L. Pastorelli, *J. Chromatogr.*, 1983, **262**, 175.
[123] D. Sybilska and T. Koscielski, *J. Chromatogr.*, 1983, **261**, 357.
[124] J. H. Abalain, D. Picart, F. Berthou, R. Ollivier, Y. Amet, J. Y. Daniel, and H. H. Floch, *J. Chromatogr.*, 1983, **272**, 305.
[125] R. H. Liu and W. W. Ku, *J. Chromatogr.*, 1983, **271**, 309.

covered in the next section.) Enantioselective GC/MS has been used to separate chiral photodimers of 2-cycloalkenones,[126] derivatized amino acids[127,128] and 3-hydroxyacids,[128] γ-vinyl-γ-aminobutyric acid as its *N*-trifluoroacetyl *O*-methyl ester,[129] and *exo*- and *endo*-brevicomin.[130] Chiral resolution holds much promise in pharmacology because of the known enantiomer-specific biological activity of many drugs. For example, the two enantiomers of γ-vinyl-γ-aminobutyric acid (an inhibitor of γ-aminobutyric acid-degrading enzyme) showed different biological half-lives.[129]

Derivatization. – Articles of general interest only are covered in this section, the remainder being discussed in the appropriate sections below.

To continue the theme of enantiomeric separations by GC/MS, an alternative strategy to the use of chiral stationary phases is the application of chiral derivatizing agents followed by conventional (achiral) GC/MS. *N*-Trifluoroacetylamino acid chlorides have proved popular for acylation of amines[131] and alcohols.[132] Following the derivatization, most of ten commonly abused amine drugs, including their enantiomers, were separated and identified by capillary-column GC/MS.[131] For the *Myrmica* ant pheromone, 3-octanol, shown to be mainly the *R* enantiomer with a small amount of the *S* analogue, the (+)-*trans*-chrysanthemoyl ester was preferred.[132] Mixtures of acyclic, isomeric isoprenoid alcohols were also resolved by GC and GC/MS following formation of chrysanthemate diastereoisomers.[133] Enantiomeric resolution of carboxylic acids as (−)-menthyl esters[134] and of 5-hydroxyhexanoic acid as its *O*-D-2-phenylpropionyl methyl ester[135] has been achieved by gas chromatography with mass-spectrometric detection. Such methodology is as yet far from its full potential, so continued development and increasing applications are expected. For forensic identification of some illicit drugs, specificity to an active enantiomer may be mandatory.

Several on-column derivatizations have been reported, such as formation of *N*-mono-trifluoroacetyl derivatives of amphetamine analogues,[136] trimethylsilyl ethers[137] and boronates[138] of aliphatic diols, and per[$^2H_3$]methyl derivatives of

[126] E. Anklam, W. A. Koenig, and P. Margaretha, *Tetrahedron Lett.*, 1983, **24**, 5851.

[127] J. Rotgans, R. Wodarz, W. Schoknecht, and K. Drysch, *Arch. Oral Biol.*, 1983, **28**, 1121.

[128] W. A. König, I. Benecke, N. Lucht, E. Schmidt, J. Schulze, and S. Sievers, *J. Chromatogr.*, 1983, **279**, 555.

[129] K. D. Haegele, J. Schoun, R. G. Alken, and N. D. Huebert, *J. Chromatogr.*, 1983, **274**, 103.

[130] V. Schurig, R. Weber, G. J. Nicholson, A. C. Oehlschlager, H. Pierce, jun., A. M. Pierce, J. H. Borden, and L. C. Ryker, *Naturwissenschaften*, 1983, **70**, 92.

[131] R. H. Liu, W. W. Ku, and M. P. Fitzgerald, *J. Assoc. Off. Anal. Chem.*, 1983, **66**, 1443.

[132] A. B. Attygalle, E. D. Morgan, R. P. Evershed, and S. J. Rowland, *J. Chromatogr.*, 1983, **260**, 411.

[133] S. J. Rowland and J. R. Maxwell, *J. Chromatogr. Sci.*, 1983, **21**, 298.

[134] K. D. Ballard, T. D. Eller, and D. R. Knapp, *J. Chromatogr.*, 1983, **275**, 161.

[135] J. P. Kamerling, M. Duran, L. Bruinvis, D. Ketting, S. K. Wadman, and J. F. G. Vliegenthart, *Clin. Chim. Acta*, 1982, **125**, 247.

[136] T. A. Brettell, *J. Chromatogr.*, 1983, **257**, 45.

[137] A. I. Mikaya, A. V. Antonova, V. G. Zaikin, N. S. Prostakov, V. Yu. Rumyantsev, E. V. Slivinskii, and S. M. Loktev, *Izv. Akad. Nauk SSSR, Ser. Khim.*, 1983, 2502.

[138] M. E. Rose, C. Longstaff, and P. D. G. Dean, *J. Chromatogr.*, 1982, **249**, 174.

nucleosides.[139] Conversion of bifunctional compounds into cyclic derivatives has several advantages such as lower mass increments and higher abundances of molecular ions than most acyclic derivatives. Whilst TMS derivatives are of general use for vicinal dihydroxy long-chain acids and provide diagnostic fragmentation, they afford few molecular ions.[140] Cyclic methane- or butaneboronate methyl esters yielded clear molecular-ion peaks but little informative fragmentation, and the boronates were unhelpful for fatty acids with two or more vicinal diol groupings.[140] Bifunctional amines can be derivatized selectively with methyldichlorophosphine and sulphur to yield stable cyclic phospholidinethiones. The method was applied to a selected-ion monitoring assay for ephedrine.[141] Dialkylsilylene derivatives of 1,2-, 1,3-, and 1,4-diols can be formed in good yield under mild conditions by use of di-isopropylsilyl and di-t-butylsilyl ditriflates.[142] The six-membered ring derivatives were stable to hydrolysis. This strategy for silylation of diols has been used only to a small extent in the past, so application of the synthetic work[142] to analysis would be interesting.

Work on the location of double bonds in long-chain compounds continues. Both methyl branch points and unsaturated centres in fatty acids were characterized by GC/MS and the use of the readily prepared picolinyl esters.[143] A further report of dimethyldisulphide adducts has appeared,[144] this time applied to determining by GC/MS the double-bond position in mono-unsaturated acetates at the nanogramme level. Multiple bonds in conjugated enyne acetates have been characterized by epoxidation followed by hydrogenation (or deuteriation) to saturated mono-hydroxy compounds that were amenable to GC/MS.[145] One novel approach to double-bond location in alkenes concerns their Diels–Alder adducts. Reaction of alkenes with ketals of tetrachlorocyclopentadiene yields derivatives that are suitable for GC/MS and whose mass spectra are diagnostic of double-bond position.[146]

Derivatization of carboyxlic acids has come under examination. Volatile organic acids (or their methyl esters) are readily lost by evaporation during work-up. Both alkylamide[147] and butyl ester[148] derivatives have the advantage of having sufficiently low volatility to prevent evaporative losses yet possessing good chromatographic properties for GC/MS analysis. In a continuing series of papers on aryldiazomethanes as esterification agents, the preparation and GC/MS properties of naphthalenemethyl[149] and phenanthrenemethyl[150] esters have been illustrated. The selectivity of various t-butyldimethylsilyl (TBDMS) and TMS

139 R. G. Teece, D. Slowikowski, and K. H. Schram, *Biomed. Mass Spectrom.*, 1983, **10**, 30.
140 E. H. Oliw, *J. Chromatogr.*, 1983, **275**, 245.
141 K. Jacob, W. Voty, C. Krauss, G. Schnabl, and M. Knedel, *Biomed. Mass Spectrom.*, 1983, **10**, 175.
142 E. J. Corey and P. B. Hopkins, *Tetrahedron Lett.*, 1982, **23**, 4871.
143 D. J. Harvey, *Biomed. Mass Spectrom.*, 1982, **9**, 33.
144 H. R. Buser, H. Arn, P. Guerin, and S. Rauscher, *Anal. Chem.*, 1983, **55**, 818.
145 J. Einhorn, H. Virelizier, and A. Guerrero, *Colloq. INRA*, 1982, **7**, 95.
146 D. A. Kidwell and K. Biemann, *Anal. Chem.*, 1982, **54**, 2462.
147 K. Y. Tserng, C. A. Gilfillan, and S. C. Kalhan, *Anal. Chem.*, 1984, **56**, 517.
148 D. G. Burke and B. Halpern, *Anal. Chem.*, 1983, **55**, 822.
149 D. L. Corina and K. Isaac, *J. Chromatogr.*, 1983, **260**, 51.
150 D. L. Corina and K. Platt, *J. Chromatogr.*, 1984, **291**, 127.

reagents towards esterification of fatty acids has been studied.[151] Application to quantitative GC/MS was illustrated, as for example with arachidonic acid from human skin.

Trimethylsilylation of secondary amines with *N,O*-bis(trimethylsilyl)trifluoroacetamide (BSTFA) in dimethylformamide (DMF) should be used with caution. It has been shown[152] that the required product is accompanied by *N*-(aminomethylene)-2,2,2-trifluoroacetamides incorporating CH atoms from DMF and a CF_3CO group from BSTFA. α-Ketoacids from urine and plasma have been purified on a hydrazide-gel column, but the well known 2,4-dinitrophenylhydrazine derivatives were not employed for analysis because of the complications resulting from the presence of *syn–anti* isomers. Instead, the TMS–quinoxalinol derivatives were utilized.[153]

Derivatization and GC/MS analysis of inorganic compounds are treated separately at the end of this chapter.

Metastable-ion Techniques. – In the previous Report[2] this topic was covered in some detail. Again, the 'GC/MS/MS' terminology has been avoided here because in some cases (see below) it is misleading. 'Metastable-ion techniques' as a term is not misleading because the techniques are the same whether metastable-ion or collisional-activation processes are being studied. It is also less likely to confuse and alienate the newcomer.

Instrumentation has been reviewed.[154] Two commendable trends are noticeable in the literature: (i) comparative studies to ascertain the relative merits of metastable-ion techniques with and without on-line chromatography and GC/MS and (ii) gas chromatography coupled to a triple-quadrupole mass spectrometer emerging as the front-runner in instrumentation for chromatography/metastable-ion techniques, as suggested in the previous Report.[2]

For quantification of polychlorinated biphenyls and tetrachlorodibenzofuran, techniques based on GC/MS (high-resolution selected-ion monitoring and metastable-peak monitoring) were favoured over a non-chromatographic metastable-ion method (MS/MS). The advantages of the latter in terms of speed were outweighed by unpredictably high results due to matrix interference and the impossibility of analysing separately different isomers.[155] A similar comparison for analysis of 1,1-dichlorodimethylsulphone in aquatic organisms came out in favour of both direct-probe and gas-chromatographic metastable-ion techniques.[156]

Direct-probe inlet of urinary samples and constant neutral scanning with a triple-quadrupole analyser in the hydroxide-ion chemical-ionization mode have permitted rapid detection of over 100 organic acids.[157] However, the method

[151] P. M. Woollard, *Biomed. Mass Spectrom.*, 1983, **10**, 143.
[152] S. K. Sethi, P. F. Crain, and J. A. McCloskey, *J. Chromatogr.*, 1983, **254**, 109.
[153] H. Todoriki, T. Hayashi, H. Naruse, and S. Ikeda, *J. Chromatogr.*, 1982, **232**, 394.
[154] R. A. Yost and D. D. Fetterolf, *Mass Spectrom. Rev.*, 1983, **2**, 1.
[155] R. D. Voyksner, J. R. Hass, G. W. Sovocool, and M. M. Bursey, *Anal. Chem.*, 1983, **55**, 744.
[156] K. Lindstroem and R. Schubert, *J. High Resolut. Chromatogr. Chromatogr. Commun.*, 1984, **7**, 68.
[157] D. F. Hunt, A. B. Giordani, G. Rhodes, and D. A. Herold, *Clin. Chem.*, 1982, **28**, 2387.

detects molecular masses rather than identifies the components. Tentative assignment of structures for the acids is based on their previous unambiguous identification by GC/MS. The method also suffers in that carboxylic acids with the same nominal mass are all detected together, so many of the quantitative estimates are higher, by as much as two orders of magnitude, than those reported for individual acids by GC/MS.[157] Hence, metabolic profiling by the non-chromatographic method appears to rely on GC/MS and yet be less reliable. Selected-reaction monitoring with capillary-column GC inlet offers far superior selectivity and good sensitivity.[158] The ease of switching between different reactions with a triple-quadrupole system makes selection of an internal standard less difficult than it is with magnetic-sector instruments, for which multiple metastable-peak monitoring is rarely reported. The large number of instrumental conditions that can be optimized for maximum sensitivity and selectivity in selected-reaction monitoring, even for the relatively simple triple-quadrupole instrument, leads to great flexibility[159] but holds also the potential for complexity and irreproducibility. Relative to GC/MS, sample preparation can be reduced or eliminated through the extra selectivity of using a triple-quadrupole analyser as a GC detector.[160] Several examples of environmental analysis appear to confirm this conclusion (but recall the caution on use of crude samples[88]). The author wisely comments that 'metastable-peak monitoring combines all the advantages of GC/MS and MS/MS techniques except for the speed of analysis [of the latter]'.[160] A comparison of response factors determined for 53 pollutants by triple- and single-quadrupole GC/MS showed that quantification with the former system was possible with precision similar to that of conventional GC/MS.[161] Quantitative assay of analytes whose structures have been determined by collisional-activation methods is clearly viable using the triple-quadrupole system.[161] Applications of the triple-quadrupole instrument coupled to a gas chromatograph include detection of 500 fg of 2,3,7,8-TCDD by monitoring the loss of $COCl^{\bullet}$ from molecular ions during capillary-column GC/MS[162] and a study of the effect of methyl substitution on the relative ease of various fragmentation routes of cyclohexanones.[163]

Selected metastable-peak monitoring has been used to determine the concentrations of 3- and 4-hydroxyphenylacetic acids (as their pentafluoropropionyl trifluoroethyl esters) in single caudate nuclei[164] and for steranes[165] and tricyclic terpanes[166] in geochemical samples. Monitoring of m/z 217 for steranes often affords complex chromatograms, but increased selectivity is available by monitor-

[158] R. Endele and M. Senn, *Int. J. Mass Spectrom. Ion Phys.*, 1983, **48**, 81.
[159] J. R. B. Slayback and P. A. Taylor, *Spectra*, 1983, **9** (3), 18.
[160] R. F. Bonner, *Int. J. Mass Spectrom. Ion Phys.*, 1983, **48**, 311.
[161] A. D. Sauter, L. D. Betowski, and J. M. Ballard, *Anal. Chem.*, 1983, **55**, 116.
[162] B. Shushan, J. E. Fulford, B. A. Thomson, W. R. Davidson, L. M. Danylewych, A. Ngo, S. Nacson, and S. D. Tanner, *Int. J. Mass Spectrom. Ion Phys.*, 1983, **46**, 225.
[163] W. J. Richter, W. Blum, and H. Schwarz, *Int. J. Mass Spectrom. Ion Phys.*, 1983, **46**, 221.
[164] D. A. Durden, *J. Neurosci. Methods*, 1983, **7**, 61.
[165] G. A. Warburton and J. E. Zumberge, *Anal. Chem.*, 1983, **55**, 123.
[166] J. M. Moldowan, W. K. Seifert, and E. J. Gallegos, *Geochim. Cosmochim. Acta*, 1983, **47**, 1531.

ing the unimolecular decompositions, m/z 372 → 217, m/z 386 → 217, and m/z 400 → 217, in one capillary-column GC/MS analysis by programmed changes in the accelerating voltage.[165] Linked scanning at constant B^2/E for the product ion at m/z 191, again during capillary-column GC/MS, was utilized to aid identification of tricyclic terpanes in petroleum.[166]

A new approach in metastable-ion techniques has begun to emerge with time-resolved magnetic dispersion.[43, 167] Information on all product ions of all precursor ions can be obtained very rapidly (certainly on the chromatographic time-scale) with a single-focusing magnetic-sector mass spectrometer modified to record momentum and velocity data simultaneously. This is accomplished through pulsing of the ion beam at the ion source and time-resolved detection, relying on the fact that a product ion has essentially the same velocity as its precursor. The approach offers excellent potential because the necessary basic instrumentation is simple and common and because in a single scan the data gained will contain the full gamut of metasable-ion mass spectra (constant precursor, constant product, and constant neutral species).[167] Improvements in sensitivity, detection, and acquisition are needed to fulfil its promise. Time-array detection and a transient recorder (capable of acquiring 25 000 spectra per second!) are currently being developed. It should be noted that, whilst the authors rightly point out the potential for very powerful GC/MS/MS and LC/MS/MS, the term 'MS/MS' is a nonsense since there are no two sequential mass analyses with their method.

Kinetic energy released on decomposition of metastable ions has been measured by mass-analysed ion-kinetic-energy spectroscopy of capillary-column GC eluents.[168, 169] The measurements can be made at nanogramme[168] or sub-nanogramme levels[169] and are a sensitive probe of structure, allowing isomeric brominated[168] and chlorinated[169] compounds to be differentiated. The method is recommended for complex mixtures of isomers that are difficult or impossible to unravel by conventional GC/MS.

3 Applications: Gas Chromatography/Mass Spectrometry

Long-chain Compounds. – A useful review surveying a wide range of current methods, including GC/MS and LC/MS, for analysis of glycerolipids has been published.[170] Complex mixtures of wax esters, steryl esters, triacylglycerols, and alkyldiacylglycerols from geochemical and biological sources have been characterized by capillary-column GC/MS.[171] Neutral lipids containing up to 65 carbon atoms were analysed intact on short SE-52 and SE-30 glass columns. Wax and steryl esters were best examined by methane CI, whilst EI mass spectra were helpful for identifying the glycerides.[171] Packed-column GC/CIMS with different

[167] J. T. Stults, C. G. Enke, and J. F. Holland, *Anal. Chem.*, 1983, **55**, 1323.
[168] J. R. Hass, Y. Tondeur, and R. D. Voyksner, *Anal. Chem.*, 1983, **55**, 295.
[169] R. D. Voyksner, J. R. Hass, M. M. Bursey, and W. R. Kenan, jun., *Anal. Chem.*, 1983, **55**, 914.
[170] A. Kuksis, L. Marai, and J. J. Myher, *J. Chromatogr.*, 1983, **273**, 43.
[171] S. G. Wakeham and N. M. Frew, *Lipids*, 1982, **17**, 831; S. G. Wakeham, *Geochim. Cosmochim. Acta*, 1982, **46**, 2239.

reactant gases has also been utilized for analysis of wax esters.[172] As well as the cyclic boronates already described,[140] TMS methyl ester derivatives have been employed for characterizing the vicinal dihydroxyacids that are hepatic metabolites of linolenic and linoleic acids.[173] Hepatic, hydroxylated metabolites of linoleic, arachidonic, and docosahexaenoic acids have been examined also as TMS methyl esters.[174] Selected-ion monitoring has been used with a deuteriated palmitate internal standard to measure ^{13}C-labelled palmitate in a biological tracer study[175] and with methyl nonadecanoate as internal standard for assay of myocardial free fatty acids down to the picogramme level.[176]

Determination of double-bond positions in long-chain compounds is still an active area. Polyunsaturated fatty acids can yield spectra diagnostic of the double-bond positions following reduction with deuteriodi-imide and conversion to pyrrolidide derivatives[177] or by formation of the methoxy derivative.[178] It is claimed that conventional EI spectra acquired during GC/MS are sufficiently diagnostic to locate double bonds in seven positional isomers of dodecenol acetate[179] and C_{12}–C_{18} acetates and alcohols[180] without derivatization. In the latter study[180] the double-bond position was determined only to within ±1 carbon atom.

Analysis of autoxidized fats by GC/MS has continued with the examination of the volatile products of thermal decomposition of hydroperoxy cyclic peroxides in a GC injector port.[181] The fatty-acid synthetase reaction has been probed by tracer studies using the labelled fatty-acid precursors [2-$^{2}H_2$]malonyl-CoA[182] and [^{2}H]propionyl-CoA.[183] The latter was the primary substrate for biosynthesis of heptadecanoic acid by *Brevibacterium ammoniagenes.* A series of studies on the metabolism of unsaturated fatty acids includes a deuterium-labelling study of the stereochemistry of the reaction catalysed by *trans*-2-enoyl-coenzyme A reductase from *E. coli.*[184] One strategy for bacterium identification and classification is to characterize the long-chain fatty-acid esters that they produce. Derivatization to TBDMS ethers followed by GC/MS has been applied to mixtures

[172] R. D. Plattner and G. F. Spencer, *Lipids*, 1983, **18**, 68.
[173] E. H. Oliw, *Biochem. Biophys. Res. Commun.*, 1983, **111**, 644.
[174] H. Hughes, C. V. Smith, E. C. Horning, and J. R. Mitchell, *Anal. Biochem.*, 1983, **130**, 431.
[175] P. F. Bougneres and D. M. Bier, *J. Lipid Res.*, 1982, **23**, 502.
[176] D. H. Hunneman and C. Schweickhardt, *J. Mol. Cell. Cardiol.*, 1982, **14**, 339.
[177] A. Kawaguchi, Y. Kobayashi, Y. Ogawa, and S. Okuda, *Chem. Pharm. Bull.*, 1983, **31**, 3228.
[178] J. Lankelma, E. Ayanoglu, and C. Djerassi, *Lipids*, 1983, **18**, 853.
[179] M. Horiike and C. Hirano, *Agric. Biol. Chem.*, 1982, **46**, 2667; M. Horiike and C. Hirano, *Biomed. Mass Spectrom.*, 1984, **11**, 145.
[180] B. A. Leonhardt, E. D. DeVilbiss, and J. A. Klun, *Org. Mass Spectrom.*, 1983, **18**, 9.
[181] E. N. Frankel, W. E. Neff, and E. Selka, *Lipids*, 1983, **18**, 353.
[182] K. Saito, A. Kawaguchi, S. Nozoe, Y. Seyama, and S. Okuda, *Biochem. Biophys. Res. Commun.*, 1982, **108**, 995.
[183] K. Arai, A. Kawaguchi, Y. Saito, N. Koike, S. Okuda, Y. Seyama, and T. Yamakawa, *J. Biochem.*, 1982, **91**, 11.
[184] M. Mizugaki, T. Nishimaki, T. Shiraishi, H. Yamanaka, A. Kawaguchi, and S. Okuda, *J. Biochem.*, 1982, **92**, 1649.

extracted from *Mycobacterium tuberculosis*,[185] and several chromatographic techniques were utilized for identifying pathogenic coryneform bacteria.[186]

Prostaglandins and Related Eicosanoids. – The value of GC/MS in this area is such that it attracts some excellent work. Recent advances have occurred, particularly in sample-preparation procedures, application of derivatization, and ionization methods. Many developments are aimed at increasing the efficiency of quantitative analysis of eicosanoids. Several review articles have appeared,[187–191] and many useful studies incorporating GC/MS methodologies have been published in a book.[192]

Arachidonic acid metabolites can be isolated rapidly by using disposable octadecylsilane reversed-phase columns.[193] For both prostaglandins and thromboxanes, TBDMS ether derivatives are useful.[194–196] In particular, the $(M - Bu^{\bullet})^+$ ion affords a large (often the base) peak that can be monitored selectively for quantitative analysis. Thromboxane B_2 (TXB_2) and 6-keto-prostaglandin $F_{1\alpha}$ (6-keto-$PGF_{1\alpha}$) in blood plasma were determined in this way by capillary-column GC/MS;[196] the detection limit was 50 pg. Whilst TBDMS derivatives may be the current best option for positive-ion mass spectrometry, the combination of a more electronegative derivative and negative-ion chemical ionization (GC/NICIMS) can afford highly sensitive and selective analytical methods.[197–203] The

185 A. I. Mallet, D. E. Minnikin, and G. Dobson, *Biomed. Mass Spectrom.*, 1984, **11**, 79.
186 M. Athalye, W. C. Noble, A. I. Mallet, and D. E. Minnikin, *J. Gen. Microbiol.*, 1984, **130**, 513.
187 C. Fischer and J. C. Froelich, *Adv. Lipid Res.*, 1982, **19**, 185.
188 I. A. Blair, *Br. Med. Bull.*, 1983, **39**, 223.
189 G. Trugnan, *Vie Med.*, 1983, **64**, 151.
190 I. W. Reimann, C. Fischer, B. Rosenkranz, and J. C. Froelich, *Prostaglandins Kidney: Biochem., Physiol., Pharmacol., Clin. Appl.*, 1983, 99.
191 H. Kindahl and E. Granstroem, *Acta Obstet. Gynecol. Scand., Suppl.*, 1983, **113**, 15.
192 'Methods in Enzymology, Vol. 86: Prostaglandins and Arachidonate Metabolites', ed. W. E. M. Lands and W. L. Smith, Academic, New York, 1982.
193 J. R. Luderer, D. L. Riley, and L. M. Demers, *J. Chromatogr.*, 1983, **273**, 402.
194 J. Mai, B. German, and J. E. Kinsella, *J. Chromatogr.*, 1983, **254**, 91.
195 H. Matsuda, M. Inui, T. Kuzuya, and M. Tada, *Koenshu - Iyo Masu Kenkyukai*, 1982, **7**, 189.
196 H. Matsuda, T. Kuzuya, M. Inui, and M. Tada, *Koenshu - Iyo Masu Kenkyukai*, 1983, **8**, 155.
197 K. A. Waddell, C. Robinson, M. A. Orchard, S. E. Barrow, C. T. Dollery, and I. A. Blair, *Int. J. Mass Spectrom. Ion Phys.*, 1983, **48**, 233.
198 K. A. Waddell, I. A. Blair, and J. Welby, *Biomed. Mass Spectrom.*, 1983, **10**, 83; K. A. Waddell, S. E. Barrow, C. Robinson, M. A. Orchard, C. T. Dollery, and I. A. Blair, *ibid.*, 1984, **11**, 68.
199 I. A. Blair, S. E. Barrow, K. A. Waddell, P. J. Lewis, and C. T. Dollery, *Adv. Prostaglandin, Thromboxane, Leukotriene Res.*, 1983, **11**, 197.
200 C. Robinson, J. R. S. Hoult, K. A. Waddell, I. A. Blair, and C. T. Dollery, *Biochem. Pharmacol.*, 1984, **33**, 395.
201 R. J. Strife and R. C. Murphy, *J. Chromatogr.*, 1984, **305**, 3; *Prostaglandins, Leukotrienes Med.*, 1984, **13**, 1.
202 H. Miyazaki, M. Ishibashi, H. Takayama, K. Yamashita, I. Suwa, and M. Katori, *J. Chromatogr.*, 1984, **289**, 249.
203 B. J. Smith, D. A. Herold, R. M. Ross, F. G. Marquis, R. L. Bertholf, C. R. Ayers, M. R. Wills, and J. Savory, *Res. Commun. Chem. Pathol. Pharmacol.*, 1983, **40**, 73.

preferred derivatives are pentafluorobenzyl esters with selected-ion monitoring of the abundant $(M - C_6F_5CH_2^{\bullet})^-$ ion during capillary-column GC/MS. Detection limits of 1–8 pg for prostanoids,[197,198] 0.5 pg ml^{-1} of 6-oxo-$PGF_{1\alpha}$ in human plasma,[199] and 200 fg of $PGF_{2\alpha}$[202] have been recorded.

Metabolism of arachidonic acid has been the subject of a great many papers. The recommended GC/MS coupling of introducing the end of a flexible capillary column directly into the ion source has distinct advantages for labile arachidonic acid metabolites, mainly because active surfaces are eliminated.[204] The method was illustrated with the identification by bonded-phase capillary-column GC/EIMS and GC/CIMS of six eicosatetraenoic acids from fibroblast cultures. The main metabolite of arachidonic acid in human-skin fibroblast cultures was shown to be PGE_2, and the cyclo-oxygenase pathway was more important than that of lipoxygenase.[205] 5,6-, 8,9-, 11,12-, and 14,15-epoxyeicosatrienoic acids have been shown to be arachidonic acid metabolites in rat liver.[206] 11,12-Epoxy-5,8,14- and 14,15-epoxy-5,8,11-eicosatrienoic acids were identified as a variety of different derivatives after incubating arachidonic acid with rabbit-liver microsomes.[207] Not surprisingly, the vicinal diols, 11,12-dihydroxy-5,8,14- and 14,15-dihydroxy-5,8,11-eicosatrienoic acids, could be shown, by selected-ion monitoring against octadeuteriated analogues, also to be produced when arachidonic acid is incubated with isolated rat hepatocytes and renal cells.[208] Application of boronate derivatives[140] in such an analysis would be interesting. The presence of lipoxygenase pathways for arachidonic acid in murine eosinophils has been established by packed-column GC/MS[209] and in mouse peritoneal macrophages by high-efficiency capillary-column GC/MS.[210] Several arachidonic acid metabolites have been determined in normal, benign, and malignant human mammary tissues. 6-Keto-$PGF_{1\alpha}$ was found to be present in higher concentrations in breast carcinomas than in normal tissues.[211]

Quantitative measurement of PGE_2 has been effected through extensive derivatization procedures.[212–214] The readily prepared 9-enol-PGE_2-methyl ester TMS ether was utilized for selected-ion monitoring of nanogramme amounts of PGE_2 in human urine, the internal standard being a tetradeuteriated analogue.[212] Simultaneous quantification of PGE_2 and PGE_3 has been accomplished after conversion to the methyl ester TMS ether derivatives of their PGB_2 and PGB_3 counterparts, respectively.[213] A linear response in the low nanogramme range

[204] H. Gleispach, B. Mayer, L. Rauter, and E. Wurz, *J. Chromatogr.*, 1983, **273**, 166.
[205] H. Gleispach, B. Mayer, R. Moser, and H. Leis, *Fresenius' Z. Anal. Chem.*, 1984, **317**, 740.
[206] N. Chacos, J. R. Falck, C. Wixtrom, and J. Capdevila, *Biochem. Biophys. Res. Commun.*, 1982, **104**, 916.
[207] E. H. Oliw, F. P. Guengerich, and J. A. Oates, *J. Biol. Chem.*, 1982, **257**, 3771.
[208] E. H. Oliw and P. Moldeus, *Biochim. Biophys. Acta*, 1982, **721**, 135.
[209] J. Turk, T. H. Rand, R. L. Maas, J. A. Lawson, A. R. Brash, L. J. Roberts, D. G. Colley, and J. A. Oates, *Biochim. Biophys. Acta*, 1983, **750**, 78.
[210] H. Rabinovitch-Chable, J. Durand, J. C. Aldigier, P. Chebroux, N. Gualde, J. L. Beneytout, and M. Rigaud, *Prostaglandins, Leukotrienes Med.*, 1984, **13**, 9.
[211] I. F. Stamford, M. A. Carroll, A. Civier, C. N. Hensby, and A. Bennett, *J. Pharm. Pharmacol.*, 1983, **35**, 48.
[212] C. Fischer, *Biomed. Mass Spectrom.*, 1984, **11**, 114.
[213] A. Ferretti, V. P. Flanagan, and J. M. Roman, *Lipids*, 1982, **17**, 825.

was observed, again with tetradeuteriated PGE_2 as internal standard. Selected-ion monitoring of the molecular ion of the same PGB_2 derivative has served to determine PGE_2 in gastrointestinal fluids.[214] Prostaglandin $F_{2\alpha}$ as its triacetyl methyl ester was also quantified in this study, along with a comparison of radioimmunological assay and the packed-column GC/MS technique.[214] A detection limit of 40 pg was achieved for determination of 16,16-dimethyl-*trans*-Δ^2-PGE, as its TBDMS ether methyl ester, against a dideuteriated internal standard.[215] 13,14-Dihydro-15-keto-PGE_2 is a major metabolite of PGE_2 in the human uterus. An assay that is sensitive to 2 ng of the metabolite in tissue samples utilized the oxime TBDMS methyl ester derivative.[216]

Urinary levels of the prostacyclin metabolites 6-keto-$PGF_{1\alpha}$ and 2,3-dinor-6-keto-$PGF_{1\alpha}$ have been examined in adults and neonates by capillary-column GC/MS. The large difference in relative proportions of the metabolites in adults and neonates indicated an inverse relationship between the activity of β-oxidation and prostacyclin formation.[217] Capillary-column GC/MS afforded assays of $PGF_{2\alpha}$ and 6-oxo-$PGF_{1\alpha}$ down to levels of 20 pg ml^{-1} of urine.[218] Ammonia chemical ionization, tetradeuteriated analogues as internal standards, and a fused-silica bonded-phase column were used. Monitoring of the long-lived urinary metabolite of $PGF_{2\alpha}$, 9α,11α-dihydroxy-15-oxo-2,3,4,5,20-pentanor-19-carboxyprostanoic acid (PGF-M), has been proposed as an indication of total body synthesis of $PGF_{2\alpha}$.[219] Unlike most studies reported here, the internal standard was a homologue, the 1-methyl-20-ethyl ester of PGF-M.

As an approach to investigating the low incidence of cardiovascular disease amongst Eskimos, whose diet is high in polyunsaturated fats such as eicosapentaenoic acid (EPA), capillary-column GC/MS analysis showed that humans on an EPA-rich diet produce PGI_3, which is anti-aggregating in blood.[220] The major urinary PGI_3 metabolite, Δ^{17}-2,3-dinor-6-keto-$PGF_{1\alpha}$, was identified by GC/MS. It is proposed that dietary changes might induce less thrombogenic states.[220] An automated solid injection system has been applied to determination of 13,14-dihydro-15-keto-PGE_2 as its oxime TBDMS ether methyl ester.[221] Eicosapolyenoic acids in blood serum were measured as their methyl esters by selected-ion monitoring of $(M + NH_4)^+$ ions formed by ammonia CI, using octadeuterioeicosatetraenoic acid as internal standard.[222] Application of capillary columns to this already rapid assay would be worthwhile. The roles of prostaglandins and thromboxane in the accumulation of exudate in rat carrageenin-induced pleurisy have been investigated by GC/MS using various dimethylisopropylsilyl ether derivatives.[223] The daily excretion rates of 2,3-dinor-thromboxane B_2 in healthy

[214] K. Bukhave, K. Green, and J. Rask-Madsen, *Biomed. Mass Spectrom.*, 1983, **10**, 265.
[215] V. Dimov, K. Green, M. Bygdeman, Y. Konishi, K. Imaki, and M. Hayashi, *Prostaglandins*, 1983, **25**, 225.
[216] R. W. Kelly and M. H. Abel, *Biomed. Mass Spectrom.*, 1983, **10**, 276.
[217] S. Fischer, B. Scherer, and P. C. Weber, *Biochim. Biophys. Acta*, 1983, **750**, 127.
[218] H. Gleispach, B. Mayer, L. Rauter, and E. Wurz, *J. Chromatogr.*, 1983, **273**, 161.
[219] A. Ferretti, V. P. Flanagan, and J. M. Roman, *Anal. Biochem.*, 1984, **136**, 217.
[220] S. Fischer and P. C. Weber, *Nature (London)*, 1984, **307**, 165.
[221] R. W. Kelly, *Anal. Proc.*, 1982, **19**, 256.
[222] M. Suzuki, M. Nishizawa, T. Miyatake, and Y. Kagawa, *J. Lipid Res.*, 1982, **23**, 363.
[223] Y. Harada, K. Tanaka, Y. Uchida, A. Ueno, S. Ohishi, K. Yamashita, M. Ishibashi, H. Miyazaki, and M. Katori, *Prostaglandins*, 1982, **23**, 881.

humans have been determined by GC/MS.[224] Capillary columns have been found to be mandatory for separation of the stable cyclo-oxygenase metabolites of arachidonic acid ($PGF_{2\alpha}$, PGE_2, PGD_2, 6-keto-$PGF_{1\alpha}$, and thromboxane B_2).[225] Rapid quantitative profiling of prostaglandins and thromboxane has been reported by addition of deuteriated analogues as internal standards to the biological samples and single-step extraction on C_{18} reversed-phase cartridges followed by derivatization and selected-ion monitoring during capillary-column GC/MS.[226]

Isoprenoid Compounds. – Application of GC/MS to terpenoid profiling is very well established. Some recent examples are listed here.

Three novel acyclic triterpenes (botryococcenes) have been found in the alga *Botryococcus braunii*,[227] and the unstable tetracyclic triterpenes eupha-7,24-diene and (20*R*)-dammara-13(17),24-diene were isolated from fresh leaves of *Polypodium* species.[228] Several fungitoxin terpenoids have been identified in tobacco leaves.[229] The terpenoid content of *Mentha spicata* was characterized by both GC/EIMS and GC/NICIMS.[230]

A major growth inhibitor of tulip bulbs has been shown by GC/MS to be abscisic acid.[231] The same acid has been quantified, against deuteriated internal standards, in plant tissues.[232,233] Determination of abscisic acid and 2-*trans*-abscisic acid in eucalyptus leaves was effected by GC/CIMS,[232] and both abscisic and indol-3-ylacetic acids were quantified simultaneously in *Zea mays* roots.[233] The changes in ratio of *cis–trans* to *trans–trans* abscisic acid during ripening of apple fruits have been studied partly by capillary-column GC/MS.[234]

Whole-body extracts of mosquitos have been examined by selected-ion monitoring for juvenile hormones.[235] Previously detected JH-I and JH-II were absent, but JH-III was detected. Analysis of anhydroretinol by GC/EIMS and GC/CIMS has served as a useful indicator of plasma retinol. The anhydroretinol is formed from retinol directly on GC columns, and optimization of conditions afforded a linear response with nanogramme amounts of injected retinol.[236] Mono-unsaturated and tri-unsaturated forms of phytanic acid have been identified, after derivatization to locate the double bonds and use of GC/EIMS and GC/

[224] Q. Vesterqvist, K. Green, F. H. Lincoln, and O. K. Sebek, *Thromb. Res.*, 1984, **33**, 39.
[225] C. Chiabrando, A. Noseda, and R. Fanelli, *J. Chromatogr.*, 1982, **250**, 100.
[226] C. Chiabrando, A. Noseda, M. N. Catagnoli, M. Romano, and R. Fanelli, *J. Chromatogr.*, 1983, **279**, 581.
[227] P. Metzger and E. Casadevall, *Tetrahedron Lett.*, 1983, **24**, 4013.
[228] Y. Arai, K. Masuda, and H. Ageta, *Chem. Pharm. Bull.*, 1982, **30**, 4219.
[229] A. Fuchs, W. Slobbe, P. C. Mol, and M. A. Posthumus, *Phytochemistry*, 1983, **22**, 1197.
[230] V. P. Papageorgiou, N. Arguriadou, S. Kokkini, and A. S. Mellidis, *Chem. Chron.*, 1983, **12**, 27.
[231] P. H. Terry, L. H. Aung, and A. A. De Hertogh, *Plant Physiol.*, 1982, **70**, 1574.
[232] A. G. Netting, B. V. Milborrow, and A. M. Duffield, *Phytochemistry*, 1982, **21**, 385.
[233] L. Rivier and P. E. Pilet in ref. 5, p. 219.
[234] F. Bangerth, *Planta*, 1982, **155**, 199.
[235] F. C. Baker, H. H. Hagedorn, D. A. Schooley, and G. Wheelock, *J. Insect Physiol.*, 1983, **29**, 465.
[236] M. E. Cullum, J. A. Olson, and S. W. Veysey, *Int. J. Vitam. Nutr. Res.*, 1984, **54**, 3; M. E. Cullum, M. H. Zile, and S. W. Veysey, *ibid.*, p. 11.

CIMS, in plasma of a patient with Refsum's disease.[237] More than 30 gibberellins and kaurenoids were detected by GC/MS in seed extracts of *Cucurbita maxima*, including some new 12α- and 12β-hydroxy derivatives.[238] Gibberellin A_9 has been identified, after methyl esterification and fused-silica capillary-column GC/MS, at levels of 1 ng g^{-1} of Norway spruce.[239]

Oxygen Heterocycles and Phenols. – Analysis of cannabinoids in biological fluids[240] forms the major activity in this section. In common with trends in prostaglandin and environmental applications, negative-ion mass spectrometry is a growth area. The pentafluoropropyl pentafluoropropionyl derivative of Δ^1-tetrahydrocannabinol-11-oic acid was utilized for a capillary-column GC/NICIMS assay of the parent compound in urine.[241] The advantage of the method over positive-ion GC/EIMS and GC/CIMS in terms of sensitivity was reported to be 200-fold. On the other hand, ammonia positive-ion CI of TMS derivatives[242] and NICI of trifluoroacetyl derivatives of Δ^1-tetrahydrocannabinol (Δ^1-THC) and two metabolites[243] both permitted quantification down to levels of about 0.2 ng ml^{-1} in plasma, blood, or urine. Other workers[244] have favoured the use of pentafluorobenzyl derivatives for GC/NICIMS determination of Δ^1-THC, cannabinol, and cannabidol in human blood and urine. Oxygenated constituents of the essential oil of *Cannabis sativa* L. have been tentatively identified by GC/NICIMS.[245] Conventional GC/EIMS of TMS derivatives of hydroxylated metabolites of Δ^6-THC suggests that 1,6-epoxidation occurs in human-liver microsomes.[246] Two radioimmunoassay techniques have been compared with GC/MS for forensic analysis of Δ^1-THC.[247]

Details of the measurement of the first mammalian lignans, enterolactone and enterodiol, in physiological fluids have been described.[248] Capillary-column GC/MS has been used to determine warfarin and its hydroxylated metabolites extracted from microsomal incubations.[249] Trimethylsilylation of isoflavones is reported to afford by-products that are isomeric with the required derivative,[250] and trimethylsilylation of hydroxylated anthraquinones gave derivatives with

[237] J. E. Evans and J. T. Dulaney, *Biochim. Biophys. Acta*, 1983, **752**, 346.
[238] S. Blechschmidt, U. Castel, P. Gaskin, P. Hedden, J. E. Graebe, and J. MacMillan, *Phytochemistry*, 1984, **23**, 553.
[239] P.-C. Oden, B. Andersson, and R. Gref, *J. Chromatogr.*, 1982, **247**, 133.
[240] 'The Analysis of Cannabinoids in Biological Fluids', ed. R. Hawks, U.S. Government Printing Office, Washington, 1982.
[241] L. Karlsson, J. Jonsson, K. Aaberg, and C. Roos, *J. Anal. Toxicol.*, 1983, **7**, 198.
[242] R. L. Foltz and B. J. Hidy, *NIDA Res. Monogr.*, 1982, **42**, 99.
[243] R. L. Foltz, K. M. McGinnis, and D. M. Chinn, *Biomed. Mass Spectrom.*, 1983, **10**, 316.
[244] H. Hattori, O. Suzuki, and M. Asano, *Koenshu - Iyo Masu Kenkyukai*, 1983, **8**, 159.
[245] H. Hendriks and A. P. Bruins, *Biomed. Mass Spectrom.*, 1983, **10**, 377.
[246] I. Yamamoto, S. Narimatsu, K. Watanabe, T. Shimonishi, H. Yoshimura, and T. Nagano, *Chem. Pharm. Bull.*, 1983, **31**, 1784.
[247] V. W. Hanson, M. H. Buonarati, R. C. Baselt, N. A. Wade, C. Yep, A. A. Biasotti, V. C. Reeve, A. S. Wong, and M. W. Orbanowski, *J. Anal. Toxicol.*, 1983, **7**, 96.
[248] K. D. R. Setchell, A. M. Lawson, L. M. McLaughlin, S. Patel, D. N. Kirk, and M. Axelson, *Biomed. Mass Spectrom.*, 1983, **10**, 227.
[249] E. D. Bush, L. K. Low, and W. F. Trager, *Biomed. Mass Spectrom.*, 1983, **10**, 395.
[250] M. D. Woodward, *Phytochemistry*, 1982, **21**, 1403.

poor chromatographic properties.[251] Reductive silylation was found to eliminate the latter problem. The degradation products of lignins from both deciduous and coniferous wood have been subjected to GC/MS as TMS derivatives. The former products were compounds of syringyl and guaicyl types and the latter of guaicyl type only.[252] Traces of syringyl compounds in *Pinus radiata* lignin were identified by pyrolysis GC/MS.[253]

Carbohydrates. – A review of mass-spectrometric strategies for determination of monosaccharides, oligosaccharides, and glycoconjugates is recommended reading,[254] and a book on glycoconjugates contains many pertinent and useful reports.[255] A review of mass-spectral methods applied to iridoids also covers GC/MS.[256]

Amino sugars have come under considerable examination. Analysis of urinary sialic acids has been based on their dissociation into pyruvate and *N*-acylmannosamines by the action of *N*-acetylneuraminate lyase, followed by GC and GC/MS of the latter products as TMS diethyl dithioacetals.[257] It was observed that cancer patients excreted greater concentrations of *N*-acetylneuraminic acid than normal subjects. Salla disease patients also excrete increased levels of *N*-acetylneuraminic acid, and abnormally large amounts were found in liver and skin fibroblasts of such patients.[258] The formation and GC/MS behaviour of per-*O*-TMS ketoxime derivatives of neuraminic acid have been described.[259] Seven novel sialic acids in bovine submandibular-gland glycoprotein have been detected following an improved isolation procedure.[260] Using *N*-[$^{2}H_3$]acetylneuraminic acid as internal standard and the TMS derivative, the natural compound was quantified down to the picogramme range.[261] Application of the method to plasma samples showed greater levels of *N*-acetylneuraminic acid in lung-cancer patients than in controls. Sialic acids can be analysed not only by GC/MS of TMS derivatives but also underivatized by HPLC/MS.[262] The permethylated alditols derived from hexosamine-containing oligosaccharides and glycosphingolipids often exhibit poor chromatographic properties inasmuch as their retention times are rather long. Trifluoroacetolysis of oligosaccharides affords permethyl-

[251] L. M. Henriksen and H. Kjøsen, *J. Chromatogr.*, 1983, **258**, 252.
[252] R. Concin, P. Burtscher, E. Burtscher, and O. Bobleter, *Int. J. Mass Spectrom. Ion Phys.*, 1983, **48**, 63.
[253] T. J. Fullerton and R. A. Franich, *Holzforschung*, 1983, **37**, 267.
[254] V. N. Reinhold and S. A. Carr, *Mass Spectrom. Rev.*, 1983, **2**, 153.
[255] 'Glycoconjugates', ed. M. A. Chester, D. Heinegard, A. Lundblad, and S. Svensson, Proceedings of the 7th International Symposium on Glycoconjugates, Secretariat, Lund, 1983.
[256] S. S. Popov and N. V. Handjieva, *Mass Spectrom. Rev.*, 1983, **2**, 481.
[257] K. Kakehi, K. Maeda, M. Teramae, S. Honda, and T. Takai, *J. Chromatogr.*, 1983, **272**, 1.
[258] M. Renlund, M. A. Chester, A. Lundblad, J. Parkkinen, and T. Krusius, *Eur. J. Biochem.*, 1983, **130**, 39.
[259] T. P. Mawhinney, M. A. Madson, R. H. Rice, M. S. Feather, and G. J. Barbero, *Carbohydr. Res.*, 1982, **104**, 169.
[260] G. Reuter, R. Pfeil, S. Stoll, R. Schauer, J. P. Kamerling, and J. F. G. Vliegenthart, *Eur. J. Biochem.*, 1983, **134**, 139.
[261] Y. Sugawara, M. Iwamori, J. Portoukalian, and Y. Nagai, *Anal. Biochem.*, 1983, **132**, 147.
[262] R. Schauer, A. K. Shukla, C. Schroeder, and E. Müller, *Pure Appl. Chem.*, 1984, **56**, 907.

ated *N*-trifluoroacetates with increased volatility. The improved procedure has been reported in detail.[263]

Quantification of glucose in clinical and nutritional science has been reviewed.[264] Derivatization of glucose to 1,2:5,6-di-*O*-isopropylidene-α-D-glucofuranose and packed-column GC/MS have been compared with the use of the 1,2:3,5-bis(butaneboronate)-6-acetate and capillary-column GC/MS.[265] Both methods employed uniformly labelled [^{13}C]glucose as internal standard for determination of serum glucose. The results agreed well, with both methods having excellent precision, but samples for the former required greater preparation.[265] Other workers[266] also favoured the diboronate-6-acetate derivative for measuring serum glucose, this time against a dideuteriated internal standard. Yet another method for glucose in serum has been reported.[267] It utilized a tetradeuteriated analogue as internal standard and fused-silica-column GC/MS. Galactose, mannose, and fructose were also determined.[267]

In a useful paper[268] the GC/MS behaviour of *O*-acetyl-*O*-methylgalactononitriles has been summarized. Several mono-*O*-methylaldoses have been resolved from common aldoses as their isopropylidene derivatives.[269] Several different derivatives, including the isopropylidenes, have been used to characterize hexosuloses and pentosuloses,[270] and TMS isopropylidene derivatives of sorbose have been examined by capillary-column GC/MS.[271] Enantiomeric tetruloses and pentuloses were characterized as their diastereoisomeric trifluoroacetylated *O*-(−)-menthyloxime derivatives by GC/MS with an OV-225 glass capillary column.[272] Chemical ionization was preferred over EI for determination of trifluoroacetylated 2- and 3-pentulose, 2- and 3-hexulose, and 2,5-hexodiulose *O*-butoximes.[273,274] A fused-silica capillary column (SP-2100) was considered to be superior to a glass capillary column (SP-2250) for resolving methoxime TMS derivatives of many monosaccharides. Quantification was effected by selected-ion monitoring and an isotope-dilution method.[275] The presence of alditols in eye lenses is well known, but examination of late-eluting components by capillary-column GC/MS demonstrated for the first time that lenses also contained a heptitol and an octitol.[276] Chemical characterization of micro-organisms

[263] B. Nilsson and D. Zopf, *Methods Enzymol.*, 1982, **83**, 46.

[264] D. E. Matthews and D. M. Bier, *Annu. Rev. Nutr.*, 1983, **3**, 309.

[265] E. White, V. M. J. Welch, T. Sun, L. T. Sniegoski, R. Schaffer, H. S. Hertz, and A. Cohen, *Biomed. Mass Spectrom.*, 1982, **9**, 395.

[266] P. Dan, P. M. Clemons, M. I. Sperling, M. J. Gelfand, I. W. Chen, M. A. Sperling, and E. J. Norman, *Anal. Lett.*, 1983, **16** (B9), 655.

[267] O. Pelletier and S. Cadieux, *Biomed. Mass Spectrom.*, 1983, **10**, 130.

[268] C. A. Stortz, M. C. Matulewicz, and A. S. Cerezo, *Carbohydr. Res.*, 1982, **111**, 31.

[269] K. H. Aamlid and S. Morgenlie, *Carbohydr. Res.*, 1983, **124**, 1.

[270] P. A. Garcia Ruiz, I. Cartagena Travesedo, and A. Soler Martinez, *Can. J. Chem.*, 1984, **62**, 870.

[271] R. Novina, *Chromatographia*, 1982, **15**, 704.

[272] H. Schweer, *Carbohydr. Res.*, 1983, **116**, 139.

[273] H. Schweer, *Carbohydr. Res.*, 1982, **111**, 1.

[274] P. Decker, H. Schweer, and R. Pohlmann, *J. Chromatogr.*, 1982, **244**, 281.

[275] O. Pelletier and S. Cadieux, *J. Chromatogr.*, 1982, **231**, 225.

[276] S. A. Barker, F. P. Fish, M. Tomana, L. C. Garner, R. L. Settine, and J. Prchal, *Biochem. Biophys. Res. Commun.*, 1983, **116**, 988.

by capillary-column GC/MS profiling of neutral and amino sugars has been assessed.[277]

Oligosaccharide sequencing by GC/MS is well established. Novel developments only are reported here. Reductive cleavage with an organosilane reagent system (hydrosilylation) can yield monosaccharides that preserve the original ring structure or a series of oligomers. Detection of the oligomers by HPLC has been improved by incorporating fluorophoric groups either before or after reductive cleavage. For complementary analytical data, such oligosaccharides can be completely reduced and characterized by GC/MS as monosaccharide acetates.[278] Traditional methylation analysis of sugars is now well supplemented by FAB mass spectrometry for sequence determination.[279] The use of GC/MS for examining oligosaccharide mixtures produced by *N*-deacetylation and deamination of glycopeptides containing sialic acid has been discussed.[280] Methanolysis of oligosialosyl compounds can give rise to anhydro compounds[281] and some de-*N*-acetylation.[282] Alditols of eight common monosaccharides have been partially methylated then acetylated. The resulting mixtures were analysed by GC/MS to serve as standards for the characterization of all possible partially methylated alditol acetates from methylation analysis of polysaccharides containing those monosaccharides.[283] Application of ammonia CI, a bonded-phase fused-silica capillary column, and selected-ion monitoring has increased the sensitivity available in methylation analysis. Glycoprotein oligosaccharides could be analysed at 100–500 pmol levels.[284] A further refinement of methylation analysis is available by analysing the alditol acetates on capillary columns coated with a mixture of OV-17 and OV-225 (37:13).[285] Methylation analysis of *N*-linked oligosaccharides is possible on intact glycoproteins (*i.e.* no pronase digestion is necessary) if the partially methylated alditol acetates are separated from non-carbohydrate material on a silica-gel column prior to GC/MS analysis.[286] Finally, difficulties in methylation analysis of glycosphingolipids have been overcome by a rapid GC/MS method for identifying oligosaccharides released by trifluoroacetolysis from glycolipids.[287]

Phospholipids. – Both field desorption[288] and FAB mass spectrometry[289] can complement GC/MS studies of phospholipids. Conversion of phospholipids of

277 M. D. Walla, P. Y. Lau, S. L. Morgan, A. Fox, and A. Broun, *J. Chromatogr.*, 1984, **288**, 399.

278 V. N. Reinhold, E. Coles, and S. A. Carr, *J. Carbohydr. Chem.*, 1983, **2**, 1.

279 E. A. Nothnagel, M. McNeil, P. Albersheim, and A. Dell, *Plant Physiol.*, 1983, **71**, 916.

280 I. Mononen, *Carbohydr. Res.*, 1982, **104**, 1.

281 M. R. Lifely and F. H. Cottee, *Carbohydr. Res.*, 1982, **107**, 187.

282 S. Inoue, G. Matsumura, and Y. Inoue, *Anal. Biochem.*, 1982, **125**, 118.

283 J. Klok, H. C. Cox, J. W. De Leeuw, and P. A. Schenck, *J. Chromatogr.*, 1982, **253**, 55.

284 R. Geyer, H. Geyer, S. Kühnhardt, W. Mink, and S. Stirm, *Anal. Biochem.*, 1983, **133**, 197.

285 P. A. J. Gorin, E. M. Giblin, G. P. Slater, and L. Hogge, *Carbohydr. Res.*, 1982, **106**, 235.

286 M. E. Lowe and B. Nilsson, *Anal. Biochem.*, 1984, **136**, 187.

287 B. Nilsson and D. Zopf, *Arch. Biochem. Biophys.*, 1983, **222**, 628.

288 J. Sugatani, M. Kino, K. Saito, T. Matsuo, H. Matsuda, and I. Katakuse, *Biomed. Mass Spectrom.*, 1982, **9**, 293.

289 R. W. Gross, *Biochemistry*, 1984, **23**, 158.

rat brain to diglycerides preceded GC/MS of their TBDMS derivatives,[288] whereas phospholipid constituents of canine myocardial sarcolemma were subjected to methanolysis prior to GC/MS analysis.[289] Various sphingoglycolipids have been identified by GC/CIMS following trimethylsilylation or permethylation,[290] and phosphatidylethanolamines and phosphatidylcholines in rat-brain myelin and synaptosomes have been determined as their TBDMS derivatives.[291] Reduction with $^{2}H_2$ of unsaturated fatty-acyl chains within diacylglycerols derived from phospholipids, followed by formation of TBDMS derivatives, has provided good resolution by GC/MS whilst preserving information on the position and degree of unsaturation of each acid.[292] Two types of platelet-activating factor have been detected in human neutrophils by selected-ion monitoring.[293] Phospholipids of a demosponge have been shown to contain saturated and unsaturated 2-methoxy fatty acids with chain lengths of 19-28 carbon atoms.[294]

Pyrimidines, Purines, Nucleosides, and Nucleotides. – The role of mass spectrometry in nucleic acid research has been reviewed.[295] *De novo* biosynthesis of pyrimidines has been studied by tracing the incorporation of [^{13}C]bicarbonate and $^{13}CO_2$ into the uracil nucleotide pool in L1210 tumours *in vitro* and in mice.[296] Identification and quantification of cyclic guanosine-3′:5′-monophosphate in maize seedlings have been accomplished by GC/MS.[297] Deuteriomethylation of dimethylxanthines has been utilized in a GC/MS method to confirm their presence in extracts of horse urine.[298]

The use of [$^{2}H_9$]trimethylanilinium hydroxide as an on-column permethylation reagent for nucleosides distinguishes native and derivative methyl groups.[139] Approaches to understanding the fragmentation of TMS derivatives of nucleosides are underway.[299] Given the importance of trimethylsilylation in producing volatile and thermally stable derivatives suitable for GC/MS from labile nucleosides, characterization of fragmentation pathways is key for further application to new nucleosides and determination of chemically and biologically incorporated isotopes. Nucleoside TMS derivatives have been examined on a short fused-silica column directly coupled to the ion source of a mass spectrometer equipped with an ion-counting detector for acquiring data on minor constituents and weak signals such as from high-resolution selected-ion monitoring and metastable-peak monitoring.[300] Sub-microgramme amounts of DNA have been analysed for 5-methylcytosine as its TBDMS derivative by an isotope-dilution GC/MS method

[290] T. Ariga, *Koenshu - Iyo Masu Kenkyukai*, 1982, **7**, 49.
[291] M. Kino, T. Matsumura, M. Gamo, and K. Saito, *Biomed. Mass Spectrom.*, 1982, **9**, 363.
[292] B. F. Dickens, C. S. Ramesha, and G. A. Thompson, jun., *Anal. Biochem.*, 1982, **127**, 37.
[293] K. Satouchi, M. Oda, K. Yasunaga, and K. Saito, *J. Biochem. (Tokyo)*, 1983, **94**, 2067.
[294] E. Ayanoglu, S. Popov, J. M. Kornprobst, A. Aboud-Bichara, and C. Djerassi, *Lipids*, 1983, **18**, 830.
[295] M. Linscheid, *Trends Anal. Chem.*, 1983, **2**, 32.
[296] J. M. Strong, L. W. Anderson, A. Monks, C. A. Chisena, and R. L. Cysyk, *Anal. Biochem.*, 1983, **132**, 243.
[297] B. Janistyn, *Planta*, 1983, **159**, 382.
[298] E. Houghton, *Biomed. Mass Spectrom.*, 1982, **9**, 103.
[299] H. Pang, K. H. Schram, D. L. Smith, S. P. Gupta, L. B. Townsend, and J. A. McCloskey, *J. Org. Chem.*, 1982, **47**, 3923.
[300] D. L. Smith and J. A. McCloskey, *Int. J. Mass Spectrom. Ion Phys.*, 1983, **48**, 237.

with a detection limit of 0.3 pmol of the base.[301] 1-Methylpseudouridine is a novel modified nucleoside found in archaebacterial tRNA, in which it replaces ribosylthymine.[302] A model system has been described for studying radiation-induced DNA base damage.[303] Irradiated samples of nucleosides and nucleotides were hydrolysed, silylated (TMS or TBDMS), and analysed by capillary-column GC/MS to identify the products.

Cytokinins are powerful cell-division factors occurring in plant tissues and micro-organisms in low concentrations. They can be analysed by GC/MS as trifluoroacetyl derivatives[304] and quantified against deuteriated analogues as internal standards.[305,306]

Steroids. – As with the previous review period,[2] the sphere of application of GC/MS includes the identification of new and uncommon steroids and metabolites, quantification of known structures in biochemical, clinical, and forensic studies, and confirmation of synthesis of isotope-labelled analogues as either internal standards or precursors for metabolic tracer studies. A book containing several articles of interest for the mass spectrometrist in this area has been published.[307] A review of the isolation, identification, and quantification of steroids in biological samples is recommended.[308] An excellent survey of mass-spectrometric methods for analysis of steroids is available,[309] and the determination of steroid hormones in human endocrine tissues has been described.[310]

A modern approach to isolation and fractionation of steroids has been described.[311] Extraction of conjugated steroids from plasma is accomplished by Amberlite XAD-2 or Sep-Pak C_{18} cartridges, then separation into unconjugated material, glucuronides, monosulphates, and disulphates is achieved on a strong anion exchanger. Hydrolysis of the conjugates and separation of the released steroids into neutral and phenolic fractions on the same anion-exchanger column precede formation of methoxime TMS derivatives and capillary-column GC/MS.[311] Oestrogens with vicinal *cis*-diol groups can be retained strongly by complexation with anion exchangers in their borate form. Hence the steroids are readily resolved from those not containing such diol functions.[312] Borate complexation also protects the labile *ortho*-dihydroxyphenyl function of catechol oestrogens from decomposition during chromatography. The use of 'Tris–HCl' buffers during work-up of saturated 3-oxosteroids must be regarded with caution

[301] P. F. Crain and J. A. McCloskey, *Anal. Biochem.*, 1983, **132**, 124.
[302] H. Pang, M. Ihara, Y. Kuchino, S. Nishimura, R. Gupta, C. R. Woese, and J. A. McCloskey, *J. Biol. Chem.*, 1982, **257**, 3589.
[303] M. Dizdaroglu, *J. Chromatogr.*, 1984, **295**, 103.
[304] M. Ludewig, K. Dörffling, and W. A. König, *J. Chromatogr.*, 1982, **243**, 93.
[305] T. Sugiyama and T. Hashizume, *Kagaku to Seibutsu*, 1982, **20**, 760.
[306] L. M. S. Palni, R. E. Summons, and D. S. Letham, *Plant Physiol.*, 1983, **72**, 858.
[307] 'Advances in Steroid Analysis', Analytical Chemistry Symposia Series, Vol. 10, ed. S. Görög, Elsevier, Amsterdam, 1982.
[308] J. Sjövall and M. Axelson, *Vitam. Horm. (N.Y.)*, 1982, **39**, 31.
[309] S. J. Gaskell, *Methods Biochem. Anal.*, 1983, **29**, 385.
[310] S. J. Gaskell, *Horm. Norm. Abnorm. Hum. Tissues*, 1983, **3**, 121.
[311] M. Axelson and B.-L. Sahlberg, *J. Steroid Biochem.*, 1983, **18**, 313.
[312] T. Fotsis and R. Heikkinen, *J. Steroid Biochem.*, 1983, **18**, 357.

following the observation of a reaction leading to imine products in 20–40% yield.[313]

Selective derivatizations can aid characterization of steroids.[314,315] Application of the microbial enzyme cholesterol oxidase as a selective oxidant for 3β-hydroxy-steroids and cyclic boronate formation with diols of appropriate configuration have been discussed.[314] Screening of brassinosteroids at nanogramme levels in plants has also benefited from methaneboronate ester formation.[316] Conditions for converting 3-ketosteroids into 3-enol-TBDMS ethers have been reported.[317] Free, unhindered hydroxyl groups also react. Hydroxyl functions that do not react can be trimethylsilylated before GC/MS analysis.[317] Upon trimethylsilylation, 3-keto-nor-19-methyl-4,5- or -1,2-epoxide steroids undergo aromatization of the A-ring both thermally in the GC injector and in the silylation vessel.[318] Following work-up by immunoadsorption,[96,97] GC/NICIMS was utilized for determining the pentafluorobenzyloxime TMS ether of testosterone and the bis-heptafluorobutyrate of oestradiol-17β with detection limits in the low picogramme range.[319] The method was used to confirm the presence of testosterone in saliva.

The review period has seen the preparation of several deuterium-labelled steroids for use as internal standards in selected-ion monitoring or in tracer studies: [$^2H_8$]oestradiol, [$^2H_7$]oestrone, [$^2H_6$]2-hydroxyoestrone, and [$^2H_6$]4-hydroxy-oestrone,[320] oestriol, 16α-hydroxyoestrone, and oestriol 16-glucuronide,[321] trideuteriated analogues of cholesterol, pregnenolone, progesterone, and 3β-hydroxyandrost-5-en-17-one,[322] [$^2H_6$]cholesterol,[323] and 5α-androstane-3,17β-diols.[324]

The measurement of daily urinary-production rates and metabolic-clearance rates of hormonal steroids in man has been reported,[325,326] using clinically acceptable deuterium-labelled precursors and selected-ion monitoring. Capillary-column GC/MS was favoured for qualitative and quantitative determination of

313 C.-G. Eriksson, L. Nordström, and P. Eneroth, *J. Steroid Biochem.*, 1983, **19**, 1199.
314 C. J. W. Brooks, W. J. Cole, T. D. V. Lawrie, J. MacLachlan, J. H. Borthwick, and G. M. Barrett, *J. Steroid Biochem.*, 1983, **19**, 189.
315 C. J. W. Brooks, G. M. Barrett, and W. J. Cole, *J. Chromatogr.*, 1984, **289**, 231.
316 N. Ikekawa, S. Takatsuto, T. Kitsuwa, H. Saito, T. Morishita, and H. Abe, *J. Chromatogr.*, 1984, **290**, 289.
317 S. H. G. Andersson and J. Sjövall, *J. Chromatogr.*, 1984, **289**, 195.
318 E. Schwarz, S. Adbel-Baky, P. W. Lequesne, and P. Vouros, *Int. J. Mass Spectrom. Ion Phys.*, 1983, **47**, 511.
319 S. J. Gaskell and P. W. Brooks, *Int. J. Mass Spectrom. Ion Phys.*, 1983, **48**, 241.
320 R. Knuppen, O. Haupt, and H.-O. Hoppen, *Steroids*, 1982, **39**, 667.
321 M. Numazawa, M. Nagoka, and M. Ogata, *Chem. Pharm. Bull.*, 1984, **32**, 618.
322 L. J. Goad, M. A. Breen, N. B. Rendell, M. E. Rose, J. N. Duncan, and A. P. Wade, *Lipids*, 1982, **17**, 982.
323 D. N. Kirk, M. J. Varley, H. L. J. Makin, and D. J. H. Trafford, *J. Chem. Soc., Perkin Trans. 1*, 1983, 2563.
324 F. Jacolot, F. Berthou, D. Picart, J. Y. Le Gall, and H. H. Floch, *J. Labelled Compd. Radiopharm.*, 1982, **19**, 553.
325 D. W. Johnson, T. J. Broom, L. W. Cox, C. D. Matthews, G. Phillipou, and R. F. Seamark, *J. Steroid Biochem.*, 1983, **19**, 203.
326 D. W. Johnson, T. J. Broom, L. W. Cox, G. Phillipou, and R. F. Seamark, *Int. J. Mass Spectrom. Ion Phys.*, 1983, **48**, 249.

daily variations in urinary steroids.[327] An off-line computer system used methylene-unit retention indices in library searching of the acquired data. 19-Nor-deoxycorticosterone, which has a potential role in the pathogenesis of low-renin essential hypertension, has been detected in normal rat serum.[328] There has been a plethora of 'definitive' GC/MS analyses of cortisol in human serum and plasma, some of which are referenced here.[329–334] The favoured strategy is ^{2}H- or ^{14}C-labelled isotope dilution with selected-ion monitoring of m/z 605 for methoxime TMS cortisol (and a close ion appropriate for the particular internal standard). Quantification in the nmol l^{-1} range is common, and a detection limit of 200 pg has been achieved.[331]

The nature of 11 non-ecdysteroid steroids identified in the haemolymph of larvae of *Sarcophaga bullata* by capillary-column GC/NICIMS suggested an active metabolism of progesterone and testosterone.[335] Analysis by GC/MS of cholesterol and desmosterol produced by silkworm larvae fed on labelled stigmasterol indicated that migration of hydrogen from C-25 to the C-24 position occurs during dealkylation, in common with the known conversion of sitosterol into cholesterol.[336] Biosynthesis of cholesterol in chick muscle cells has been examined by studying the effect of 20,25-diazacholesterol.[337]

As examples of the use of GC/MS in identifying sterols in marine invertebrates, the free sterols of the gorgonian *Lephogorgia subcompressa* as well as their acetates and TMS ethers have been characterized using a fused-silica capillary column,[338] and three novel steroids with highly branched side chains, 25-methylfucosterol, 24-ethyl-24-methylcholesterol, and axinyssasterol, have been identified in a *Pseudoaxinyssa* sponge.[339] A scheme was proposed for a biomethylation sequence that could lead to the observed steroids.[339] Detection and quantification of 19-norandrosterone in the urine of competing athletes serve as a means of detecting doping with the anabolic steroid 19-nortestosterone decanoate up to six weeks after administration.[340] Oestrogens have been quantified in serum

[327] J. J. Vrbanac, C. C. Sweeley, and J. D. Pinkston, *Biomed. Mass Spectrom.*, 1983, **10**, 155.
[328] Y. Kobayashi, T. Ogihara, Y. Yamamura, F. Watanabe, T. Kiguchi, I. Ninomiya, and Y. Kumahara, *Endocrinol. Jpn.*, 1984, **31**, 49.
[329] J. A. Jonckheere and A. P. De Leenheer, *Biomed. Mass Spectrom.*, 1983, **10**, 197.
[330] D. G. Patterson, M. B. Patterson, P. H. Culbreth, D. M. Fast, J. S. Holler, E. J. Sampson, and D. D. Bayse, *Clin. Chem.*, 1984, **30**, 619.
[331] L. Siekmann and H. Breuer, *J. Clin. Chem. Clin. Biochem.*, 1982, **20**, 883.
[332] S. J. Gaskell, C. J. Collins, G. C. Thorne, and G. V. Groom, *Clin. Chem.*, 1983, **29**, 862.
[333] L. Oest, O. Falk, O. Lantto, and I. Bjoerkhem, *Scand. J. Clin. Lab. Invest.*, 1982, **42**. 181.
[334] O. Lantto, B. Lindback, A. Aakvaag, M. Damkjaer-Nielsen, U. M. Pomoell, and I. Bjoerkhem, *Scand. J. Clin. Lab. Invest.*, 1983, **43**, 433.
[335] D. De Clerck, H. Diederik, and A. De Loof, *Insect Biochem.*, 1984, **14**, 199.
[336] Y. Fujimoto, M. Kimura, A. Takasu, F. A. M. Khalifa, M. Morisaki, and N. Ikekawa, *Tetrahedron Lett.*, 1984, **25**, 1501.
[337] G. T. Emmons, E. R. Rosenblum, J. N. Peace, J. M. Malloy, D. L. Doerfler, I. R. McManus, and I. M. Campbell, *Biomed. Mass Spectrom.*, 1982, **9**, 278.
[338] J. Rovirosa, O. Munoz, A. San-Martin, A. M. Seldes, and E. G. Gros, *Lipids*, 1983, **18**, 570.
[339] X. Li and C. Djerassi, *Tetrahedron Lett.*, 1983, **24**, 665.
[340] I. Björkhem and H. Ek, *J. Steroid Biochem.*, 1982, **17**, 447.

from pregnant women[341] and seminal plasma of man, bull, boar, and stallion.[342] In the former study[341] oestrone, oestradiol, and oestriol were subjected to extractive alkylation then acetylation. The derivatives are more hydrolytically stable and afford less fragmentation than the commoner TMS ethers. Detection limits of 10 pg of oestrone and oestradiol and 200 pg of oestriol were reported.[341] 2-, 4-, 6-, or 16-hydroxylation of oestrogens by human follicles has been established by GC/MS.[343]

Measurement of vitamin D and its metabolites by GC/MS has been reviewed.[344] Human saliva possibly contains 25-hydroxyvitamin D_3 at very low levels but not 25-hydroxyvitamin D_2. Earlier reports of the presence of hydroxyvitamin D species are thought to be erroneous on the basis of the poor selectivity of the competitive protein-binding assays used.[345]

Complementing FAB studies of intact species, GC/MS has confirmed as phosphates the ecdysteroid conjugates in newly laid eggs of the desert locust. Following hydrolysis, the TMS derivative of phosphate was identified by GC/MS. The compounds were identified as the 22-phosphate esters of ecdysone, 2-deoxy-, 20-hydroxy-, and 2-deoxy-20-hydroxyecdysone.[346] Similar 22-phosphate conjugates have been identified in developing eggs of the grasshopper, the steroids being identified by hydrolysis, silylation, and GC/MS.[347,348] These workers claim that the 22-N^6-(isopentenyl)adenosine monophosphoric ester of ecdysone occurs in eggs of the grasshopper.[349] Ecdysteroids in adults of the nematode *Dirofilaria immitis* have been quantified as their TMS derivatives by selected-ion monitoring.[350]

Studies of bile acids continue unabated. Reviews[351–353] containing good sections on isolation and purification procedures[351] and on comparisons of radioimmunoassay and GC/MS[353] have appeared. Continuing the clinical characterization of the rare inherited liver disease cerebrotendinous xanthomatosis (CTX), analysis of the bile alcohols in urine has been shown to provide a diagnosis of CTX and an evaluation of the effect of chenodeoxycholic acid therapy.[354] Bile acids appear not to have a causative role in the pruritus of cholestatic liver disease,[355]

[341] K.-Y. Tserng, R. K. Danish, and J. S. Bendt, *J. Chromatogr.*, 1983, **272**, 233.
[342] A. Reiffsteck, L. Dehennin, and R. Scholler, *J. Steroid Biochem.*, 1982, **17**, 567.
[343] L. Dehennin, C. Blacker, A. Reiffsteck, and R. Scholler, *J. Steroid Biochem.*, 1984, **20**, 465.
[344] H. L. J. Makin and D. J. H. Trafford, *Vitam. D: Basic Clin. Aspects*, 1984, 497.
[345] D. J. H. Trafford and H. L. J. Makin, *Clin. Chim. Acta*, 1983, **129**, 19.
[346] R. E. Isaac, M. E. Rose, H. H. Rees, and T. W. Goodwin, *Biochem. J.*, 1983, **213**, 533.
[347] G. Tsoupras, B. Luu, and J. A. Hoffmann, *Steroids*, 1982, **40**, 551.
[348] G. Tsoupras, C. Hetru, B. Luu, E. Constantin, M. Lagueux, and J. A. Hoffmann, *Tetrahedron*, 1983, **39**, 1789.
[349] G. Tsoupras, B. Luu, and J. A. Hoffman, *Science*, 1983, **220**, 507.
[350] A. H. W. Mendis, M. E. Rose, H. H. Rees, and T. W. Goodwin, *Mol. Biochem. Parasitol.*, 1983, **9**, 209.
[351] K. D. R. Setchell and A. Matsui, *Clin. Chim. Acta*, 1983, **127**, 1.
[352] J. Sjövall, *Koenshu – Iyo Masu Kenkyukai*, 1983, **8**, 29.
[353] J. M. Street, D. J. H. Trafford, and H. L. J. Makin, *J. Lipid Res.*, 1983, **24**, 491.
[354] B. G. Wolthers, M. Volmer, J. van der Molen, B. J. Koopman, A. E. J. de Jager, and R. J. Waterreus, *Clin. Chim. Acta*, 1983, **131**, 53.
[355] T. C. Bartholomew, J. A. Summerfield, B. H. Billing, A. M. Lawson, and K. D. R. Setchell, *Clin. Sci.*, 1982, **63**, 65.

but patients do excrete two pentahydroxycholestanes in their urine.[356] Their presence may indicate alternative pathways of cholic acid synthesis in liver disease. Norcholic, $3\alpha,7\beta,12\alpha$-trihydroxy-5β-cholanoic, hyocholic, $1\beta,3\alpha,12\alpha$-trihydroxy-5β-cholanoic, and allocholic acids have been identified in urine of patients with liver cirrhosis.[357]

General methods for obtaining profiles of bile acids and related compounds in faeces have been described.[358] The article is strong on isolation and fractionation techniques. The monohydroxylated fraction of bile acids of human meconium has been analysed by GC/MS on capillary columns of SE-30 and SE-54.[359] Four etianic acids were identified. Quantitative determination of the unusual bile acids 3β-hydroxy- and $3\beta,12\alpha$-dihydroxy-5-cholenoic acids has been reported by GC/MS of TMS methyl ester derivatives.[360] The latter bile acid was detected in serum of patients with cholestasis and of infants with biliary atresia. Selected-ion monitoring forms a basis for accurate quantification of cholic, chenodeoxycholic, and deoxycholic acids in serum, using deuteriated standards.[361] The method was applied to healthy individuals in the fasting state. Fasting levels of monoketonic bile acids in human peripheral and portal blood have also been reported.[362] The problem of long retention times of ketonic bile acids relative to hydroxy bile acids was eliminated by reduction with sodium borodeuteride. The resulting mixtures of hydroxy bile acids and $^{2}H_{1}$-labelled hydroxy bile acids were analysed as TMS methyl esters, and the ratio of deuteriated to non-deuteriated forms was determined by GC/MS.[362] Bile acid kinetics have been investigated with [24-^{13}C]-chenodeoxycholic acid as a precursor.[363] Several free and sulphate and glucuronide conjugated bile acids have been fractionated on an anion-exchange column and quantified after enzymic hydrolysis as hexafluoroisopropyl trifluoroacetyl derivatives.[364] Application of GC/NICIMS to this procedure may prove useful. The identity of a commonly used internal standard, a commercial sample of labelled 23-nordeoxycholic acid, has been called into question.[365] It was in fact a monoacetate.

Amines. – Mass spectrometry has an important role in this field, particularly for biogenic amines. This fact is reflected in the large number of reviews of the

[356] H. Ludwig-Kohn, H. V. Henning, A. Sziedat, D. Matthaei, G. Spiteller, J. Reiner, and H. J. Egger, *Eur. J. Clin. Invest.*, 1983, **13**, 91.

[357] Y. Amuro, E. Hayashi, T. Endo, K. Higashino, and S. Kishimoto, *Clin. Chim. Acta*, 1983, **127**, 61.

[358] K. D. R. Setchell, A. M. Lawson, N. Tanida, and J. Sjoevall, *J. Lipid Res.*, 1983, **24**, 1085.

[359] J. St. Pyrek, R. Lester, E. W. Adcock, and A. T. Sanghvi, *J. Steroid Biochem.*, 1983, **18**, 341.

[360] M. Tohma, H. Takeshita, R. Mahara, and I. Makino, *Koenshu – Iyo Masu Kenkyukai*, 1983, **8**, 77.

[361] I. Bjoerkhem and O. Falk, *Scand. J. Clin. Lab. Invest.*, 1983, **43**, 163.

[362] I. Björkhem, B. Angelin, K. Einarsson, and S. Ewerth, *J. Lipid Res.*, 1982, **23**, 1020.

[363] F. Stellaard, R. Schubert, and G. Paumartner, *Biomed. Mass Spectrom.*, 1983, **10**, 187.

[364] H. Takikawa, H. Otsuka, T. Beppu, Y. Seyama, and T. Yamakawa, *J. Biochem. (Tokyo)*, 1982, **92**, 985.

[365] A. F. Attill, M. Angelico, L. Capocaccia, V. Pieche, and A. Catanfora, *J. Lipid Res.*, 1982, **23**, 211.

topic in the last two years. A book[366] contains chapters on GC/MS and selected-ion monitoring of biogenic amines and their metabolites and on quantitative high-resolution mass spectrometry applied to biogenic amines. Analysis of catecholamines and their metabolites by GC/MS has been surveyed extensively and authoritatively in another book.[367] Quantification of biogenic amines was also discussed in the book[367] and elsewhere.[368] As in other areas reported (*e.g.* prostaglandins, cannabinoids, and environmental studies), GC/NICIMS is making welcome inroads into biogenic amine applications. Much of its success in enhancing sensitivity and selectivity relies on proper choice of (electronegative) derivatives. This point has been emphasized in a thorough review of GC/NICIMS with fused-silica capillary columns for quantitative analysis of small-molecule neurotransmitters and related substances.[369] Qualitative and quantitative analyses of tetrahydro-β-carbolines,[370] catecholamines and metabolites,[371] and polyamines and polyamine conjugates[372] have been reviewed.

The occurrence, or not, of tetrahydro-β-carbolines (THBCs, or tryptolines) in mammalian systems has been[2] and still is a contentious issue. Pictet–Spengler condensation between indoleamines and aldehydes accounts for THBCs, but the reaction may occur either *in vitro* or *in vivo*.[373–375] The presence of formaldehyde in both biological media and solvents can result in artifactual formation of THBCs,[373,374] so incorporation of aldehyde-trapping reagents into work-up procedures appears to be obligatory for unambiguous detection of THBCs. The compounds may be involved in alcohol toxicity as a result of a reaction between metabolically produced acetaldehyde and biogenic amines. Both 1-methyl-THBC and 6-hydroxy-1-methyl-THBC are normal constituents of human urine,[376] hence providing further evidence[2] against their role in ethanol intake, but during intoxication alcoholics excrete larger amounts of 6-hydroxy-1-methyl-THBC than control subjects.[377] Of five THBCs considered by GC/NICIMS only THBC (tryptoline) itself occurred in rat brain.[375] The use of chloroform as extraction solvent for amines also seems fraught with difficulties. Various reactions between amines and chloroform or its common contaminants have been reported,[378,379] and dichloromethane would seem to be a preferable solvent.

[366] 'Evaluation of Analytical Methods in Biological Systems. Part A. Analysis of Biogenic Amines', ed. G. B. Baker and R. T. Coutts, Elsevier, Amsterdam, 1982.
[367] 'Methods in Biogenic Amine Research', ed. S. Parvez, T. Nagatsu, and I. Nagatsu, Elsevier, Amsterdam, 1983.
[368] F. Karoum and H. H. Neff, *Mod. Methods Pharmacol.*, 1982, **1**, 39.
[369] K. F. Faull and J. D. Barchas, *Methods Biochem. Anal.*, 1983, **29**, 325.
[370] S. A. Barker, *Prog. Clin. Biol. Res.*, 1982, **90**, 113.
[371] M. R. Holdiness, *Am. Lab.*, 1983, **15**, 34.
[372] G. D. Daves, jun., R. G. Smith, and C. A. Valkenburg, *Methods Enzymol.*, 1983, **94**, 48.
[373] T. R. Bosin, B. Holmstedt, A. Lundman, and O. Beck, *Anal. Biochem.*, 1983, **128**, 287.
[374] T. R. Bosin, B. Holmstedt, A. Lundman, and O. Beck, *Prog. Clin. Biol. Res.*, 1982, **90**, 15.
[375] K. F. Faull, R. B. Holman, G. R. Elliott, and J. D. Barchas, *Prog. Clin. Biol. Res.*, 1982, **90**, 135.
[376] O. Beck, T. R. Bosin, B. Holmstedt, and A. Lundman, *Prog. Clin. Biol. Res.*, 1982, **90**, 29.
[377] O. Beck, T. R. Bosin, A. Lundman, and S. Borg, *Biochem. Pharmacol.*, 1982, **31**, 2517.
[378] A. T. Kacprowicz, *J. Chromatogr.*, 1983, **269**, 61.
[379] E. J. Cone, W. F. Buchwald, and W. D. Darwin, *Drug Metab. Dispos.*, 1982, **10**, 561.

Reaction of catecholamines with methyl chloroformate in aqueous media produces quantitative yields of formates. Following formation of TBDMS ethers, the *N*-formate derivatives are suitable for quantitative GC/MS analysis at pg–ng levels.[380] The isothiocyanate derivative of β-phenylethylamine can be prepared *in situ* from the biological sample to which CS_2 is added.[381] The isothiocyanate derivative is amenable to selected-ion monitoring of the amine in human urine and rat brain. In an uncommon study to profile amines in normal and uremic urine for the purposes of clinical diagnosis, phenolic and aliphatic amines were monitored during packed-column GC/MS.[382] Deuteriated analogues of various biogenic amines have been used as internal standards for quantifying the natural compounds in regions of human brain.[383] Application of fused-silica capillary columns and GC/NICIMS to the assay of catecholamines and their *O*-methylated metabolites (as pentafluoropropionyl and TBDMS pentafluorobenzyl derivatives) has been strongly advocated.[384]

As suggested,[2] utilization of GC/NICIMS with an electronegative derivative enhances the sensitivity and reliability of analysis of histamine.[385,386] Both papers concern tris(pentafluorobenzyl)histamine, monitoring the ion at m/z 430, and report a detection limit in the range 100–500 fg. Another group of workers[387,388] also describes a selective and sensitive method for histamine, this time using positive-ion (ammonia) chemical ionization. The method involves the heptafluorobutyroyl derivative and has improved accuracy and precision over, and similar sensitivity (detection limit of 40 fmol) to, radioenzymatic assay.

Communication of results in the field of biogenic amine metabolism is hindered rather than helped by the inconsistent nomenclature applied; standardization of trivial names would be welcome. The concentrations of octopamine, tyramine, and their α-methylated analogues in regions of rat brain have been determined before and after administration of amphetamine.[389] The method utilizes pentafluoropropionyl derivatives, deuteriated internal standards, and GC/NICIMS and is capable of assays in the femtogramme range. The same derivative was employed for various phenylethylamine and phenylacetic acid metabolites of dopamine in a study that compared EI, CI, and NICI methods.[390] The use of the negative-ion technique resulted in considerably enhanced sensi-

[380] A. P. J. M. De Jong and C. A. Cramers, *J. Chromatogr.*, 1983, **276**, 267.
[381] O. Suzuki and H. Hattori, *Biomed. Mass Spectrom.*, 1983, **10**, 430.
[382] T. Ohki, A. Saito, K. Ohta, T. Niwa, K. Maeda, and J. Sakakibara, *J. Chromatogr.*, 1982, **233**, 1.
[383] J. Herranen, A. Huhtikangas, H. Tirronen, T. Halonen, M. Huuskonen, K. Reinikainen, and P. Reikkinen, *J. Chromatogr.*, 1984, **307**, 241.
[384] J. T. Martin, J. D. Barchas, and K. F. Faull, *Anal. Chem.*, 1982, **54**, 1806.
[385] L. J. Roberts, *J. Chromatogr.*, 1984, **287**, 155.
[386] L. J. Roberts and J. A. Oates, *Anal. Biochem.*, 1984, **136**, 258.
[387] J. J. Keyzer, H. Breukelman, H. Elzinga, B. J. Koopman, B. G. Wolthers, and A. P. Bruins, *Biomed. Mass Spectrom.*, 1983, **10**, 480.
[388] J. J. Keyzer, B. G. Wolthers, F. A. J. Muskiet, H. Breukelman, H. F. Kauffman, and K. de Vries, *Anal. Biochem.*, 1984, **139**, 474.
[389] P. H. Duffield, D. F. H. Dougan, D. N. Wade, G. K. C. Low, and A. M. Duffield, *Spectrosc.: Int. J.*, 1983, **2**, 311.
[390] P. L. Wood, *Biomed. Mass Spectrom.*, 1982, **9**, 302.

tivities. Conjugated and unconjugated dopamine in cerebrospinal fluid of monkey has been assayed by GC/MS using the trifluoroacetyl derivatives.[391] A systematic comparison of HPLC, fluorimetry, and GC/MS for analysis of homovanillic and 5-hydroxyindoleacetic acids in cerebrospinal fluid has been presented.[392] Homovanillic and 4-hydroxy-3-methoxymandelic (vanilmandelic) acids in urine, serum, plasma, and cerebrospinal fluid have been determined simultaneously as dimethylthiophosphinyl methyl esters by fused-silica bonded-phase capillary-column GC/MS.[393] Volumes of biological fluids of only 0.1–1.0 ml[393] or, using a different and simple procedure for sample preparation, 50 μl[394] are required for GC/MS analysis of these two acidic metabolites. 2-Hydroxy-2-phenylpropanoic acid and [$^{2}H_6$]mandelic acid have been compared as internal standards for assay of mandelic acid in blood plasma, using TMS derivatives.[395] Plasma concentrations of 3- and 4-hydroxyphenylacetic and phenylacetic acids have been measured after fasting and consumption of deuterium-labelled amine precursors.[396] High-resolution selected-ion monitoring of pentafluoropropionyl trifluoroethyl esters was used to determine the effects of fasting, age, and sex on deamination metabolism. 3-Hydroxyphenylglycol, as its tris(pentafluoropropionate), has been determined in mammalian urine.[397]

In the area of tryptophan metabolites, several reports have described the quantification of tryptamine compounds.[398–401] Capillary columns were used for quantifying tryptamine in brain tissue.[398,399] The failure to detect [^{13}C, ^{15}N]-*N*-methyltryptamine in urine after administration of [^{13}C, ^{15}N]tryptamine to schizophrenic patients does not support the indoleamine transmethylation hypothesis of schizophrenia.[400] The GC/MS method for 5-hydroxytryptamine in plasma has been improved.[401] Negative-ion CI techniques have also been employed effectively in this category.[402,403] Serotonin, as its trifluoroacetate, has been measured in blood and urine by capillary-column GC/NICIMS,[402] and melatonin, as its pentafluoropropionate, was determined down to the femtogramme range in plasma.[403]

[391] M. A. Elchisak, K. H. Powers, and M. H. Ebert, *J. Neurochem.*, 1982, **39**, 726.
[392] G. M. Anderson, M. B. Bowers, R. H. Roth, G. J. Young, C. C. Hrbek, and D. J. Cohen, *J. Chromatogr.*, 1983, **277**, 282.
[393] K. Jacob, W. Vogt, and C. Schwertfeger, *J. Chromatogr.*, 1984, **290**, 331.
[394] D. H. Hunneman, *Clin. Chim. Acta*, 1983, **135**, 169.
[395] H. Luthe, H. Ludwig-Koehn, and U. Langenbeck, *Biomed. Mass Spectrom.*, 1983, **10**, 183.
[396] B. A. Davies, D. A. Durden, and A. A. Boulton, *J. Chromatogr.*, 1982, **230**, 219.
[397] J. R. Crowley, M. W. Couch, C. M. Williams, M. I. James, K. E. Ibrahim, and J. M. Midgley, *Biomed. Mass Spectrom.*, 1982, **9**, 146.
[398] F. Artigas, C. Sunol, J. M. Tusell, E. Martinez, and E. Gelpi, *Biomed. Mass Spectrom.*, 1984, **11**, 142.
[399] O. Beck and F. Flodberg, *Biomed. Mass Spectrom.*, 1984, **11**, 155.
[400] R. W. Walker, L. R. Mandel, L. Delisi, R. J. Wyatt, and W. J. A. Vandenheuvel, *J. Chromatogr.*, 1984, **289**, 223.
[401] S. Baba, M. Utoh, M. Horie, and Y. Mori, *J. Chromatogr.*, 1984, **307**, 1.
[402] T. Hayashi, S. Kamada, H. Naruse, and Y. Iida, *Koenshu - Iyo Masu Kenkyukai*, 1983, **8**, 165.
[403] D. J. Skene, R. M. Leone, I. M. Young, and R. E. Silman, *Biomed. Mass Spectrom.*, 1983, **10**, 655.

Previously reported GC/MS procedures could not distinguish choline from *N*-amino-*N*,*N*-dimethylaminoethanol and acetyl choline from *O*-acetyl-*N*-amino-*N*,*N*-dimethylaminoethanol. This deficiency has been corrected by employing a deuterium-labelling method.[404] Traces of aza-arenes (detection limits 0.02–0.5 ng) in sea water have been determined by selected-ion monitoring,[405] and qualitative and quantitative analysis of naturally occurring and synthetic pteridines as TMS derivatives has been reported.[406] Steroidal, indole, and piperidinyldipyridine alkaloids in skin of a poisonous frog,[407] erythrina alkaloids in seeds and leaves,[408] and alkaloids of a toxic lupin[409] have been identified by GC/MS.

Amino Acids and Peptides. – Optimum analytical conditions for GC/MS of amino acids are largely established. Most of the literature concerns application rather than development of techniques, so the discussion here is brief. Examination of peptides is not so mature and methods are still evolving.

Following oral administration of deuterium-labelled tyrosine to healthy humans, plasma levels of the amino acid and its metabolites could be measured by GC/NICIMS of pentafluorobenzoyl TMS derivatives.[410,411] Perfluoroacyl alkyl esters remain the favoured derivatives for GC/MS analysis, as with the heptafluorobutyroyl methyl ester of 3,5-di-iodotyrosine used to determine the urinary excretion of the parent compound in humans of differing thyroidal status,[412] *N*-trifluoroacetyl isobutyl esters of the five-carbon β-, γ-, and δ-amino alkanoic acids,[413] the trifluoroacetyl methyl ester of thyroxine,[414] and *N*-pentafluoropropionyl hexafluoroisopropyl esters of putative amino acid neurotransmitters.[415] Detection limits of the neurotransmitter candidates, already 50 fmol to 1 pmol, may well be improved by use of capillary columns and GC/NICIMS. Methylimidazole-4-acetic acids[416,417] and indole-3-acetic acid[418] have been determined in human body fluids and Rhizobium culture supernatants, respectively. For example, 1-methylimidazole-4-acetic acid in cerebrospinal fluid can be identified and quantified as its heptafluorobutyroyl methyl ester against a deuteriated internal standard.[416] A detection limit of 200 pg ml^{-1} was recorded.

[404] M. W. Newton, B. Ringdahl, and D. J. Jenden, *Anal. Biochem.*, 1983, **130**, 88.

[405] R. Shinohara, A. Kido, Y. Okamoto, and R. Takeshita, *J. Chromatogr.*, 1983, **256**, 81.

[406] T. Kuster, A. Matasovic, and A. Niederwieser, *J. Chromatogr.*, 1984, **290**, 303; T. Kuster and A. Niederwieser, *J. Chromatogr.*, 1983, **278**, 245.

[407] T. Tokuyama and J. W. Daly, *Tetrahedron*, 1983, **39**, 41.

[408] A. S. Chawla, A. H. Jackson, and P. Ludgate, *J. Chem. Soc., Perkin Trans. 1*, 1982, 2903.

[409] I.-C. Kim. M. F. Balandrin, and A. D. Kinghorn, *J. Agric. Food Chem.*, 1982, **30**, 796.

[410] S. Kamada, T. Hayashi, H. Naruse, Y. Iida, and S. Daishima, *Koenshu – Iyo Masu Kenkyukai*, 1982, **7**, 195.

[411] T. Hayashi, S. Kamada, H. Naruse, and Y. Iida, *Koenshu – Iyo Masu Kenkyukai*, 1982, **7**, 199.

[412] M. N. Mohammed, M. Farmer, and D. B. Ramsden, *Biomed. Mass Spectrom.*, 1983, **10**, 507.

[413] J. R. Cronin, G. U. Yuen, and S. Pizzarello, *Anal. Biochem.*, 1982, **124**, 139.

[414] B. Moeller, L. Bjoerkhem, O. Falk, O. Lantto, and A. Larsson, *J. Clin. Endocrinol. Metab.*, 1983, **56**, 30.

[415] M. Wolfensberger and U. Amsler, *J. Neurochem.*, 1982, **38**, 451.

[416] C. G. Swahn and G. Sedvall, *J. Neurochem.*, 1983, **40**, 688.

Amino acid racemization in heated proteins, in stored and heated milk powders, and in alkali-treated casein has been examined by deuterium-labelling and application of the chiral Chirasil-Val stationary phase to differentiate between the D-isomers already present in the protein and those formed during treatment.[419] This elegant method was described, with some reservations, in the previous Report.[2] Selected-ion monitoring was used to detect the amino acids, and it was concluded that racemization partly accounted for nutritional losses induced by alkali treatment.[419] Studies on the important process of alkylation of amino acid residues in proteins continue.[420,421] *In vivo* exposure to alkylating agents ethylene oxide and propylene oxide has been monitored through GC/MS analysis of *S*-methylcysteine[420] and 3′-(2-hydroxyethyl)-[420] and 3′-(2-hydroxy-propyl)-histidine[420,421] in haemoglobin. *N*-Heptafluorobutyroyl alkyl ester derivatives and capillary columns with both achiral and chiral liquid phases were employed. The power of GC/MS for investigating uncommon and novel amino acids is clear.[2] For instance, γ-carboxylation of a glutamic acid residue in a tetrapeptide has been studied, again relying on deuterium labelling.[422]

Metabolic studies in the amino acid field have been, and still are, based on incubations with ^{15}N-labelled tracers and GC/MS to detect and measure the isotopic content of the resulting amino acids. Incorporation of $^{15}NH_3$ into a wide range of amino acids can be investigated readily by GC/MS.[423,424] Specifically, NH_4Cl is a major precursor of glutamic acid in *Brevibacterium lactofermentum*,[425] $(^{15}NH_4)_2SO_4$ in *Brevibacterium flavum* gives rise to L-[^{15}N]lysine,[426] and the labelled nitrogen of NH_4Cl in adult human plasma appears in the guanidino nitrogens of arginine.[427] By GC/MS, as little as 0.15 mol% excess of ^{15}N could be detected in 50 μl of plasma.[427] [^{15}N]Amino acids have also been used as precursors.[428,429] In this way it has been shown that much glutamic acid and glutamine in organotypic cerebellar explants is derived from leucine.[428] Kinetic aspects (turnover rate and pool size) of the metabolism of glycine in rabbits can be elucidated by GC/MS after a single dosage with [^{15}N]glycine.[429] Effects of insulin and glucagon on the metabolism were evaluated.

417 J. K. Khandelwal, L. B. Hough, B. Pazhenchevsky, A. M. Morrishow, and J. P. Green, *J. Biol. Chem.*, 1982, **257**, 12 815.
418 J. Badenoch-Jones, R. E. Summons, M. A. Djordjevic, J. Shine, D. S. Letham, and B. G. Summons, *Appl. Environ. Microbiol.*, 1982, **44**, 275.
419 R. Liardon and R. F. Hurrell, *J. Agric. Food Chem.*, 1983, **31**, 432.
420 E. Bailey, P. B. Farmer, and J. B. Campbell, *Anal. Proc.*, 1982, **19**, 239.
421 P. B. Farmer, S. M. Gorf, and E. Bailey, *Biomed. Mass Spectrom.*, 1982, 9, 69.
422 A. I. Burgess, M. P. Esnouf, K. Rose, and R. E. Offord, *Biochem. J.*, 1983, **215**, 75.
423 K. Samukawa, *Radioisotopes*, 1982, **31**, 166.
424 M. Yudkoff, I. Nissim, S. U. Kim, D. Pleasure, and S. Segal, *J. Neurochem.*, 1984, **42**, 283.
425 Z. E. Kahana and A. Lapidot, *Anal. Biochem.*, 1983, **132**, 160.
426 C. S. Irving, C. L. Cooney, L. T. Brown, D. Gold, J. Gordon, and P. D. Klein, *Anal. Biochem.*, 1983, **131**, 93.
427 I. Nissim, M. Yudkoff, T. Terwilliger, and S. Segal, *Anal. Biochem.*, 1983, **131**, 75.
428 M. Yudkoff, I. Nissim, S. Kim, D. Pleasure, K. Hummeler, and S. Segal, *Biochem. Biophys. Res. Commun.*, 1983, **115**, 174.
429 I. Nissim and A. Lapidot, *Biochem. Med.*, 1984, **31**, 185.

Sequencing of peptides and proteins continues, with the synergistic combination of GC/MS and FAB approaches being highly favoured.[430–433] As an example,[434] the structure of a toxic octapeptide of a sawfly larva was determined largely by FAB mass spectrometry, but the configurations of the residues were assigned after hydrolysis, formation of *N*-trifluoroacetyl isopropyl esters, and GC/MS on a column of Chirasil-Val. Half of the amino acids had the D-configuration.[434]

Isobutane, ammonia, and methanol have been compared as reactant gases for GC/CIMS analysis of *N*-acetyl and *N*-[$^2H_3$]acetyl methyl esters of urinary dipeptides in patients suffering from immunopeptiduria, and the use of methanol was advocated.[41] Application of capillary columns and perfluoroacyl derivatives would be worthwhile in such a study. As with methanol, a mixture of isobutane and argon has been reported[435] to produce prominent protonated molecules and fragment (sequence) ions from small peptides, this time as their *O*-TMS-polyamino alcohol derivatives. Diastereoisomeric dipeptides have been resolved as their TMS derivatives using GC/MS.[436] On the achiral capillary column used, individual enantiomers were not separated. Two tetrapeptides that have a role in the inflammatory process in skin disorders have been determined by GC/MS (EI and NICI). They were examined as trifluoroacetyl TMS derivatives of the corresponding amino alcohols.[437] The crosslinked amino acids of food proteins have been determined by GC/MS after alkali treatment.[438] The trifluoroacetyl n-butyl esters of Orn-Ala and Lys-Ala were compared. *N*-Trifluorodideuterioethyl-*O*-TMS polyamino alcohol derivatives from partial hydrolysis of glucagon have been shown to afford much more sequence information than the corresponding *N*-trifluoroacetyl-*N*, *O*-permethyl derivatives by GC/MS.[439] Yields of the latter were lower because of by-product formation. The permethyl derivatives were more prone to losses in the GC/MS system (adsorption and/or decomposition) but the derivatization is simpler and their mass spectra are simpler too. Nevertheless, the polyamino alcohol procedure is the preferred one. Some workers stand by the permethylation method for GC/MS.[440–443] Permethylation of acylated peptides, again from glucagon, has been accomplished at the 2–10 nmol level by an iodomethane/dimethylsulphinyl anion procedure.[440] The

[430] K. Biemann, *Int. J. Mass Spectrom. Ion Phys.*, 1982, **45**, 183.
[431] K. L. Rinehart, jun., *et al.*, *Pure Appl. Chem.*, 1982, **54**, 2409.
[432] 'Methods of Protein Sequence Analysis', Proceedings of the 4th International Conference, ed. M. Elzinga, Humana, Clifton, N.J., 1982.
[433] H. Matsuda, *Kagaku to Seibutsu*, 1982, **20**, 242.
[434] D. H. Williams, S. Santikarn, P. B. Oelrichs, F. De Angelis, J. K. MacLeod, and R. J. Smith, *J. Chem. Soc., Chem. Commun.*, 1982, 1394.
[435] J. R. Yates and R. J. Anderegg, *Biomed. Mass Spectrom.*, 1983, **10**, 567.
[436] M. Dizdaroglu and M. G. Simic, *J. Chromatogr.*, 1982, **244**, 293.
[437] A. I. Mallet, K. Rose, and J. D. Priddle, *Biomed. Mass Spectrom.*, 1983, **10**, 120.
[438] K. Hasegawa and S. Iwata, *Agric. Biol. Chem.*, 1982, **46**, 2513.
[439] C. T. Pederson and P. Roepstorff, *Int. J. Mass Spectrom. Ion Phys.*, 1983, **48**, 193.
[440] K. Rose, M. G. Simona, and R. E. Offord, *Biochem. J.*, 1983, **215**, 261.
[441] K. Rose, M. G. Simona, R. E. Offord, C. P. Prior, B. Otto, and D. R. Thatcher, *Biochem. J.*, 1983, **215**, 273.
[442] K. Rose, A. Bairoch, and R. E. Offord, *J. Chromatogr.*, 1983, **268**, 197.
[443] K. Rose, J. Gladstone, and R. E. Offord, *Biochem. J.*, 1984, **220**, 189.

strategy has been applied to bacterially produced γ-interferon[441] and methionine-enkephalin.[442] Capillary-column GC/MS is recommended in these studies.[442] Incorporation of ^{18}O into amino acid residues from $H_2{}^{18}O$-enriched media, followed by GC/MS, serves as an additional tool for studying biochemical and chemical transformations of polypeptides in solution.[441,443] The use of dipeptidyl peptidases in conjunction with GC/MS continues to be applied to sequence studies. The method is particularly advantageous for polypeptides with blocked N-termini.[444] Dipeptidyl carboxypeptidase and GC/MS have been applied successfully to several light chains from anti-β-(1,6)-D-galactan-binding monoclonal antibodies.[445]

Conventional Edman-degradation methodology has been supplemented with GC/MS analysis to determine the sequence of osteocalcin,[446] and the primary structure of the anti-tumour antibiotic macromycin was deduced by the same two techniques with further corroboration from FAB studies.[447] This paper[447] shows very well the complementary nature of the three techniques, with much of the sequence being determined by GC/MS of the mixture generated by a single partial acid hydrolysis.

Insect Pheromones and Other Secretions. – A brief methodological review of GC/MS has emphasized application to insect pheromones,[448] and 13 papers constituting a 'symposium in print' on animal chemical defences by many leading workers have appeared.[449] Subjects included marine invertebrates, chrysomelid larvae and adults, termites, ants, ladybug, and striped skunk.

Many non-volatile components of venoms are well characterized, but the volatile constituents have not been subjected to the same level of scrutiny. One paper[450] has reported that the venom gland of the European hornet contains traces of volatile semiterpenoid esters and alcohols. Sesquiterpenes from termite soldiers include a new $5\beta,7\beta,10\beta$-eudesmane, amiteol (1),[451] and the defensive secretions of daddy long-legs (harvestmen) contain *N,N*-dimethyl-2-phenylethylamine as a major component along with nicotine, bornyl acetate, bornyl propion-

HO H

(1)

[444] H. C. Krutsch, *Methods Enzymol.*, 1983, **91**, 511.

[445] M. Pawlita, M. Potter, and S. Rudikoff, *J. Immunol.*, 1982, **129**, 615.

[446] P. V. Hauschka, S. A. Carr, and K. Biemann, *Biochemistry*, 1982, **21**, 638.

[447] T. S. Samy, K.-S. Hahm, E. J. Modest, G. W. Lampman, H. T. Keutmann, H. Umezawa, W. C. Herlihy, B. W. Bigson, S. A. Carr, and K. Biemann, *J. Biol. Chem.*, 1983, **258**, 183.

[448] L. R. Hogge and D. J. H. Olson, *Ind. Res. Dev.*, 1982, **24**, 144.

[449] 'Animal Defense Mechanisms', ed. J. Meinwald, *Tetrahedron*, 1982, **38**, 1855–1970.

[450] J. W. Wheeler, M. T. Shamim, P. Brown, and R. M. Duffield, *Tetrahedron Lett.*, 1983, **24**, 5811.

[451] Y. Naya, G. D. Prestwich, and S. G. Spanton, *Tetrahedron Lett.*, 1982, **23**, 3047.

ate, camphene, and limonene,[452] very different from the quinones and phenols secreted by other laniatorids. The defensive secretion of the pentatomid bug *Aspavia brunna* has been compared to and contrasted with those of various Heteropteras using GC/EIMS and GC/CIMS.[453] The chemotaxonomic significance of the occurrence of four β-hydroxyalkanoic acids and 6-octenoic acid in the abdominal tips of the water beetle has been discussed.[454]

The role of enantiomeric separations for full characterization of pheromones has been well illustrated for the case of myrmica ants,[132] and the identity of the sex pheromone of the w-marked cutworm was confirmed by using a GC/MS technique with two parallel fused-silica capillary columns.[46] In a capillary-column study that departs from straightforward structure elucidation, biosynthesis of bombykol in the pheromone gland of the silkworm pupa was investigated by use of deuterium-labelled (*Z*)-11-hexadecenoic acid as a tracer.[455] By conventional GC/MS analyses, (*E*,*Z*)-2,13-octadecadien-1-ol acetate was confirmed as a new pheromone for sesiid moths,[456] structure (2) was deduced to be the aggregation

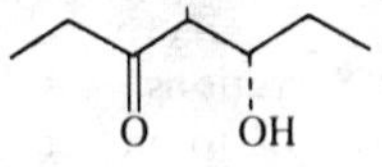

O OH

(2)

pheromone of rice and maize weevils,[457] the first cyclopropane derivative, *trans*-1-vinyl-2-(1*E*,3*Z*-hexadienyl)cyclopropane, found to act as a hormone in sexual reproduction was determined in a *Hormosira* seaweed,[458] (*Z*3,*Z*6,*Z*9)-1,3,6,9-nonadecatetraene was assigned as the sex pheromone of the winter moth[459] and (*Z*,*Z*,*Z*)-1,3,6,9-heneicosatetraene and (*Z*,*Z*)-6,9-heneicosadiene as pheromones of an arctiid moth,[460] 7-*exo*-ethyl-5-methyl-6,8-dioxabicyclo-[3.2.1]-3-octene was identified as a mouse pheromone or pheromone adjuvant,[461] and 2-aminoacetophenone was shown to be a component of the sex pheromone of the web-spinning larch sawfly.[462] Such publications are typical of the use of

[452] O. Ekpa, J. W. Wheeler, J. C. Cockendolpher, and R. M. Duffield, *Tetrahedron Lett.*, 1984, **25**, 1315.
[453] M. I. Akpata and T. O. Olagbemiro, *Z. Naturforsch., Teil B*, 1982, **37**, 935.
[454] H. Schildknecht, B. Weber, and K. Dettner, *Z. Naturforsch., Teil B*, 1983, **38**, 1678.
[455] R. Yamaoka, Y. Taniguchi, and K. Hayashiya, *Experientia*, 1984, **40**, 80.
[456] M. Schwarz, J. A. Klun, B. A. Leonhardt, and D. T. Johnson, *Tetrahedron Lett.*, 1983, **24**, 1007.
[457] N. R. Schmuff, J. K. Phillips, W. E. Burkholder, H. M. Fales, C.-W. Chen, P. P. Roller, and M. Ma, *Tetrahedron Lett.*, 1984, **25**, 1533.
[458] D. G. Müller, M. N. Clayton, G. Gassmann, W. Boland, F.-J. Marner, and L. Jaenicke, *Experientia*, 1984, **40**, 211.
[459] H. J. Bestmann, T. Brosche, K. H. Koschatzky, K. Michaelis, H. Platz, K. Roth, J. Süss, O. Vostrowsky, and W. Knauf, *Tetrahedron Lett.*, 1982, **23**, 4007.
[460] S. C. Jain, D. E. Dussourd, W. E. Conner, T. Eisner, A. Guerrero, and J. Meinwald, *J. Org. Chem.*, 1983, **48**, 2266.
[461] M. Novotny, F. J. Schwende, D. Wiesler, J. W. Jorgenson, and M. Carmack, *Experientia*, 1984, **40**, 217.
[462] R. Baker, C. Longhurst, D. Selwood, and D. Billany, *Experientia*, 1983, **39**, 993.

GC/MS in this area. Selected-ion monitoring has also been utilized for the determination of pheromone alkenyl acetates of moths.[463,464] Use of ammonia CI provides enhanced sensitivity and selectivity, the detection limits being in the mid-picogramme range.[464]

Clinical and Metabolic Studies. – The frenetic pace of publishing in this category has been checked somewhat. New developments and key applications only are covered here. The analysis of drugs and their metabolites is treated in the next chapter. An extensive review, covering modes of usage of instruments, recent applications in clinical chemistry, and current developments likely to have an impact on the clinical-chemistry laboratory of the future, is strongly recommended.[465] Three books have described GC/MS applications in microbiology,[466] pharmacology,[467] and metabolic research involving labelled tracers.[468] Reviews have covered identification of drugs in acute drug intoxications,[469] drug bioavailability,[470] determination of acidic metabolites in blood serum and urine for clinical diagnosis,[471] stable isotopes in biomedical[472] and nutritional research,[473] quantification in biochemistry and biomedicine,[81,82] and high-resolution mass spectrometry for analysis of endogenous and exogenous compounds in tumour tissue.[474] One of these articles[471] contains a very useful table collating a large number of organic acids that have been identified in urine and serum, with references for all acids in the list.

Mixtures of organic acids are still the prime samples for the clinical chemist. Alkylamide,[147] butyl ester,[184,475] and oxime TMS derivatives[476] are amongst those recently reported for GC/MS analysis of urinary acids. Some improvement of performance of the stationary phase OV-1701 over OV-17 was noted for oxime TMS derivatives.[476] Krebs cycle and related acids were quantified simultaneously as n-butyl esters against their n-[$^{2}H_{9}$]butyl ester counterparts as internal standards.[475] The method was satisfactory for acids at the pg level in perchloric acid extracts of tissues. Profiles of organic acids in human milk have been determined by GC/MS of TMS derivatives. Identification and quantification

[463] M. H. Benn, R. A. Galbreath, V. A. Holt, H. Young, G. Down, and E. Priesner, *Z. Naturforsch., Teil C*, 1982, **37**, 1130.
[464] C. Loefstedt and G. Odham, *Biomed. Mass Spectrom.*, 1984, **11**, 106.
[465] R. E. Hill and D. T. Whelan, *Clin. Chim. Acta*, 1984, **139**, 231.
[466] 'Gas Chromatography/Mass Spectrometry Applications in Microbiology', ed. G. Odham, L. Larsson, and P.-A. Mardh, Plenum, New York, 1984.
[467] 'Pharmaceutical Analysis: Modern Methods, Part A', ed. J. W. Munson, Marcel Dekker, Basel, 1982.
[468] 'Laboratory and Research Methods in Biology and Medicine, Vol. 9: Tracers in Metabolic Research: Radioisotope and Stable Isotope/Mass Spectrometry Methods', ed. R. R. Wolfe, Liss, New York, 1984.
[469] J. E. Pettersen and B. Skuterud, *Anal. Chem. Symp. Ser.*, 1983, **13**, 111.
[470] J. Vink, *Mass Spectrom. Rev.*, 1982, **1**, 349.
[471] H. M. Liebich, *J. High Resolut. Chromatogr. Chromatogr. Commun.*, 1983, **6**, 640.
[472] N. J. Haskins, *Biomed. Mass Spectrom.*, 1982, **9**, 269.
[473] D. E. Matthews and D. M. Bier, *Ann. Rev. Nutr.*, 1983, **3**, 309.
[474] S. J. Gaskell, H. M. Leith, and B. G. Brownsey, *Anal. Proc.*, 1984, **21**, 29.
[475] L. Marai and A. Kuksis, *J. Chromatogr.*, 1983, **268**, 447.
[476] A. P. J. M. De Jong, *J. Chromatogr.*, 1982, **233**, 297.

of several acids in colostrum and mature milk were achieved.[477] Both methyl and isobutyl esters of several microbial acids (*e.g.* tuberculostearic, muramic, and 2,6-diaminopimelic acids) in artificial culture and in body fluids of patients with infectious diseases have been examined by capillary-column GC/EIMS and GC/CIMS.[478] The metabolic changes in organic acids in renal tissue under induced ischaemic conditions have been studied. Several acids increased in concentration at first and then decreased, whilst 3-deoxyaldonic acids simply decreased in renal-tissue biopsy.[479]

Evidence for a defect in general acyl-CoA dehydrogenase (C_6–C_{10}-dicarboxylic aciduria) is accumulating.[480,481] For example, selected-ion monitoring of TMS derivatives indicated that 5-hydroxyhexanoic acid and hexanoglycine were excreted in excessive amounts.[481] Urinary organic-acid profiles have been used to characterize dihydrolipoyl dehydrogenase deficiency.[482] Other evaluations of enzyme activities by GC/MS include various steroid hydroxylases.[483–485] Urinary steroid profiles of normal controls and patients with 17α- or 21-hydroxylase deficiency have been compared by use of alkoxime TMS derivatives and some new hydroxylated pregnanes identified for the first time.[483] Selected-ion monitoring of 7α-hydroxycholesterol forms a basis for studying cholesterol 7α-hydroxylase activity and of mevalonate for hydroxymethylglutaryl-CoA reductase activity.[484] A suspected 17α-hydroxylase block in a patient with delayed puberty has been confirmed by observing steroid profiles by GC/MS since no 17-hydroxyl steroids were present.[485] On the topic of steroid profiling, capillary-column GC/MS has been used to observe quantitative alterations of urinary steroids associated with diabetes mellitus,[486] a disease that has been shown to be characterized by elevated urinary excretion of 4-heptanone.[487] In the urine of propionic acidemia sufferers, levels of 3-ethyl-3-hydroxyglutaric acid[488,489] and a number of other acids[490] have been shown to be elevated. 2,3-Dimethyl-3-hydroxyglutaric acid was found

[477] T. Masui, M. Fujishima, Y. Mori, S. Tamaoki, T. Shinka, and I. Matsumoto, *Koenshu - Iyo Masu Kenkyukai*, 1983, **8**, 271; T. Masui, M. Fujishima, Y. Mori, T. Shinka and I. Matsumoto, *Int. J. Mass Spectrom. Ion Phys.*, 1983, **48**, 225.
[478] G. Odham, A. Tunlid, L. Larsson, and P. A. Mardh, *Chromatographia*, 1982, **16**, 83.
[479] T. Niwa, N. Yamamoto, H. Asada, A. Kawanishi, M. Yokoyama, K. Maeda, and T. Ohki, *J. Chromatogr.*, 1983, **272**, 227.
[480] S. Kølvraa, N. Gregersen, E. Christensen, and N. Hobolth, *Clin. Chim. Acta*, 1982, **126**, 53.
[481] N. Gregersen, S. Kølvraa, K. Rusmussen, P. B. Mortensen, P. Divry, M. David, and N. Hobolth, *Clin. Chim. Acta*, 1983, **132**, 181.
[482] T. Kuhara, T. Shinka, Y. Inoue, M. Matsumoto, M. Yoshino, Y. Sakaguchi, and I. Matsumoto, *Clin. Chim. Acta*, 1983, **133**, 133.
[483] T. Mizumoto, *Kurume Igakkai Zasshi*, 1982, **45**, 904.
[484] M. Galli Kienle, G. Galli, E. Bosisio, G. Cighetti, and R. Pocoletti, *J. Chromatogr.*, 1984, **289**, 267.
[485] P. V. Fennessey, P. G. Marsh, E. R. Orr, P. Burnstein, and G. Betz, *Clin. Chim. Acta*, 1983, **129**, 1.
[486] M. Alasandro, D. Wiesler, G. Rhodes, and M. Novotny, *Clin. Chim. Acta*, 1982, **126**, 243.
[487] H. M. Liebich, *J. Chromatogr.*, 1983, **273**, 67.
[488] T. Kuhara, Y. Inoue, T. Shinka, I. Matsumoto, and M. Matsuo, *Biomed. Mass Spectrom.*, 1983, **10**, 629.
[489] R. J. Pollitt, *Biomed. Mass Spectrom.*, 1983, **10**, 253.
[490] T. Kuhara, T. Shinka, M. Matsuo, and I. Matsumoto, *Clin. Chim. Acta*, 1982, **123**, 101.

in urine of a patient with β-ketothiolase deficiency.[489] A new organic aciduria, fumaric aciduria, associated with mental retardation and speech defects is characterized by a net secretion of fumaric acid by the renal tubules.[491]

Examination of α-ketoacids in urine, plasma, and cerebrospinal fluid can be effected through GC/MS of TMS-quinoxalinol derivatives.[153,492] α-Ketoglutaric acid concentrations, measured by GC/CIMS with benzoylformic acid as internal standard, have been correlated with physiological variations in paediatrics.[492] Acid-catalysed n-butyl esterification of three α-ketoacids was accompanied by undesirable formation of variable amounts of n-butyl ketals as shown by GC/MS,[493] so this commonly used procedure should be regarded with caution.

The unusual acid *cis*-4-hydroxycyclohexanecarboxylic acid, identified by GC/MS as its TMS derivative in urine of a child, is thought to be a by-product of intestinal bacterial metabolism and not endogenous to mammalian metabolism.[494] The significance of the occurrence of 5-hydroxy-6-methoxy- and 6-hydroxy-5-methoxyindolyl-2-carboxylic acids and 5-hydroxy-6-methoxyindole in a pigmented-melanoma cell-culture supernatant has been discussed.[495] Determination of mycolic acids from corynebacteria is helping to characterize the pathogenic corynebacteria[186,496] such as those associated with leprosy.[496] Patients afflicted with the genetic disease Hawkinsinuria excrete hawkinsin, *cis*- and *trans*-4-hydroxycyclohexylacetic acid, quinolacetic acid, and pyroglutamic acid along with 4-hydroxyphenolic acids.[497] Striking abnormalities have been noted in long-chain dicarboxylic acids in serum[498] and aromatic compounds in urine[499] of patients with Reye's and Reye's-like syndrome. Quantitative analysis of amniotic fluid by GC/CIMS has shown elevated concentrations of 2-methyl-3-hydroxybutyric and methylmalonic acids when foetuses were affected with methylmalonic acidemia,[500] of 3-hydroxyisovaleric acid in the case of a carboxylase deficiency (an inherited disorder of leucine metabolism),[501] and of glutaric acid with a foetus affected with glutaric aciduria type II.[502] Such methods involving amniotic fluid have great potential for prenatal diagnosis of inherited diseases. Hereditary tyrosinemia is characterized by the presence in urine of 4,6-dioxoheptanoic, 3,5-dioxo-octanedioic, and 4-oxo-6-hydroxyheptanoic

[491] D. T. Whelan, R. E. Hill, and S. McClorry, *Clin. Chim. Acta*, 1983, **132**, 301.
[492] F. Rocchiccioli, J. P. Leroux, and P. H. Cartier, *Biomed. Mass Spectrom.*, 1984, **11**, 24.
[493] L. Marai and A. Kuksis, *J. Chromatogr.*, 1982, **249**, 359.
[494] J. B. Kronik, O. A. Mamer, J. Montgomery, and C. R. Scriver, *Clin. Chim. Acta*, 1983, **132**, 205.
[495] S. Pavel, F. A. J. Muskiet, L. De Ley, T. H. The, and W. Van der Slik, *J. Cancer Res. Clin. Oncol.*, 1983, **105**, 275.
[496] C. Gailly, P. Sandra, M. Verzele, and C. Cocito, *Eur. J. Biochem.*, 1982, **125**, 83.
[497] C. H. Hocart, B. Halpern, L. A. Hick, C. D. Wong, J. W. Hammond, and B. Wilcken, *J. Chromatogr.*, 1983, **275**, 237.
[498] K. J. Ng, B. D. Andresen, M. D. Hilty, and J. R. Bianchine, *J. Chromatogr.*, 1983, **276**, 1.
[499] F. Rocchiccioli, P. H. Cartier, and P. F. Bougneres, *Biomed. Mass Spectrom.*, 1984, **11**, 127.
[500] C. Jakobs, L. Sweetman, and W. L. Nyhan, *Clin. Chim. Acta*, 1984, **140**, 157.
[501] C. Jakobs, L. Sweetman, W. L. Nyhan, and S. Packman, *J. Inherited Metab. Dis.*, 1984, **7**, 15.
[502] C. Jakobs, L. Sweetman, S. K. Wadman, M. Duran, J. M. Saudubray, and W. L. Nyhan, *Eur. J. Pediatr.*, 1984, **141**, 153.

acids.[503] Infusion of the tracer [$^{13}C_3$]lactic acid into humans has served as a basis for studying lactate metabolism by application of GC/MS.[504]

New deoxyalditols and inositol isomers have been discovered in both normal and uraemic urine.[505] The serum and urine levels of several isomers varied significantly between normal subjects and uremic patients, suggesting altered metabolism of chiroinositol and 1-deoxyglucose (1,5-anhydroglucitol) in uremia. 4-*O*-β-D-Mannopyranosyl-2-acetamido-2-deoxy-D-glucose is excreted in larger amounts than normal in a patient with mucolipidosis II.[506] The metabolic origins of the sulphur and two nitrogen atoms of biotin from *E. coli* have been investigated by tracer studies with $^{34}SO_4^{2-}$, L-[sulphane-^{34}S]thiocystine, and [^{15}N]methionine.[507] The key steps are incorporation of the N atom of methionine into 7-keto-8-aminopelargonic acid and of the S atom of cysteine into dethiobiotin during biosynthesis of biotin.

Food and Agricultural Chemistry. – Again this field is clearly divided into determination of flavour and odour components and of toxic compounds in food sources. A very large number of applications of GC/MS, most of which are routine, fall particularly into the latter category, so a few representative examples are given in this section. The proceedings of the 1981 conference on developments in food analysis have much to interest the mass spectrometrist,[508] and there have been several reviews of foods, beverages, and flavours.[509–513] The role of mass spectrometry for regulatory analysis of pesticides and industrial chemicals in foodstuffs has been reviewed.[514]

Volatile constituents of jackfruits have been analysed on two different packed columns of widely different separating characteristics. A computer matched the mass spectra of the two sets of eluents and assigned two retention indices to each component. Mass-spectral library searching then used a retention-index 'window' as a filter for fast and reliable identifications.[515] Both capillary-column GC/MS and GC/FTIR have been used for analysis of cherimoya fruit pulp, and in total 208 volatile components were identified.[516] For quantification of vanillin

[503] B. Lindblad and G. Steen, *Biomed. Mass Spectrom.*, 1982, **9**, 419.
[504] R. A. Neese, E. W. Gertz, J. A. Wisneski, L. D. Gruenke, and J. C. Craig, *Biomed. Mass Spectrom.*, 1983, **10**, 458.
[505] T. Niwa, N. Yamamoto, K. Maeda, K. Yamada, T. Ohki, and M. Mori, *J. Chromatogr.*, 1983, **277**, 25.
[506] M. Lemonnier, C. Derappe, L. Poenaru, M. A. Chester, A. Lundblad, S. Svensson, and P.-A. Öckerman, *FEBS Lett.*, 1982, **141**, 263.
[507] E. DeMoll, R. H. White, and W. Shive, *Biochemistry*, 1984, **23**, 558.
[508] 'Recent Developments in Food Analysis', ed. W. Baltes, P. B. Czedik-Eysenberg, and W. Pfannhauser, Verlag Chemie, Weinheim, 1982.
[509] C. Merritt, jun. and D. H. Robertson, *Dev. Food Sci.*, 1982, **3A**, 49.
[510] D. A. Cronin, *Dev. Food Sci.*, 1982, **3A**, 15.
[511] M. C. Petrini and E. Fedeli, *Riv. Ital. Sostanze Grasse*, 1982, **59**, 251.
[512] A. K. Foltz, J. A. Yeransian, and K. G. Sloman, *Anal. Chem.*, 1983, **55**, 164R.
[513] I. Horman in 'Analysis of Foods and Beverages, Modern Techniques', ed. G. Charalambous, Academic Press, New York, 1984.
[514] T. Cairns, E. G. Siegmund, R. A. Jacobson, T. Barry, G. Petzinger, W. Morris, and D. Heikes, *Biomed. Mass Spectrom.*, 1983, **10**, 301.
[515] P. Rasmussen, *Anal. Chem.*, 1983, **55**, 1331.
[516] H. Idstein, W. Herres, and P. Schreier, *J. Agric. Food Chem.*, 1984, **32**, 383.

in vanilla beans, GC/MS was preferred to MS/MS (on the basis of reproducibility) and HPLC (because of column contamination).[517] The new indole glucosinolates 4-hydroxy- and 4-methoxy-3-indolylmethylglucosinolate were identified in cabbage and other vegetables partly by GC/MS of silylated desulphoglucosinolates.[518] The main quinoxalines derived from various homoglucans have been identified following silylation.[519] The alkaline *o*-phenylenediamine method yields information on linkages within the glucans; for instance, the ratio of 1,4- to 1,3-linkages in barley glucan was estimated to be 1.6.[519]

The presence and amounts of 6-hydroxy-1-methyl-1,2,3,4-tetrahydro-β-carboline in beer, wine, fruits, tomatoes, and cheese have been determined by capillary-column GC/MS,[520] complementing the same group's studies in the biological area.[373,374,376,377] A new class of hop constituents, tricyclic sesquiterpenoids, has been characterized and its biogenic relationship to germacrenes demonstrated.[521] Headspace concentration and 'heart-cutting' have been used in analysing the aroma-related compounds of various types of coffee.[522] New devices for extracting orange juice from Satsuma mandarin fruit have been tested by GC/MS evaluation of the juices' flavour qualities.[523]

A study of the natural products (other than cannabinoids) produced by *Cannabis* has revealed a new series of compounds, typified by cannabispiradienone (3). The biosynthesis of the spirans was compared with the processes that lead to cannabinoids.[524] Many differences between the basic fractions of marihuana and tobacco smoke condensates have been found by capillary-column GC/MS.[525] From the low-resolution mass spectra of phenol TMS ethers

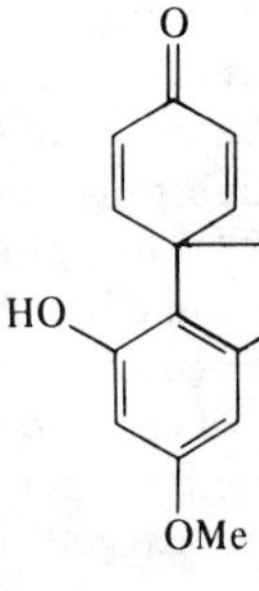

O

HO

OMe

(3)

[517] D. Fraisse, F. Maquin, D. Stahl, K. Suon, and J. C. Tabet, *Analusis*, 1984, **12**, 63.
[518] R. J. W. Truscott, D. G. Burke, and I. R. Minchinton, *Biochem. Biophys. Res. Commun.*, 1982, **107**, 1258; R. J. W. Truscott, I. R. Minchinton, D. G. Burke, and J. P. Sang, *ibid.*, p. 1368.
[519] N. Morita, K. Hayashi, M. Takagi, and K. Miyano, *Agric. Biol. Chem.*, 1983, **47**, 757.
[520] O. Beck, T. R. Bosin, and A. Lundman, *J. Agric. Food Chem.*, 1983, **31**, 288; T. R. Bosin, A. Lundman, and O. Beck, *ibid.*, p. 444.
[521] R. Tressl, K.-H. Engel, M. Kossa, and H. Köppler, *J. Agric. Food Chem.*, 1983, **31**, 892.
[522] T. H. Wang, H. Shanfield, and A. Zlatkis, *Chromatographia*, 1983, **17**, 411.
[523] H. Ohta, K. Tonohara, A. Watanabe, K. Iino, and S. Kimura, *Agric. Biol. Chem.*, 1982, **46**, 1385.
[524] L. Crombie and W. M. L. Crombie, *J. Chem. Soc., Perkin Trans. 1*, 1982, 1455.
[525] M. Novotny, F. Merli, D. Wiesler, and T. Saeed, *Chromatographia*, 1982, **15**, 564.

derived from smoke, accurate mass measurement of ions was achieved by bleeding perfluorotributylamine as a calibration compound into the ion source throughout the GC/MS analysis.[526] New components of cigarette smoke are still being discovered by application of modern techniques of GC/MS.[527,528]

Toxicological aspects of food chemistry are dominated by nitrosamines, pesticide residues, mycotoxins, and contamination from plastic packaging. Analysis of nitrosamines by GC/MS and LC/MS has been reviewed.[529] Of the many publications on nitrosamines, three typical ones are mentioned here. The presence of *N*-nitrosopyrrolidine and *N*-nitrosodimethylamine in bacon, beer, and malt is considered confirmed by appropriate responses during selected monitoring of three m/z values for each compound before photolysis with UV light and a lack of response after the irradiation.[530] Most, but not all, of the *N*-nitrosothiazolidine found by GC/MS in extracts of fried bacon was produced artifactually as a result of analysis in the presence of residual nitrite,[531] and the determination of *N*-nitroso-3-hydroxypyrrolidine in fried foods has been described in detail.[532] In a study utilizing GC, LC, GC/MS, and LC/MS[533] the hydrolytic degradation products of the organophosphorothioate insecticide chlorpyrifos were identified. The study was aimed at providing a kinetic expression to define the disappearance of chlorpyrifos in aquatic ecosystems. Several methods were also employed to observe components of the chlorinated insecticide toxaphene, namely GC/EIMS, GC/CIMS, and GC/NICIMS.[534] The complex mixture contained at least 202 compounds.

Selected-ion monitoring at medium resolution and with a bonded-phase fused-silica capillary column has been developed into an assay for aflatoxins B_1 and B_2 in peanuts at levels over 0.1 p.p.b.[535] Several workers[536-538] have investigated trichothecene mycotoxins in foodstuffs. The complexity of the chemical matrix of fungal-contaminated food can cause erroneous identification of toxins when structures are assigned by GC retention times alone, as with deoxynivalenol in cultures of *Fusarium solani* on corn.[536] The unknown metabolite of the fungus that co-eluted with deoxynivalenol could be differentiated from it by mass spectrometry. Several trichothecene mycotoxins, including deoxynivalenol down to 100 fg, can be determined in corn, wheat, and mixed feeds by GC/NICIMS,[537] whilst T-2 toxin, HT-2 toxin, diacetoxyscirpenol, and zearalenone have been confirmed and quantified in corn by packed-column GC/MS of TMS deriva-

[526] R. Wittowski, M. Kellert, and W. Baltes, *Int. J. Mass Spectrom. Ion Phys.*, 1983, **48**, 339.
[527] Y. Saint-Jalm, P. Moree-Testa, and A. Testa, *Analusia*, 1983, **11**, 12.
[528] P. Moree-Testa, Y. Saint-Jalm, and A. Testa, *J. Chromatogr.*, 1984, **290**, 263.
[529] K. S. Webb, L. M. Libbey, J. H. Hotchkiss, R. A. Scanlan, T. Fazio, and M. Castegnaro, *IARC Sci. Publ.*, 1983, **45**, 449.
[530] W. I. Kimoto and W. Fiddler, *J. Assoc. Off. Anal. Chem.*, 1982, **65**, 1162.
[531] W. I. Kimoto, J. W. Pensabene, and W. Fiddler, *J. Agric. Food Chem.*, 1982, **30**, 757.
[532] J. S. Lee, L. M. Libbey, and R. A. Scanlan, *IARC Sci. Publ.*, 1983, **45**, 411.
[533] D. L. Macalady and N. L. Wolfe, *J. Agric. Food Chem.*, 1983, **31**, 1139.
[534] M. A. Saleh, *J. Agric. Food Chem.*, 1983, **31**, 748.
[535] R. T. Rosen, J. D. Rosen, and V. P. DiProssimo, *J. Agric. Food Chem.*, 1984, **32**, 276.
[536] A. Visconti and F. Palmisano, *J. Chromatogr.*, 1982, **252**, 305.
[537] J. M. Rothberg, J. L. MacDonald, and J. C. Swims, *ACS Symp. Ser.*, 1983, **234**, 271.

tives.[538] Confirmation of identity as well as quantification was achieved at levels above about 25 p.p.b. Sep-Pak cartridges are proving useful in providing a fast and convenient extraction method for toxins.[535,538] Plastic packing of foods had been shown to contaminate the food within. Several volatile impurities in polystyrene cups are extracted into hot water in just a few minutes,[539] and styrene[540] and butadiene[541] have been determined in plastics and foods by automatic headspace GC/MS analysis. Sterilization of polyethylene film can be effected by electron-beam irradiation, but this process produces several odorous volatile compounds.[542]

Environmental Science and Toxicology. – In the search for even better sensitivity and selectivity for detecting pollutants there has been a noticeable shift towards GC/NICIMS, particularly for polyhalogenated compounds and polycyclic aromatic hydrocarbons (PAHs). This brief review highlights such recent developments rather than lists numerous applications of established methods. A number of relevant books describe the following areas in more detail: general environmental chemistry,[543–545] carcinogens and hazardous substances,[546] hydrocarbons including PAHs,[547–549] chlorinated dioxins and related compounds,[550,551] water pollutants,[552] and drugs of abuse.[553] Numerous review articles have appeared that cover to some extent GC/MS applied to environmental problems: fused-silica capillary-column GC/MS and NICI,[554] air pollution,[555,556] water pollution

538 R. T. Rosen and J. D. Rosen, *J. Chromatogr.*, 1984, **283**, 223.
539 C. A. Eiceman and M. Carpen, *Anal. Lett.*, 1982, **15**, 1169.
540 J. Gilbert and J. R. Startin, *J. Sci. Food Agric.*, 1983, **34**, 647.
541 J. R. Startin and J. Gilbert, *J. Chromatogr.*, 1984, **294**, 427.
542 K. Azuma, T. Hirata, T. Ishitani, Y. Tanaka, and H. Tsunoda, *Agric. Biol. Chem.*, 1983, **47**, 855.
543 'Analytical Techniques in Environmental Chemistry', Vol. 2, ed. J. Albaiges, Pergamon, Oxford, 1982.
544 I. L. Marr and M. S. Cresser, 'Environmental Chemical Analysis', International Textbook Co., New York, 1983.
545 'Analytical Aspects of Environmental Chemistry', ed. D. F. S. Natusch and P. K. Hopke, Wiley, New York, 1983.
546 'Handbook of Carcinogens and Hazardous Substances: Chemical and Trace Analysis', ed. M. C. Bowman, Dekker, New York, 1982.
547 'Polynuclear Aromatic Hydrocarbons: Physical and Biological Chemistry', ed. M. Cooke, A. J. Dennis, and L. Gerald, Battelle Press, Columbus, 1982.
548 'Current Topics in Environmental and Toxicological Chemistry, Vol. 5: Chemistry and Analysis of Hydrocarbons in the Environment', ed. J. Albaiges, R. W. Frei, and E. Merian, Gordon and Breach, New York, 1983.
549 'Handbook of Polynuclear Aromatic Hydrocarbons', ed. A. Bjørseth, Marcel Dekker, New York, 1983.
550 'Human and Environmental Risks of Chlorinated Dioxins and Related Compounds', ed. R. E. Tucker *et al.*, [*Environ. Sci. Res.*, 1983, **26**], Plenum, New York, 1983.
551 'Chlorinated Dioxins and Dibenzofurans in the Total Environment', ed. G. Choudhary, L. H. Keith, and C. Rappe, Butterworth, Boston, 1983.
552 'Analysis of Organic Micropollutants in Water', ed. A. Bjørseth and G. Angeletti, Reidel, Dordrecht, 1982.
553 L. Fishbein, 'Chromatography of Environmental Hazards: Drugs of Abuse', Elsevier, New York, 1982.
554 D. J. Dixon, *Comm. Eur. Communities, [Rep.] EUR*, 1982, **EUR 7137**, 139.
555 J. Freudenthal, *Int. J. Mass Spectrom. Ion Phys.*, 1982, **45**, 247.
556 D. L. Fox and H. E. Jeffries, *Anal. Chem.*, 1983, **55**, 233R.

and effluents,[557–560] pesticides, chlorinated dioxins, and carcinogens,[561–564] and forensic science.[565]

Sample Preparation and Papers of General Interest. A one-step normal-phase HPLC separation for analysis of organic compounds on flyash has been developed.[566] All polychlorinated dibenzodioxins (PCDDs) elute in one fraction for a very convenient subsequent analysis of those compounds by capillary-column GC/MS. The use of fused-silica capillary columns (usually directly inserted into the ion source) for general environmental applications is now very popular, with SE-54 as liquid phase perhaps having been scrutinized most.[567, 568] However, nitrophenols could not be eluted successfully from SE-54 columns, and loss of efficiency was noted when a single column was used to tackle successively acidic and basic analytes.[567] Coupling of capillary columns to FTIR spectrometers has been used alongside the more established GC/MS method. In at least one study (of hazardous wastes), the GC/FTIR method identified fully more components than did GC/MS,[569] although the sensitivity of the latter is considerably better.[570] Of course, the two methods should not be regarded as competitive but complementary.[22, 23] Given the stated trend towards negative-ion CI methods, it is not surprising that pulsed positive-ion negative-ion chemical ionization has been applied to fused-silica capillary-column GC/MS analysis of hazardous aromatic compounds found at dump sites.[571]

Air and Airborne Particulate Pollution. An HPLC method has been devised that affords a PAH fraction suitable for capillary-column GC/MS without further purification,[572] and a two-step column chromatographic separation, also of crude extracts from airborne particulate matter, isolates oxygenated PAHs with high selectivity.[573] Ketones, quinones, anhydrides, coumarins, and aldehydes were characterized by GC and GC/MS. Alkyl homologues of PAHs and oxygenated PAHs on diesel particulate matter have been examined as a function of engine conditions. Increasing cylinder temperature led to lower concentrations of the

[557] S. P. Scott, N. Sutherland, and R. J. Vincent, *Anal. Proc.*, 1984, **21**, 179.
[558] M. J. Fishman, D. E. Erdmann, and J. R. Garbarino, *Anal. Chem.*, 1983, **55**, 102R.
[559] D. E. Wells and A. A. Cowan, *Anal. Proc.*, 1982, **19**, 242.
[560] F. Karrenbrock and K. Haberer, *Vom Wasser*, 1983, **60**, 237.
[561] H. R. Buser, *Trends Anal. Chem.*, 1982, **1**, 318.
[562] J. Sherma and G. Zweig, *Anal. Chem.*, 1983, **55**. 57R.
[563] W. B. Crummet, *Chemosphere*, 1983, **12**, 429.
[564] M. L. Gross, *IARC Sci. Publ.*, 1983, **39**, 443.
[565] T. A. Brettell and R. Saferstein, *Anal. Chem.*, 1983, **55**, 19R.
[566] H. Y. Tong, D. L. Shore, F. W. Karasek, P. Helland, and E. Jellum, *J. Chromatogr.*, 1984, **285**, 423.
[567] J. R. Dahlgran and L. Abrams, *J. High Resolut. Chromatogr. Chromatogr. Commun.*, 1982, **5**, 656.
[568] G. T. Hunt and M. P. Hoyt, *J. High Resolut. Chromatogr. Chromatogr. Commun.*, 1982, **5**, 291.
[569] K. H. Shafer, T. L. Hayes, J. W. Brasch, and R. J. Jakobsen, *Anal. Chem.*, 1984, **56**, 237.
[570] D. F. Gurka, M. Hiatt, and R. Titus, *Anal. Chem.*, 1984, **56**, 1102.
[571] L. D. Betowski, H. M. Webb, and A. D. Sauter, *Biomed. Mass Spectrom.*, 1983, **10**, 369.
[572] T. Romanowski, W. Funcke, J. Koenig, and E. Balfanz, *Anal. Chem.*, 1982, **54**, 1285.
[573] J. Koenig, E. Balfanz, W. Funcke, and T. Romanowski, *Anal. Chem.*, 1983, **55**, 599.

PAH derivatives.[574] Synthetic fuel doped with PAHs has been used to study the relationships between diesel fuel aromaticity, PAH content, and the mutagenic activity of the associated soot.[575] It was shown by GC/MS and MS/MS that exhaust emissions of PAHs and nitro-PAHs from diesel engines were influenced markedly by incomplete combustion of PAHs through soot-forming mechanisms.

Negative-ion techniques have become powerful tools for airborne trace analysis. Negative-ion CI discriminates in favour of compounds with mutagenic or carcinogenic properties (*e.g.* PAHs, halogenated compounds) and so is used to advantage in this area. Amounts in the region of 100 pg can be identified by full scanning, and detection limits of about 50–100 fg are not uncommon.[576] The methane/nitrous oxide NICI spectra can differentiate many PAH[577, 578] and polychlorinated isomers[577] in both standards and air particulate extracts. Quantitative determination of chlorinated compounds at levels of pg m^{-3} in Arctic air can be achieved by the same method.[579] Substituted PAHs in aerosols have been extracted with liquid carbon dioxide, to minimize losses of reactive species, pre-fractionated by HPLC, and then analysed by GC/NICIMS in a method showing good selectivity.[580] Benzo[a]pyrene in engine exhaust and airborne particles has been determined by a fast selected-ion-monitoring NICI method.[581] Much effort has been expended to analyse airborne nitrated PAHs, and negative-ion mass spectrometry appears to be the strongest method.[582–585] The combination of SE-54 fused-silica capillary-column GC and NI atmospheric-pressure ionization was capable of separating and analysing four mononitro-PAH isomers,[583] whilst a DB-5 fused-silica capillary column was suitable for GC/NICIMS separation of 15 nitrated PAHs in a synthetic mixture.[584] Application of the latter to analysis of urban air and a detection limit of 1 pg for single-ion monitoring of 2-methyl-1-nitronaphthalene were demonstrated.[584] The common feature of all of these negative-ion studies[576–585] is the use of capillary columns, mostly manufactured from fused silica. No matter what is the method of ionization, such columns are recommended for compounds as labile as nitro-PAHs.[586, 587] Thermal decomposition during GC of 1-nitropyrene is minimized also by using cool on-column injection.[586] The product of on-column decomposition may be aminopyrene.[587]

[574] T. E. Jensen and R. A. Hites, *Anal. Chem.*, 1983, **55**, 594.
[575] T. R. Henderson, J. D. Sun, A. P. Li, R. L. Hanson, W. E. Bechtold, T. M. Harvey, J. Shabanowitz, and D. F. Hunt, *Environ. Sci. Technol.*, 1984, **18**, 428.
[576] M. Oehme, *Int. J. Mass Spectrom. Ion Phys.*, 1983, **48**, 287.
[577] M. Oehme, S. Manoe, and H. Stray, *J. Chromatogr.*, 1983, **279**, 649.
[578] M. Oehme, *Anal. Chem.*, 1983, **55**, 2290.
[579] M. Oehme and H. Stray, *Fresenius' Z. Anal. Chem.*, **1982**, **311**, 665.
[580] H. Stray, S. Manoe, A. Mikalsen, and M. Oehme, *J. High Resolut. Chromatogr. Chromarogr. Commun.*, 1984, **7**, 74.
[581] S. Daishima, A. Shibata, Y. Iida, and K. Furuya, *Bunseki Kagaku*, 1983, **32**, 761.
[582] M. Oehme, S. Manø, and H. Stray, *J. High Resolut. Chromatogr. Chromatogr. Commun.*, 1982, **5**, 417.
[583] W. A. Korfmacher, P. P. Fu, M. Chou, and R. K. Mitchum, *J. High Resolut. Chromatogr. Chromatogr. Commun.*, 1984, **7**, 41.
[584] T. Ramdahl and K. Urdal, *Anal. Chem.*, 1982, **54**, 2256.
[585] T. Ramdahl, G. Becher, and A. Bjørseth, *Environ. Sci. Technol.*, 1982, **16**, 861.
[586] H. Y. Tong, J. A. Sweetman, and F. W. Karasek, *J. Chromatogr.*, 1983, **264**, 231.
[587] J. A. Sweetman, F. W. Karasek, and D. Schuetzle, *J. Chromatogr.*, 1982, **247**, 245.

For accurate quantification of 1-nitropyrene in airborne particulate matter and fly ash, a deuteriated analogue as internal standard was added to compensate for the lability of the analyte.[587] The well known predilection of nitro compounds to be reduced to amines during mass spectrometry has been publicized again.[588] Apparently, reduction is most pronounced during LC/CIMS of nitro-PAHs.

Trace levels of other toxicants in air have also been measured by GC/MS. For example, 2,4-toluene di-isocyanate used in the polyurethane industry was determined after hydrolysis and derivatization of the resulting diamine,[589] and detection limits of 100–200 ng m^{-3} were recorded for seven volatile nitrosamines.[590]

Water Pollution, Effluents, and Fuel Spills. A wide range of organic compounds in water has been determined by SP-2250 packed-column and SE-54 capillary-column GC/MS.[591] At the low p.p.b. level both systems gave mean relative standard deviations of about 20%, but mean recoveries were better for the latter method. Identification of contaminants derived from chlorinated phenols in well water has been achieved not only by GC/MS but also by field desorption, direct CI, LC/MS, and nuclear-magnetic-resonance methods. Such multifarious approaches are to be encouraged.[592] In another comparative study a long-term evaluation of an automated system for extracting spectra and library searching to identify components of waste water has appeared.[593] The samples were surveyed for all compounds rather than just those on a target list. An overall reliability of 71% was recorded when differentiation of specific isomers was not needed, with the reliability of lower-quality matches, but not higher-quality ones, being enhanced markedly by considering retention data as well as mass spectra.[593] The products of PAHs on reaction with chlorine have been elucidated. The results have implications for disinfection of drinking water by chlorination.[594] A strategy for determining the source of organic pollutants in rivers consisted of GC/MS profiling of river-water samples and candidate industrial discharges, followed by a comparison of retention indices on two different columns, response factors of selective detectors, and mass spectra.[595] Organic profiles and selective detection methods in water analysis,[596] determination of phthalate esters in

[588] M. A. Quilliam, F. Messier, P. A. D'Agostino, B. E. McCarry, and M. S. Lant, *Spectrosc.: Int. J.*, 1984, **3**, 33.

[589] A. DePascale, L. Cobelli, R. Paladino, L. Pastorello, A. Frigerio, and C. Sala, *J. Chromatogr.*, 1983, **256**, 352.

[590] R. S. Marano, W. S. Updegrove, and R. C. Machen, *Anal. Chem.*, 1982, **54**, 1947.

[591] J. W. Eichelberger, E. H. Kerns, P. Olynyk, and W. L. Budde, *Anal. Chem.*, 1983, **55**, 1471.

[592] R. F. Bonner, M. G. Foster, D. E. Games, and O. Meresz, *Int. J. Mass Spectrom. Ion Phys.*, 1983, **48**, 315.

[593] W. M. Shackleford, D. M. Cline, L. Faas, and G. Kurth, *Anal. Chim. Acta*, 1983, **146**, 15.

[594] A. R. Oyler, R. J. Liukkonen, M. K. Lukasewycz, D. A. Cox, D. A. Peake, and R. M. Carlson, *Environ. Health Perspect.*, 1982, **46**, 73.

[595] J. C. Peterson, G. Guiochon, C. Democrate, and M. Dutang, *Int. J. Environ. Anal. Chem.*, 1983, **14**, 23.

[596] P. Burchill, A. A. Herod, K. M. Marsh, C. A. Pirt, and E. Pritchard, *Water Res.*, 1983, **17**, 1891, 1905.

marine samples,[597] minimization of weathering effects and use of biomarkers in identifying spilled crude oils,[598] and quantitative analysis of oil pollution around North Sea production platforms[599] have been discussed. Studies on the uptake of oil by shellfish show preferential accumulation of the more toxic aromatic hydrocarbons.[599]

Pesticides and Halogenated Residues. The advantages and limitations of a GC/MS system equipped with a pulsed positive-ion negative-ion CI accessory for determining pesticide residues have been outlined.[600] Organochlorine compounds generally are good candidates for analysis by GC/NICIMS. On-column injection and capillary-column GC/NICIMS of chlorinated paraffin mixtures have been discussed,[601] and many organochlorine pesticides were detected more sensitively by negative-ion than positive-ion CI methods.[602] Molecular anions are mainly generated for polychlorinated aromatic compounds, and for all types of organochlorine pesticides examined fragment ions consisted primarily of $(M-H_mCl_n)^-$ and Cl^- ions, where $m \geqslant 0$ and $n \geqslant 1$. The high selectivity of capillary-column GC/NICIMS permits detection of pesticides down to 0.2–50 pg amounts in river water, soil, and sediment without prior separation of polychlorinated biphenyls (PCBs) or phthalate esters.[602]

Approximately 96% of waste-transformer oils with their interferences has been removed by an HPLC technique, leaving a fraction clean enough to be analysed for PCBs down to 100 p.p.b. levels by GC/MS.[603] Negative-ion CI methods are also being applied for PCB analysis. Markedly increased selectivity towards PCBs was noted when GC/EIMS was replaced by GC/NICIMS.[604] Reactant-gas mixtures of CH_4/O_2, Ar/O_2, CH_4/H_2O, and CH_4 were compared for analysis of PCBs in marine sediments, and the water-containing reactant was suitable for monitoring of the PCBs.[604] A fused-silica capillary column was also used in a GC/NICIMS assay of PCBs in commercial silica gel.[605] Sediment samples can be checked rapidly for the presence of PCB residues at p.p.m. levels by pyrolytic desorption and GC/MS.[606] No lengthy chemical manipulations are required, so the method may be useful when the extent of contamination needs to be known urgently. As a less satisfactory alternative to GC/NICIMS for improving the selectivity of PCB determinations, conventional GC/MS can be used with limited mass range scanning.[607] Monitoring of the molecular-ion

[597] M. J. Waldock, *Chem. Ecol.*, 1983, **1**, 261.
[598] J. Shen, *Anal. Chem.*, 1984, **56**, 214.
[599] R. J. Law, *Anal. Proc.*, 1982, **19**, 248.
[600] S. J. Stout and W. A. Steller, *Biomed. Mass Spectrom.*, 1984, **11**, 207.
[601] M. D. Müller and P. P. Schmid, *J. High Resolut. Chromatogr. Chromatogr. Commun.*, 1984, **7**, 33.
[602] S. Daishima, Y. Iida, and T. Kajiki, *Nippon Kagaku Kaishi*, 1983, 1271; *ibid.*, 1984, p. 739.
[603] V. P. Nero and R. D. Hudson, *Anal. Chem.*, 1984, **56**, 1041.
[604] E. Lewis and W. D. Jamieson, *Int. J. Mass Spectrom. Ion Phys.*, 1983, **48**, 303.
[605] A. Bergman, L. Reutergardh, and M. Ahlman, *J. Chromatogr.*, 1984, **291**, 392.
[606] K. McMurtrey, N. J. Wildman, and H. Tai, *Bull. Environ. Contam. Toxicol.*, 1983, **31**, 734.
[607] R. B. Westerberg, S. L. Alibrando, and F. J. van Lenten, *J. Chromatogr.*, 1984, **284**, 447.

cluster serves as a surer basis than single-ion monitoring for qualitative analysis of PCBs whilst maintaining reasonable sensitivity compared to full scanning. Negative-ion mass spectrometry has also been applied to polybrominated aromatic compounds, their behaviour with respect to formation of molecular and halogen anions as a function of ionization conditions having been reported.[608] Differentiation of polybrominated[609] and polychlorinated biphenyl[610] isomers based on their mass spectra alone relies on the occurrence of *ortho* effects for appropriately substituted isomers. Of course, GC retention behaviour provides a further criterion for structural distinction.

The preferred column for GC/MS analysis of the acetate esters of chlorinated phenols was reported to be a 50 m SE-30 fused-silica capillary column, and mass-spectral fragmentation can distinguish some structural isomers.[611] Chlorinated phenoxyphenols in a technical chlorophenol formulation have been analysed,[612] and polychlorophenols have been examined by GC/NICIMS of their pentafluorobenzyl ethers and pentafluorobenzoate esters.[613] These derivatives yield abundant negative ions incorporating the chlorophenoxy portion of the molecules and hence are suitable for identification of chlorophenols at trace levels. The GC/NICIMS method, with methane as reactant gas, was applied to samples of starch adhesive and water collected near an industrial dump.[613]

The existence of many positional isomers of polychlorinated dibenzo-*p*-dioxins (PCDDs) presents a considerable challenge for the analyst who wishes to measure individual isomers.[2] A useful tabulation of relative retention times for 39 PCDDs has been published[614] following high-resolution GC with a fused-silica SP-2100 column directly inserted into an API mass spectrometer. Preparation, HRGC/MS analysis, and identification of all PCDD isomers containing four or more chlorine substituents have been described.[615] Analytical conditions for a Silar 10c column were established to resolve and detect isomers with the 2,3,7,8-tetrachloro substitution pattern that seems to equate with highest toxicity. Clearly, capillary columns are obligatory in this area as exemplified by the isomer-specific determination by high-resolution selected-ion monitoring of p.p.b. levels of PCDDs in particulate emissions resulting from wood combustion.[616] [^{13}C]Analogues were used as internal standards. The lack of reference standards for the 75 isomers of PCDDs makes their quantification difficult. An assay that relies on relative response factors for each isomeric group has been developed to circumvent the problem.[617] Fly ash from a high-temperature incinerator, generally considered to be free of organic compounds, has been shown to contain PCDDs

[608] J. Greaves, J. G. Bekesi, and J. Roboz, *Biomed. Mass Spectrom.*, 1982, **9**, 406.
[609] G. W. Sovocool and N. K. Wilson, *J. Org. Chem.*, 1982, **47**, 4032.
[610] L. G. M. T. Tuinstra and W. A. Traag, *J. Assoc. Off. Anal. Chem.*, 1983, **66**, 708.
[611] I. O. O. Korhonen and J. Knuutinen, *J. Chromatogr.*, 1983, **256**, 135.
[612] T. Humppi, R. Laitinen, E. Kantolahti, J. Knuutinen, J. Passivirta, J. Tarhanen, M. Lahtiperä, and L. Virkki, *J. Chromatogr.*, 1984, **291**, 135.
[613] S.-Z. Sha and A. M. Duffield, *J. Chromatogr.*, 1984, **284**, 157.
[614] W. A. Korfmacher and R. K. Mitchum, *J. High Resolut. Chromatogr. Chromatogr. Commun.*, 1982, **5**, 681.
[615] H. R. Buser and C. Rappe, *Anal. Chem.*, 1984, **56**, 442.
[616] T. J. Nestrick and L. L. Lamparski, *Anal. Chem.*, 1982, **54**, 2292.
[617] A. C. Viau and F. W. Karasek, *J. Chromatogr.*, 1983, **270**, 235.

(but not tetra- or penta-chloro isomers) along with chlorinated cyclohexanes, PCBs, and other compounds.[618a] The presence of PCDDs in emissions of waste incinerators may be the result of combustion of PVC.[618b] A survey has been published[619] of the PCDDs and polychlorinated dibenzofurans (PCDFs) identified in incinerator emissions by both GC/EIMS and GC/NICIMS. Formation of chloro-substituted 1,2-benzoquinone anions under API conditions with an oxygen-rich plasma yields structural information and has been used as the basis of an isomer-specific method for 2,3,7,8-TCDD in the low p.p.t. range by capillary-column GC/MS.[620] The first report of a (hydroxylated) microbial metabolite of TCDD has appeared,[621] and 1- and 8-methoxy-2,3,7,8-TCDDs were confirmed by GC/MS as metabolites of 2,3,7,8-TCDD in isolated rat hepatocytes.[622]

Synthesis of the 38 tetrachlorodibenzofuran isomers has been reported.[623] For capillary-column GC/MS of the isomers, liquid phases of SP-2330 and SE-54 are recommended. Many co-eluting TCDFs on one column can be resolved with the alternative one, forming a valuable technique for identifying TCDF isomers.[623] A detection limit of 0.5 pg has been obtained for 2,3,7,8-TCDF using GC/APIMS, but all the TCDF isomers that were examined afforded identical mass spectra under the conditions used.[624] Hydroxy-PCDFs in technical pentachlorophenol have been analysed as alkyl ethers by GC/NICIMS on a fused-silica SE-54 column.[625]

Toxicology and Forensic Science. Biological tissues have been examined extensively for residues of drugs, pesticides, and pollutants by GC and GC/MS, usually using selected-ion monitoring for quantitative assay. Such usage of GC/MS can be considered as routine. Typical of the genre, components of chlordane were detected and identified in fish by GC/MS in EI and negative-ion modes,[626] and organosulphur compounds detected in oysters and mussels by GC/MS were indicative of oil pollution at sea.[627]

Mercapturic acids are end products of an important metabolic pathway for xenobiotic substances (conjugation to glutathione) and detoxification of reactive electrophiles. Ten synthetic mercapturic acids have been resolved on an OV-101 column, but GC/MS confirmed that a number of the compounds had

[618] (*a*) F. W. Karasek and A. C. Viau, *J. Chromatogr.*, 1983, **265**, 79; (*b*) F. W. Karasek, A. C. Viau, G. Guichon, and M. F. Gonnord, *J. Chromatogr.*, 1983, **270**, 227.
[619] A. Cavallaro, L. Luciani, G. Ceroni, I. Rocchi, G. Invernizzi, and A. Gorni, *Chemosphere*, 1982, **11**, 859.
[620] R. K. Mitchum, W. A. Korfmacher, and G. F. Moler, *Int. J. Mass Spectrom. Ion Phys.*, 1983, **48**, 307.
[621] M. Philippi, J. Schmid, H. K. Wipf, and R. Hütter, *Experientia*, 1982, **38**, 659.
[622] T. Sawahata, J. R. Olson, and R. A. Neal, *Biochem. Biophys. Res. Commun.*, 1982, **105**, 341.
[623] T. Mazer, F. D. Hileman, R. W. Noble, and J. J. Brooks, *Anal. Chem.*, 1983, **55**, 104.
[624] W. A. Korfmacher, R. K. Mitchum, F. D. Hileman, and T. Mazer, *Chemosphere*, 1983, **12**, 1243.
[625] M. Deinzer, D. Griffin, T. Miller, J. Lamberton, P. Freeman, and V. Jonas, *Biomed. Mass Spectrom.*, 1982, **9**, 85.
[626] M. Ribick and J. Zajicek, *Chemosphere*, 1983, **12**, 1229.
[627] M. Ogata and K. Fujisawa, *J. Chromatogr. Sci.*, 1983, **21**, 420.

decomposed thermally in the injection port.[628] An acetylation procedure has been reported[629] that affords *S*-acetates of the prosthetic moieties of glutathione-conjugated metabolites in the mercapturic acid pathway. The acetates were characterized by GC/MS. A detailed description of a device for vacuum distillation of volatile compounds from fish-tissue samples and examples of its use have been reported.[630] The first report of trihydroxylated metabolites of 3-methylcholanthrene has appeared.[631] The 13 triol metabolites of mouse-liver microsomes were identified by GC/MS as their TMS derivatives. Livers of sole exposed to fuel oil have been examined for metabolites of 2- and 3-ring aromatic hydrocarbons by qualitative and quantitative GC/MS.[632]

It is sad that there is a need for analysts to develop techniques for detecting agents of chemical and biological warfare. Four *Fusarium* mycotoxins and poly(ethylene glycol), a suspected carrier medium, were found to occur in the purported chemical-warfare agent 'yellow rain' by GC/MS analysis of TMS derivatives.[633] Samples of leaves, water, cereal grains, and soil and of blood, urine, and body tissues of alleged Asian chemical-warfare victims were also shown to contain trichothecenes not indigenous to that region.[634] Allegations that such results are unreliable have been refuted.[635] Tear-gas ingredients and their residues in clothing can be analysed by GC/MS,[636] and lachrymator residues from exploding bank security devices (also tear gas) have been examined by GC/EIMS, GC/CIMS, and GC/NICIMS.[637] Both mass chromatograms[638] and selected-ion monitoring[639] have assumed a role in arson analysis. In a comparative study, capillary-column GC/EIMS, GC/CIMS, and GC/NICIMS have been evaluated for analysis of 2,4,6-trinitrotoluene and related nitro-aromatic species.[640] Human hair seems to have become a suitable sample for forensic analysis by GC/MS. Methamphetamine,[641,642] amphetamine,[642] chloroquine,[643] and mono-desethylchloroquine[643] have been determined in hair of drug users. Selected-ion monitoring in the CI mode was sufficiently sensitive to detect methamphetamine

628 W. Onkenhout, G. F. Guijt, H. J. De Jong, and N. P. E. Vermeulen, *J. Chromatogr.*, 1982, **243**, 362.
629 J. E. Bakke, *Biomed. Mass Spectrom.*, 1982, **9**, 74.
630 M. H. Hiatt, *Anal. Chem.*, 1983, **55**, 506.
631 K. Hashimoto, Y. Suzuki, K. Kinoshita, G. Takahashi, and K. Yasuhira, *J. Chromatogr.*, 1983, **260**, 429.
632 M. M. Krahn and D. C. Malins, *J. Chromatogr.*, 1982, **248**, 99.
633 R. T. Rosen and J. D. Rosen, *Biomed. Mass Spectrom.*, 1982, **9**, 443.
634 C. J. Mirocha, R. A. Pawlosky, K. Chatterjee, S. Watson, and W. Hayes, *J. Assoc. Off. Anal. Chem.*, 1983, **66**, 1485.
635 R. Stevenson, *Chem. Brit.*, 1984, **20**, 593; J. D. Rosen, *ibid.*, p. 986.
636 J. Nowicki, *J. Forensic Sci.*, 1982, **27**, 704.
637 R. M. Martz, D. J. Reutter, and L. D. Lasswell, *J. Forensic Sci.*, 1983, **28**, 200.
638 R. M. Smith, *Anal. Chem.*, 1982, **54**, 1399A; *J. Forensic Sci.*, 1983, **28**, 318.
639 M. A. Trimpe and R. Tye, *Arson Anal. Newsl.*, 1983, **7**, 26.
640 D. S. Weinburg and J. P. Hsu, *J. High Resolut. Chromatogr. Chromatogr. Commun.*, 1983, **6**, 404.
641 S. Suzuki, T. Inoue, T. Yasuda, T. Niwaguchi, H. Hori, and S. Inayama, *Eisei Kagaku*, 1984, **30**, 23.
642 O. Suzuki, H. Hattori, and M. Asano, *J. Forensic Sci.*, 1984, **29**, 611.
643 A. Viala, E. Deturmeny, C. Aubert, M. Estadieu, A. Durand, J. P. Cano, and J. Delmont, *J. Forensic Sci.*, 1983, **28**, 922.

and amphetamine as trifluoroacetates in a single human hair.[642] Analysis of castor-oil triglycerides in cosmetic smudges has been accomplished by transesterification and GC/MS of the resulting methyl esters.[644] Finally, administration of synthetic corticosteroids to horses can be confirmed by GC/NICIMS analysis of methoxime TMS derivatives.[645] The use of high-boiling solvents in conjunction with on-column injection has increased the speed of such analyses.[646]

Organic Geochemistry and Fuel. – Some books[647–649] with many chapters on GC/MS in organic geochemistry and reviews involving mass spectrometry of petroleum products,[650,651] pyrolysis methods in geochemistry,[652,653] determination of biogenic and anthropogenic organic matter in the environment,[654] the geological fate of steroids,[655] and analysis of oil shales[656] have been published. One issue of a journal of limited circulation has provided a useful introduction and review for the newcomer to geochemical mass spectrometry.[657] Some important new work with organometallic geochemicals and GC/MS is covered in the last section of this chapter. A typical procedure for analysis of geochemical samples involves extraction by liquid–liquid partition and/or liquid chromatographic methods (HPLC being favoured), followed by capillary-column GC/EIMS, often using selected-ion monitoring. The more selective metastable-peak monitoring has been employed for steranes[165] and terpanes,[166] but alternative methods of ionization have had little impact in the review period.

A large number of papers has considered the roles of sulphur and particularly nitrogen compounds in petroleum products. Specific detection of sulphur compounds in gasoline was effected with GC/NICIMS with ammonia as reactant gas.[658] The main advantage of this rare use of NICI in geochemistry clearly concerns selectivity: hydrocarbons hardly contribute to the total-ion-current trace, having very low ionization efficiencies in this mode. A homologous series of novel biomarkers, bicyclic and tetracyclic terpenoid sulphoxides and sulphides, have been identified in a bitumen.[659] The identity of basic nitrogen compounds

[644] R. L. Keagy, *J. Forensic Sci.*, 1983, **28**, 623.
[645] E. Houghton, P. Teale, M. C. Dumasia, and J. K. Wellby, *Biomed. Mass Spectrom.*, 1982, **9**, 459.
[646] E. Houghton, P. Teale, and M. C. Dumasia, *Analyst*, 1984, **109**, 273.
[647] 'Advances in Organic Geochemistry 1979', ed. A. G. Douglas and J. R. Maxwell, Pergamon, Oxford, 1982.
[648] 'Soil Analysis: Instrumental Techniques and Related Procedures', ed. K. A. Smith, Dekker, New York, 1983.
[649] 'Advances in Organic Geochemistry. Proceedings of the 10th International Meeting', ed. M. Bjoroey, Wiley, Chichester, 1983.
[650] T. A. Norris and V. P. Nero, *Lubrication*, 1982, **68**, 25.
[651] R. E. Terrell, *Anal. Chem.*, 1983, **55**, 245R.
[652] S. R. Larter and A. G. Douglas, *J. Anal. Appl. Pyrolysis*, 1982, **4**, 1.
[653] R. P. Philp, *Trends Anal. Chem.*, 1982, **1**, 237.
[654] B. R. T. Simoneit, *Int. J. Environ. Anal. Chem.*, 1982, **12**, 177.
[655] A. S. MacKenzie, S. C. Brassell, G. Eglinton, and J. R. Maxwell, *Science*, 1982, **217**, 491.
[656] P. F. V. Williams, *Fuel*, 1983, **62**, 756.
[657] E. J. Gallegos (ed.), *Spectra*, 1982, **8** (2/3), 3–60 (publ. Finnigan Corporation).
[658] J. Guieze, G. Devant, and D. Loyaux, *Int. J. Mass Spectrom. Ion Phys.*, 1983, **46**, 313.
[659] J. D. Payzant, D. S. Montgomery, and O. P. Strausz, *Tetrahedron Lett.*, 1983, **24**, 651.

in three coal-liquefaction products was aided by selected methylation of primary amines with dimethylformamide-dimethylacetal prior to GC/MS.[660] Separation of the potent microbial mutagens amino-PAHs from other nitrogen-containing polycyclic aromatic compounds in coal-derived liquids relied upon forming their pentafluoropropionyl derivatives, gel-permeation chromatography, and hydrolysis back to amino-PAHs. Fused-silica capillary-column GC/MS was used to identify the amines.[661] Several different extraction methods for isolating basic nitrogen compounds from coal tar have been assessed by GC/MS of the fractions obtained.[662] Aqueous-acid extraction and cation-exchange chromatography were particularly selective, whilst the latter and a promising method utilizing liquid chromatography on a polar silica afforded the highest recoveries. Nitrogen compounds in coal tar have been investigated without prior fractionation[663] and after a two-step partition process.[664] Whilst on the topic of coal tars, it is noted here that volatile compounds and volatile pyrolysis products[665] and large PAHs[666] in coal tars have been analysed by GC/MS. The latter study necessitated the use of a short thermostable (bonded-phase) capillary column. Isolation of aza-arenes from sediments, air particulate matter, and sewage effluents involved Sephadex LH-20 chromatography,[667] while nitrogen compounds in hydrated shale oils were determined by preparative HPLC and GC/MS.[668] Azanaphthalenes in crude oil[669] and aza-arenes in Arabian light crude oil[670,671] have been identified by capillary-column GC/MS.

Excellent resolution and good chromatographic behaviour of the wide range of compounds found in coal liquids (nitrogen bases, neutral aromatic and non-aromatic substances, and hydroxy aromatic compounds) have been achieved with a single 15 m SE-54 fused-silica capillary column for GC/MS studies.[672] The H-coal process liquefies coal to produce refinery feedstock. Compositional analysis of H-coal by GC/MS has been reported.[673,674] The SRC-11 coal-liquefaction product has also been examined,[674] in particular for phenols.[675] Coal-derived

[660] P. Burchill, A. A. Herod, and E. Pritchard, *J. Chromatogr.*, 1982, **246**, 271.
[661] D. W. Later, T. G. Andros, and M. L. Lee, *Anal. Chem.*, 1983, **55**, 2126.
[662] P. Burchill, A. A. Herod, J. P. Mahon, and E. Pritchard, *J. Chromatogr.*, 1983, **265**, 223.
[663] P. Burchill, A. A. Herod, and E. Pritchard, *Fuel*, 1983, **62**, 11.
[664] M. Novotny, D. Wiesler, and F. Merli, *Chromatographia*, 1982, **15**, 374.
[665] J. G. Moncur and W. G. Bradshaw, *J. High Resolut. Chromatogr. Chromatogr. Commun.*, 1983, **6**, 595.
[666] T. Romanowski, W. Funcke, I. Grossmann, J. Koenig, and E. Balfanz, *Anal. Chem.*, 1983, **55**, 1030.
[667] E. T. Furlong and R. Carpenter, *Geochim. Cosmochim. Acta*, 1982, **46**, 1385.
[668] L. Chan, J. Ellis, and P. T. Crisp, *J. Chromatogr.*, 1984, **292**, 355.
[669] J. M. Schmitter, I. Ignatiadis, and P. J. Arpino, *Geochim. Cosmochim. Acta*, 1983, **47**, 1975.
[670] G. Grimmer, J. Jacob, and K. W. Naujack, *Anal. Chem.*, 1983, **55**, 2398.
[671] G. Grimmer, J. Jacob, K. W. Naujack, and D. Schneider, *Ber. – Dtsch. Ges. Mineraloelwiss. Kohlechem.*, 1983, Project 325, p. 1–49.
[672] D. M. Parees and A. Z. Kamzelski, *J. Chromatogr. Sci.*, 1982, **20**, 441.
[673] J. L. Wong and C. M. Gladstone, *J. Chromatogr.*, 1983, **267**, 303.
[674] L. J. S. Young, N. C. Li, and D. Hardy, *Fuel*, 1983, **62**, 718.
[675] C. M. White and N. C. Li, *Anal. Chem.*, 1982, **54**, 1570.

liquids have been analysed by GC/FIMS[676] and, following fractionation on silica gel, by GC/EIMS for phenols and indanols using a Curie-point flash-vaporization technique.[677]

Crude- and heavy-oil fractions have been analysed for polycyclic aromatic hydrocarbons and related compounds.[678,679] The presence of over 200 PAHs and isosteric O- and S-containing analogues in crude oils has been reported,[678] and, after removal of polar material by acid/base extraction, PAHs of bitumen and heavy oils were examined by HPLC and capillary-column GC/MS.[679] Spilled crude oils of similar origin have been differentiated by a GC/MS method that is virtually independent of weathering effects.[598] Correlating the compositions of crude oils by location of unique biomarkers has been described.[598,680] A range of petroleum samples has been found to contain 4β(H)-eudesmane from higher-plant precursors and 8β(H)-drimane probably derived from microbial sources.[681]

For a wide range of largely non-polar organic compounds in aquatic sediments, Soxhlet extraction with CH_2Cl_2 was generally more efficient than ultrasonic extraction methods. For the most polar compounds, ultrasonic extraction with CH_2Cl_2/MeOH gave better recoveries.[682] The configuration of acyclic isoprenoid acids in sediments of various maturity may be examined by GC/MS of their diastereoisomeric (−)-menthyl esters.[683] The effect of maturation on configurational isomerization was studied. Recent sediments have been shown to contain tightly bound β-hydroxy C_{10}–C_{28} acids of a bacterial source.[684] Distributions and identities of hydroxy acids in recent and ancient sediments, determined by GC/MS of TMS ether methyl esters, indicate both higher-plant and microbial sources.[685] A previously misidentified unusual C_{27} triterpane in North Sea sediments is now assigned to 17α(H),18α(H),21β(H)-25,28,30-trisnorhopane.[686] Two other triterpanes and their possible origins were described. The unusual 4-methylsterols in recent lacustrine sediments are thought to be due to *in situ* bacterial biosynthesis,[687] whilst those in Black Sea sediments may be derived from free-living dinoflagellates.[688] Other classes of compounds found in sediments

[676] D. D. Whitehurst, S. E. Butrill, jun., F. J. Derbyshire, M. Farcasiu, G. A. Odoerfer, and L. R. Rudnik, *Fuel*, 1982, **61**, 994.
[677] W. H. McClannan, H. L. C. Meuzelaar, G. S. Metcalf, and G. R. Hill, *Fuel*, 1983, **62**, 1422.
[678] G. Grimmer, J. Jacob, and K.-W. Naujack, *Fresenius' Z. Anal. Chem.*, 1983, **314**, 29.
[679] M. A. Poirier and B. S. Das, *Fuel*, 1984, **63**, 361.
[680] D. A. Flory, H. A. Lichtenstein, K. Biemann, J. E. Biller, and C. Barker, *Oil Gas J.*, 1983, **81**, 91.
[681] R. Alexander, R. Kagi, and R. Noble, *J. Chem. Soc., Chem. Commun.*, 1983, 226.
[682] S. Sporstøl, N. Gjøs, and G. E. Carlberg, *Anal. Chim. Acta*, 1983, **151**, 231.
[683] A. S. MacKenzie, R. L. Patience, D. A. Yon, jun., and J. R. Maxwell, *Geochim. Cosmochim. Acta*, 1982, **46**, 783.
[684] K. Kawamura and R. Ishiwatari, *Nature (London)*, 1982, **297**, 144.
[685] J. N. Cardoso and G. Eglinton, *Geochim. Cosmochim. Acta*, 1983, **47**, 723.
[686] J. K. Volkman, R. Alexander, R. I. Kagi, and J. Rullkoetter, *Geochim. Cosmochim. Acta*, 1983, **47**, 1033.
[687] F. Mermoud, F. O. Gülacar, S. Siles, B. Chassaing, and A. Bucks, *Chemosphere*, 1982, **11**, 557.
[688] J. W. De Leeuw, W. I. C. Rijpstra, P. A. Schenck, and J. K. Volkman, *Geochim. Cosmochim. Acta*, 1983, **47**, 455.

include the novel A-nor-steranes,[689] phthalate esters,[690] and aromatic hydrocarbons.[691]

A new degradative pathway is believed to be responsible for de-A-steroid ketones and de-A-aromatic steroid hydrocarbons in Cretaceous black shale.[692] Study of oil shales and shale oils from the Green River and Rundle Formations has revealed over 1000 compounds,[693] and Condor (Australia) oil shale yielded over 600 compounds,[694] from a wide range of chemical classes. Accurate mass chromatograms at the nominal m/z value of 121 can be made characteristic of homologous n-alkylpyridines, some terpenes, and methylated phenols in Paraho shale oil,[695] and tricyclic terpenoid carboxylic acids and their parent alkanes have been determined in Alberta oil sands.[696]

Characterization of humic substances from aquatic[697,698] and terrestrial[698,699] sources using a variety of mass-spectrometric techniques has been reported, and results continue to emerge of GC/MS analysis of the Murchison meteorite.[700,701] Its distribution of nitrogen heterocycles is similar to that after catalytic reaction of ammonia with simple aldehydes,[700] and a series of partially racemic amino acids was found in interior fragments of the meteorite.[701]

Pyrolysis-GC/MS. – The value of examining natural and man-made macromolecules by GC/MS of their pyrolysis products has been reviewed.[702–706] A compendium and atlas of pyrolysis mass spectrometry of recent and fossil biomaterials[707] and a lengthy guide to analytical pyrolysis[708] are available.

689 G. Van Graas, F. De Lange, J. W. De Leeuw, and P. A. Schenck, *Nature (London)*, 1982, **296**, 59.
690 J. C. Peterson and D. H. Freeman, *Int. J. Environ. Anal. Chem.*, 1982, **12**, 277.
691 S. Sportol, N. Gjos, R. G. Lichtenthaler, K. O. Gustavsen, K. Urdal, F. Oreld, and J. Skei, *Environ. Sci. Technol.*, 1983, **17**, 282.
692 G. Van Graas, F. De Lange, J. W. De Leeuw, and P. A. Schenk, *Nature (London)*, 1982, **299**, 437.
693 L. L. Ingram, J. Ellis, P. T. Crisp, and A. C. Cook, *Chem. Geol.*, 1983, **38**, 185.
694 C. E. Rovere, P. T. Crisp, J. Ellis, and P. D. Bolton, *Fuel*, 1983, **62**, 1274.
695 E. J. Gallegos, *Anal. Chem.*, 1984, **56**, 701.
696 T. D. Cyr and O. P. Strausz, *J. Chem. Soc., Chem. Commun.*, 1983, 1028.
697 W. Liao, R. F. Christman, J. D. Johnson, D. S. Millington, and J. R. Hass, *Environ. Sci. Technol.*, 1982, **16**, 403.
698 M. A. Wilson, R. P. Philp, A. H. Gillam, T. D. Gilbert, and K. R. Tate, *Geochim. Cosmochim. Acta*, 1983, **47**, 497.
699 J. Abbott, T. I. Balkas, O. Basturk, and A. F. Gaines, *Int. J. Mass Spectrom. Ion Phys.*, 1983, **48**, 145.
700 P. G. Stoks and A. W. Schwartz, *Geochim. Cosmochim. Acta*, 1982, **46**, 309.
701 M. H. Engel and B. Nagy, *Nature (London)*, 1982, **296**, 837.
702 S. A. Liebman and E. J. Levy, *Adv. Chem. Ser.*, 1983, **203** (Polym. Charact.), 617.
703 E. Reiner and T. F. Moran, *Adv. Chem. Ser.*, 1983, **203** (Polym. Charact.), 705.
704 S. A. Liebman and E. J. Levy, *J. Chromatogr. Sci.*, 1983, **21**, 1.
705 J. Chiu, *Anal. Calorim.*, 1984, **5**, 197.
706 H. K. Yuen and G. W. Mappes, *Thermochim. Acta*, 1983, **70**, 269.
707 H. L. C. Meuzelaar, J. Haverkamp, and F. D. Hileman, 'Pyrolysis Mass Spectrometry of Recent and Fossil Biomaterials', Elsevier, Amsterdam, 1982.
708 W. J. Irwin, 'Analytical Pyrolysis. A Comprehensive Guide', Marcel Dekker, Basel, 1982.

Rotary drying of aqueous soil suspensions in the pyrolysis probe ensures uniform distribution of soil samples and thus improves reproducibility of the technique.[709]

In the polymer industry, the method of pyrolysis-GC/MS has been used to examine the polymers themselves and also their additives or impurities. Particles of poly(methylmethacrylate) in tissue, originating from wear of nearby endoprotheses, have been detected rapidly and sensitively,[710] the bonded chains present in reversed-phase HPLC stationary phases can be determined,[711] and the volatile compounds formed by heating an epoxy resin, composed of tetraglycidyl methylene dianiline and diamino diphenyl sulphone, have been identified by pyrolysis-GC/MS.[712] The antioxidant 2,6-di-t-butyl-*p*-cresol in butadiene/styrene block copolymers was identified and quantified by pyrolysis-GC/MS,[713] and the pyrolysis products of poly(1,3-phenylene isophthalamide) and poly(1,4-phenylene terephthalamide) have been investigated.[714]

Thermal degradation of carbohydrates has been studied by pyrolysis-GC/MS methods,[715,716] and the effects of inorganic salts in the pyrolysis matrix have been shown to be marked.[716] The leading references in the geochemical area were covered in the previous section, but it is noted here that estuarine colloid material, when pyrolysed, was shown[717] to produce components typical for thermal decomposition of both carbohydrates (furans, lactones, *etc.*) and proteins (pyrroles, amides, *etc.*), suggesting that the particulate organic matter in surface waters was derived mainly from aquatic micro-organisms rather than plant debris. Also, various lignins and soft brown coal woods have been compared and contrasted by pyrolysis-GC/MS.[718]

Gases evolved from differential thermal analysers, a thermobalance, or a micro-furnace have been analysed quantitatively by on-line GC/MS. The methods were applied to the thermal behaviour of calcium propanoate.[719] Measurement of the isotopic content of oxygen of potassium chlorate and bromate[720] and the identification of aniline dyes in environmental samples[721] have been accomplished by pyrolysis-GC/MS.

[709] M. Gassiot-Matas, J. M. Alcaniz-Baldellou, and J. Andress-Canadell, *J. Anal. Appl. Pyrolysis*, 1982, **4**, 241.
[710] G. Buchhorn, I. Lüderwald, K.-E. Müller, and H. G. Willert, *Fresenius' Z. Anal. Chem.*, 1982, **312**, 539.
[711] P. Mussche and M. Verzele, *J. Anal. Appl. Pyrolysis*, 1983, **4**, 273.
[712] J. G. Moncur, A. B. Campa, and P. C. Pinoli, *J. High Resolut. Chromatogr. Chromatogr. Commun.*, 1982, **5**, 322.
[713] N. Lichtenstein and K. Quellmalz, *Fresenius' Z. Anal. Chem.*, 1983, **316**, 268.
[714] J. R. Brown and A. J. Power, *Polym. Degrad. Stabil.*, 1982, **4**, 379.
[715] T. Yoneya, *J. SCCJ*, 1982, **16**, 57 (*Chem. Abstr.*, 1982, **97**, 164 869).
[716] A. Van der Kaaden, J. Haverkamp, J. J. Boon, and J. W. De Leeuw, *J. Anal. Appl. Pyrolysis*, 1983, **5**, 199.
[717] A. C. Sigleo, T. C. Hoering, and G. R. Helz, *Geochim. Cosmochim. Acta*, 1982, **46**, 1619.
[718] R. P. Philp, N. J. Russell, T. D. Gilbert, and J. M. Friedrich, *J. Anal. Appl. Pyrolysis*, 1982, **4**, 143.
[719] P. A. Barnes, G. Stephenson, and S. B. Warrington, *J. Therm. Anal.*, 1982, **25**, 299.
[720] E. Larsen, H. Egsgaard, and N. Bjerre, *J. Trace Microprobe Tech.*, 1983, **1**, 387.
[721] L. E. Abbey, J. P. Gould, and T. F. Moran, *J. – Water Pollut. Control Fed.*, 1982, **54**, 474.

Miscellaneous. – Concurring with other workers' finding of formaldehyde in biological tissues,[373–375] it has been shown that derivatization with pentafluorophenylhydrazine followed by selected-ion monitoring with $^{2}H_2{}^{13}CO$ as internal standard is a useful procedure for quantifying formaldehyde in tissue samples as small as 20 mg wet weight.[722] Characterization of formaldehyde oligomers in H_2O/MeOH solutions has been accomplished by formation of TMS derivatives and capillary-column GC/CIMS.[723] Ethylene/propylene copolymer and PVC samples have been examined for volatile components by dynamic headspace sampling, collection and concentration in a cold trap, then capillary-column GC/MS,[724] and mixtures of linear and cyclic poly(dimethylsiloxanes) have been studied.[725]

Fluoroacetic acid, a highly toxic compound produced by some tropical plants including tea, can be determined by pentafluorobenzylation and selected-ion monitoring. At low ionization energy, the detection limit was 5 ng g^{-1} of plant sample.[726] Volatile excretions of plants have also been studied.[727,728] In one of these investigations,[728] a chemically bonded thick-film (1 μm) non-polar capillary column of fused silica that remained unaffected by oxygen and water was employed for GC/MS.

Measurement of the exchange rate of carbonyl oxygen in α-ketoacids, free in solution or bound in enzyme complexes, has been achieved by an ^{18}O incorporation procedure.[729] Airborne hydroxyl radicals can be trapped with α-4-pyridyl-*N*-t-butylnitrone-α-1-oxide and the resulting adduct converted to its TMS ether for GC/MS analysis.[730] Finally, a series of carbonates and diglycol carbonates has been analysed by GC/MS.[731]

4 High-performance Liquid Chromatography/Mass Spectrometry

In this section, reports of genuine HPLC/MS are reviewed. Studies that utilize the moving-belt interface, for example, as a rapid and automatic direct-insertion probe of pure samples are omitted. Also, papers describing simple home-made direct-liquid-introduction (DLI) interfaces are not covered unless they offer distinct advantages. In particular, devices reported to be unsuitable for non-volatile compounds are excluded.

[722] H. d'A. Heck, E. L. White, and M. Casanova-Schmitz, *Biomed. Mass Spectrom.*, 1982, **9**, 347.
[723] D. F. Utterback, D. S. Millington, and A. Gold, *Anal. Chem.*, 1984, **56**, 470.
[724] S. Jacobsson, *J. High Resolut. Chromatogr. Chromatogr. Commun.*, 1984, **7**, 185.
[725] L. J. Bogunovic and M. D. Dragojevic, *Glas. Hem. Drus. Beograd*, 1982, **47**, 667.
[726] T. Vartiainen and P. Kauranen, *Anal. Chim. Acta*, 1984, **157**, 91.
[727] V. A. Isidorov, I. G. Zenkevich, and B. V. Ioffe, *Dokl. Akad. Nauk SSSR*, 1982, **263**, 893.
[728] W. F. Burns and D. T. Tingey, *J. Chromatogr. Sci.*, 1983, **21**, 341.
[729] T. S. Viswanathan, C. E. Hignite, and H. F. Fisher, *Anal. Biochem.*, 1982, **123**, 295.
[730] T. Watanabe, M. Yoshida, S. Fujiwara, K. Abe, A. Onoe, M. Hirota, and S. Igarashi, *Anal. Chem.*, 1982, **54**, 2470; K. Abe, H. Suezawa, M. Hirota, and T. Ishii, *J. Chem. Soc., Perkin Trans. 2*, 1984, 29.
[731] S. P. Griff, *Appl. Spectrosc.*, 1983, **37**, 354.

Reviews of LC/MS have appeared on average at the prodigious rate of once a month. Editors may wish to consider if this publication rate is warranted. Relevant reviews include a collection of articles forming a useful introduction for the newcomer,[732] general LC/MS,[9,10,733–741] HPLC detectors including mass spectrometers,[742] LC/MS applied to natural products,[743–745] pesticides,[743] water samples,[746] glycerolipids,[170] and steroids and terpenes,[747] microbore liquid-chromatography columns,[748] microbore LC/MS,[749–752] ion emission from liquids,[753] and thermospray LC/MS.[754,755]

The LC/MS combination is a technique caught midway between the experimental stage and the 'routine use' stage. Of course, some interface designs are further advanced than others but none has yet arrived at the latter stage. Also, it is still true that the choice of interface depends on the class of compound and type of investigation being addressed;[746] for example, thermospray LC/MS is a clear leader for ionic compounds but not for non-polar substances. Thus, choice of purchase is difficult for the user who has many different analytical problems to solve. In both DLI and thermospray methods, a portion or the whole of the LC effluent is passed directly into the ion source. In the former, use of a small diaphragm to effect nebulization, a desolvation chamber, and a water-cooled probe has extended the range of compounds amenable to DLI to polar ones of low volatility. In the ion source the LC solute becomes the CI reactant gas. The

[732] D. E. Games (ed.), *Spectra*, 1983, **9** (1), 3–28 (publ. Finnigan MAT).
[733] R. C. Willoughby and R. F. Browner in 'Trace Analysis', ed. J. F. Lawrence, Academic Press, 1982, Vol. 2, p. 69.
[734] W. H. McFadden, *Anal. Proc.*, 1982, **19**, 258.
[735] D. A. Yorke, *Proc. Inst. Pet., London*, 1982, 159.
[736] Z. F. Curry, *J. Liq. Chromatogr.*, 1982, **5** (Suppl. 2), 257.
[737] P. J. Arpino in 'Liquid Chromatography Detectors', ed. T. M. Vickrey, Marcel Dekker, New York, 1983, p. 243.
[738] G. Guiochon and P. J. Arpino, *J. Chromatogr.*, 1983, **271**, 13.
[739] D. E. Games, *Adv. Chromatogr.*, 1983, **21**, 1.
[740] K. Levsen, *Nachr. Chem., Tech. Lab.*, 1983, **31**, 782.
[741] C. G. Edmonds, J. A. McCloskey, and V. A. Edmonds, *Biomed. Mass Spectrom.*, 1983, **10**, 237.
[742] J. H. Knox, *Anal. Proc.*, 1982, **19**, 166.
[743] D. E. Games, C. Eckers, M. S. Lant, E. Lewis, N. C. A. Weerasinghe, and S. A. Westwood, *Anal. Proc.*, 1982, **19**, 253.
[744] D. J. Dixon, *Analusis*, 1982, **10**, 343.
[745] D. E. Games, *Pharm. Weekbl.*, 1984, **119**, 30.
[746] D. E. Games, M. G. Foster, and O. Meresz, *Anal. Proc.*, 1984, **21**, 174.
[747] M. E. Rose, *Biochem. Soc. Trans.*, 1983, **11**, 561.
[748] R. P. W. Scott, 'Small Bore Liquid Chromatography Columns', Wiley, Chichester, 1984; R. P. W. Scott, *Adv. Chromatogr.*, 1983, **22**, 247; F. J. Yang, *J. High Resolut. Chromatogr. Chromatogr. Commun.*, 1983, **6**, 348; D. Ishii and T. Takeuchi, *Adv. Chromatogr.*, 1983, **21**, 131.
[749] D. E. Games and S. A. Westwood, *Eur. Spectrosc. News*, 1983, **48**, 14.
[750] J. Henion, D. Skrabalak, E. Dewey, and G. Maylin, *Drug Metab. Rev.*, 1983, **14**, 961.
[751] D. E. Games, N. J. Alcock, and M. A. McDowall, *Anal. Proc.*, 1984, **21**, 24.
[752] J. Henion, *J. Chromatogr. Libr.*, 1984, **28**, 260.
[753] M. L. Vestal, *Springer Ser. Chem. Phys.*, 1983, **25** (Ion Form. Org. Solids), 246.
[754] M. L. Vestal, *Mass Spectrom. Rev.*, 1983, **2**, 447.
[755] R. Schubert, *GIT Fachz. Lab.*, 1984, **28**, 323.

thermospray technique utilizes rapid heating of the effluent, adiabatic expansion of the vapour into a vacuum, and charge exchange between salt ions and the eluent during evaporation of the droplets. Detection limits are in the low pg range in favourable cases, and large flows of aqueous solutions are tolerated. The method has been used mainly with quadrupole instruments but, with precautions, can be interfaced to magnetic-sector mass spectrometers. A device that is more suitable for magnetic-sector instruments is the electrospray interface,[756] which is currently under development. To some extent, interest in the belt interface has been rekindled with the overcoming of the problem of irregular sample deposition by use of spray deposition, and application of FAB ionization directly to eluents on the belt. With spray deposition the need for heating is removed, and with application of FAB the belt system has become the most versatile interface in terms of compatible ionization modes (EI, CI, SIMS, laser desorption, and FAB).

Thermospray LC/MS. – Both of the suggestions made in the previous Report[2] regarding this interface (incorporation of electrical heating and commercial exploitation) were fulfilled shortly after writing and before publication. The current commercial system (Vestec Corporation) takes the LC effluent through a capillary tube at up to 400 °C directly into the ion source, fitted with one extra mechanical pump, where an expanding aerosol jet of sample and solvent molecules is formed. As long as the sample is polar and the solvent contains a salt such as ammonium acetate, ions of the sample are formed in the spray without external means of ionization. Weakly ionized eluents require conventional CI with the LC solvent acting as reactant gas. The (quadrupole) mass spectrometer for analysis of those ions is at right angles to the jet stream. The method results in little fragmentation and no apparent thermal decomposition of analytes. Its transfer efficiency depends on a number of factors, particularly the flow rate and temperature of the various parts of the interface,[757] although the method has been applied successfully to widely different structures without changing the operating conditions.[758] The electrically heated vaporizer has improved the stability and reproducibility of effluent vaporization, and ionization at flow rates up to 2 ml min^{-1} of aqueous mobile phase is possible.[757] Despite these advances, day-to-day reproducibility of ionization efficiency and fragmentation pathways is still a limitation. Optimization of the configuration of the vaporizer and ion sampling may not yet be established, and a firmer grasp of the detailed mechanism of ionization would help to overcome the remaining problems. As recommended in the last Report,[2] the mechanism of ionization has been studied.[759] The vaporizer produces a superheated mist carried in a supersonic jet of vapour. The droplets preferentially retain non-volatile molecules and are charged positively or negatively according to statistical expectations for random sampling of a neutral fluid containing discrete positive and negative charges. Under the influence of the large local electric fields on the drops, molecular ions with clustered solvent evaporate from the droplets and the clusters equilibrate

[756] M. Dole, L. L. Mack, and R. C. Hines, *J. Chem. Phys.*, 1968, **49**, 2240.
[757] C. R. Blakley and M. L. Vestal, *Anal. Chem.*, 1983, **55**, 750.
[758] I. A. S. Lewis, VG Analytical Insight, 1983, No. 13.
[759] M. L. Vestal, *Int. J. Mass Spectrom. Ion Phys.*, 1983, **46**, 193.

with vapour in the ion source. Some ion/molecule reactions may occur. Finally, ions diffuse to the sampling aperture and are then analysed.[757, 759] A conventional direct-insertion-probe inlet has formed the basis of a new dual-purpose LC/MS interface.[760] It provides both standard DLI LC/MS and, by heating the copper vaporizer, thermospray ionization. Given that LC/MS users currently need to use different types of interface depending on the analytical problem being pursued, this DLI/thermospray combination is to be welcomed. Detection limits of about 100 ng in the thermospray mode need some improvement. Post-column addition of buffer prior to thermospray LC/MS could eliminate the necessity to incorporate a volatile salt into the LC solvent. A co-axial tee piece has been shown[761] to be useful for post-column addition of buffers during gradient LC and thermospray mass spectrometry of carbamate and urea pesticides.

The thermospray method has been applied to analysis of glucuronides[762] and amino acids and peptides.[763–765] Aliquots containing 10 ng of simple glucuronides were sufficient for LC/MS and identification by acquisition of full spectra.[762] Both negative-ion and positive-ion spectra were reported with abundant $(M - H)^-$, $(M + H)^+$, and $(M + NH_4)^+$ ions. Under some conditions,[763] peptides yield protonated molecules without accompanying fragment ions. By adjusting the pH of the effluent, singly, doubly, triply, and quadruply charged pre-formed ions, $(M + nH)^{n+}$, can be observed. Given an analyte with a sufficient number of ionizable side chains, the technique permits examination of peptides with mass greater than the upper scan limit of a quadrupole mass filter (*e.g.* glucagon with relative molar mass of 3483), but no sequence information is obtained. Coupling to a magnetic-sector instrument provides a better means of overcoming both upper-mass and resolution limitations of a quadrupole system. An interesting approach to peptide sequencing involves hydrolysis with immobilized carboxypeptidase Y and on-line analysis of hydrolysates with thermospray LC/MS.[764] For example, the first ten C-terminal residues of RNase were readily assigned. Mixtures of carnitine, acetylcarnitine, and propionylcarnitine have been analysed at nmol levels against their stable-isotope analogues as internal standards.[765]

Direct Liquid Introduction. – Many non-volatile compounds (amino acids, peptides, carbohydrates, nucleosides, *etc.*) can be analysed by DLI LC/MS when a nebulizing sample-introduction system is employed.[766, 767] Chemical ionization with the LC solvent as reactant gas and API[766] are compatible mass-spectrometric techniques, but it would also be interesting to consider applying FAB ionization to the vaporized droplets.[36] Flow rates of aqueous solvents up to 0.15 ml min^{-1}

[760] T. Covey and J. Henion, *Anal. Chem.*, 1983, **55**, 2275.
[761] R. D. Voyksner, J. T. Bursey, and E. D. Pellizzari, *Anal. Chem.*, 1984, **56**, 1507.
[762] D. J. Liberato, C. C. Fenselau, M. L. Vestal, and A. L. Yergey, *Anal. Chem.*, 1983, **55**, 1741.
[763] D. Pilosof, H. Y. Kim, D. F. Dyckes, and M. L. Vestal, *Anal. Chem.*, 1984, **56**, 1236.
[764] H. Y. Kim, D. Pilosof, M. L. Vestal, D. F. Dyckes, J. P. Kitchell, and E. Dvorin, *Pept.: Struct. Funct., Proc. Am. Pept. Symp., 8th*, 1983, 719.
[765] A. L. Yergey, D. J. Liberato, and D. S. Millington, *Anal. Biochem.*, 1984, **139**, 278.
[766] M. Sakairi and H. Kambara, *Shitsuryo Bunseki*, 1983, **31**, 87.
[767] J. A. Apffel, U. A. T. Brinkman, R. W. Frei, and E. A. I. M. Evers, *Anal. Chem.*, 1983, **55**, 2280.

have been shown to be compatible with a DLI interface for magnetic-sector mass spectrometers.[768] The tip of the nozzle in the DLI device described is heated electrically to provide rapid vaporization to a fast beam of solvated solute molecules (and solvent vapour). A large part of the vapour is pumped away at a jet-separator stage, while solvated sample molecules traverse the separator and enter the source through a fine orifice. It is unfortunate that, apart from adenosine and leucinylalanine, more realistic analytes were not chosen to test the system. Ions have been extracted from a micro-LC effluent by the spraying action of a non-uniform electric field and the collision of the droplets with air molecules.[769] The molecular ions move supersonically through a nozzle into a region of low vacuum and thence through a diaphragm into the high-vacuum analyser. Protonated and cationized molecules only were observed for the dansyl derivative of arginine.[769]

Several commendable studies have considered basic aspects of DLI LC/MS, in particular, optimization of vacuum nebulizers and desolvation chambers, the theory of evaporation of liquids, and understanding the nature of the ionization process(es). In one DLI system[770] droplets from the nebulizer are passed through a heated desolvation chamber at sonic velocities prior to introduction into the CI ion source, but complete desolvation is avoided as undesirable. The negative-ion spectrum of vitamin B_{12} obtained by this system is impressive.[770] Continuing a series of papers on optimization of DLI operating parameters, the design of the desolvation chamber has been discussed.[771] Its shape focuses the electrically charged droplets from the diaphragm nebulizer into the CI ion source. The voltage applied to the drift tube influenced the ionization efficiency, and a stable but unsustained Townsend discharge occurred at over 140 V.[771] There are clear advantages in using microbore columns with DLI LC/MS[772–775] because it is easier to match the LC effluent to the mass spectrometer than *vice versa*. Applicability of a micro-LC/MS system was improved markedly by incorporating a water cooling jacket and a bubble saturator for the nebulizing gas. The most advantageous temperature profile around the nebulizer tip was also discussed.[772] A DLI interface for microbore LC has been constructed from fused-silica capillary tubing that protrudes into the CI ion source.[773] Transfer of heat from the source block to the tip of the capillary tube was necessary to support evaporation of the solution. Strong dependence on temperature was again noted for the rate of vaporization within the transfer line.[773] Solute enrichment within a desolvation

[768] J. R. Chapman, E. H. Harden, S. Evans, and L. E. Moore, *Int. J. Mass Spectrom. Ion Phys.*, 1983, **46**, 201.

[769] M. L. Aleksandrov, L. N. Gall, N. V. Krasnov, V. I. Nikolaev, V. A. Pavlenko, V. A. Shkurov, G. I. Baram, M. A. Grachev, V. D. Knorre, and Yu. S. Kusner, *Bioorg. Khim.*, 1984, **10**, 710 (*Chem. Abstr.*, 1984, **101**, 65 234).

[770] M. Dedieu, C. Juin, P. J. Arpino, and G. Guichon, *Anal. Chem.*, 1982, **54**, 2372.

[771] P. J. Arpino, J. P. Bounine, M. Dedieu, and G. Guiochon, *J. Chromatogr.*, 1983, **271**, 43.

[772] H. Yoshida, K. Matsumoto, K. Itoh, S. Tsuge, Y. Hirata, K. Mochizuki, N. Kokubun, and Y. Yoshida, *Fresenius' Z. Anal. Chem.*, 1982, **311**, 674.

[773] A. P. Bruins and B. F. H. Drenth, *J. Chromatogr.*, 1983, **271**, 71.

[774] F. R. Sugnaux, D. S. Skrabalak, and J. D. Henion, *J. Chromatogr.*, 1983, **264**, 357.

[775] Y. Iida, S. Daishima, and S. Okada, *Kenkyu Hokoku – Asahi Garasu Kogyo Gijutsu Shoreikai*, 1982, **40**, 73 (*Chem. Abstr.*, 1983, **99**, 113 405).

chamber for LC/MS of entire micro-LC effluents has been described and applied to quantitative analysis of ng amounts of corticosteroids.[774] Nebulizers using the Babington principle[776] are almost impossible to block and have been much used in atomic spectroscopy, but not in LC/MS. In such nebulizers the gas emerges from the smaller orifice and the sample is fed over it as a film. Where blockage problems occur with conventional devices, Babington-type nebulizers may be worth investigation. For LC/MS, solvents are usually chosen for their chromatographic properties rather than their efficacies as CI reactant gases. Using common LC solvent systems like methanol/water, acetonitrile, and acetonitrile/water, the resulting CI plasmas are not well characterized. Three studies in this area represent useful first steps in understanding the precise nature of ion formation in DLI LC/MS.[777-779] Collisional-activation studies were used to study the reactant-gas cluster ions.[777] The difficulty of maintaining optimal operating conditions is compounded during gradient elution, since response varies with ion-source pressure,[778,779] and this in turn varies with solvent composition.[778] Changes in solvent composition appeared to affect the degree of fragmentation more than the abundance of protonated molecules. The extent of fragmentation was shown to depend also on temperature, and optimization of the DLI probe position was discussed.[778]

The DLI approach to LC/MS has been utilized for studies of steroid glucuronides by NICI,[780] cardiac glycosides in digitalis,[780] vitamin D_3 and its metabolites,[780] sequencing of underivatized peptides,[781] nucleosides,[782] quantitative assay of triacylglycerols[783] and diacylglycerol moieties of glycerophospholipids,[784] steroids and antibiotics in equine urine,[785] explosives,[786-788] drugs,[787,789] biogenic catecholamine ion pairs,[790] and phenylureas.[791] Both negative-ion and positive-ion modes were employed for sequencing the enkephalins and α-amanitin,[781] and positive-ion CI allowed about 1 μg of explosives injected on to the column (10 ng entering the ion source after 99:1 splitting) to be analysed,[788] whereas NICI allowed 100 ng of TNT on column (1 ng after splitting) to be detected.[786] The loss of sensitivity incurred by effluent splitting is eliminated by utilizing the whole

776 R. F. Browner and A. W. Boorn, *Anal. Chem.*, 1984, **56**, 875A.
777 R. D. Voyksner, J. R. Hass, and M. M. Bursey, *Anal. Chem.*, 1982, **54**, 2465.
778 R. D. Voyksner, C. E. Parker, R. J. Hass, and M. M. Bursey, *Anal. Chem.*, 1982, **54**, 2583.
779 J. Yinon and A. Cohen, *Org. Mass Spectrom.*, 1983, **18**, 47.
780 C. N. Kenyon, P. C. Goodley, D. J. Dixon, J. O. Whitney, K. R. Faull, and J. D. Barchas, *Int. Lab.*, 1983, **13** (4), 60 [*Am. Lab.*, 1983, **15** (1), 38].
781 C. N. Kenyon, *Biomed. Mass Spectrom.*, 1983, **10**, 535.
782 E. L. Esmans, Y. Luyten, and F. C. Alderweireldt, *Biomed. Mass Spectrom.*, 1983, **10**, 347.
783 L. Marai, J. J. Myher, and A. Kuksis, *Can. J. Biochem. Cell Biol.*, 1983, **61**, 840; J. J. Myher, A. Kuksis, L. Marai, and F. Manganaro, *J. Chromatogr.*, 1984, **283**, 289.
784 S. Pind, A. Kuksis, J. J. Myher, and L. Marai, *Can. J. Biochem. Cell Biol.*, 1984, **62**, 301.
785 C. Eckers, J. D. Henion, G. A. Maylin, D. S. Skrabalak, J. Vessman, A. M. Tivert, and J. C. Greenfield, *Int. J. Mass Spectrom. Ion Phys.*, 1983, **46**, 205.
786 C. E. Parker, R. D. Voyksner, Y. Tondeur, J. D. Henion, D. J. Harvan, J. R. Hass, and J. Yinon, *J. Forensic Sci.*, 1982, **27**, 495.
787 J. Yinon, *Int. J. Mass Spectrom. Ion Phys.*, 1983, **48**, 253.
788 J. Yinon and D.-G. Hwang, *J. Chromatogr.*, 1983, **268**, 45.

effluent from microbore columns, as with analyses at low ng levels of corticosteroid metabolites and antibiotics[785] and drugs such as trichlormethiazide.[789]

Moving-belt Interfaces. – Given the popularity of FAB mass spectrometry for analysis of polar compounds and its potential compatibility with the fairly well established moving-belt technique, it is surprising that only two papers have reported on-line LC/FABMS.[792,793] Even so, along with spray deposition on to the belt and utilization of microbore columns, the prospect of LC/FABMS is responsible for renewed interest in the method.

A commercial moving-belt interface was modified such that the belt extends into the ion source, enabling not only direct EI and CI but also FAB mass spectrometry by directing a beam of argon atoms on to the tip of the belt.[792] Using ionization directly from the solid phase, FAB spectra of saccharides and peptides were obtained. An order of magnitude increase in sensitivity was attendant on coating the belt with glycerol as liquid matrix.[792] Belt cleaning (preferably by washing rather than heating) on the 'return journey' is necessary to prevent memory effects.

A marked improvement in both normal-phase and reversed-phase LC/MS efficiency occurs when effluent from the column is forced through a jet that deposits the sample onto the belt as a dry spray.[794–797] All infrared heating can be removed from the interface when this technique is used. Detection limits around 40 pg were noted for PAHs, and mobile phases with high water content are permissible.[795] One sprayer was best used with a nebulizing gas and held at an angle of 45° to the plane of the belt.[794] Spray systems can be applied to gradient-elution LC/MS with microbore columns.[796] Another method of ionization that is compatible with the moving-belt interface is laser desorption.[797] Amino acids, antibiotics, nucleosides, and peptides were deposited on the belt with a thermospray vaporizer then ionized and desorbed by a Q-switched laser beam.

Glass-lined stainless-steel microbore columns provide performances comparable with conventional columns, but the lower flow rate of mobile phase improves LC/MS sensitivity since the entire effluent is compatible with the belt interface

[789] C. Eckers, D. S. Skrabalak, and J. Henion, *Clin. Chem.*, 1982, **28**, 1882.
[790] H. Milon and H. Bur, *J. Chromatogr.*, 1983, **271**, 83.
[791] K. Levsen, K. H. Schäfer, and J. Freudenthal, *J. Chromatogr.*, 1983, **271**, 51.
[792] P. Dobberstein, E. Korte, G. Meyerhoff, and R. Pesch, *Int. J. Mass Spectrom. Ion Phys.*, 1983, **46**, 185.
[793] I. A. S. Lewis and P. W. Brooks, Abstracts of 31st Ann. Conf. Mass Spectrom. and Allied Topics, Boston, 1983, p. 850.
[794] E. P. Lankmayr, M. J. Hayes, B. L. Karger, P. Vouros, and J. M. McGuire, *Int. J. Mass Spectrom. Ion Phys.*, 1983, **46**, 177.
[795] M. J. Hayes, E. P. Lankmayer, P. Vouros, B. L. Karger, and J. M. McGuire, *Anal. Chem.*, 1983, **55**, 1745.
[796] M. J. Hayes, H. E. Schwartz, P. Vouros, B. L. Karger, A. D. Thruston, jun., and J. M. McGuire, *Anal. Chem.*, 1984, **56**, 1229.
[797] E. D. Hardin, T. P. Fan, C. R. Blakley, and M. L. Vestal, *Anal. Chem.*, 1984, **56**, 2.

and background due to solvent impurities is reduced.[798] The system tolerates gradient elution, flow programming, and high-percentage aqueous solutions.[798] Samples analysed by the microbore LC/MS technique include carbamate pesticides,[798] β-lactam antibiotics,[799] and pseudomonic acids.[799]

Band broadening other than in the column is negligible in capillary GC/MS at least with well designed interfaces. However, in LC/MS extra-column band broadening leads to a significant increase in height equivalent to a theoretical plate. Until now, investigations of the chromatographic behaviour of moving-belt systems have been qualitative at best. It is to be applauded that quantitative studies aiming to improve understanding of the factors contributing to band broadening are now being undertaken.[794, 800] Determination of peak area, variance, and skew formed the basis of studying the effects of belt speed, mobile-phase flow rate, and spray conditions on resolution.[794] Again by considering system variance, another group of workers has concluded that a mass spectrometer is a detector of low time constant and low effective dead volume, suitable for use with microbore HPLC.[800]

Moving-belt interfaces have been evaluated with or applied to mono- to tetra-saccharides, nucleotides, bile acids, riboflavin, gibberellin A_3, various glycosides, riboflavin phosphate, and a quaternary ammonium drug,[801] a test-well sample from a landfill site,[802] the pesticide coumaphos and its oxygen analogue in eggs and milk,[803] thermally labile carbamate and urea pesticides,[804] dexamethazone and betamethasone,[805] steroids and drugs,[806] sphingoid bases,[807] kepone and mirex,[808] amygdalin,[809] peptide sequencing by use of *N*-acetyl-*N*,*O*-permethylated derivatives[810] and examination of *C*-methylation artifacts from the permethylation reaction of peptides,[811] polynuclear aromatic and heterocyclic compounds

798 M. S. Lant, D. E. Games, S. A. Westwood, and B. J. Woodhall, *Int. J. Mass Spectrom. Ion Phys.*, 1983, **46**, 189; N. J. Alcock, L. Corbelli, D. E. Games, M. S. Lant, and S. A. Westwood, *Biomed. Mass Spectrom.*, 1982, **9**, 499.
799 M. A. McDowall, D.E. Games, and J. L. Gower, *Int. J. Mass Spectrom. Ion Phys.*, 1983, **48**, 157.
800 D. E. Games, M. J. Hewlins, S. A. Westwood, and D. J. Morgan, *J. Chromatogr.*, 1982, **250**, 62.
801 D. E. Games, M. A. McDowall, K. Levsen, K. H. Schafer, P. Dobberstein, and J. L. Gower, *Biomed. Mass Spectrom.*, 1984, **11**, 87.
802 M. G. Foster, O. Meresz, D. E. Games, M. S. Lant, and S. A. Westwood, *Biomed. Mass Spectrom.*, 1983, **10**, 338.
803 K. D. White, Z. Min, W. C. Brumley, R. T. Krause, and J. A. Sphon, *J. Assoc. Off. Anal. Chem.*, 1983, **66**, 1358.
804 T. Cairns, E. G. Siegmund, and G. M. Doose, *Biomed. Mass Spectrom.*, 1983, **10**, 24.
805 T. Cairns, E. G. Siegmund, J. J. Stamp, and J. P. Skelly, *Biomed. Mass Spectrom.*, 1983, **10**, 203.
806 J. D. Baty and R. G. Willis, *Anal. Proc.*, 1982, **19**, 251.
807 F. B. Jungalwala, J. E. Evans, H. Kadowaki, and R. H. McCluer, *J. Lipid Res.*, 1984, **25**, 209.
808 J. N. Huckins, D. L. Stalling, J. D. Petty, D. R. Buckler, and B. T. Johnson, *J. Agric. Food Chem.*, 1982, **30**, 1020.
809 T. Cairns and E. G. Siegmund, *Biomed. Mass Spectrom.*, 1982, **9**, 307.
810 P. Roepstorff, M. A. McDowall, M. P. L. Games, and D. E. Games, *Int. J. Mass Spectrom. Ion Phys.*, 1983, **48**, 197.
811 T. J. Yu, B. L. Karger, and P. Vouros, *Biomed. Mass Spectrom.*, 1983, **10**, 633.

in petroleum,[812] pepper and capsicum oleoresins,[813] and coumarins.[814] A glass-lined stainless-steel microbore column with gradient elution was utilized along with capillary-column GC/MS for the acidic fraction of a test-well sample.[802] When compared with GC/MS for analysis of coumaphos, LC/MS provided a better detection limit but poorer reproducibility and analysis time.[803] A marked degree of thermal decomposition of some steroidal drugs on a moving belt was noted especially at low concentration,[805] whilst others[806] found that 250 ng of steroids injected on to the LC column was sufficient for identification by LC/MS. An HPLC/MS method capable of sequencing peptides with up to seven residues[810] requires development before it becomes a practicable sequencing technique. Comparison of LC/MS and MS/MS for analysis of natural oxygen heterocyclic compounds showed that major coumarins could be located by the latter method but, unlike the former, isomeric compounds could not be resolved thereby.[814]

Miscellaneous Interfaces. – The on-line combination of HPLC and ^{252}Cf fission-fragment-induced desorption mass spectrometry has been improved.[815–817] Twelve discrete samples are collected as thin films on foils mounted on a rotatable disc and analysed at the rate of one per minute.[815–817] The previous spectrum acquisition time was 10 min.[2] The system has been tested with anti-tumour and anti-arrhythmic drugs.[815]

Of pertinence to LC/MS is the liquid ion-evaporation technique.[162, 818, 819] Since the liquid flow rate for atmospheric-pressure ion-evaporation mass spectrometry (APIEMS) is about 1 ml min^{-1}, eluents from HPLC columns could be fed directly to the ion source. The method may complement other LC/MS methods since the solvent needs to be predominantly aqueous.[818, 819] The process is applicable to most positive and negative monovalent ions (both inorganic and organic) and some neutral organic species that evaporate, clustered with ions. Only molecular or quasi-molecular ions are likely to be produced,[818] but collisional activation could be used to provide fragmentation information.[162, 819] Detection limits down to 200 pg, depending on the sample, are possible.[818]

The interface that employs a resistance-heated stationary wire to concentrate the liquid stream of effluent[2] has been improved by addition of an ultrasonic spraying device to overcome the difficulties of introducing aqueous solvents into

[812] R. W. Smith, D. E. Games, and S. F. Noel, *Int. J. Mass Spectrom. Ion Phys.*, 1983, **48**, 327.

[813] D. E. Games, N. J. Alcock, J. van der Greef, L. M. Nyssen, H. Maarse, and M. C. ten Noever de Brauw, *J. Chromatogr.*, 1984, **294**, 269.

[814] N. J. Alcock, W. Kuhnz, and D. E. Games, *Int. J. Mass Spectrom. Ion Phys.*, 1983, **48**, 153.

[815] H. Jungclas, H. Danigel, L. Schmidt, and J. Dellbrügge, *Org. Mass Spectrom.*, 1982, **17**, 499; L. Schmidt, H. Danigel, and H. Jungclas, *Nucl. Instrum. Methods Phys. Res.*, 1982, **198**, 165.

[816] H. Jungclas, H. Danigel, and L. Schmidt, *Int. J. Mass Spectrom. Ion Phys.*, 1983, **46**, 197.

[817] H. Jungclas, H. Danigel, and L. Schmidt, *J. Chromatogr.*, 1983, **271**, 35.

[818] J. V. Iribarne, P. J. Dziedzic, and B. A. Thomson, *Int. J. Mass Spectrom. Ion Phys.*, 1983, **50**, 331.

[819] B. A. Thomson, J. V. Iribarne, and P. J. Dziedzic, *Anal. Chem.*, 1982, **54**, 2219.

the ion source.[820] Finally, a simple mechanical system comprising a screw rod driven by a motor has been described for collecting fractions of microbore LC effluent at flow rates up to 0.5 ml min^{-1}.[821] Up to 30 fractions can be monitored subsequently by laser mass spectrometry in this mechanized off-line combination.

5 Inorganic and Organometallic Compounds

Independently, chromatography and mass spectrometry are making inroads into instrumental analysis of inorganic species in solution (*e.g.* ion chromatography and plasma mass spectrometry), traditionally the domain of optical and electrochemical methods. It is reasonable to expect this trend to continue and to include combined chromatography/mass spectrometry. At present, application of GC/MS and LC/MS to inorganic and organometallic compounds falls far short of its potential, and this overview seeks to make more widely known the possibilities of such analyses.

The most obvious attribute of LC/MS for this field is its lack of dependence on volatility. Various ion (liquid) chromatographic techniques are rapidly growing in popularity for speciation of inorganic species, so coupling to mass spectrometers seems a logical and desirable extension (*e.g.* for isotope studies) but has not yet been reported. Both GC and GC/MS have been employed for inorganic and organometallic compounds that are volatile or that can be made so by derivatization/complexation. A catalogue of experimental GC methods for organometallic compounds is available,[822] and metal-chelate gas chromatography has been reviewed.[49,823–825] Trace elements have been determined particularly as β-diketonate[824] or dithiocarbamate complexes[825] by packed- and capillary-column GC. With respect to established atomic-spectroscopic techniques, the main advantage of combining chromatographic and mass spectrometry is that the detailed *form* in which the metal occurs can be ascertained. Also covered in this section is GC/MS of inorganic anions, the derivatization of which has been well reviewed.[826]

Up to 1980, major methods based on GC for analysis of organometallic compounds were (*a*) pyrolysis GC, (*b*) chemical conversion to organic compounds followed by GC, and (*c*) direct GC analysis of simple intact organometallic compounds. Mass spectrometry was used mainly to show that substances had, or often had not, retained their molecular integrity during chromatography. With modern methods of (on-column) injection and deactivation of columns and interfaces, analysis of a broader range of organometallic species is possible

[820] R. G. Christensen, E. White, S. Meiselman, and H. S. Hertz, *J. Chromatogr.*, 1983, **271**, 61.
[821] J. F. K. Huber, T. Dzido, and F. Heresch, *J. Chromatogr.*, 1983, **271**, 27.
[822] T. R. Compton, 'Gas Chromatography of Organometallic Compounds', Plenum, New York, 1982.
[823] P. Mushak in 'Handbook of Derivatives for Chromatography', ed. K. Blau and G. King, Heyden, London, 1978, p. 433.
[824] P. M. Patni and S. P. Pande, *Chem. Era*, 1982, **18**, 1.
[825] R. Neeb, *Pure Appl. Chem.*, 1982, **54**, 847.
[826] W. C. Butts in 'Handbook of Derivatives for Chromatography', ed. K. Blau and G. King, Heyden, London, 1978, p. 411.

without decomposition, so such applications should become more widespread. Formation of metal chelates (particularly dithiocarbamates and fluorinated dithiocarbamates) is now rapid and quantitative even at trace levels, and their extraction into organic solvents is straightforward, so GC/MS analysis of metal ions is also to be encouraged.

The analysis of organometallic compounds in the geosphere is a key area of application of GC/MS because of their importance in emerging synthetic fuel processes and their impact on the environment,[827] even at trace levels. The presence of methyl- and phenyl-arsonic acids in oil shale from Green River formation has been established[827,828] in part by capillary-column GC/MS using their five-co-ordinate 3-methylcatecholates (4). In the same study the occurrence

(4) R = Me or Ph

of arsenate ion (AsO_4^{3-}) was verified through its tris(TMS) derivative.[827] In another investigation in which GC/MS complemented HPLC/atomic-absorption spectroscopy, residues of organoarsenic compounds in vegetables grown in soil treated with arsenic acid were analysed.[829] Methyldi-iodoarsine was characterized by packed-column GC/MS. Inorganic arsenic and methylarsenic compounds in soil, plants, urine, and river water have been determined by selected-ion monitoring after borohydride reduction to arsine and mono-, di-, and tri-methylarsines and isolation in a heptane cold trap.[830] Detection limits of about 200–400 pg ml^{-1} and standard deviations of 2–5% at the p.p.b. level were reported. The method was used to assess biomethylation of arsenic compounds in animals, and trimethylarsine was identified as a new urinary metabolite.[830] Microbial methylation of inorganic tin(IV) *in vitro* by a strain of *Pseudomonas* species from Chesapeake Bay was investigated by GC with a tin-selective detector and confirmed by GC/MS.[831] In a novel application of capillary-column GC and GC/MS, metal(III) and metal(IV) porphyrin complexes have been analysed.[832–834] Silicon(IV), aluminium(III), gallium(III), and rhodium(III) alkyl-substituted porphyrins were characterized, mainly as their TMS derivatives, and interestingly it was noted[832] that their retention indices depended on the oxidation state of the central metal. The bis(TMS), silicon(IV) derivatives proved to be well suited

[827] R. H. Fish, R. S. Tannous, W. Walker, C. S. Weiss, and F. E. Brinckman, *J. Chem. Soc., Chem. Commun.*, 1983, 490.
[828] R. H. Fish and F. E. Brinckman, *Prepr. – Am. Chem. Soc., Div. Pet. Chem.*, 1983, **28**, 177.
[829] R. A. Pyles and E. A. Woolson, *J. Agric. Food Chem.*, 1982, **30**, 866.
[830] Y. Odanaka, N. Tsuchiya, O. Matano, and S. Goto, *Anal. Chem.*, 1983, **55**, 929.
[831] J.-A. A. Jackson, W. R. Blair, F. E. Brinckman, and W. P. Iverson, *Environ. Sci. Technol.*, 1982, **16**, 110.

to GC/MS, affording diagnostic EI spectra,[832,833] and peak shapes on a 20 m CPSil 5 capillary column were generally excellent. When applied to petroporphyrins from Boscan crude oil,[834] the less volatile metallo and free-base alkyl porphyrins were analysed with a 6 m OV-1 column.

Detection limits of 1 $\mu g\ l^{-1}$ have been recorded for determination by GC/MS of tributyltin chloride and dibutyltin dichloride in water following derivatization by pentylation[835] and of 0.5 p.p.b. for tributyltin species in water and sediments after methylation to tributylmethyltin.[836] Swiss river and lake waters were shown to contain 1–15 $ng\ l^{-1}$ of Bu_3Sn species.[836] Metal ions in lyophilized samples of various tissues in a case of fatal poisoning with $NiSO_4$ were extracted as diethyldithiocarbamate complexes, and the presence of nickel (and absence of copper) was confirmed by GC/MS of the extracted chelate.[837]

To elute successfully through a GC column, a metal ion must be made volatile by encapsulation within a chelate as with the diethyldithiocarbamate of nickel above.[837] A general problem then becomes decomposition during chromatography, particularly with β-diketones and packed columns. Degradation seems to be associated mainly with adsorption on and interaction with the support material, the injector cavity, and glass-wool plugs. Hence, utilization of capillary columns improves greatly the applicability of GC/MS methods. Independent of column conditions, some metal chelates and especially β-diketones are prone to on-column decomposition. This is due to their involatility: the column temperatures required to elute them cause thermal degradation. The major breakthrough that made GC of metal chelates a useful analytical tool was the introduction of fluorinated ligands, which impart greater volatility on the resulting chelates.[824,825] Currently, di(trifluoroethyl)dithiocarbamates are one of the most useful types of derivatives for determination of traces of metals by GC and GC/MS.[825,838] In lengthy studies of fluorinated and non-fluorinated dialkyldithiocarbamate chelates[838,839] it was shown that chelates of metals with small ionic radii are more volatile than those with larger radii and that volatility generally increases as the oxidation state of the co-ordinated metal decreases. Successful GC, as assessed by GC/MS, was associated with symmetrical chelate structures with low dipole moments. Fragmentations of the chelates were also proposed.[838,839] The high volatility and thermal stability of fluorinated β-diketonate chelates are also advantageous in GC/MS, as with the determination of chromium in sea water by isotope dilution using tris(1,1,1-trifluoropentane-2,4-diono)chromium(III).[840] Even using a packed column, precisions of 5% were achieved at p.p.b. and sub-p.p.b. levels

[832] P. J. Marriott, J. P. Gill, and G. Eglinton, *J. Chromatogr.*, 1982, **249**, 291.
[833] P. J. Marriott and G. Eglinton, *J. Chromatogr.*, 1982, **249**, 311.
[834] P. J. Marriott, J. P. Gill, R. P. Evershed, G. Eglinton, and J. R. Maxwell, *Chromatographia*, 1982, **16**, 304,
[835] H. Ochi, Y. Shinozaki, and H. Hayashi, *Ehime-ken Kogai Gijutsu Senta Shoho*, 1983, 14 (*Chem. Abstr.*, 1984, **101**, 59 857).
[836] M. D. Mueller, *Fresenius' Z. Anal. Chem.*, 1984, **317**, 32.
[837] S. C. Szathmary and T. Daldrup, *Fresenius' Z. Anal. Chem.*, 1982, **313**, 48.
[838] M. L. Riekkola, *Ann. Acad. Sci. Fenn., Ser. A2*, 1983, **199**, p. 1–55.
[839] M. L. Riekkola, *Acta Chem. Scand., Ser. A.*, 1983, **37**, 691.
[840] K. W. M. Siu, M. E. Bedna, and S. S. Berman, *Anal. Chem.*, 1983, **55**, 473.

by selected-ion monitoring. A mixture of mixed-ligand Cr^{III} complexes synthesized from Cr^{III}, 1,1,1,5,5,5-hexafluoropentane-2,4-dione, 1,1,1-trifluoropentane-2,4-dione, and 2,2-dimethyl-6,6,7,7,8,8,8-heptafluoro-octane-3,5-dione has been investigated by GC/MS with an SE-54 fused-silica capillary column. Of the 25 possible products, 24 were chromatographically resolved, and their gas-phase stabilities were discussed.[841] Following the improved chromatographic behaviour of chelates through use of fluorinated ligands and fused-silica capillary columns, application of GC/MS to mixtures of metal ions should expand since it offers good sensitivity for trace analysis, great separating power, and the opportunity to perform various isotope analyses.

Several reports of GC/MS analysis of anions have also been published in the review period.[842–845] The common oxyanions have been known to be amenable to GC and GC/MS as their TMS derivatives for more than a decade, but some, particularly bis(TMS)sulphate, are prone to decomposition in solution and during chromatography. Such instability has been overcome by use of t-butyldimethylsilyl derivatives.[842] The TBDMS esters of carbonate, sulphite, sulphate, selenite, selenate, borate, phosphite, orthophosphate, vanadate, arsenite, arsenate, and pyrophosphate anions have been reported to be stable and suitable for GC. Each ester produced a simple mass spectrum dominated, as expected, by $(M - Bu^{\bullet})^+$ ions.[842] As with trimethylsilylation, the derivatization succeeds only with the ammonium salts of the oxyanions or the free acids. Despite this definite improvement over previously reported methods, the TMS derivatives are still used, for example in the verification of the presence of arsenate ion, as $(TMS)_3AsO_4$, in oil shale by GC/MS.[827] The importance of measuring nitrite ion in environmental and biological media resides largely in its probable role in forming carcinogenic nitrosamines by reaction with amines. Nitrite has been determined by GC and GC/MS as its pentafluorobenzyl derivative.[843,844] A GC method using electron-capture detection was shown to be specific and sensitive, and in good quantitative agreement with spectrophotometry.[843] Alkylation with pentafluorobenzyl bromide also yields derivatives of CN^-, SCN^-, and I^- that are compatible with GC/MS, but several other common anions failed to yield useful products.[844] Thiocyanate did not produce the expected product but bis(pentafluorobenzyl)-sulphide. Since electron-capture detection is favourable for these derivatives, expansion to GC/NICIMS would be expected to enhance further selectivity and sensitivity. The flux of nitrite in blood of mice has been established by oral dosage with $^{15}NO_2^-$, derivatization to (5) on reaction with 1-hydrazinophthalazine, then selected-ion monitoring.[845] This interesting method was selective to nitrite and the detection limit was 100 ng ml^{-1}.

[841] R. E. Sievers and K. C. Brooks, *Int. J. Mass Spectrom. Ion Phys.*, 1983, **47**, 527.

[842] T. P. Mawhinnery, *Anal. Lett.*, 1983, **16** (A2), 159; *J. Chromatogr.*, 1983, **257**, 37.

[843] H.-L. Wu, S.-H. Chen, K. Funazo, M. Tanaka, and T. Shono, *J. Chromatogr.*, 1984, **291**, 409.

[844] H.-L. Wu, S.-H. Chen, S.-J. Lin, W.-R. Hwang, K. Funazo, M. Tanaka, and T. Shono, *J. Chromatogr.*, 1983, **269**, 183.

[845] A. Tanaka, N. Nose, H. Masaki, and H. Iwasaki, *J. Chromatogr.*, 1984, **306**, 51.

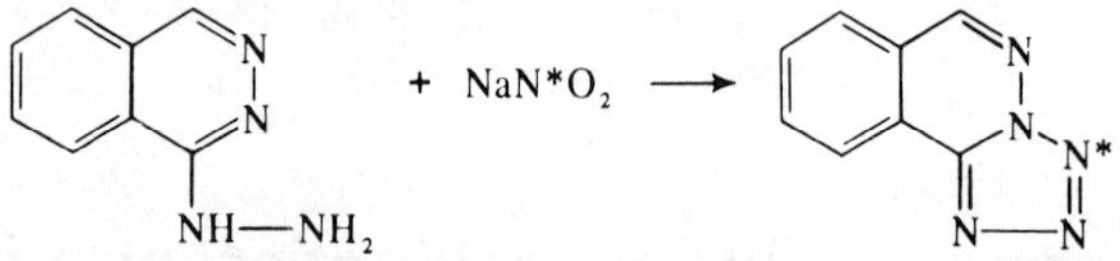

+ $NaN^{*}O_2$ →

NH—NH_2

($N^{*} = {}^{15}N$)

(5)

6 Concluding Remarks

Along with the newer areas of application of combined chromatography/mass spectrometry, such as analysis of racemic mixtures and inorganic and organometallic compounds, several methodologies are also expanding. In GC/MS, the GC/FTIR/MS combination has great potential for complex mixtures, whilst cheap specialized mass-spectometric detectors for GC could find a wide market for undemanding yet effective GC/MS work as in clinical and teaching laboratories for instance. Improvements in HPLC/MS have continued slowly but surely, the combination with FAB mass spectrometry and the thermospray approach being particularly attractive. One technique that could compete with LC/MS for analysing the more difficult samples is supercritical fluid chromatography in combination with mass spectrometry. Further development of SFC/MS promises rich rewards.

10

Drug Metabolism, Pharmacokinetics, and Toxicity

BY D. J. HARVEY

1 Introduction

This review covers the period from July 1982 until June 1984 and is concerned with the applications of mass spectrometry to the identification and detection of drugs and their metabolites, measurements of drugs in body fluids, studies on mechanisms of drug toxicity, pharmacokinetics, and forensic aspects of drugs of abuse. Wider aspects of drug research such as measurements of altered hormone or neurotransmitter concentrations are beyond the scope of this review.

Books of interest that have been published during the review period include two further volumes of Florey's 'Analytical Profiles of Drug Substances',[1] two books presenting collections of spectra as the eight largest peaks,[2,3] and a two-volume set containing mass- and other spectrometric data on 600 drugs and drug-related compounds.[4] General reviews containing information on the mass spectrometry of drugs include the fundamental review in *Analytical Chemistry*,[5] three reviews on the analysis of drugs of abuse,[6–8] chapters in two books on drug toxicity,[9,10] and a review on applications in clinical chemistry.[11] The latter review contains a list of 64 drug assays published between 1980 and 1983. A large number of full spectra of drugs and their metabolites has been published by Maurer and Pfleger in a series of papers on drug screening. These include butyrophenone and bisfluorophenyl neuroleptics,[12] anti-inflammatory anal-

[1] (*a*) 'Analytical Profiles of Drug Substances', ed. K. Florey, Academic Press, New York, 1982, Vol. 11; (*b*) *ibid*., 1983, Vol. 12.

[2] 'Handbook of Mass Spectra of Drugs', ed. I. Sunshine, CRC Press, Boca Raton, FL, 1981.

[3] 'An Eight Peak Index of Mass Spectra of Compounds of Forensic Interest', ed. R. E. Ardrey, C. Brown, A. R. Allan, T. S. Bal, and A. C. Moffat, The Forensic Science Society, Scottish Academic Press, Edinburgh, 1983.

[4] (*a*) 'Instrumental Data for Drug Analysis', ed. T. Mills, tert., W. N. Price, P. T. Price, and J. C. Roberson, Elsevier, Amsterdam, 1982, Vol. 1; (*b*) *ibid*., 1984, Vol. 2.

[5] A. L. Burlingame, J. O. Whitney, and D. H. Russell, *Anal. Chem*., 1984, **56**, 417R.

[6] T. A. Gough and P. B. Baker, *J. Chromatogr. Sci*., 1982, **20**, 289.

[7] T. A. Gough and P. B. Baker, *J. Chromatogr. Sci*., 1983, **21**, 145.

[8] P. B. Baker and G. F. Phillips, *Analyst (London)*, 1983, **108**, 777.

[9] 'Drug Metabolism and Drug Toxicity', ed. J. R. Mitchell and M. G. Horning, Raven Press, New York, 1984.

[10] 'Topics in Forensic and Analytical Toxicology', Anal. Chem. Symp. Ser., Vol. 20, ed. R. A. A. Maes, Elsevier, Amsterdam, 1984.

[11] R. E. Hill and D. T. Whelan, *Clin. Chim. Acta*, 1984, **139**, 231.

[12] H. Maurer and K. Pfleger, *J. Chromatogr*., 1983, **272**, 75.

gesics,[13] opioids and other analgesics,[14] phenothiazines and other neuroleptics,[15] and anti-depressants.[16] A number of recent meetings also contained material of interest. These include the annual symposia organized by the Italian Group for Mass Spectrometry in Biochemistry and Medicine,[17,18] the Fourth International Symposium on Quantitative Mass Spectrometry in the Life Sciences,[19] the Ninth Annual Mass Spectrometry Conference (Vienna),[20,21] the Second Workshop on LC/MS and MS/MS (Montreux),[22] and 'Biochemische Analytik '82'.[23] Other more specific reviews are included below in the relevant sections.

2 Current Trends

Ionization Techniques. – With the advent of fast-atom-bombardment (FAB) ionization it has become possible to analyse a large number of antibiotics, drug conjugates, and ionic species not previously amenable to mass-spectrometric analysis. Two general reviews containing applications to antibiotics have appeared.[24,25] A method for direct examination of spots from TLC plates has been published;[26] the spots were removed onto the probe tip with double-sided masking tape, and the resulting spectra were free from ions produced from the TLC absorbant and tape. Glucuronide conjugates have received considerable attention as described below. FAB mass spectrometry has enabled the first acyl-linked mercapturic acid metabolite to be identified in man; clofibrate mercapturate was detected in urine, and it was proposed to be formed from the glucuronide by electrophilic attack on the SH group.[27] A rapid FD technique for the analysis of drugs and their metabolites in urine involving simple ethyl acetate extraction and direct FD examination of the extracts has been used to identify nine metabolites from 3,5-dinitro-2-hydroxytoluene.[28]

Negative-ion mass spectrometry, particularly electron-capture negative-ion mass spectrometry, is becoming increasingly popular, as described below, for measurement of drugs in body fluids at high dilution. The technique has recently

[13] H. Maurer and K. Pfleger, *Fresenius' Z. Anal. Chem.*, 1983, **314**, 586.

[14] H. Maurer and K. Pfleger, *Fresenius' Z. Anal. Chem.*, 1984, **317**, 42.

[15] H. Maurer and K. Pfleger, *J. Chromatogr.*, 1984, **306**, 125.

[16] H. Maurer and K. Pfleger, *J. Chromatogr.*, 1984, **305**, 309.

[17] 'Recent Developments in Mass Spectrometry in Biochemistry, Medicine and Environmental Research', Anal. Chem. Symp. Ser., Vol. 12, ed. A. Frigerio, Elsevier, Amsterdam, 1982.

[18] 'Chromatography and Mass Spectrometry in Biomedical Sciences', Anal. Chem. Symp. Ser., Vol. 14, ed. A. Frigerio, Elsevier, Amsterdam, 1983.

[19] *Biomed. Mass Spectrom.*, 1983, **10(3)**, 113.

[20] (*a*) *Int. J. Mass Spectrom. Ion Phys.*, 1982, **45**; (*b*) *ibid.*, 1983, **46**; (*c*) *ibid.*, 1983, **47**.

[21] 'Mass Spectrometry Advances', ed. E. R. Schmid, K. Varmuza, and I. Fogy, Elsevier, Amsterdam, 1982.

[22] *J. Chromatogr.*, 1983, **271**.

[23] *Fresenius' Z. Anal. Chem.*, 1982, **311**.

[24] K. L. Busch and R. G. Cooks, *Science*, 1982, **218**, 247.

[25] K. L. Rinehart, jun., *Science*, 1982, **218**, 254.

[26] T. T. Chang, J. O. Lay, jun., and R. J. Francel, *Anal. Chem.*, 1984, **56**, 111.

[27] M. Stogniew and C. Fenselau, *Drug Metab. Dispos.*, 1982, **10**, 609.

[28] J. van der Greef and D. C. Leegwater, *Biomed. Mass Spectrom.*, 1983, **10**, 1.

been briefly reviewed.[29] Alternate positive-ion EI/negative-ion CI spectra have been reported to give better complementary information for methadone and a hydroxylated metabolite of pentobarbitone than corresponding spectra recorded in the positive-ion CI mode.[30]

Liquid Chromatography/Mass Spectrometry. – LC/MS coupling for use in forensic analysis has been reviewed,[31] and applications to forensic analysis of drugs such as barbiturates have been discussed.[32] Use of the thermospray and related direct-introduction interfaces has received increasing attention for drug analyses over the last two years at the expense of the belt interface and has shown promise for high-sensitivity assays. Thus, β-hydroxyethyltheophylline has been detected by using selected-ion monitoring at the level of 0.005 ng with a S/N ratio of 2,[33] and spectra of drugs such as caffeine, betamethasone, and reserpine have been recorded at the 100 ng level after chromatography on a 2 mm microbore column running at a flow rate of 0.15 ml min^{-1}.[34] Microbore LC/MS has also been reviewed.[35] By use of microbore LC/MS with direct introduction, detection of drugs such as steroids, antibiotics, and trichlormethazine together with their metabolites has been achieved in the low nanogram range.[36,37] Trichlormethazine was detected more sensitively in the negative- than in the positive-ion mode.[36] Nebulizer interfaces have been used to record spectra from erythromycin A,[38] clobazam,[39] and ranitidine.[39] A method for recording ^{252}Cf-fission fragment-induced desorption spectra has been published.[40–42] It consists of a 12-sector rotating disc incorporating an aluminium/plastic foil on to which the column effluent is deposited with a nebulizer. Analysis used a time-of-flight mass spectrometer, and spectra were recorded from the antitumour drugs etoposide and teniposide. An ultrasonic spray interface has been used[43] to measure valproic acid in human serum after direct injection of the serum on to the column. The standard was injected separately, and the ion at *m/z* 145 was monitored.

[29] H. Brandenberger, *Fresenius' Z. Anal. Chem.*, 1984, **317**, 634.
[30] H. Brandenberger, *Int. J. Mass Spectrom. Ion Phys.*, 1983, **47**, 213.
[31] 'High Performance Liquid Chromatography in Forensic Chemistry', ed. I. S. Lurie and J. D. Wittwer, jun., Marcel Dekker, New York, 1983.
[32] J. Yinon, *Int. J. Mass Spectrom. Ion Phys.*, 1983, **48**, 253.
[33] C. R. Blakley and M. L. Vestal, *Anal. Chem.*, 1983, **55**, 750.
[34] T. Covey and J. Henion, *Anal. Chem.*, 1983, **55**, 2275.
[35] D. E. Games, N. J. Alcock, and M. A. McDowall, *Anal. Proc.*, 1984, **21**, 24.
[36] C. Eckers, D. S. Skrabalak, and J. Henion, *Clin. Chem.*, 1982, **28**, 1882.
[37] C. Eckers, J. D. Henion, G. A. Maylin, D. S. Skrabalak, J. Vessman, A. M. Tivert, and J. C. Greenfield, *Int. J. Mass Spectrom. Ion Phys.*, 1983, **46**, 205.
[38] M. Dedieu, C. Juin, P. J. Arpino, and G. Guiochon, *Anal. Chem.*, 1982, **54**, 2372.
[39] J. A. Apffel, U. A. Th. Brinkman, R. W. Frei, and E. A. I. M. Evers, *Anal. Chem.*, 1983, **55**, 2280.
[40] H. Jungclas, H. Danigel, L. Schmidt, and J. Dellbrugge, *Org. Mass Spectrom.*, 1982, **17**, 499.
[41] H. Jungclas, H. Danigel, and L. Schmidt, *Int. J. Mass Spectrom. Ion Phys.*, 1983, **46**, 197.
[42] H. Jungclas, H. Danigel, and L. Schmidt, *J. Chromatogr.*, 1983, **271**, 35.
[43] R. G. Christensen, E. White V, S. Meiselman, and H. S. Hertz, *J. Chromatogr.*, 1983, **271**, 61.

Gas Chromatography/Mass Spectrometry. – GC/MS remains a standard technique for drug analysis, and many examples are given below. Fused-silica capillary columns are now used extensively, and their advantages of increased precision and sensitivity have been reviewed.[44] Quantitative aspects of capillary-column GC/MS have been discussed[45] in relation to the analysis of timolol. The use of computer-generated selected-ion chromatogram plots has been a routine technique in GC/MS analysis for some time. A program has now been produced[46] to enable ion-cluster chromatograms to be plotted, and it can be used to locate isotopically labelled species in chromatograms of drug metabolites.

Mass Spectrometry/Mass Spectrometry. – Tandem mass spectrometry of pharmaceuticals has been reviewed.[47] Triple-quadrupole instruments are now in use in several laboratories for the rapid identification of drugs in body fluids and, because of their selectivity, enable simple extraction methods to be used. A typical technique is to locate a drug using a parent-ion- or neutral-loss scan and then to obtain more information by recording the spectrum in the daughter-ion mode.[48–50] Anti-convulsant drugs in tissues have been analysed with a triple quadrupole incorporating laser desorption;[51] the $[M + H]^+$ ion was selected with the first quadrupole, then collisionally activated, and the fragments were analysed in the third stage. A soft-ionization technique involving desorption of ions from the liquid phase followed by MS/MS analysis has been described for the analysis of drugs such as opiates, phenazone, and methaqualone.[52] The sensitivity obtained with MS/MS techniques for the analysis of drugs in biological media is generally excellent because of the high selectivity involved. Caffeine, for example, has been measured in urine to the low picogram range with a simple extraction requiring only two minutes,[53] and an analysis for the deacylated metabolite of metipranolol was reported to be ten times as sensitive as a conventional GC/MS assay.[54] The antiparasitic drug ivermectin has been detected in cattle tissue by use of MS/MS techniques.[55] Although analyses of drugs directly from biological matrices has been reported, isolation by, for example, TLC is regarded as essential by some workers.[56]

[44] J. Settlage and H. Jaeger, *J. Chromatogr. Sci.*, 1984, **22**, 192.

[45] W. J. A. VandenHeuvel, J. R. Carlin, and R. W. Walker, *J. Chromatogr. Sci.*, 1983, **21**, 119.

[46] R. J. Anderegg, *J. Chromatogr.*, 1983, **275**, 154.

[47] W. J. Richter, W. Blum, U. P. Schlunegger, and M. Senn in 'Tandem Mass Spectrometry', ed. F. W. McLafferty, Wiley, New York, 1983, p. 417.

[48] R. J. Perchalski, R. A. Yost, and B. J. Wilder, *Anal. Chem.*, 1982, **54**, 1466.

[49] R. A. Yost, H. O. Brotherton, and R. J. Perchalski, *Int. J. Mass Spectrom. Ion Phys.*, 1983, **48**, 77.

[50] H. O. Brotherton and R. A. Yost, *Anal. Chem.*, 1983, **55**, 549.

[51] R. J. Perchalski, R. A. Yost, and B. J. Wilder, *Anal. Chem.*, 1983, **55**, 2002.

[52] B. A. Thomson, J. V. Iribarne, and P. J. Dziedzic, *Anal. Chem.*, 1982, **54**, 2219.

[53] B. Shushan, J. E. Fulford, B. A. Thomson, W. R. Davidson, L. M. Danylewych, A. Ngo, S. Nacson, and S. D. Tanner, *Int. J. Mass Spectrom. Ion Phys.*, 1983, **46**, 225.

[54] R. Endele and M. Senn, *Int. J. Mass Spectrom. Ion Phys.*, 1983, **48**, 81.

[55] P. C. Tway, G. V. Downing, J. R. B. Slayback, G. S. Rahn, and R. K. Isensee, *Biomed. Mass Spectrom.*, 1984, **11**, 172.

[56] J. Henion, G. A. Maylin, and B. A. Thomson, *J. Chromatogr.*, 1983, **271**, 107.

Quantitative Aspects. – There has been a trend over the past few years for HPLC-based methods to be preferred to gas-phase methods for the routine quantitative analysis of drugs in body fluids. However, GC/MS still remains the most sensitive technique and is widely used for assays of drugs present in low concentration. Various methods for increasing sensitivity above that provided by conventional selected-ion monitoring are in use. These methods mainly depend on selective-ionization techniques or on increasing the selectivity of the mass spectrometer to reduce the number of interfering ions produced by material co-extracted with the drug, and they can be illustrated by the MS/MS experiments cited above. The related technique of metastable-peak monitoring on a conventional dual-sector instrument has been used to measure warfarin in as little as 4 μl of plasma.[57] Ion monitoring under high-resolution conditions also results in an increase in selectivity as illustrated by a recent assay for salbutamol in plasma.[58] The drug was measured by monitoring the ion at *m/z* 369.1738 using a resolution of 5000 in order to separate it from the ion at *m/z* 369.3476 derived from cholesterol. Nicotine has been measured at a resolution of 10 000 to remove interference from $C_5H_8O^{+\bullet}$ ions at *m/z* 84.[59]

The most sensitive drug assays reported to date are all based on electron-capture ionization of fluorinated derivatives. A limit of 500 fg ml^{-1} has been recorded for both guanfacine (1)[60] and 4-hydroxydebrisoquine (2),[61] using

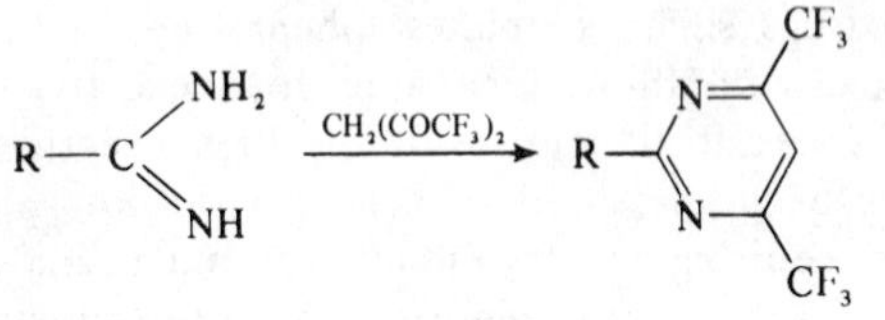

R—C(NH$_2$)=NH $\xrightarrow{CH_2(COCF_3)_2}$ R–(4,6-bis-CF_3-pyrimidin-2-yl)

(1) guanfacine, R = 2,6-$Cl_2C_6H_3$—CH_2—CO—NH—

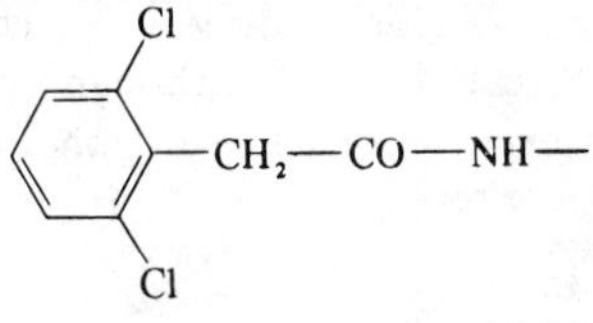

(2) 4-hydroxydebrisoquine, R = (4-hydroxy-1,2,3,4-tetrahydroisoquinolin-2-yl)

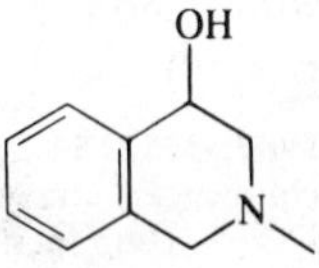

Scheme 1

57 N. W. Davies, J. C. Bignall, and M. S. Roberts, *Biomed. Mass Spectrom.*, 1983, **10**, 646.

58 R. J. N. Tanner, L. E. Martin, and J. Oxford, *Anal. Proc.*, 1983, **20**, 38.

59 D. Jones, M. Curvall, L. Abrahamsson, E. Kazemi-Vala, and C. Enzell, *Biomed. Mass Spectrom.*, 1982, **9**, 539.

60 C. Julien-Larose, C. Lange, D. Lavene, J. R. Kiechel, and J. J. Basselier, *Int. J. Mass Spectrom. Ion Phys.*, 1983, **48**, 221.

61 S. Murray, G. C. Kahn, A. R. Boobis, and D. S. Davies, *Int. J. Mass Spectrom. Ion Phys.*, 1983, **48**, 89.

bis(trifluoromethyl)pyrimidinyl derivatives formed by condensation of the drugs with hexafluoroacetylacetone (Scheme 1). Reagents for the further derivatization of the 4-hydroxy group in the latter compound have also been investigated;[62] perfluoroacyl derivatives produced extensive fragmentation, but trimethylsilyl (TMS) derivatives gave responses similar to those of the parent compound. Other assays employing negative-ion CI are described below.

The relative merits of stable-isotope, analogue, or homologue standards have been discussed in earlier reviews and will not be repeated here. Most of the assays reported during the present review period have used stable-isotope or analogue standards requiring multiple-ion monitoring even though greater instrumental precision can be obtained with single-ion monitoring. Greater precision and reproducibility in the earlier stages of an assay can, however, be obtained with stable-isotope standards, but problems can arise at the measurement stage because of interference by unwanted isotopes. Polynomial regression[63] and the use of either linear or non-linear models[64] have recently been proposed as a means for overcoming this problem, and Moler *et al.*[65] have discussed methods for assessing confidence limits in assays of this type.

As in the last review period, there has been little use made of carrier effects in drug monitoring at low levels. Vink *et al.*[66] have, however, recommended the use of stable-isotope standards to reduce peak tailing in an assay for the serotonin re-uptake blocker ORG-6582 in plasma.

3 General Applications and Techniques

Drug Screening. – In a recent survey[67] of 484 fatally injured drivers in Ontario, various techniques, including GC/MS, were used to indicate the presence of alcohol in 57% of all cases. Other drugs were present in 26% of the group and were mainly accounted for by diazepam and cannabis. Deaths attributable to drugs in Los Angeles county, California, in 1975–76 were mainly caused by alcohol and phencyclidine.[68] Studies in Germany with 2000 cases of acute poisoning have shown the most frequently encountered drugs to be benzodiazepines (455), barbiturates (324), pyrazolones (170), diphenhydramine (117), tricyclic antidepressants (97), and phenothiazines (82).[69] The most appropriate method for mass-spectrometric analysis varied with the drug. Simple screening methods such as spot tests are appropriate in most cases, but GC/MS has been

[62] S. Murray and K. A. Waddell, *Biomed. Mass Spectrom.*, 1982, **9**, 466.
[63] J. A. Jonckheere and A. P. De Leenheer, *Anal. Chem.*, 1983, **55**, 153.
[64] W. T. Yap, R. Schaffer, H. S. Hertz, E. White V, and M. J. Welch, *Biomed. Mass Spectrom.*, 1983, **10**, 262.
[65] G. F. Moler, R. R. Delongchamp, W. A. Korfmacher, B. A. Pearce, and R. K. Mitchum, *Anal. Chem.*, 1983, **55**, 835.
[66] J. Vink, H. J. M. van Hal, and G. L. M. van der Laar, *Biomed. Mass Spectrom.*, 1982, **9**, 370.
[67] G. Cimbura, D. M. Lucas, R. C. Bennett, R. A. Warren, and H. M. Simpson, *J. Forensic Sci.*, 1982, **27**, 855.
[68] R. D. Budd, D. M. Lindstrom, E. C. Griesemer, and T. T. Noguchi, *Bull. Narcotics*, 1983, **35**(i), 41.
[69] C. Koppel and J. Tenczer, *Int. J. Mass Spectrom. Ion Phys.*, 1983, **48**, 213.

reported to give the best results for unknown drugs.[70] A GC/MS method for characterization of amine drugs such as amphetamine, mescaline, and caffeine and their enantiomers has been reported to give resolution of all compounds, including the enantiomers.[71] GC/MS studies of human hair have shown that basic drugs such as antidepressants and nicotine are found only in sections of hair grown after administration of the drug.[72] It was therefore proposed that hair analysis could be used to determine the chronology of intoxication.

Analytical Artefacts. – The recognition of artefacts produced by contaminants, decomposition products, or unwanted side-reactions is essential for successful drug analysis. A review on contaminants in mass spectrometry has appeared,[73] and several cases of artefact formation by interaction of drugs with reagents or solvents have been reported. Thus, exposure to aged solutions of HCl in methanol has been shown to cause a reduction in extraction yield of chlorpromazine, amitriptyline, codeine, and methadone because of their reaction with dimethyl ether and peroxides formed in the reagent.[74] Phosgene and ethyl chloroformate in chloroform form carbamoyl chlorides and carbamates from drugs with amine substituents,[75] and chloroform itself undergoes reaction with mexiletine on KOH/Carbowax 20M GLC columns.[76] Contaminants such as plasticizers and preservatives and their metabolites are frequently present in biological fluids or are introduced during work-up. Several papers report detection or mass-spectral characteristics of compounds of this type. Thus 2-mercaptobenzothiazole, a rubber-vulcanization agent, has been found in drugs packed in syringes with rubber closures,[77] and the FAB mass spectra of disodium pamoate, a pharmaceutical additive, have been studied.[78]

As metabolites of plasticizers and food additives are frequently encountered in drug-metabolism studies, several metabolic investigations of these compounds have been carried out. Metabolic studies on the phthalate plasticizers bis-(9-decenyl)phthalate and bis-(2-ethylhexyl)phthalate have shown hydrolysis, hydroxylation, and β-oxidation.[79–81] In metabolic studies with antioxidants it has been shown that glutathione,[82] glucuronide,[82] and cysteine conjugates[83] are formed by butylated hydroxytoluene and that the intermediate for the sulphur conjugates is probably a toxic quinone methide.[83] The 2,4,6-tri-t-butylphenoxy

[70] J. E. Pettersen and B. Skuterud, *Anal. Chem. Symp. Ser.*, 1983, **13**, 111.
[71] R. H. Liu, W. W. Ku, and M. P. Fitzgerald, *J. Assoc. Off. Anal. Chem.*, 1983, **66**, 1443.
[72] I. Ishiyama, T. Nagai, and S. Toshida, *J. Forensic Sci.*, 1983, **28**, 380.
[73] M. Ende and G. Spiteller, *Mass Spectrom. Rev.*, 1982, **1**, 29.
[74] W. H. Swallow, *J. Forensic Sci. Soc.*, 1982, **22**, 297.
[75] E. J. Cone, W. F. Buchwald, and W. D. Darwin, *Drug Metab. Dispos.*, 1982, **10**, 561.
[76] A. T. Kacprowicz, *J. Chromatogr.*, 1983, **269**, 61.
[77] J. C. Reepmeyer and Y. H. Juhl, *J. Pharm. Sci.*, 1983, **72**, 1302.
[78] D. V. Bowen, L. A. Broad, T. Norris, M. Barber, R. S. Bordoli, G. J. Elliott, R. D. Sedgwick, and A. N. Tyler, *Analyst (London)*, 1984, **109**, 353.
[79] P. W. Albro, J. T. Corbett, D. Marbury, and C. Parker, *Xenobiotica*, 1984, **14**, 389.
[80] P. W. Albro, S. T. Jordan, J. L. Schroeder, and J. T. Corbett, *J. Chromatogr.*, 1982, **244**, 65.
[81] E. J. Draviam, K. H. Pearson, and J. Kerkay, *Anal. Lett.*, 1982, **15**, 1729.
[82] K. Tajima, K. Yamamoto, and T. Mizutani, *Chem. Pharm. Bull.*, 1983, **31**, 3671.
[83] Y. Nakagawa, K. Hiraga, and T. Suga, *Biochem. Pharmacol.*, 1983, **32**, 1417.

radical has been identified by mass spectrometry as a metabolite of the haemorrhagic antioxidant 2,4,6-tri-t-butylphenol.[84] Other preservatives whose metabolism has been studied include butylated hydroxyanisole,[85] methylparaben,[86] and 2-t-butyl-4-methoxyphenol.[87]

The analytical utility of multiple peaks produced by drugs or their TMS derivatives in GC/MS studies has been discussed;[88] 46 of the 116 drugs examined showed this behaviour. Tauroline has been reported to undergo thermal decomposition on GLC columns,[89] as have several benzodiazepines as discussed below; nifedipine, however, in contrast to earlier reports, does not.[90]

Derivatization. – Several new reactions and fragmentation mechanisms of TMS and related derivatives pertinent to drug studies have been reported. Thus the reaction of several secondary amines with *N,O*-bis(trimethylsilyl)trifluoroacetamide (BSTFA) in the presence of DMF leads to the formation of a fluorine-containing derivative rather than to a TMS derivative (Scheme 2).[91] Chlorine migration has been detected in the mass spectra of t-butyldimethylsilyl derivatives of chloro-alcohols (Scheme 3).[92] Triethylsilyl derivatives have been prepared

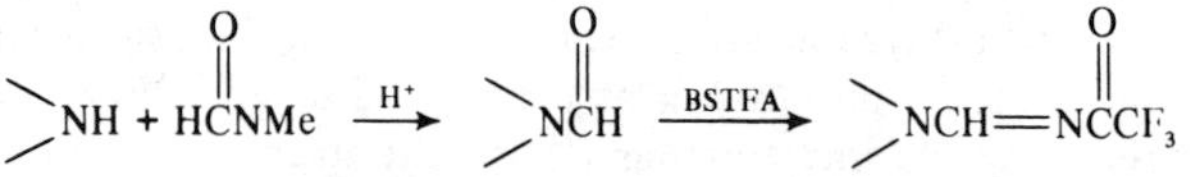

$$>NH + HC(=O)NMe \xrightarrow{H^+} >NC(=O)H \xrightarrow{BSTFA} >NCH{=}NC(=O)CF_3$$

Scheme 2

$$[\text{t-Bu}(Me_2)Si{-}O{-}CH_2CH_2CH_2CH_2{-}Cl]^{+\bullet} \xrightarrow{-Bu^\bullet} \text{cyclic } {-}Si{=}O^+ \cdots Cl \rightarrow Cl{-}Si{-}\overset{+}{O}{=}CH_2 \rightarrow Cl{-}Si^+$$

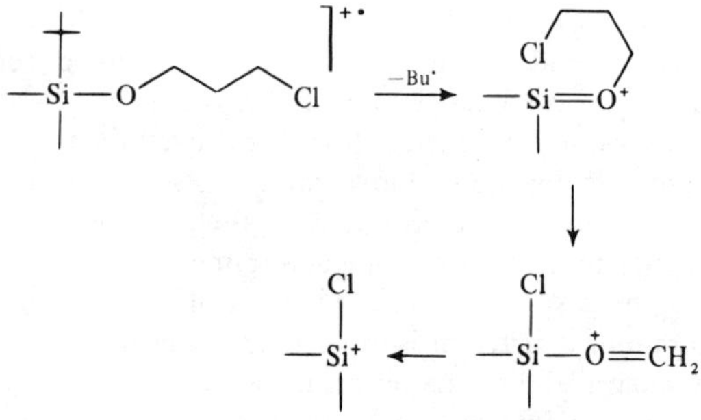

Scheme 3

[84] O. Takahashi and K. Hiraga, *Xenobiotica*, 1983, **13**, 319.
[85] R. El-Rashidy and S. Niazi, *Biopharm. Drug Dispos.*, 1983, **4**, 389.
[86] K. W. Hindmarsh, E. John, L. A. Asali, J. N. French, G. L. Williams, and W. G. McBride, *J. Pharm. Sci.*, 1983, **72**, 1039.
[87] A. Guarna, L. D. Corte, M. G. Giovannini, F. De Sarlo, and G. Sgaragli, *Drug. Metab. Dispos.*, 1983, **11**, 581.
[88] M. C. Dutt, *J. Chromatogr.*, 1982, **248**, 115.
[89] F. Erb, M. Imbenotte, J. P. Huvenne, M. Vankemmel, P. Scherpereel, and R. W. Pfirrmann, *Eur. J. Drug Metab. Pharmacokinet.*, 1983, **8**, 163.
[90] L. J. Lesko, A. K. Miller, R. L. Yeager, and D. C. Chatterji, *J. Chromatogr. Sci.*, 1983, **21**, 415.
[91] S. K. Sethi, P. F. Crain, and J. A. McCloskey, *J. Chromatogr.*, 1983, **254**, 109.
[92] S. T. Hill and M. Mokotoff, *J. Org. Chem.*, 1984, **49**, 1441.

with 2-methyl-1-triethylsilyloxy-1-methoxypropene; primary alcohols reacted at room temperature, hindered hydroxy groups reacted in the presence of a trace of trifluoromethanesulphonic acid, whereas phenols required heating.[93]

Use of Stable Isotopes. – In addition to their use as internal standards for quantitative work, drugs labelled with stable isotopes have been used in many studies of drug uptake, metabolism, pharmacokinetics, and toxicity.[94,95] Many applications can be found below, but these can be classified into several types of experiment. Thus, co-administration of equal proportions of a drug and a labelled analogue to animals or humans and examination of the spectra of the resulting metabolites for doublets, the so-called 'twin-ion' or 'isotope-doublet' technique, can be used in tracer studies to identify drug-derived compounds in the presence of endogenous material. Metabolic replacement of deuterium can give information on the site of hydroxylation, although the primary kinetic isotope effect frequently results in inhibition of metabolism at the labelled site and may produce metabolic switching. Lack of an isotope effect, as with many aromatic hydroxylations, indicates mechanisms other than direct oxygen insertion such as the intermediacy of an arene oxide. In these cases, displacement of deuterium to adjacent carbon atoms can occur, a mechanism termed the NIH shift. Mass-spectrometric monitoring of a labelled drug given as a substitute dose to patients receiving chronic treatment with the drug, a technique known as 'pulse labelling', enables the pharmacokinetics of a single dose of the drug to be studied under chronic administration conditions without disrupting the treatment. The technique can also be used for studies of drugs of abuse where drug residues from previous illicit use may be present. Drug metabolites having structures identical to those of endogenous compounds are easily differentiated by the incorporation of an isotopic label, and the fate of hydrogen and acetyl groups removed from drugs during metabolism can be studied by measuring their incorporation into endogenous biochemicals. Pseudo-racemates of racemic drugs, in which one enantiomer contains a label, enable pharmacokinetic and metabolic studies of both isomers to be determined simultaneously. The administration of a drug by a specific route or in a specific formulation concomitantly with a dose containing a different isotope given intravenously yields information on uptake and bioavailability. Experiments with stable isotopes on the determination of metabolic intermediates are numerous, and some are described below. The relative contribution of alternative metabolic pathways can be determined by measurement of isotopic ratio in metabolites, as was recently reported for the formation of 3-bromocinnamic acid (5) from the antiepileptic 3-bromo-*N*-ethylcinnamamine (3) (Scheme 4).[96] Formation of TMS molecular adduct ions in DCI spectra using tetramethylsilane and its deuteriated analogue has been reported as a method for determining molecular weights.[97]

[93] E. Yoshii and K. Takeda, *Chem. Pharm. Bull.*, 1983, **31**, 4586.
[94] M. Eichelbaum, G. E. von Unruh, and A. Somogyi, *Clin. Pharmacokinet.*, 1982, **7**, 490.
[95] D. Halliday and M. J. Rennie, *Clin. Sci.*, 1982, **63**, 485.
[96] H.-S. Lin, R. H. Levy, E. A. Lane, and W. P. Gordon, *J. Pharm. Sci.*, 1984, **73**, 285.
[97] R. N. Stillwell, D. I. Carroll, J. G. Nowlin, and E. C. Horning, *Anal. Chem.*, 1983, **55**, 1313.

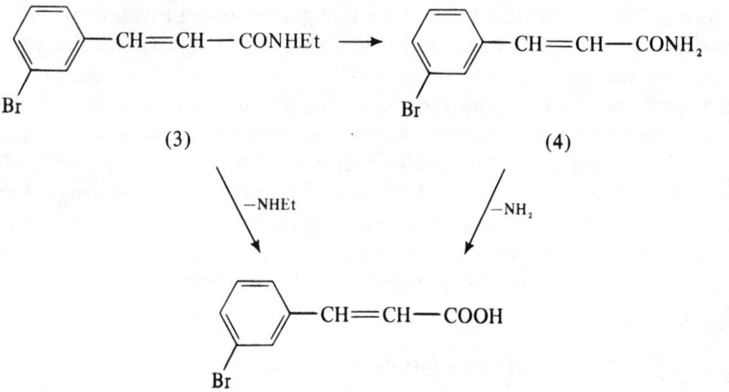

(3) 3-Br-C6H4-CH=CH-CONHEt → (4) 3-Br-C6H4-CH=CH-CONH$_2$

(3) —NHEt→ (5); (4) —NH$_2$→ (5)

(5) 3-Br-C6H4-CH=CH-COOH

Scheme 4

Glucuronide Conjugates. – With the advent of the newer ionization techniques, considerable advances have been made in the analysis of drug conjugates, particularly glucuronides. Several quaternary ammonium-linked glucuronides have been shown to give clear molecular ions in the positive-ion mode[98,99] and to afford diagnostic fragment ions by cleavage at the glycosidic linkage.[98] Negative-ion FAB has been used to characterize glucuronides of *N*-hydroxyamides in the underivatized state.[100] A survey[101] of the FAB spectra of thirty glucuronides containing seven glycosidic linkages has shown that the negative-ion spectra give the best molecular-weight information and fragmentation except for quaternary ammonium, carbinolamine, and epoxide-linked glucuronides, where positive-ion spectra were preferred. Diagnostic $[M - 177]^-$ ions were abundant in the negative-ion spectra and yielded structural information on the aglycone. FAB studies on zomepirac glucuronide have revealed intramolecular acyl migration at physiological pH, yielding four compounds that could not be cleaved by β-glucuronidase.[102] It is possible that this rearrangement may be common to other glucuronides and could be of considerable importance, as enzymic cleavage prior to mass-spectrometric detection is a standard technique in glucuronide analysis. FAB has also been used to identify a glucuronide metabolite of hydroxyclofazimine.[103]

Studies of seven oxygen-linked glucuronides by laser desorption have shown the presence of abundant molecular and $[M + X]^+$ ions ($X = Na$ or K)[104] and

[98] J. P. Lehman and C. Fenselau, *Drug Metab. Dispos.*, 1982, **10**, 446.
[99] F. W. Janssen, S. K. Kirkman, C. Fenselau, M. Stogniew, B. R. Hofmann, E. M. Young, and H. W. Ruelius, *Drug Metab. Dispos.*, 1982, **10**, 599.
[100] P. C. C. Feng, C. Fenselau, M. E. Colvin, and J. A. Hinson, *Drug Metab. Dispos.*, 1983, **11**, 103.
[101] C. Fenselau, L. Yelle, M. Stogniew, D. Liberato, J. Lehman, P. Feng, and M. Colvin, jun., *Int. J. Mass Spectrom. Ion Phys.*, 1983, **46**, 411.
[102] J. Hasegawa, P. C. Smith, and L. Z. Benet, *Drug Metab. Dispos.*, 1982, **10**, 469.
[103] P. C. C. Feng, C. C. Fenselau, and R. R. Jacobson, *Drug Metab. Dispos.*, 1982, **10**, 286.
[104] R. B. Van Breemen, J.-C. Tabet, and R. J. Cotter, *Biomed. Mass Spectrom.*, 1984, **11**, 278.

have resulted in the identification of an acid glucuronide of flufenamic acid.[105] A comparison of the positive-ion and negative-ion ammonia CI spectra of several glucuronides has shown more characteristic fragmentation in the positive-ion mode with *m/z* 194 being diagnostic of the glucuronide function.[106,107] Both studies employed LC/MS sample introduction using thermospray[106] and belt interfaces,[107] respectively. The mass spectra of TMS derivatives of dihydrodiol glucuronides have been discussed,[108] and, in addition to the drugs discussed below, glucuronides have been identified by mass spectrometry for ipomeanol,[109] trimetozine,[110] *S*-carboxymethyl-L-cysteine,[111] disulfiram,[112] bromocriptine,[113] bromperidol,[114] amrinone,[115] ciramadol,[116] and lofexidine.[117]

4 Studies on Specific Groups of Drugs

Model Compounds. – Several metabolic pathways have been investigated with model compounds. Thus chain elongation of benzoic acid to 3-hydroxy- and 3-keto-3-phenylpropionic acids has been detected in the horse and was thought to involve participation of benzoyl Co-A.[118] Experiments with ^{18}O have shown air to be the source of oxygen in the hydroxymethyl intermediate involved in demethylation of *N*-methylcarbazole[119] and in the formation of carbon monoxide from 1,3-benzodioxoles.[120] In the latter experiment, ^{13}C-labelling identified the methylene group as the source of the eliminated carbon atom. New metabolic routes include the first report of an *N*-acetylornithine conjugate (of 3-phenoxybenzoic acid) in animals,[121] the identification of amidoximes as metabolites of

105 R. J. Cotter and J.-C. Tabet, *Int. J. Mass Spectrom. Ion Phys.*, 1983, **53**, 151.
106 D. J. Liberato, C. C. Fenselau, M. L. Vestal, and A. L. Yergey, *Anal. Chem.*, 1983, **55**, 1741.
107 T. Cairns and E. G. Siegmund, *Anal. Chem.*, 1982, **54**, 2456.
108 J. E. Bakke, V. J. Feil, and C. Struble, *Biomed. Mass Spectrom.*, 1982, **9**, 246.
109 C. N. Statham, J. S. Dutcher, S. H. Kim, and M. R. Boyd, *Drug. Metab. Dispos.*, 1982, **10**, 264.
110 K. Kawahara and T. Ofuji, *J. Chromatogr.*, 1982, **231**, 333.
111 R. H. Waring, S. C. Mitchell, and R. R. Shah, *Xenobiotica*, 1983, **13**, 311.
112 D. I. Eneanya, B. D. Andresen, N. Gerber, and J. R. Bianchine, *Res. Commun. Chem. Pathol. Pharmacol.*, 1983, **41**, 441.
113 G. Maurer, E. Schreier, S. Delaborde, H. R. Loosli, R. Nufer, and A. P. Shukla, *Eur. J. Drug Metab. Pharmacokinet.*, 1982, **7**, 281.
114 F. A. Wong, C. P. Bateman, C. J. Shaw, and J. E. Patrick, *Drug Metab. Dispos.*, 1983, **11**, 301.
115 J. F. Baker, B. W. Chalecki, D. P. Benziger, P. E. O'Melia, S. D. Clemans, and J. Edelson, *Drug Metab. Dispos.*, 1982, **10**, 168.
116 S. F. Sisenwine, C. O. Tio, and H. W. Ruelius, *Drug Metab. Dispos.*, 1982, **10**, 161.
117 I. Midgley, A. G. Fowkes, L. F. Chasseaud, D. R. Hawkins, R. Girkin, and K. Kesselring, *Xenobiotica*, 1983, **13**, 87.
118 M. V. Marsh, J. Caldwell, A. J. Hutt, R. L. Smith, M. W. Horner, E. Houghton, and M. S. Moss, *Biochem. Pharmacol.*, 1982, **31**, 3225.
119 G. L. Kedderis, L. A. Dwyer, D. E. Rickert, and P. F. Hollenberg, *Mol. Pharmacol.*, 1983, **23**, 758.
120 M. W. Anders, J. M. Sunram, and C. F. Wilkinson, *Biochem. Pharmacol.*, 1984, **33**, 577.
121 K. R. Huckle, G. Stoydin, D. H. Hutson, and P. Millburn, *Drug Metab. Dispos.*, 1982, **10**, 523.

benzamidines,[122] thiomethylation and thiohydroxylation of heterocyclic compounds,[123] and the metabolism of a stable *N*-hydroxymethyl metabolite of *N*-methylbenzamide to benzamide.[124] Compounds containing *N*-hydroxymethyl groups are proposed intermediates in *N*-demethylation reactions, and factors affecting their stability have been examined.[125,126] The identification by GC/MS of thioethers and mercapturic acid conjugates of styrene oxide,[127] 4-ethenylmethylbenzenes,[128] and vinyltoluene[129] has given more information on sulphur conjugation reactions occurring as a result of metabolism through epoxide formation. Oxides are also involved in aromatic hydroxylations; studies with deuteriated bromobenzenes have shown that formation of 2-bromophenol from bromobenzene involves participation of a 2,3- rather than a 1,2-arene oxide.[130] Not all aromatic hydroxylations, however, involve arene oxide intermediates, as recently shown for 3-hydroxylation of 2,2′,5,5′-tetrachlorobiphenyl.[131] The cysteine conjugate of bromobenzene is hydrolysed by intestinal micro-organisms, demonstrating for the first time the contribution of these organisms to the formation of thiol compounds from cysteine conjugates.[132]

Halogenated Hydrocarbons. – Many of these compounds such as halothane are used as anaesthetics, and others find various uses as solvents. Most show toxic properties related to metabolic transformations and covalent binding to endogenous compounds, as illustrated by the recent identification of a complex between the trichloromethyl radical derived from carbon tetrachloride and cholesterol.[133] Dichloromethylcarbene has been identified as a metabolite of carbon tetrachloride by trapping with 2,3-dimethyl-2-butene to give 1,1-dichloro-2,2,3,3-tetramethylcyclopropane (6) (Scheme 5), which was identified by GC/MS.[134] Electrophilic chlorine from carbon tetrachloride has been trapped with dimethylphenol and also identified by GC/MS;[135,136] use of ^{35}Cl-labelled CCl_4 and $Na^{37}Cl$ showed that all of the trapped chlorine originated from the

[122] B. Clement, *Xenobiotica*, 1983, **13**, 467.
[123] R. Pal and G. Spiteller, *Xenobiotica*, 1982, **12**, 813.
[124] D. Ross, P. B. Farmer, A. Gescher, J. A. Hickman, and M. D. Threadgill, *Life Sci.*, 1983, **32**, 597.
[125] D. Ross, P. B. Farmer, A. Gescher, J. A. Hickman, and M. D. Threadgill, *Biochem. Pharmacol.*, 1982, **31**, 3621.
[126] D. Ross, P. B. Farmer, A. Gescher, J. A. Hickman, and M. D. Threadgill, *Biochem. Pharmacol.*, 1983, **32**, 1773.
[127] K. Nakatsu, S. Hugenroth, L.-S. Sheng, E. C. Horning, and M. G. Horning, *Drug Metab. Dispos.*, 1983, **11**, 463.
[128] T. Kuhler, *Xenobiotica*, 1984, **14**, 417.
[129] T. H. H. Heinonen, *Biochem. Pharmacol.*, 1984, **33**, 1585.
[130] T. J. Monks, S. S. Lau, L. R. Pohl, and J. R. Gillette, *Drug Metab. Dispos.*, 1984, **12**, 193.
[131] B. D. Preston, J. A. Miller, and E. C. Miller, *J. Biol. Chem.*, 1983, **258**, 8304.
[132] S. Suzuki, H. Tomisawa, S. Khihara, H. Fukazawa, and M. Tateishi, *Biochem. Pharmacol.*, 1982, **31**, 2137.
[133] G. A. S. Ansari, M. T. Moslen, and E. S. Reynolds, *Biochem. Pharmacol.*, 1982, **31**, 3509.
[134] L. R. Pohl and J. W. George, *Biochem. Biophys. Res. Commun.*, 1983, **117**, 367.
[135] B. A. Mico and L. R. Pohl, *Biochem. Biophys. Res. Commun.*, 1982, **107**, 27.
[136] B. A. Mico, R. V. Branchflower, and L. R. Pohl, *Biochem. Pharmacol.*, 1983, **32**, 2357.

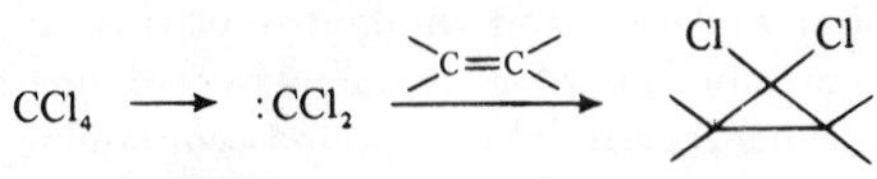

$$CCl_4 \longrightarrow :CCl_2 \xrightarrow{>C=C<} \text{(6)}$$

(6)

Scheme 5

carbon tetrachloride. Likewise, phosgene from metabolism of chloroform has been trapped with 1-cysteine, to give 2-oxothiazolidine-4-carboxylate,[137] and with glutathione.[138] 2,2,2-Trifluoroethanol has been identified and implicated in the toxicity of fluorinated ether anaesthetics,[139] and epoxides have been detected as metabolites of 1,1-[140] and 1,2-dichloroethylenes.[141] Although conjugation with glutathione is usually associated with detoxification of reactive intermediates of this type and has been identified for several halogenated hydrocarbons,[142, 143] there is evidence that some sulphur conjugates such as the ethylene-*S*-glutathionylepisulphonium ion formed from 1,2-dihaloethanes are toxic.[144]

Anticancer Drugs. – As these compounds are frequently rather involatile, they are good candidates for examination by the newer ionization techniques. Several anticancer drugs have been examined using FAB. The bisguanosine adduct of cisplatin has been characterized using FAB with collision-induced dissociation,[145] and platinum has been detected by laser-microprobe mass spectrometry in renal proximal tubule walls after intravenous cisplatin administration;[146] nephrotoxicity is the most important dose-limiting factor for this drug. Przybylski[147] has compared FAB and FD spectra of several anticancer drugs and has concluded that, whereas FD can be used for quantitative measurements, FAB could not. FD mass spectrometry has been used successfully to characterize glutathione and other thiol conjugates of the anticancer drug 4′-(9-acridinylamino)methanesulphon-*m*-anisidide[148] and the oligoglutamyl conjugates of methotrexate.[149]

[137] M. B. Bailie, J. H. Smith, J. F. Newton, and J. B. Hook, *Toxicol. Appl. Pharmacol.*, 1984, **74**, 285.

[138] R. V. Branchflower, D. S. Nunn, R. J. Highet, J. H. Smith, J. B. Hook, and L. R. Pohl, *Toxicol. Appl. Pharmacol.*, 1984, **72**, 159.

[139] M. J. Murphy, D. A. Dunbar, and L. S. Kaminsky, *Toxicol. Appl. Pharmacol.*, 1983, **71**, 84.

[140] A. K. Costa and K. M. Ivanetich, *Biochem. Pharmacol.*, 1982, **31**, 2083.

[141] A. K. Costa and K. M. Ivanetich, *Biochem. Pharmacol.*, 1982, **31**, 2093.

[142] J. A. Nash, L. J. King, E. A. Lock, and T. Green, *Toxicol. Appl. Pharmacol.*, 1984, **73**, 124.

[143] T. D. Landry, T. S. Gushow, P. W. Langvardt, J. M. Wall, and M. J. McKenna, *Toxicol. Appl. Pharmacol.*, 1983, **68**, 473.

[144] J. C. Livesey, M. W. Anders, P. W. Langvardt, C. L. Putzig, and R. H. Reitz, *Drug Metab. Dispos.*, 1982, **10**, 201.

[145] G. Puzo, J. C. Prome, J. P. Macquet, and I. A. S. Lewis, *Biomed. Mass Spectrom.*, 1982, **9**, 552.

[146] A. H. Verbueken, R. E. Van Grieken, G. J. Paulus, G. A. Verpooten, and M. E. DeBroe, *Biomed. Mass Spectrom.*, 1984, **11**, 159.

[147] M. Przybylski, *Fresenius' Z. Anal. Chem.*, 1983, **315**, 402.

[148] K. Gaudich and M. Przybylski, *Biomed. Mass Spectrom.*, 1983, **10**, 292.

[149] M. Przybylski, *Arzneim.-Forsch.*, 1982, **32**, 995.

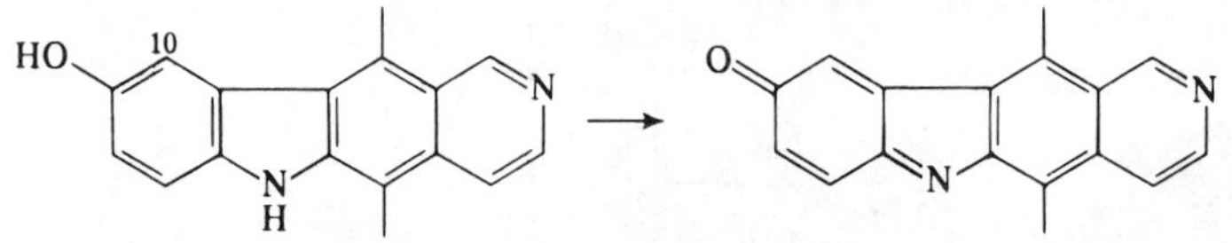

(7) (8)

Scheme 6

Autoxidation of 9-hydroxyellipticine (7) to a substituted quinoneimine (8) (Scheme 6) has been detected by ammonia DCI techniques[150] and is thought to be related to the compound's cytotoxicity. A similar quinoneimine has been identified following peroxidase-catalysed oxidation of N^2-methyl-9-hydroxyellipticine and is involved in covalent binding to protein.[151] A quinoneimine is also an intermediate in formation of the 10-*S*-glutathione conjugate of hydroxyellipticinium acetate after metabolic hydroxylation.[152] The 9-*O*-glucuronide of N^2-methyl-9-hydroxyellipticinium acetate has also been identified by DCI and FD techniques[153] together with an *o*-quinone (9).[154]

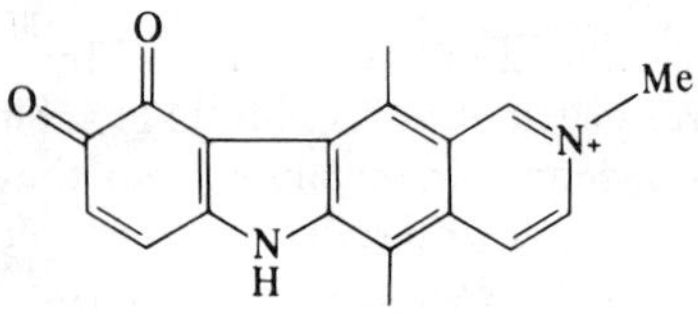

(9)

Studies on the mechanism of action of the nitrosoureas by EI and FAB techniques have shown that *N*-(2-fluoroethyl)-*N'*-cyclohexyl-*N*-nitrosourea reacts with DNA to give O^6-(2-fluoroethyl)guanosine by attack of the 2-haloethyl group on the 6-position of guanine.[155] This rearranges to 2-hydroxyethyl-substituted guanosine. The nitrosourea PCNU [1-(2-chloroethyl)-3-(2,6-dioxo-3-piperidyl)-1-nitrosourea] has been measured in plasma to the 1 ng ml^{-1} level by DCI techniques, and its concentration was shown to fall from 1000 ng ml^{-1} to 1 ng ml^{-1} in 240 minutes.[156]

The electron-impact-induced fragmentation of cyclophosphamide and related compounds has been described,[157] and the reactive iminocyclophosphamide has

[150] C. Auclair, K. Hyland, and C. Paoletti, *J. Med. Chem.*, 1983, **26**, 1438.
[151] C. Auclair, B. Meunier, and C. Paoletti, *Biochem. Pharmacol.*, 1983, **32**, 3883.
[152] M. Maftouh, B. Monsarrat, R. C. Rao, B. Meunier, and C. Paoletti, *Drug Metab. Dispos.*, 1984, **12**, 111.
[153] M. Maftouh, G. Meunier, B. Dugue, B. Monsarrat, B. Meunier, and C. Paoletti, *Xenobiotica*, 1983, **13**, 303.
[154] J. Bernadou, G. Meunier, C. Paoletti, and B. Meunier, *J. Med. Chem.*, 1983, **26**, 574.
[155] W. P. Tong, M. C. Kirk, and D. B. Ludlum, *Biochem. Pharmacol.*, 1983, **32**, 2011.
[156] R. G. Smith, L. K. Cheung, L. G. Feun, and T. L. Loo, *Biomed. Mass Spectrom.*, 1983, **10**, 404.
[157] H. I. Kenttamaa, P. Savolahti, and J. Koskikallio, *Int. J. Mass Spectrom. Ion Phys.*, 1983, **47**, 463.

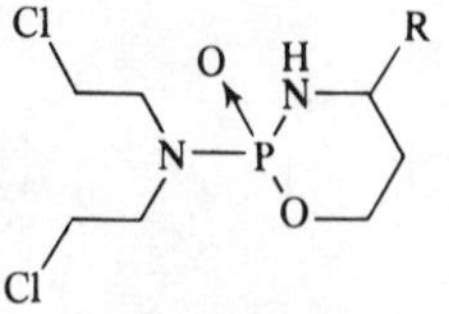

(10) R = CN
(11) R = OH
(12) R = ^2H

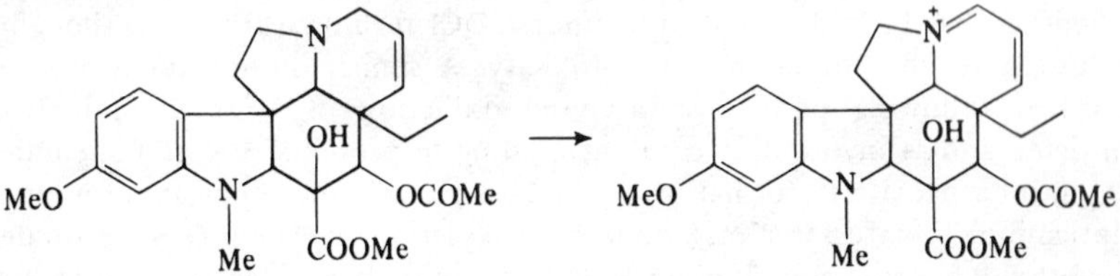

(13) → (14)

been trapped as its cyanide adduct (10) and identified by EI mass spectrometry.[158] The intermediate 4-hydroxycyclophosphamide (11) inhibits cytochrome P450 and delays formation of 7-deoxyadriamycinol from the anticancer drug adriamycin.[159] Reduction of the imine with sodium borodeuteride gives an α-deuterio derivative (12), which can also be used to characterize the intermediate.[158] A corresponding reaction has been used to characterize an imine intermediate (14) of the alkaloid 16-*O*-acetylvindoline (13).[160] Metabolic studies using EI and GC/MS methods have also been reported for these naturally occurring anticancer compounds[161,162] and for the naturally occurring steroid derivative withaferin-A;[163] hydroxylations and oxidations were prominent.

GC/MS studies of 1-(4-carboxamidophenyl)-[164] and 1-(4-acetylphenyl)-3,3-dimethyltriazene[165] have shown demethylation and, in the latter compound, keto reduction and formation of 4-aminoacetophenone. Covalent binding of hexamethylmelamine to DNA involves a hydroxymethyl metabolite as shown by GC/MS analysis of pentamethylmelamine obtained from hydrolysis of the

[158] C. Fenselau, J. P. Lehman, A. Myles, J. Brandt, G. S. Yost, O. M. Friedman, and O. M. Colvin, *Drug Metab. Dispos.*, 1982, **10**, 636.
[159] P. Dodion, C. E. Riggs, jun., S. R. Akman, J. M. Tamburini, O. M. Colvin, and N. R. Bachur, *J. Pharmacol. Exp. Ther.*, 1984, **229**, 51.
[160] F. S. Sariaslani, F. M. Eckenrode, J. M. Beale, jun., and J. P. Rosazza, *J. Med. Chem.*, 1984, **27**, 749.
[161] F. Eckenrode, W. Peczynska-Czoch, and J. P. Rosazza, *J. Pharm. Sci.*, 1982, **71**, 1246.
[162] B. C. Mayo, S. R. Biggs, D. R. Hawkins, L. F. Chasseaud, A. Darragh, G. A. Baldock, and B. R. Whitby, *J. Pharmacobiodynam.*, 1982, **5**, 951.
[163] J. Fuska, J. Prousek, J. Rosazza, and M. Budesinsky, *Steroids*, 1982, **40**, 157.
[164] G. Sava, T. Giraldi, L. Lassiani, C. Nisi, and P. B. Farmer, *Biochem. Pharmacol.*, 1982, **31**, 3629.
[165] P. Farina, E. Benfenati, R. Reginato, L. Torti, M. D'Incalci, M. D. Threadgill, and A. Gescher, *Biomed. Mass Spectrom.*, 1983, **10**, 485.

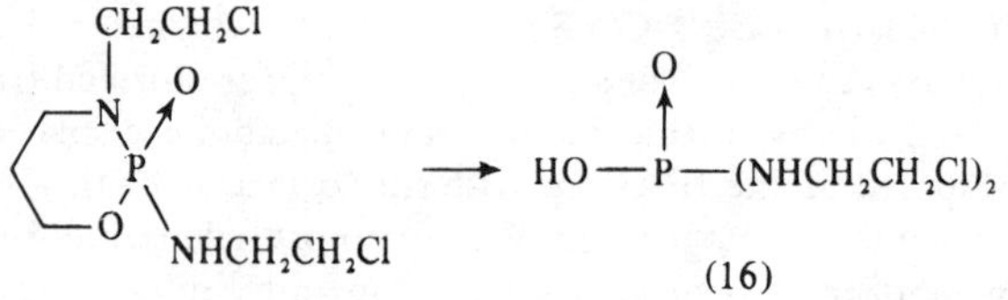

$$\text{(15)} \longrightarrow HO{-}P(\rightarrow O){-}(NHCH_2CH_2Cl)_2 \; \text{(16)}$$

(15): cyclic N-(CH_2CH_2Cl), P=O, $NHCH_2CH_2Cl$

Scheme 7

adduct.[166] Other metabolic studies using direct-insertion techniques include those of aclarubicin[167] and ifosfamide (15);[168] six metabolites including products of oxidation, ring opening (Scheme 7), and loss of a chloroethyl group were identified from the latter compound.

Nucleosides. – Compounds of this type are used both as anticancer and antiviral agents. FD and EI mass spectrometry have been used to show the presence of 9β-D-arabinofuranosyl-2-fluorohypoxanthine as a metabolite of 9β-D-arabinofuranosyl-2-fluoroadenine in blood and, additionally, of the toxic 2-fluoroadenine in urine.[169] Urapidil, 6-{3-[4-(2-methoxyphenyl)piperazinyl]propylamino}-1,3-dimethyluracil, is metabolized by hydroxylation and *O*-demethylation,[170] and 5-(2-bromo-*E*-ethenyl)-2′-deoxyuridine gives 5-(2-bromo-*E*-ethenyl)uracil.[171] A similar metabolic pathway exists for 5-ethyl-2′-deoxyuridine with additional 1-hydroxylation of the ethyl group.[172] 1-(Tetrahydro-2-furanyl)-5-fluorouracil (Ftorafur), a pro-drug of 5-fluorouracil (5-FU), is metabolized by hydroxylation;[173] fragment ions in the EI spectrum formed by migration of labile hydrogen were particularly helpful in structural determination of this metabolite. The carbohydrate ring of this drug is degraded to γ-butyrolactone and succinaldehyde.[174] Quantification of 5-FU has been achieved by GC/MS using 5-Br-FU[175] and ^{15}N[176] standards, and a Br-containing standard has also been used for measurement of 5′-deoxy-5-fluorouridine in humans;[177] CI resulted in a more sensitive

[166] M. M. Ames, M. E. Sanders, and W. S. Tiede, *Cancer Res.*, 1983, **43**, 500.
[167] R. G. Smith, A. A. Miller, J. A. Benvenuto, M. Valdivieso, and T. L. Loo, *Cancer Treatment Rep.*, 1983, **67**, 351.
[168] R. F. Struck, D. J. Dykes, T. H. Corbett, W. J. Suling, and M. W. Trader, *Br. J. Cancer*, 1983, **47**, 15.
[169] R. F. Struck, A. T. Shortnacy, M. C. Kirk, M. C. Thorpe, R. W. Brockman, D. L. Hill, S. M. El Dareer, and J. A. Montgomery, *Biochem. Pharmacol.*, 1982, **31**, 1975.
[170] E. Sturm and K. Zech, *Biomed. Mass Spectrom.*, 1984, **11**, 211.
[171] Y. Robinson, N. Gerry, R. D. Brownsill, and C. W. Vose, *Biomed. Mass Spectrom.*, 1984, **11**, 199.
[172] R. Kaul, K. Keppeler, G. Kiefer, B. Hempel, and P. Fischer, *Chemosphere*, 1982, **11**, 539.
[173] T. Marunaka, *Biomed. Mass Spectrom.*, 1982, **9**, 381.
[174] Y. M. El Sayed and W. Sadee, *Biochem. Pharmacol.*, 1982, **31**, 3006.
[175] C. Aubert, J. P. Sommadossi, Ph. Coassolo, J. P. Cano, and J. P. Rigault, *Biomed. Mass Spectrom.*, 1982, **9**, 336.
[176] T. Marunaka and Y. Umeno, *Chem. Pharm. Bull.*, 1982, **30**, 1868.
[177] J.-P. Sommadossi, C. Aubert, J.-P. Cano, J. Gouveia, P. Ribaud, and G. Mathe, *Cancer Res.*, 1983, **43**, 930.

assay than EI in the latter case. GC/MS studies have shown that cyclic metabolites containing either oxygen (18)[178] or sulphur (22)[179,180] are formed from 1-propyl- (17) and 1-allyl-3,5-diethyl-6-chlorouracil (19); metabolic displacement of the chlorine was proposed as the first step with the formation of thiols and methylthiols.[181] The selenium analogue, 2β-D-ribofuranosylselenazole-4-carboximide, of the antitumour drug tiazofurin has been shown by negative-ion FAB to be

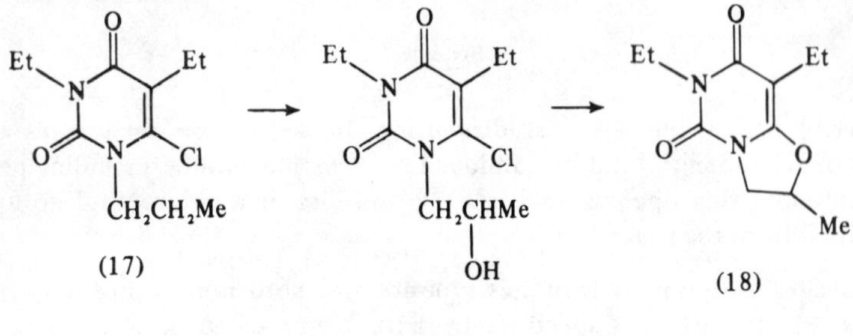

(17) (18)

Scheme 8

(19) (20) (21) (22)

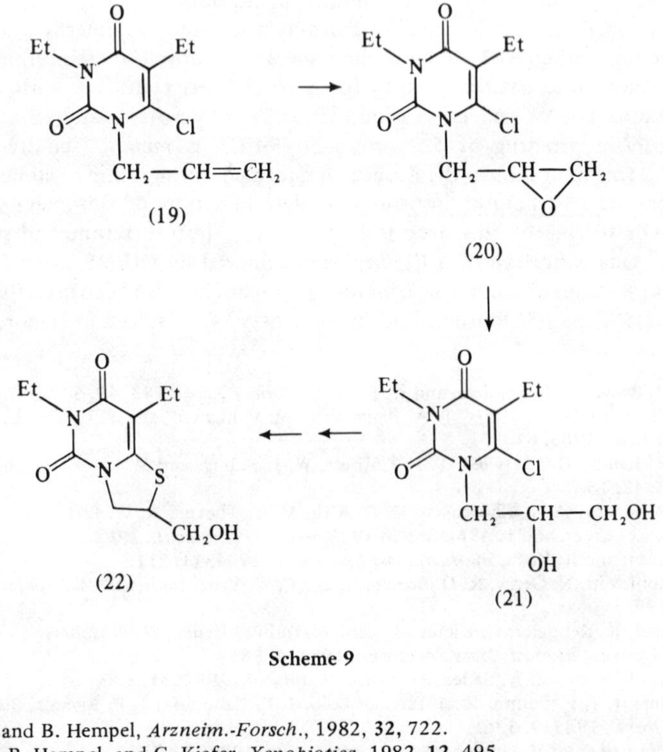

Scheme 9

[178] R. Kaul and B. Hempel, *Arzneim.-Forsch.*, 1982, **32**, 722.
[179] R. Kaul, B. Hempel, and G. Kiefer, *Xenobiotica*, 1982, **12**, 495.
[180] R. Kaul, G. Kiefer, and B. Hempel, *Arzneim.-Forsch.*, 1982, **32**, 610.
[181] R. Kaul, B. Hempel, and G. Kiefer, *J. Pharm. Sci.*, 1982, **71**, 897.

metabolized to an analogue of NAD in a manner analagous to that of the sulphur analogue.[182]

Xanthines. – Theophylline (23), caffeine (24), and three demethylated metabolites have been measured to 0.5–10 ng ml^{-1} levels in biological fluids from pre-term infants treated with theophylline. Measurement was by GC/MS as their *N*-propyl derivatives.[183] 1,7-Dimethylxanthine has been shown to interfere with an earlier assay for theophylline.[184] A ratio of 0.57 has been measured for caffeine and theophylline at steady state in infants after theophylline administration,[185] and the absence of caffeine as a metabolite of theophylline in adults has been attributed, following experiments with deuteriated analogues, to rapid metabolism rather than to a lack of formation.[186] Pulse labelling with [^{15}N,^{13}C]-theophylline during chronic theophylline treatment has shown a reduced plasma clearance in the presence of cimetidine.[187, 188] 5-Acetylamino-6-amino-3-methyluracil (25) has been identified by both EI and CI mass spectrometry as a human urinary metabolite of caffeine;[189] metabolism of the [1,3-$^{15}N_2$,8-^{13}C] analogue showed that C-8 was lost during the biotransformation and thus the acetyl group did not originate from caffeine. Its formyl derivative, 5-acetylamino-6-formylamino-3-methyluracil (26), has been identified after chloroform extraction, but it is probable that the formyl group originates artefactually from formaldehyde present in the chloroform.[190] Crown ethers have been shown to give adduct ions [crown ether + drug + H]$^+$, in the CI spectra of caffeine.[191] Among studies with synthetic xanthines, it has been found that the stimulant ethimizol undergoes

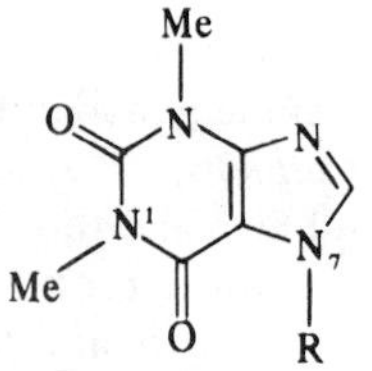

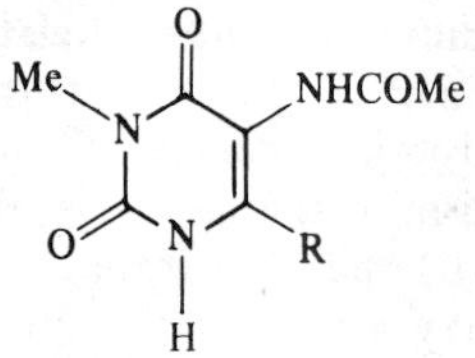

(23) R = H
(24) R = Me

(25) R = NH_2
(26) R = NHCHO

182 H. N. Jayaram, G. S. Ahluwalia, R. L. Dion, G. Gebeyehu, V. E. Marquez, J. A. Kelley, R. K. Robins, D. A. Cooney, and D. G. Johns, *Biochem. Pharmacol.*, 1983, **32**, 2633.
183 K.-Y. Tserng, *J. Pharm. Sci.*, 1983, **72**, 508.
184 J. H. G. Jonkman, R. A. de Zeeuw, and R. Schoenmaker, *Clin. Chem.*, 1982, **28**, 1988.
185 G. Lonnerholm, B. Lindstrom, L. Paalzow, and G. Sedin, *Eur. J. Clin. Pharmacol.*, 1983, **24**, 371.
186 K.-Y. Tserng, F. N. Takieddine, and K. C. King, *Clin. Pharmacol. Ther.*, 1983, **33**, 522.
187 R. E. Vestal, K. E. Thummel, and B. Musser, *Br. J. Clin. Pharmacol.*, 1983, **15**, 411.
188 R. E. Vestal, K. E. Thummel, B. Musser, S. G. Jue, G. Mercer, and W. N. Howald, *Biomed. Mass Spectrom.*, 1982, **9**, 340.
189 A. R. Branfman, M. F. McComish, R. J. Bruni, M. M. Callahan, R. Robertson, and D. W. Yesair, *Drug Metab. Dispos.*, 1983, **11**, 206.
190 B. K. Tang, D. M. Grant, and W. Kalow, *Drug Metab. Dispos.*, 1983, **11**, 218.
191 A. K. Bose, O. Prakash, G. Y. Hu, and J. Edasery, *J. Org. Chem.*, 1983, **48**, 1780.

both demethylation and hydroxylation reactions[192,193] whereas *erythro*-9-(2-hydroxy-3-nonyl)hypoxanthine is metabolized predominantly by hydroxylation and β-oxidation of the aliphatic chain.[194]

Barbiturates. – 4′-Hydroxy metabolites of butobarbitone and pentobarbitone have been measured in urine by GC/MS.[195] The pharmacokinetics of pentobarbitone have been studied by GC/MS in dogs, and a half-life of 8.2 ± 2.2 hours has been found.[196] The pharmacokinetics of hexobarbitone have been reported in humans.[197] Chloromethyldimethylsilyl derivatives have been used as GLC shift reagents for hexobarbitone metabolites and for labelling them with chlorine.[198]

Hydantoins and Other Related Anticonvulsants. – Aromatic hydroxylation, mainly in the 4′-position, is the major route for 5,5-diphenylhydantoin (DPH) metabolism in the mouse[199] and Leopard frog (*Rana pipiens*).[200] Studies on the metabolism of the [4′-^{2}H] analogue of DPH in rat and man and the observation of an NIH shift implicate an arene oxide intermediate in the formation of the 4′-hydroxy metabolite.[201] Studies on deuteriated analogues have also shown that 5-(3′-hydroxyphenyl)-5-phenylhydantoin is formed from a 3′,4′-arene oxide intermediate and not from a 2′,3′-arene oxide.[202] A method for measurement of several hydroxy metabolites of DPH in biological specimens, sensitive to 20 ng ml^{-1} of wet tissue, has been reported;[203] [$^2H_3$]methylation was used at the derivatization stage so that the 3′-methoxy metabolite could be measured. Nine metabolites of aminoglutethimide [2-(4-aminophenyl)-2-ethylglutarimide] have been identified in the rat by means of direct-insertion techniques.[204] These metabolites include products of acetylation, hydroxylation, sulphate conjugation, and ring opening. Oxidation of the amino group in this drug to a nitro group has been reported,[205] and an *N*-hydroxy metabolite has been observed to disproportionate into nitrosogluthimide and aminogluthimide in the mass spectrometer.[206]

192 L. Soltes, V. Mlynarik, and V. Mihalov, *Xenobiotica*, 1983, **13**, 683.

193 L. Soltes, S. Bezek, T. Trnovec, M. Stefek, and Z. Kallay, *Pharmacology*, 1983, **26**, 198.

194 E. H. Pfadenhauer, C. S. Bankert, J. Jensen, C. E. Jones, E. E. Jenkins, and J. A. McCloskey, *Drug Metab. Dispos.*, 1984, **12**, 280.

195 M. A. Al Sharifi, J. N. T. Gilbert, and J. W. Powell, *Xenobiotica*, 1983, **13**, 179.

196 M. C. Frederiksen, T. K. Henthorn, T. I. Ruo, and A. J. Atkinson, jun., *J. Pharmacol. Exp. Ther.*, 1983, **225**, 355.

197 N. P. E. Vermeulen, C. T. Rietueld, and D. D. Breimer, *Br. J. Clin. Pharmacol.*, 1983, **15**, 459.

198 M. Desage, J. L. Brazier, and R. Guilluy, *Int. J. Mass Spectrom. Ion Phys.*, 1983, **48**, 85.

199 S. A. Chow and L. J. Fischer, *Drug Metab. Dispos.*, 1982, **10**, 156.

200 S. W. Johnson, *Drug Metab. Dispos.*, 1982, **10**, 510.

201 M. Claesen, M. A. A. Moustafa, J. Adline, D. Vandervorst, and J. H. Poupaert, *Drug Metab. Dispos.*, 1982, **10**, 667.

202 M. A. A. Moustafa, M. Claesen, J. Adline, D. Vandervorst, and J. H. Poupaert, *Drug Metab. Dispos.*, 1983, **11**, 574.

203 M. Arboix and C. Pantarotto, *Chromatographia*, 1982, **15**, 509.

204 H. Egger, F. Bartlett, W. Itterly, R. Rodebaugh, and C. Shimanskas, *Drug Metab. Dispos.* 1982, **10**, 405.

205 R. C. Coombes, A. B. Foster, S. J. Harland, M. Jarman, and E. C. Nice, *Br. J. Cancer* 1982, **46**, 340.

206 M. Jarman, A. B. Foster, P. E. Goss, L. J. Griggs, I. Howe, and R. C. Coombes, *Biomed Mass Spectrom.*, 1983, **10**, 620.

4-Hydroxyprimidone has been detected as its *N*-propyl derivative by GC/MS as a minor urinary metabolite of primidone in the rat and man.[207]

Cannabinoids. – The proceedings of the 1982 Louisville symposium on cannabinoids have been published;[208] they contain a number of papers on mass-spectrometric applications. The first tetrahydroxycannabinoid, cannabitetrol (27), has been isolated from *Cannabis sativa*,[209] and capillary-column GC/MS has been used to identify 96 compounds in marihuana condensates.[210] The unnatural isomer Δ^6-tetrahydrocannabinol (Δ^6-THC) has been shown to be an isolation artefact formed during decarboxylation of cannabinoid acids.[211]

(27)

(28) R = Me
(29) R = COOH

The most significant advance in research into cannabinoid metabolism made during the last two years has been the identification of metabolites in man. Eleven metabolites of Δ^1-THC (28) have been identified by GC/MS[212] in human liver and shown to be mainly mono- and di-hydroxy derivatives. The fourteen metabolites identified in urine, however, were all acidic and consisted of Δ^1-THC-7-oic acid (29) and its side-chain hydroxylated derivatives and products of β-oxidation of the side chain.[213–215] Δ^1-THC glucuronide is also excreted in urine and has been identified as its TMS derivative by GC/MS.[216] Compounds isolated following metabolism of Δ^6-THC by human-liver microsomes were monohydroxy derivatives and $1\alpha,6\beta$-hexahydrocannabinol.[217] The presence of

[207] W. D. Hooper, A. M. Treston, N. W. Jacobsen, R. G. Dickinson, and M. J. Eadie, *Drug Metab. Dispos.*, 1983, **11**, 607.
[208] 'The Cannabinoids: Chemical, Pharmacologic, and Therapeutic Aspects', ed. S. Agurell, W. L. Dewey, and R. E. Willette, Academic Press, Orlando, 1984.
[209] H. N. ElSohly, E. G. Boeren, C. E. Turner, and M. A. ElSohly in ref. 208, p. 89.
[210] M. Novotny, F. Merli, D. Wiesler, and T. Saeed, *Chromatographia*, 1982, **15**, 564.
[211] A. Heitrich and M. Binder, *Experientia*, 1982, **38**, 898.
[212] M. M. Halldin, M. Widman, C. v. Bahr, J.-E. Lindgren, and B. R. Martin, *Drug Metab. Dispos.*, 1982, **10**, 297.
[213] M. M. Halldin, S. Carlsson, S. L. Kanter, M. Widman, and S. Agurell, *Arzneim.-Forsch.*, 1982, **32**, 764.
[214] M. M. Halldin, L. K. R. Andersson, M. Widman, and L. E. Hollister, *Arzneim.-Forsch.*, 1982, **32**, 1135.
[215] M. M. Halldin, M. Widman, S. Agurell, L. E. Hollister, and S. L. Kanter in ref. 208, p. 211.
[216] M. M. Halldin and M. Widman, *Arzneim.-Forsch.*, 1983, **33**, 177.
[217] I. Yamamoto, S. Narimatsu, K. Watanabe, T. Shimonishi, H. Yoshimura, and T. Nagano, *Chem. Pharm. Bull.*, 1983, **31**, 1784.

this compound suggests the involvement of an intermediate epoxide of the type found in mouse-liver preparations.[218,219] A comparison of Δ^1-THC metabolism by the livers of rat and guinea pig has shown major differences, and in particular the lungs produce more aliphatic hydroxylations than the livers in both species.[220]

Quantitative aspects of cannabinoid analysis by GC/MS have been reviewed.[221] Emphasis was on sensitive detection methods because the concentrations of Δ^1-THC encountered in body fluids is very low during intoxication and rapidly falls to levels too low to be measured by existing analytical methods. CI mass spectrometry with a glass capillary column has been used to measure the drug and its two major metabolites, 7-hydroxy-Δ^1-THC and Δ^1-THC-7-oic acid, to 0.2 ng ml^{-1};[222] H_2/NH_3 as reactant gas gave better CI spectra than CH_4/NH_3. Improved sensitivity was obtained with negative-ion CI of the TFA derivatives; the acid metabolite could be detected to 0.1 ng ml^{-1}.[223] The most sensitive method for THC analysis is still the previously reported technique of metastable-peak monitoring, which is capable of measuring the drug to 0.005 ng ml^{-1} in rabbit plasma. Applications of the method to the measurement of THC in the eye and erythrocyte membrane and to the determination of THC pharmacokinetics in rabbits and mice have appeared.[224] Pharmacokinetic studies in man by GC/MS have given values in the 20–56 hour range for the half-life,[225,226] but these are thought to be underestimates as the drug cannot be measured for long enough to obtain the true value. Cannabidiol[227] and Δ^1-THC[228] pharmacokinetics have been studied using deuterium-labelled analogues to avoid problems encountered by residual levels from previous exposure; results for cannabidiol were similar to those obtained for Δ^1-THC. THC has been detected by GC/MS in human milk,[229] in the urine of non-smokers as the result of passive inhalation,[230] in breath,[231] and on the hands of people handling cannabis.[232]

[218] I. Yamamoto, S. Narimatsu, K. Watanabe, and H. Yoshimura, *Biochem. Biophys. Res. Commun.*, 1982, **109**, 922.

[219] S. Narimatsu, K. Watanabe, I. Yamamoto, and H. Yoshimura, *Xenobiotica*, 1982, **12**, 561.

[220] M. M. Halldin, H. Isaac, M. Widman, E. Nilsson, and A. Ryrfeldt, *Xenobiotica*, 1984, **14**, 277.

[221] R. Foltz, *Adv. Anal. Toxicol.*, 1984, **1**, 125.

[222] R. L. Foltz and B. J. Hidy in Res. Monogr. Ser., Vol. 42, ed. R. L. Hawks, NIDA, Rockville, 1982, p. 99.

[223] R. L. Foltz, K. M. McGinnis, and D. M. Chinn, *Biomed. Mass Spectrom.*, 1983, **10**, 316.

[224] D. J. Harvey, J. T. A. Leuschner, D. R. Wing, and W. D. M. Paton in ref. 208, p. 291.

[225] B. M. Salder, M. E. Wall, and M. Perez-Reyes in ref. 208, p. 227.

[226] M. E. Wall, B. M. Sadler, D. Brine, H. Taylor, and M. Perez-Reyes in ref. 208, p. 185.

[227] A. Ohlsson, J. E. Lindgren, S. Andersson, S. Agurell, H. Gillespie, and L. E. Hollister in ref. 208, p. 219.

[228] S. Agurell, J.-E. Lindgren, A. Ohlsson, H. K. Gillespie, and L. E. Hollister in ref. 208, p. 165.

[229] M. Perez-Reyes and M. E. Wall, *New Engl. J. Med.*, 1982, **307**, 819.

[230] M. Perez-Reyes, S. DiGiuseppi, A. P. Mason, and K. H. Davis, *Clin. Pharmacol. Ther.*, 1983, **34**, 36.

[231] A. Manolis, L. J. McBurney, and B. A. Bobbie, *Clin. Biochem.*, 1983, **16**, 229.

[232] R. Thibault, W. J. Stall, R. G. Master, and R. R. Gravier, *J. Forensic Sci.*, 1983, **28**, 15.

Measurement of the acidic metabolite (29) is preferred for forensic detection as it is excreted in relatively high concentration in urine. A negative-ion CI technique claimed to be 200 times more sensitive than positive-ion EI or CI and using the pentafluoropropyl/pentafluoropropionyl derivative has been used for measurements of this metabolite in urine.[233] TLC methods can also be used, and a combined TLC/GC/MS method has been reported for forensic use.[234] Radioimmunoassay is the most widely used method for initial screening, but results should be confirmed by GC/MS.[235,236] Stability of THC in stored blood samples is critical for forensic work, but one recent report suggests that concentrations measured by GC/MS drop after about 17 weeks, probably because of decomposition of the protein to which the drug is bound.[237]

Phencyclidine. – Metabolic opening of the piperidine ring of phencyclidine (30) yields a carbinolamine that is oxidized to the corresponding aldehyde (34).[238] This has been trapped as the methyloxime and identified by GC/MS, with the [phenyl-$^2H_5$] analogue of phencyclidine being used to identify the

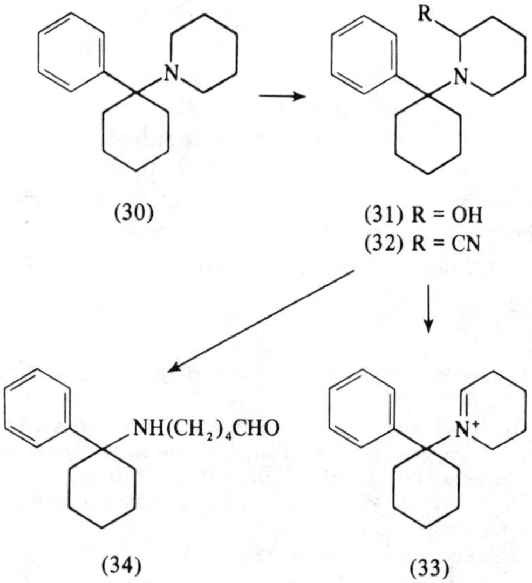

N

(30)

R

N

(31) R = OH
(32) R = CN

$NH(CH_2)_4CHO$

(34)

N^+

(33)

Scheme 10

[233] L. Karlsson, J. Jonsson, K. A. Berg, and C. Roos, *J. Anal. Toxicol.*, 1983, **7**, 198.
[234] G. R. Nakamura, W. J. Stall, V. A. Folen, and R. G. Masters, *J. Chromatogr.*, 1983, **264**, 336.
[235] M. A. Peat, B. S. Finkle, and M. E. Deyman in Res. Monogr. Ser., Vol. 42, ed. R. L. Hawks, NIDA, Rockville, 1982, p. 85.
[236] V. W. Hanson, M. H. Buonarati, R. C. Baselt, N. A. Wade, C. Yep, A. A. Biasotti, V. C. Reeve, A. S. Wong, and M. W. Orbanowsky, *J. Anal. Toxicol.*, 1983, **7**, 96.
[237] A. S. Wong, M. W. Orbanosky, V. C. Reeve, and J. D. Beede in Res. Monogr. Ser., Vol. 42, ed. R. L. Hawks, NIDA, Rockville, 1982, p. 119.
[238] G. Hallstrom, R. C. Kammerer, C. H. Nguyen, D. A. Schmitz, E. W. DiStefano, and A. K. Cho, *Drug Metab. Dispos.*, 1983, **11**, 47.

metabolite by the isotope-doublet technique. An intermediate iminium ion (33), formed from the hydroxypiperidine metabolite (31) (Scheme 10) and thought to be involved in the drug's toxicity, has been trapped as its cyanide adduct (32) and identified by CI mass spectrometry.[239] Aromatic *p*-hydroxylation is also a major metabolic route for this drug,[240] and the product has been measured to 1 ng ml^{-1} in urine by EI GC/MS.[241] Smoking of phencyclidine results in the formation of 1-phenylcyclohexene, which is metabolized mainly by allylic hydroxylation.[242,243] 1-Phenylcyclohexene is also formed as a decomposition product of phencyclidine in GLC columns containing active sites.[244] Thienyl analogues of phencyclidine, which frequently appear as contaminants in forensic samples, also decompose.[245,246]

Opiates. – Forensic analysis has been briefly reviewed;[247] analysis of acetylation by-products and other impurities is a potential method for tracing the source of illicit heroin.[248] Morphine has been detected by selected-ion monitoring in poppy seed and poppy-seed cake, indicating that its subsequent presence in urine may not indicate illicit use.[249] The alkaloids morphine, codeine, normorphine, norcodeine, and noscapine, but not thebaine, paraverine, or oripavine, have been detected in the urine of a cancer patient who had ingested 1 g day^{-1} of the 'dross' from an opium pipe.[250] Perfluoroacylation at carbon sites has been reported for certain unsaturated opioids, *e.g.* (35), in the presence of base, and the resulting derivatives allowed detection at the sub-picogram level by electron-capture GLC.[251] An assay giving a sensitivity in the 10 ng ml^{-1} range and based on GC/CIMS has been reported for morphine, codeine, and ten potential metabolites in urine,[252] and morphine pharmacokinetics have been studied by GC/MS using both analogue (nalorphine)[253] and deuteriated standards.[254] The latter method

[239] D. P. Ward, A. J. Trevor, A. Kalir, J. D. Adams, T. A. Baillie, and N. Castagnoli, jun., *Drug Metab. Dispos.*, 1982, **10**, 690.
[240] R. C. Kammerer, D. A. Schmitz, and A. K. Cho, *Xenobiotica*, 1984, **14**, 475.
[241] B. R. Kuhnert, B. S. Bagby, and N. L. Golden, *J. Chromatogr.*, 1983, **276**, 433.
[242] C. E. Cook, D. R. Brine, and C. R. Tallent, *Drug Metab. Dispos.*, 1984, **12**, 186.
[243] B. R. Martin, B. B. Bailey, H. Awaya, E. L. May, and N. Narasimhachari, *Drug Metab. Dispos.*, 1982, **10**, 685.
[244] D. Legault, *J. Chromatogr. Sci.*, 1982, **20**, 228.
[245] K. G. Rao and S. V. Soni, *J. Assoc. Off. Anal. Chem.*, 1983, **66**, 1186.
[246] R. C. Kelly and D. S. Christmore, *J. Forensic Sci.*, 1982, **27**, 827.
[247] V. Navaratnam and H. K. Fei, *Bull. Narcotics*, 1984, **36**(i), 15.
[248] B. Law, J. R. Joyce, T. S. Bal, C. P. Goddard, M. Japp, and L. J. Humphreys, *Anal. Proc.*, 1983, **20**, 611.
[249] K. Bjerver, J. Jonsson, A. Nilsson, J. Schuberth, and J. Schuberth, *J. Pharm. Pharmacol.*, 1982, **34**, 798.
[250] E. J. Cone, C. W. Gorodetzky, S. Y. Yeh, W. D. Darwin, and W. F. Buchwald, *J. Chromatogr.*, 1982, **230**, 57.
[251] J. M. Moore, A. C. Allen, and D. A. Cooper, *Anal. Chem.*, 1984, **56**, 642.
[252] E. J. Cone, W. R. Darwin, and W. F. Buchwald, *J. Chromatogr.*, 1983, **275**, 307.
[253] D. Westerling, S. Lindahl, K. E. Andersson, and A. Andersson, *Eur. J. Clin. Pharmacol.*, 1982, **23**, 59.
[254] L. L. Gustafsson, S. Friberg-Nielsen, M. Garle, A. Mohall, A. Rane, B. Schildt, and T. Symreng, *Br. J. Anaesthesiol.*, 1982, **54**, 1167.

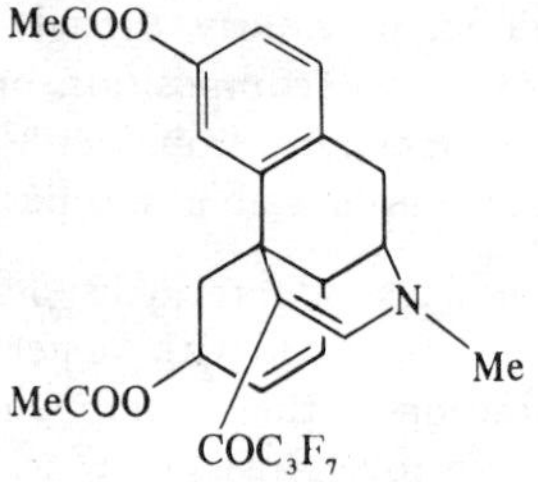

MeCOO MeCOO COC_3F_7 N Me

(35)

HO O OH N Me O

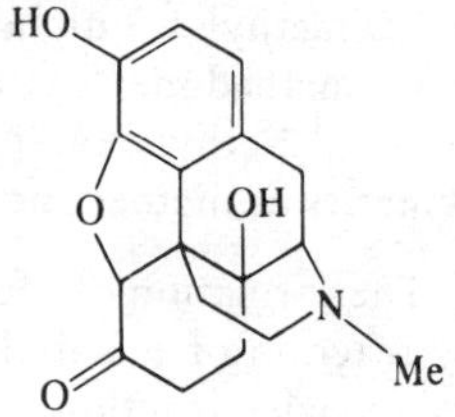

(36)

has been used to study morphine glucuronidation in the Rhesus monkey.[255] Other assays reporting simultaneous measurement of several opiates are those for morphine, codeine, and hydromorphone[256] and for morphine and codeine in biological samples,[257] both involving GC/MS. A terminal half-life of 10 days has been reported for morphine in rats, with considerable consequences for drug accumulation.[258] Long-chain fatty-acid conjugates of codeine have been identified, further demonstrating the generality of this newly discovered biotransformation route.[259]

In work with synthetic opiates, oxomorphone (36) has been shown by GC/MS to be reduced to its 6β-hydroxy analogue in all of five species in which it was examined and to the 6α-hydroxy isomer in three species;[260] in addition codorphone is metabolized by dealkylation,[261] and a negative-ion CI assay sensitive to 1 ng ml^{-1} has been reported for levorphanol in plasma.[262]

Methadone. – Pulse-labelling experiments with [$^2H_3$]methadone have been performed in patients on long-term treatment with the drug[263] and have shown a shorter half-life in cases where treatment was not successful.[264] Liver disease does not appear to affect the excretion of the drug, as shown using a quantitative method based on direct-insertion CI mass spectrometry,[265] but urinary pH

[255] A. Rane, J. Sawe, B. Lindberg, J.-O. Svensson, M. Garle, R. Erwald, and H. Jorulf, *J. Pharmacol. Exp. Ther.*, 1984, **229**, 571.
[256] J. J. Saady, N. Narasimhachari, and R. V. Blanke, *J. Anal. Toxicol.*, 1982, **6**, 235.
[257] N. B. Wu Chen, M. I. Schaffer, R.-L. Lin, and R. J. Stein, *J. Anal. Toxicol.*, 1982, **6**, 231.
[258] A. W. Jones, A. Neri, and E. Anggard, *J. Pharmacol. Exp. Ther.*, 1983, **224**, 419.
[259] E. G. Leighty and A. F. Fentiman, jun., *J. Pharm. Pharmacol.*, 1983, **35**, 260.
[260] E. J. Cone, W. D. Darwin, W. F. Buchwald, and C. W. Gorodetzky, *Drug Metab. Dispos.*, 1983, **11**, 446.
[261] J. L. Leeling, J. V. Evans, R. J. Helms, and B. A. Ryerson, *Drug Metab. Dispos.*, 1982, **10**, 649.
[262] B. H. Min and W. A. Garland, *J. Chromatogr.*, 1982, **231**, 194.
[263] M.-I. Nilsson, E. Anggard, J. Holmstrand, and L.-M. Gunne, *Eur. J. Clin. Pharmacol.*, 1982, **22**, 343.
[264] M.-I. Nilsson, L. Gronbladh, E. Widerlov, and E. Anggard, *Eur. J. Clin. Pharmacol.*, 1983, **25**, 497.
[265] M. J. Kreek, F. A. Bencsath, A. Fanizza, and F. H. Field, *Biomed. Mass Spectrom.*, 1983, **10**, 544.

does.[266] 1,5-Dimethyl-3,3-diphenyl-2-pyrrolidone, previously thought to be a metabolite of methadone, has been shown to be a decomposition product of the metabolite 2-ethylidene-1,5-dimethyl-3,3-diphenylpyrrolidine.[267] Clinical pharmacokinetics of methadone have been studied using stable isotopes.[268]

Alcohol. – The formation of tetrahydroisoquinolines and tetrahydro-β-carbolines (THBC) by reaction of metabolically derived acetaldehyde with biogenic amines in a Pictet–Spengler reaction and the contribution of these compounds to the pharmacological profile of alcohol are still a controversial issue.[269] In addition, a related reaction involving aldehydes present in chlorinated extraction solvents used in GC/MS assays has raised doubts as to whether tetrahydroisoquinolines and THBCs are natural products at all.[269,270] Higher concentrations of 6-hydroxy-1-methyl-1,2,3,4-THBC in the urine of alcoholics during intoxication than in that of controls have been measured by GC/MS,[271] but the compound is also present as a natural constituent.[272] The natural occurrence of THBCs has also been demonstrated in other laboratories,[273] and tryptoline has been measured at 0.37 ng g^{-1} in rat brain by capillary-column GC/NICIMS.[274] To show that 1-methyl-THBCs are not formed in urine from endogenous aldehydes if suitable precautions are taken, [$^{2}H_4$]tryptamine has been added to the urine and the absence of the labelled Pictet–Spengler product demonstrated by GC/MS.[275] Present results indicate that THBCs are normal constituents of urine and are not related to alcohol intake. Metabolism of THBC in rats leads to phenols and 1,2-dihydro-β-carboline-1-one.[276]

In other metabolic studies with alcohol it has been shown that deuterium from [1,1-$^{2}H_2$]ethanol is incorporated into the glycerol moiety of phospholipids[277] and into ether-containing phospholipids[278] *via* NADH formed during ethanol metabolism.

Benzodiazepines. – Mass-spectral data on benzodiazepines have been published in a comprehensive handbook.[279] A study of thermal stability during GC/MS analysis has shown several decomposition pathways related to structure.[280] Thus

[266] M.-I. Nilsson, E. Widerlov, U. Meresaar, and E. Anggard, *Eur. J. Clin. Pharmacol.*, 1982, **22**, 337.
[267] G. I. Kang and F. S. Abbott, *J. Chromatogr.*, 1982, **231**, 311.
[268] M.-I. Nilsson, *Acta Pharm. Suec.*, 1982, **19**, 472.
[269] T. R. Bosin and B. Holmstedt, *Prog. Clin. Biol. Res.*, 1982, **90**, 16.
[270] T. R. Bosin, B. Holmstedt, A. Lundmann, and O. Beck, *Anal. Biochem.*, 1983, **128**, 287.
[271] O. Beck, T. R. Bosin, A. Lundman, and S. Borg, *Biochem. Pharmacol.*, 1982, **31**, 2517.
[272] O. Beck and A. Lundman, *Biochem. Pharmacol.*, 1983, **32**, 1507.
[273] S. A. Barker, *Prog. Clin. Biol. Res.*, 1982, **90**, 113.
[274] K. F. Faull, R. B. Holman, G. R. Elliott, and J. D. Barchas, *Prog. Clin. Biol. Res.*, 1982, **90**, 135.
[275] O. Beck, T. R. Bosin, B. Holmsted, and A. Lundman, *Prog. Clin. Biol. Res.*, 1982, **90**, 29.
[276] B. Greiner and H. Rommelspacher, *Naunyn-Schmiedberg's Arch. Pharmacol.*, 1984, **325**, 349.
[277] T. Curstedt, *Biochim. Biophys. Acta*, 1982, **713**, 589.
[278] T. Curstedt, *Biochim. Biophys. Acta*, 1982, **713**, 602.
[279] H. Schultz, 'Benzodiazepines', Springer-Verlag, Berlin, 1982.
[280] J. R. Joyce, T. S. Bal, R. E. Ardrey, H. M. Stevens, and A. C. Moffat, *Biomed. Mass Spectrom.*, 1984, **11**, 284.

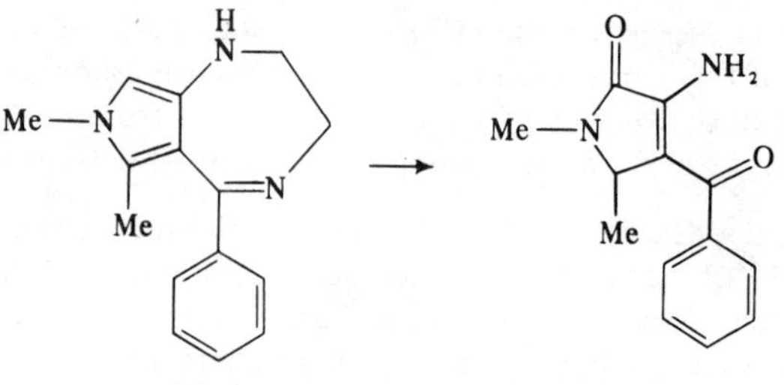

H
N
Me—N
=N
Me
(37)

→

O
NH$_2$
Me—N
O
Me
(38)

Scheme 11

ketazolam decomposes to diazepam, an oxygen radical is lost from N^4-oxides such as chlordiazepoxides, aromatic 7-nitro compounds such as nitrazepam are partially reduced to amines, α-hydroxyketones such as lorazepam lose water, and N-methyl-α-hydroxyketones lose H_2. Several benzodiazepines such as chlordiazepoxide, norchlordiazepoxide, and nordiazepam are unstable in stored blood samples,[281] and extraction methods for benzodiazepines from biological samples have been discussed with respect to the information required.[282]

Results from metabolic studies have shown that premazepam (37) undergoes ring cleavage to the ketone (38) (Scheme 11),[283] together with demethylation and hydroxylation,[284] loprazolam undergoes demethylation, N-oxidation, and reduction of the nitro group,[285] flurazepam is oxidatively deaminated to an aldehyde,[286] and imidazobenzodiazepine-3-carboxamide is extensively metabolized by hydroxylation and glucuronide conjugation.[287] Among quantitative GC/MS methods published during the review period are sensitive assays for diazepam in serum (detection limit 10 ng ml^{-1}),[288] imidazobenzodiazepine-3-carboxamide in plasma (0.1 ng ml^{-1} by negative-ion CI),[289] and midazolam and two metabolites in plasma (1.0 ng ml^{-1} by negative-ion CI).[290] A comparison of GLC and GC/MS methods for measurement of triazolam has shown good correlation over the range 0.2–20 ng ml^{-1}.[291] A pulse-labelled study with [^{15}N]-carbamazepine in four chronically treated epileptic patients has shown an increased plasma clearance indicating that the drug induces its own metabolism.[292]

281 B. Levine, R. V. Blanke, and J. C. Valentour, *J. Forensic Sci.*, 1983, **28**, 102.
282 J. A. F. de Silva, *J. Chromatogr.*, 1983, **273**, 19.
283 B. Vitiello, G. Buniva, A. Bernareggi, A. Assandri, A. Perazzi, L. M. Fuccella, and R. Palumbo, *Int. J. Clin. Pharmacol. Ther. Toxicol.*, 1984, **22**, 273.
284 A. Assandri, D. Barone, P. Ferrari, A. Perazzi, A. Ripamonti, G. Tuan, and L. F. Zerilli, *Drug Metab. Dispos.*, 1984, **12**, 257.
285 H. P. A. Illing, K. V. Watson, and J. Chamberlain, *Xenobiotica*, 1983, **13**, 531.
286 W. A. Garland, B. J. Miwa, W. Dairman, B. Kappell, M. C. C. Chiueh, M. Divoll, and D. J. Greenblatt, *Drug Metab. Dispos.*, 1983, **11**, 70.
287 S. J. Kolis, E. J. Postma, T. H. Williams, G. J. Sasso, and M. A. Schwartz, *Drug Metab. Dispos.*, 1983, **11**, 324.
288 C. Signorini, S. Tosoni, R. Ballerini, and A. Liguori, *Drugs Exp. Clin. Res.*, 1982, **8**, 185.
289 F. Rubio, B. J. Miwa, and W. A. Garland, *J. Chromatogr.*, 1982, **233**, 167.
290 F. Rubio, B. J. Miwa, and W. A. Garland, *J. Chromatogr.*, 1982, **233**, 157.
291 Ph. Coassolo, C. Aubert, and J. P. Cano, *J. Chromatogr.*, 1983, **274**, 161.
292 M. Eichelbaum, K. W. Kothe, F. Hoffmann, and G. E. von Unruh, *Eur. J. Clin. Pharmacol.*, 1982, **23**, 241.

The $[M - 1]^-$ ion in the negative-ion CI spectra of diazepam and nordiazepam is caused by traces of oxygen in the ion source and has been shown by deuterium labelling to be produced by loss of the amide hydrogen from nordiazepam and the hydrogen adjacent to the amide carbonyl in the case of diazepam.[293]

Sympathomimetics. – GC/MS studies of *N*-monotrifluoroacyl and pentafluorobenzoyl derivatives of amphetamines have been reported,[294,295] and ephedrine (39) enantiomers have been resolved as oxazolidines after reaction with naphthaldehyde.[296] The presence of impurities in illicit amphetamine (40) related to the Leuckart synthesis has found forensic application in identifying the source of the drug.[297] Among metabolic studies utilizing both direct-insertion and GC/MS techniques are reports on the excretion of *p*-hydroxymethamphetamine (41) and its glucuronide as metabolites of methamphetamine (42) in rats,[298] identification of 5-hydroxy-2-methoxyamphetamine (43), 2-methoxyphenylacetone, and 5-hydroxy-2-methoxyphenylacetone as new metabolites of methoxyphenamine

R^1, R^2, R^3, R^4 (ring substituents); —CH(R^5)—C(R^6)(R^7)—N(R^8)(R^9)

(39) $R^1 = R^2 = R^3 = R^4 = R^7 = R^8 = H$, $R^5 = OH$, $R^6 = R^9 = Me$
(40) $R^1 = R^2 = R^3 = R^4 = R^5 = R^7 = R^8 = R^9 = H$, $R^6 = Me$
(41) $R^1 = R^3 = R^4 = R^5 = R^7 = R^8 = H$, $R^2 = OH$, $R^6 = R^9 = Me$
(42) $R^1 = R^2 = R^3 = R^4 = R^5 = R^7 = R^8 = H$, $R^6 = R^9 = Me$
(43) $R^1 = OH$, $R^2 = R^3 = R^5 = R^7 = R^8 = R^9 = H$, $R^4 = OMe$, $R^6 = Me$
(44) $R^1 = R^2 = R^3 = R^5 = R^7 = R^8 = H$, $R^4 = OMe$, $R^6 = R^9 = Me$
(45) $R^1 = R^2 = R^3 = R^4 = R^5 = R^7 = H$, $R^6 = R^9 = Me$, $R^8 = CH_2Ph$
(46) $R^1 = R^2 = R^4 = R^6 = R^7 = R^8 = H$, $R^3 = R^5 = OH$, $R^9 = Me$
(47) $R^1 = CF_3$, $R^2 = R^3 = R^4 = R^5 = R^7 = R^8 = H$, $R^6 = Me$, $R^9 = Et$
(48) $R^1 = R^2 = OH$, $R^3 = R^4 = R^5 = R^8 = R^9 = H$, $R^6 = Me$, $R^7 = COOH$
(49) $R^1 = R^2 = R^5 = OH$, $R^3 = R^4 = R^6 = R^7 = R^8 = H$, $R^9 = Me$
(50) $R^1 = Me$, $R^2 = R^5 = OH$, $R^3 = R^4 = R^6 = R^7 = R^8 = H$, $R^9 = Me$
(51) $R^1 = R^2 = R^3 = R^4 = R^5 = R^8 = R^9 = H$, $R^6 = R^7 = Me$
(52) $R^1 = R^2 = R^3 = R^6 = R^7 = R^8 = H$, $R^4 = Cl$, $R^5 = OH$, $R^9 = Bu^t$
(53) $R^1 = NHCHO$, $R^2 = R^5 = OH$, $R^3 = R^4 = R^6 = R^7 = R^8 = H$, $R^9 = CH(Me)—CH_2—C_6H_4OMe$
(54) $R^1 = R^3 = R^5 = OH$, $R^2 = R^4 = R^6 = R^7 = R^8 = H$, $R^9 = Pr^i$
(55) $R^1 = CONH_2$, $R^2 = R^5 = OH$, $R^3 = R^4 = R^6 = R^7 = R^8 = H$, $R^9 = CH(Me)—C_2H_4—Ph$
(56) $R^1 = CH_2OH$, $R^2 = R^5 = OH$, $R^3 = R^4 = R^6 = R^7 = R^8 = H$, $R^9 = Bu^t$
(57) $R^1 = R^3 = R^5 = OH$, $R^2 = R^4 = R^6 = R^7 = R^8 = H$, $R^9 = Bu^t$
(58) $R^1 = CH_2OH$, $R^2 = R^5 = OH$, $R^3 = R^4 = R^6 = R^7 = R^8 = H$, $R^9 = Bu^t$

[293] W. A. Garland and B. J. Miwa, *Biomed. Mass Spectrom.*, 1983, **10**, 126.
[294] T. A. Brettell, *J. Chromatogr.*, 1983, **257**, 45.
[295] F. T. Delbeke and M. Debackere, *J. Chromatogr.*, 1983, **273**, 141.
[296] I. W. Wainer, T. D. Doyle, Z. Hamidzadeh, and M. Aldridge, *J. Chromatogr.*, 1983, **261**, 123.
[297] M. Lambrechts and K. E. Rasmussen, *Bull. Narcotics*, 1984, **36** (1), 47.
[298] T. Sakai, T. Niwaguchi, and T. Murata, *Xenobiotica*, 1982, **12**, 233.

[2-methoxyamphetamine (44)] in man,[299] identification of a nitrone[300] and several hydroxy and *N*-dealkyl metabolites of benzphetamine [*N*-benzylamphetamine (45)] in rats,[301,302] characterization of conjugates and other metabolites of *o*-[303] and *m*-synephrine [hydroxyamphetamine (46)] in man,[304] and identification of 1-(3-trifluoromethylphenyl)-2-propanol and 1-(3-trifluoromethylphenyl)-1,2-propanediol as new metabolites of fenfluramine [*N*-ethyl-3-trifluoromethylamphetamine (47)] in man.[305] MethylDopa (48) has been shown to be rapidly produced by hydrolysis of its pro-drug, the pivaloyloxyethyl ester, in a further GC/MS study.[306] Deuterium labelling has been used in studies of 3-hydroxyamphetamine metabolism to identify metabolically produced adrenaline (49) and metanephrine (50) in the presence of the endogenous compounds.[307] Cytochrome P450 has been shown to catalyse *N*-oxidation of phentermine (51) in a reconstituted system.[308]

Drugs Acting at β-Adrenergic Receptors. – Propranolol (59) is extensively metabolized with profiles showing considerable species differences.[309] New phenol metabolites have been reported,[310,311] and 3-(4-hydroxy-1-naphthoxy)-propane-1,2-diol has been identified by GC/MS as its TMS derivative in dog, rat, and man.[312] An intermediate in oxidative deamination, 3-(1-naphthoxy)-2-hydroxypropionaldehyde, has been trapped as its methyloxime and identified by CI mass spectrometry.[313] 4-Hydroxypropranolol sulphate has been isolated by ion-pair extraction[314] and its structure confirmed by FAB mass spectrometry.[315] A second conjugate was identified as the sulphate conjugate of the glycol metabolite.[315] FAB spectra of the glucuronide conjugates of (*R*)- and (*S*)-propranolol have also been recorded.[316] Propranolol enantiomers have been

299 G. McKay, J. K. Cooper, E. M. Hawes, S. D. Roy, and K. K. Midha, *Xenobiotica*, 1983, **13**, 257.

300 E. H. Jeffery and G. J. Mannering, *Mol. Pharmacol.*, 1983, **23**, 748.

301 T. Niwaguchi, T. Inoue, and S. Suzuki, *Xenobiotica*, 1982, **12**, 617.

302 T. Inoue, S. Suzuki, and T. Niwaguchi, *Xenobiotica*, 1983, **13**, 241.

303 M. I. James, J. M. Midgley, and C. M. Williams, *J. Pharm. Pharmacol.*, 1983, **35**, 559.

304 K. E. Ibrahim, J. M. Midgley, J. R. Crowley, and C. M. Williams, *J. Pharm. Pharmacol.*, 1983, **35**, 144.

305 K. K. Midha, E. M. Hawes, J. K. Cooper, J. W. Hubbard, K. Bailey, and I. J. McGilveray, *Xenobiotica*, 1983, **13**, 31.

306 S. Vickers, C. A. H. Duncan, H. G. Ramjit, M. R. Dobrinska, C. T. Dollery, H. J. Gomez, H. L. Leidy, and W. C. Vincek, *Drug Metab. Dispos.*, 1984, **12**, 242.

307 J. R. Crowley, C. M. Williams, and M. J. Fregly, *J. Pharm. Pharmacol.*, 1983, **35**, 264.

308 J. D. Duncan and A. K. Cho, *Mol. Pharmacol.*, 1982, **22**, 235.

309 E. M. Bargar, U. K. Walle, S. A. Bai, and T. Walle, *Drug Metab. Dispos.*, 1983, **11**, 266.

310 K. D. Ballard, D. R. Knapp, J. E. Oatis, jun., and T. Walle, *J. Chromatogr.*, 1983, **277**, 333.

311 T. Walle, J. E. Oatis, jun., U. K. Walle, and D. R. Knapp, *Drug. Metab. Dispos.*, 1982, **10**, 122.

312 S. M. Gupte, M. J. Bartels, B. M. Kerr, S. Laganiere, B. M. Silber, and W. L. Nelson, *Res. Commun. Chem. Pathol. Pharmacol.*, 1983, **42**, 235.

313 C.-H. Chen and W. L. Nelson, *Drug Metab. Dispos.*, 1982, **10**, 277.

314 K. H. Wingstrand and T. Walle, *J. Chromatogr.*, 1984, **305**, 250.

315 T. Walle, U. K. Walle, D. R. Knapp, E. C. Conradi, and E. M. Bargar, *Drug Metab. Dispos.*, 1983, **11**, 344.

316 J. E. Oatis, jun., J. P. Baker, J. R. McCarthy, and D. R. Knapp, *J. Med. Chem.*, 1983, **26**, 1687.

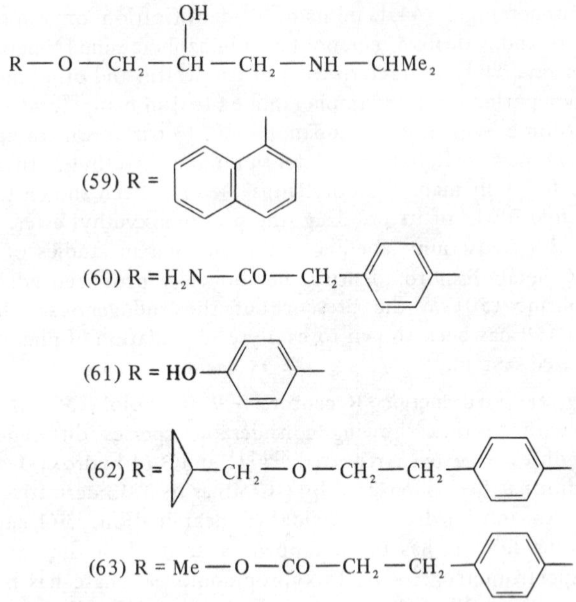

$$R-O-CH_2-CH(OH)-CH_2-NH-CHMe_2$$

(59) R = 1-naphthyl

(60) R = $H_2N-CO-CH_2-C_6H_4-$

(61) R = $HO-C_6H_4-$

(62) R = cyclopropyl$-CH_2-O-CH_2-CH_2-C_6H_4-$

(63) R = $Me-O-CO-CH_2-CH_2-C_6H_4-$

resolved as cyclic 2-oxazolidone derivatives after reaction with phosgene[317] and as adducts with *O*-[(−)-menthyl]-*N*,*N'*-di-isopropylisourea.[318] Studies with pseudo-racemates of propranolol have shown stereoselectivity in aromatic hydroxylation,[319] *N*-dealkylation,[320] glucuronide formation,[321] plasma clearance,[322] bioavailability,[323] and binding to plasma proteins.[324,325] Phenobarbitone has been shown to decrease bioavailability of propranolol,[326] and exercise has been shown to raise the concentration of the drug in plasma.[327] Cimetidine, which decreases hepatic blood flow and inhibits cytochrome P450, increases the peak plasma concentration of propranolol but does not affect the excretion of hydrophilic beta-blockers such as nadolol, which are excreted by the kidney.[328]

317 I. W. Wainer, T. D. Doyle, K. H. Donn, and J. R. Powell, *J. Chromatogr.*, 1984, **306**, 405.
318 K. D. Ballard, T. D. Eller, and D. R. Knapp, *J. Chromatogr.*, 1983, **275**, 161.
319 W. L. Nelson and M. J. Bartels, *Drug Metab. Dispos.*, 1984, **12**, 382.
320 W. L. Nelson and M. J. Bartels, *Drug Metab. Dispos.*, 1984, **12**, 345.
321 T. Walle, M. J. Wilson, K. Walle, and S. A. Bai, *Drug Metab. Dispos.*, 1983, **11**, 544.
322 L. S. Olanoff, T. Walle, U. K. Walle, T. D. Cowart, and T. E. Gaffney, *Clin. Pharmacol. Ther.*, 1984, **35**, 755.
323 S. A. Bai, M. J. Wilson, U. K. Walle, and T. Walle, *J. Pharmacol. Exp. Ther.*, 1983, **227**, 360.
324 U. K. Walle, T. Walle, S. A. Bai, and L. S. Olanoff, *Clin. Pharmacol. Ther.*, 1983, **34**, 718.
325 S. A. Bai, U. K. Walle, M. J. Wilson, and T. Walle, *Drug Metab. Dispos.*, 1983, **11**, 394.
326 V. T. Vu, S. A. Bai, and F. P. Abramson, *J. Pharmacol. Exp. Ther.*, 1983, **224**, 55.
327 G. A. Hurwitz, J. G. Webb, T. Walle, S. A. Bai, H. B. Daniell, L. Gourley, C. B. Loadholt, and T. E. Gaffney, *Br. J. Clin. Pharmacol.*, 1983, **16**, 599.
328 K. L. Duchin, M. A. Stern, D. A. Willard, and D. N. McKinstry, *Br. J. Clin. Pharmacol.*, 1984, **17**, 486.

An on-column trimethylsilylation method has been described for propranolol and other beta-blockers, but the reproducibility was not as good as with pre-column techniques.[329]

Metabolic studies with other β-adrenergic drugs have shown that atenolol (60) is oxidized to a substituted glyoxylic acid,[330] the β-agonist trimetoquinol cyclizes to a substituted tetrahydroprotoberberine,[331] tulobuterol (52) is metabolized by aromatic hydroxylation,[332] and prenalterol (61) forms β-(4-hydroxyphenoxy)-lactic acid.[333] The bis(chloromethyldimethylsilyl) ether of this compound forms a cyclic derivative involving the amino group by HCl elimination.[333] FAB mass spectrometry has been used to identify a glucuronide conjugate of formoterol (53)[334] and a sulphate of metaproterenol (54).[335] Labetalol (55) diastereoisomers have been separated as bis-cyclic butaneboronates,[336] and the presence of a small peak in HPLC chromatograms of nadolol having the same mass spectrum as the drug has been attributed to a product formed by intermolecular bonding.[337]

Quantitative GC/MS assays have been reported for tulobuterol (52),[338] betaxolol (62),[339] albuterol (56),[340] and for both terbutaline (57) and salbutamol (58) in a simultaneous assay.[341] The assay for tulobuterol (52) was by EI, whereas the other three employed positive-ion CI. Pharmacokinetic studies involving measurement by GC/MS have been reported for esmolol (63),[342] metipranolol,[343] metoprolol,[344] prenalterol,[345] and bufuralol.[346]

Salicylates. – Aspirin is metabolized by man to salicyluric acid (56–68%), salicylic and gentisic acids, and both acyl and phenolic glucuronides of salicylic acid.[347] Di- and tri-hydroxybenzoic or gentisuric acids, however, do not appear

[329] A. S. Christophersen and K. E. Rasmussen, *J. Chromatogr.*, 1982, **246**, 57.
[330] Y. Matsuki, T. Ito, S. Komatsu, and T. Nambara, *Chem. Pharm. Bull.*, 1982, **30**, 196.
[331] D. A. Williams, T. J. Maher, and D. C. Zaveri, *Biochem. Pharmacol.*, 1983, **32**, 1447.
[332] K. Matsumura, O. Kubo, T. Sakashita, H. Kato, K. Watanabe, and M. Hirobe, *Drug Metab. Dispos.*, 1982, **10**, 537.
[333] K.-J. Hoffmann, A. Arfwidsson, and K. O. Borg, *Drug Metab. Dispos.*, 1982, **10**, 173.
[334] H. Sasaki, H. Kamimura, Y. Shiobara, Y. Esumi, M. Takaichi, and T. Yokoshima, *Xenobiotica*, 1982, **12**, 803.
[335] T. R. Macgregor, L. Nastasi, P. R. Farina, and J. J. Keirns, *Drug Metab. Dispos.*, 1983, **11**, 568.
[336] T. Goromaru, Y. Matsuki, H. Matsuura, and S. Baba, *Yakugaku Zasshi*, 1983, **103**, 974.
[337] J. Kirschbaum, S. Perlman, and R. B. Poet, *J. Chromatogr. Sci.*, 1982, **20**, 336.
[338] K. Matsumura, O. Kubo, T. Sakashita, H. Kato, K. Watanabe, and M. Hirobe, *J. Pharm. Sci.*, 1983, **72**, 570.
[339] Ph. Hermann, J. Fraisse, J. Allen, P. L. Morselli, and J. P. Thenot, *Biomed. Mass Spectrom.*, 1984, **11**, 29.
[340] M. Weisberger, J. E. Patrick, and M. L. Powell, *Biomed. Mass Spectrom.*, 1983, **10**, 556.
[341] C. Lindberg and S. Jonsson, *Biomed. Mass Spectrom.*, 1982, **9**, 493.
[342] C. Y. Sum, A. Yacobi, R. Kartzinel, H. Stampfli, C. S. Davis, and C.-M. Lai, *Clin. Pharmacol. Ther.*, 1983, **34**, 427.
[343] C. Planz, H. Haass, and B. Lamberts, *Int. J. Clin. Pharmacol. Ther. Toxicol.*, 1982, **20**, 469.
[344] C. G. Regardh, S. Landahl, M. Larsson, P. Lundborg, B. Steen, K.-J. Hoffmann, and P.-O. Lagerstrom, *Eur. J. Clin. Pharmacol.*, 1983, **24**, 221.
[345] A. P. M. Greefhorst and C. L. A. Van Herwaarden, *Eur. J. Clin. Pharmacol.*, 1983, **24**, 173.
[346] R. J. Francis, P. B. East, and J. Larman, *Eur. J. Clin. Pharmacol.*, 1982, **23**, 529.
[347] A. J. Hutt, J. Caldwell, and R. L. Smith, *Xenobiotica*, 1982, **12**, 601.

to be formed.[347] The Schiff base salicylidene benzylamine (a pro-drug for salicylic acid) has been shown by GC/MS to be metabolized to salicylic acid and salicyl alcohol in rats and dogs,[348] and the anti-inflammatory drug eterylate [2-(4-acetamidophenoxy)ethyl-2-acetoxybenzoate] is hydrolysed to salicylic acid and paracetamol.[349]

Paracetamol and Related Drugs. – GC/MS properties of *N*-acetyl-*p*-benzoquinone-imine, the putative toxic metabolite of paracetamol, have been reported.[350,351] Some decomposition to paracetamol occurred on the GLC column, but a molecular-ion peak was seen at an electron-beam energy of 25 eV. Reaction with H_2O also occurred in the ion source to give $[M + 2]^+$ ions under EI conditions and $[M + 3]^+$ ions under ammonia CI conditions.[351] Alcohol has been reported to reduce the hepatotoxic effects of paracetamol; a reduction in the concentration of the mercapturic acid conjugate of the drug, as shown by CI mass spectrometry, following prior treatment with alcohol probably reflects reduced production of the quinoneimine.[352] 3-Thiomethyl adducts of paracetamol have been identified by direct-insertion mass spectrometry in rat and mouse urine,[353] and *N*-hydroxy-phenacetin *O*-glucuronide is produced by hamster-liver enzymes.[354] Metabolic deacetylation and reacetylation of paracetamol and phenacetin have been demonstrated in man by administration of trideuterioacetyl analogues of the drugs and examination of the deuterium content of the metabolites by GC/MS.[355]

Diethylstilbestrol and Structurally Related Drugs. – Diethylstilbestrol quinone (65) (Scheme 12), the postulated toxic metabolite of diethylstilbestrol (DES) (64), has been identified by GC/MS.[356] It is unstable in protic solvents, decomposing to *Z,Z*-dienestrol (66). Its formation appears to involve peroxidases.[357] The tetrafluoro analogue of dienestrol has been identified by GC/MS as a metabolite of *E*-3′,3′,5′,5′-tetrafluorodiethylstilbestrol, again suggesting formation of a quinone intermediate as this would not, in this case, be blocked by the presence of fluorine.[358] GC/MS-based screening methods for DES in meat have been published,[359,360] and the FAB spectra of the sulphate and glucuronide

[348] D. Kirkpatrick, D. R. Hawkins, L. F. Chasseaud, M. R. Al-Ani, and A. F. Al-Sayyab, *Xenobiotica*, 1983, **13**, 53.
[349] S. G. Wood, B. A. John, L. F. Chasseaud, I. Johnstone, S. R. Biggs, D. R. Hawkins, J. G. Priego, A. Darragh, and R. F. Lambe, *Xenobiotica*, 1983, **13**, 731.
[350] A. Huggett and I. A. Blair, *J. Chromatogr. Sci.*, 1983, **21**, 254.
[351] D. C. Dahlin and S. D. Nelson, *J. Med. Chem.*, 1982, **25**, 885.
[352] P. W. Banda and B. D. Quart, *Res. Commun. Chem. Pathol. Pharmacol.*, 1982, **38**, 57.
[353] S. J. Hart, K. Healey, M. C. Smail, and I. C. Calder, *Xenobiotica*, 1982, **12**, 381.
[354] A.-M. Camus, M. Friesen, A. Croisy, and H. Bartsch, *Cancer Res.*, 1982, **42**, 3201.
[355] J. D. Baty, R. G. Willis, and Y. K. Koh, *Anal. Proc.*, 1984, **21**, 14.
[356] J. G. Liehr, B. B. DaGue, A. M. Ballatore, and J. Hankin, *Biochem. Pharmacol.*, 1983, **32**, 3711.
[357] G. H. Degen, A. Wong, T. E. Eling, J. C. Barrett, and J. A. McLachlan, *Cancer Res.*, 1983, **43**, 992.
[358] J. G. Liehr, A. M. Ballatore, J. A. McLachlan, and D. A. Sirbasku, *Cancer Res.*, 1983, **43**, 2678.
[359] H. W. Durbeck and I. Bucker, *Fresenius' Z. Anal. Chem.*, 1983, **315**, 479.
[360] H. J. G. M. Derks, J. Freudenthal, J. L. M. Litjens, R. Klaassen, L. G. Gramberg, and V. Borrias-Van Tongeren, *Biomed. Mass Spectrom.*, 1983, **10**, 209.

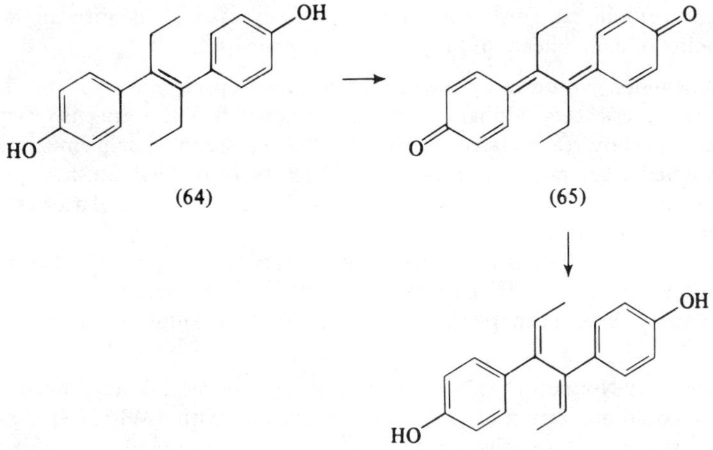

HO OH (64) → O O (65) → HO OH (66)

Scheme 12

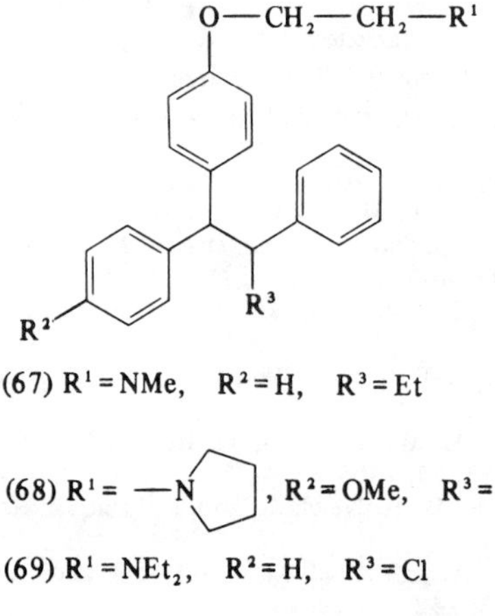

O—CH_2—CH_2—R^1 R^2 R^3

(67) R^1 = NMe, R^2 = H, R^3 = Et

(68) R^1 = —N(pyrrolidinyl), R^2 = OMe, R^3 = NO_2

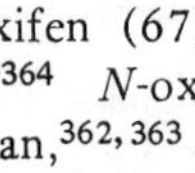

(69) R^1 = NEt_2, R^2 = H, R^3 = Cl

conjugates have been shown to contain abundant ions corresponding to $[M + Na]^+$ and $[M + K]^+$.[361]

GC/MS studies of tamoxifen (67) metabolism have shown aliphatic and aromatic hydroxylation,[362–364] *N*-oxidation,[364] *N*-demethylation,[362,364] and oxidative deamination in man,[362,363] direct-insertion studies have led to the identification of benzophenone and diarylacetophenone metabolites of nitro-

[361] J. G. Liehr, C. F. Beckner, A. M. Ballatore, and R. M. Caprioli, *Steroids,* 1982, **39**, 599.

[362] R. R. Bain and V. C. Jordan, *Biochem. Pharmacol.*, 1983, **32**, 373.

[363] J. V. Kemp, H. K. Adam, A. E. Wakeling, and R. Slater, *Biochem. Pharmacol.*, 1983, **32**, 2045.

[364] D. J. Bates, A. B. Foster, L. J. Griggs, M. Jarman, G. Leclercq, and N. Devleeschouwer, *Biochem. Pharmacol.*, 1982, **31**, 2823.

miphene (68) in rat gut,[365] and FAB has been used to identify an *N*-oxide metabolite of clomiphene (69) in rat-liver microsomes.[366]

Sulphonamides. – Studies by direct-insertion mass spectrometry of metabolites of N^4-trideuterioacetylsulphamerazine have shown that the drug is deacylated and subsequently reacetylated in man.[367,368] Acetylation of sulphanilamide has been studied in several unusual species;[369] the Tasmanian devil, brushtail possum, and pademelon (all marsupials) were shown to be capable of acetylation reactions whereas the barred bandicoot and echidna were not. Acetylation of this drug has previously been observed in sixteen other species. Screening methods involving CID/MIKES techniques,[370] GC/MS,[371,372] FD,[373] and direct-insertion EI mass spectrometry have been published for detection of sulphonamides in swine tissue.[374]

Steroids. – 19-Normethyl-1,2- or -4,5-epoxy-3-oxo-steroids have been shown to undergo aromatization of the A ring by reaction with TMS reagents in GLC injector systems.[375] Negative-ion DCI spectra of digitoxin and related cardenolides using OH^- as the reactant ion have been published,[376] and negative-ion spectra of these compounds have also been recorded using LC/MS with a moving-belt interface.[377] 16β-Hydroxydigitoxin has been identified by EI as a metabolite of digitoxin.[378] FAB and FD spectra of the quaternary ammonium-substituted steroid ORG NC45 have been recorded.[379]

Capillary-column GC/MS methods have been published for the analysis of anabolic steroids[380,381] and corticosteroid metabolites[381,382] under both positive-

365 P. C. Ruenitz and J. R. Bagley, *Life Sci.*, 1983, **33**, 1051.
366 P. C. Ruenitz, J. R. Bagley, and C. M. Mokler, *Biochem. Pharmacol.*, 1983, **32**, 2941.
367 T. B. Vree, C. A. Hekster, M. Baakman, T. Janssen, M. Oosterbaan, E. Termond, and M. Tijhuis, *Biopharm. Drug Dispos.*, 1983, **4**, 271.
368 T. B. Vree, M. W. Tijhuis, M. Baakman, and C. A. Hekster, *Biomed. Mass Spectrom.*, 1983, **10**, 114.
369 S. McLean, H. Galloway, S. Butler, D. Whittle, and S. C. Nicol, *Xenobiotica*, 1983, **13**, 81.
370 W. C. Brumley, Z. Min, J. E. Matusik, J. A. G. Roach, C. J. Barnes, J. A. Sphon, and T. Fazio, *Anal. Chem.*, 1983, **55**, 1405.
371 J. E. Matusik, C. J. Barnes, D. R. Newkirk, and T. Fazio, *J. Assoc. Off. Anal. Chem.*, 1982, **65**, 828.
372 S. J. Stout, W. A. Steller, A. J. Manuel, M. O. Poeppel, and A. R. DaCunha, *J. Assoc. Off. Anal. Chem.*, 1984, **67**, 142.
373 D. D. Giera, R. F. Abdulla, J. L. Occolowitz, D. E. Dorman, J. L. Mertz, and R. F. Sieck, *J. Agric. Food Chem.*, 1982, **30**, 260.
374 C. Friis, N. Gyrd-Hansen, P. Nielsen, C.-E. Olsen, and F. Rasmussen, *Acta Pharmacol. Toxicol.*, 1984, **54**, 321.
375 E. Schwarz, S. Abdel-Baky, P. W. Lequesne, and P. Vouros, *Int. J. Mass Spectrom. Ion Phys.*, 1983, **47**, 511.
376 A. P. Bruins, *Int. J. Mass Spectrom. Ion Phys.*, 1983, **48**, 185.
377 K. Levsen, K. H. Schafer, and P. Dobberstein, *Biomed. Mass Spectrom.*, 1984, **11**, 308.
378 L. N. Kadima, G. Lhoest, and M. Lesne, *Eur. J. Drug Metab. Pharmacokinet.*, 1982, **7**, 111.
379 W. D. Lehmann and F. M. Kaspersen, *J. Labelled Comp. Radiopharm.*, 1984, **21**, 455.
380 G. P. Cartoni, M. Ciardi, A. Giarrusso, and F. Rosati, *J. Chromatogr.*, 1983, **279**, 515.
381 E. Houghton, P. Teale, and M. C. Dumasia, *Analyst (London)*, 1984, **109**, 273.
382 E. Houghton, P. Teale, M. C. Dumasia, and J. K. Wellby, *Biomed. Mass Spectrom.*, 1982, **9**, 459.

ion EI and negative-ion CI conditions. A complication for analysis of certain anabolic steroids in horses is the identification of natural C_{18} neutral steroids (19-desmethylandrostanes) in stallion urine.[383] Mono- and di-hydroxy human-urinary metabolites of the anabolic steroid 5-chloromethandienone (4-chloro-17α-methyl-17β-hydroxy-1,4-androstadien-3-one) have been characterized as TMS derivatives by GC/MS,[384] and sulphate conjugates of 1-dehydrotestosterone have been identified using positive- and negative-ion FAB.[385] Assays have been described for the anabolic steroids ethylestrenol,[386] 19-norandrosterone,[387] 3α-hydroxy-1-methylen-5α-androstan-17-one, a metabolite of methenolone acetate,[388] and 17β-hydroxy-1α-methyl-17α-propyl-5α-androstan-3-one.[389] The latter steroid was measured in plasma to the 1 ng ml^{-1} level by single-ion monitoring (*m/z* 303) using a homologue standard. The antiandrogenic steroid oxendolone (16-ethyl-17-hydroxy-19-norandrost-5-en-3-one) is metabolized in man by reduction of the double bond and ketone group and by oxidation of the hydroxy group.[390]

Among studies with oestrogenic steroids are reports on the characterization of hydroxy metabolites of 17α-ethynylestradiol in the rat[391] and of moxestrol [11β-methoxy-19-nor-17α-pregna-1,3,5(10)-triene-20-yne-3,17β-diol] in humans,[392] the measurement of pharmacokinetic parameters of moxestrol[393] and megestrol acetate,[394] and the development of a stable-isotope assay for equiline in humans, sensitive to 0.0015–0.05 ng ml^{-1} in serum.[395] The hydrazone formation between norethindrone and isoniazid, previously reported in the rat, has now been found in the minipig,[396] and cytochrome P450-dependent oxidation of the 17α-ethynyl group in certain estrogenic steroids has been shown to lead to d-homoannulation of the D ring with incorporation of one of the ethynyl carbon atoms (Scheme 13).[392,397]

[383] E. Houghton, J. Copsey, M. C. Dumasia, P. E. Haywood, M. S. Moss, and P. Teale, *Biomed. Mass Spectrom.*, 1984, **11**, 96.
[384] H. W. Durbeck, I. Buker, B. Scheulen, and B. Telin, *J. Chromatogr. Sci.*, 1983, **21**, 405.
[385] M. C. Dumasia, E. Houghton, C. V. Bradley, and D. H. Williams, *Biomed. Mass Spectrom.*, 1983, **10**, 434.
[386] I. Bjorkhem, H. Ek, and O. Lantto, *J. Chromatogr.*, 1982, **232**, 154.
[387] I. Bjorkhem and H. Ek, *J. Steroid Biochem.*, 1982, **17**, 447.
[388] I. Bjorkhem and H. Ek, *J. Steroid Biochem.*, 1983, **18**, 481.
[389] W. Krause and U. Jakobs, *J. Pharm. Sci.*, 1984, **73**, 563.
[390] I. Midgley, A. G. Fowkes, A. Darragh, R. Lambe, L. F. Chasseaud, and T. Taylor, *Steroids*, 1983, **41**, 521.
[391] J. L. Maggs, P. S. Grabowski, M. E. Rose, and B. K. Park, *Xenobiotica*, 1982, **12**, 657.
[392] J. Salmon, D. Coussediere, C. Cousty, and J. P. Raynaud, *J. Steroid Biochem.*, 1983, **18**, 565.
[393] J. Salmon, D. Coussediere, C. Cousty, and J. P. Raynaud, *J. Steroid Biochem.*, 1983, **19**, 1223.
[394] H. Adlercreutz, P. B. Eriksen, and M. S. Christensen, *J. Pharm. Biomed. Anal.*, 1983, **1**, 153.
[395] L. Siekmann, A. Siekmann, H. Breuer, and L. Dehennin, *Biomed. Mass Spectrom.*, 1983, **10**, 168.
[396] J. C. K. Loo, N. Jordan, J. A. Menzies, and H. Watanabe, *Biopharm. Drug Dispos.*, 1983, **4**, 145.
[397] S. E. Schmid, W. Y. A. Au, D. E. Hill, F. F. Kudlubar, and W. Slikker, jun., *Drug Metab. Dispos.*, 1983, **11**, 531.

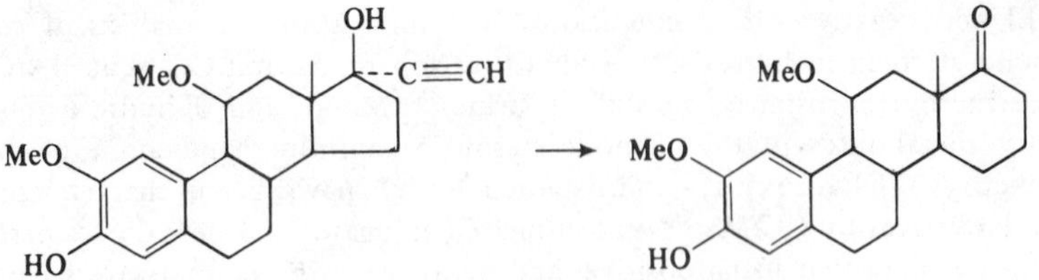

Scheme 13

The corticosteroids dexamethasone and betamethasone undergo partial thermal decomposition during LC/MS analysis with a belt interface.[398] 6β-Hydroxybudesonide is metabolized to 6β-hydroxybudesonide and 16α-prednisolone in man,[399] and the related anti-inflammatory steroid deflazacort undergoes both hydroxylation and deacylation in cynomolgus monkeys.[400] A stable-isotope dilution assay for dexamethasone has enabled the steroid to be measured for 10 hours in human plasma.[401]

Anti-inflammatory Drugs. – This large group of drugs has recently received considerable attention with respect to side-effects, some of which have proved fatal and have resulted in withdrawal of the drugs. Possible mechanisms leading to adverse reactions include the reported incorporation of some of these drugs into pathways of lipid metabolism, and some interesting applications of mass spectrometry are to be expected in this area in the future. Reports on metabolic transformations appearing during the review period include the identification by FD of the acid glucuronide of benoxaprofen (70) as a major human-urinary metabolite,[402] cyclopropane-ring hydroxylation and keto reduction in loxoprofen {2-[4-(2-oxocyclopentylmethyl)phenyl]propionate},[403] reductions, *N*-demethylation, *N*-oxidation, and *N*-glucuronide formation in ketotifen (71),[404,405] and β-oxidation, hydroxylation, reduction, and *O*-demethylation of nabumetone [2-(3-oxobutyl)-6-methoxynaphthalene].[406] Piroprofen {2-[3-chloro-4-(3-pyrrolin-1-yl)phenyl]propionic acid} is metabolized in several species by the epoxide-diol pathway and oxidation to the pyrrole analogue.[407] However, the latter reaction

[398] T. Cairns, E. G. Siegmund, J. J. Stamp, and J. P. Skelly, *Biomed. Mass Spectrom.*, 1983, **10**, 203.
[399] S. Edsbacker, S. Jonsson, C. Lindberg, A. Ryrfeldt, and A. Thalen, *Drug Metab. Dispos.*, 1983, **11**, 590.
[400] A. Assandri, P. Ferrari, A. Perazzi, A. Ripamonti, G. Tuan, and L. Zerilli, *Xenobiotica*, 1983, **13**, 181.
[401] Y. Kasuya, J. R. Althaus, J. P. Freeman, R. K. Mitchum, and J. P. Skelly, *J. Pharm. Sci.*, 1984, **73**, 446.
[402] J. Okamoto, K. Fujimoto, H. Fujitomo, and E. Hirai, *Yakugaku Zasshi*, 1983, **103**, 54.
[403] S. Naruto, Y. Tanaka, R. Hayashi, and A. Terada, *Chem. Pharm. Bull.*, 1984, **32**, 258.
[404] J. M. Begue, J. F. Le Bigot, C. Guguen-Guillouzo, J. R. Kiechel, and A. Guillouzo, *Biochem. Pharmacol.*, 1983, **32**, 1643.
[405] J. F. Le Bigot, T. Cresteil, J. R. Kiechel, and P. Beaune, *Drug Metab. Dispos.*, 1983, **11**, 585.
[406] R. E. Haddock, D. J. Jeffery, J. A. Lloyd, and A. R. Thawley, *Xenobiotica*, 1984, **14**, 327.
[407] H. Egger, F. Bartlett, H.-P. Yuan, and J. Karliner, *Drug Metab. Dispos.*, 1982, **10**, 529.

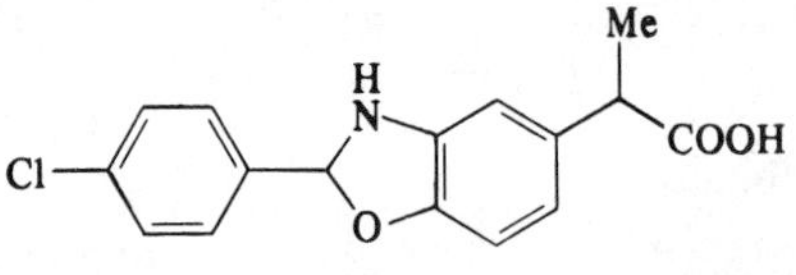

(70)

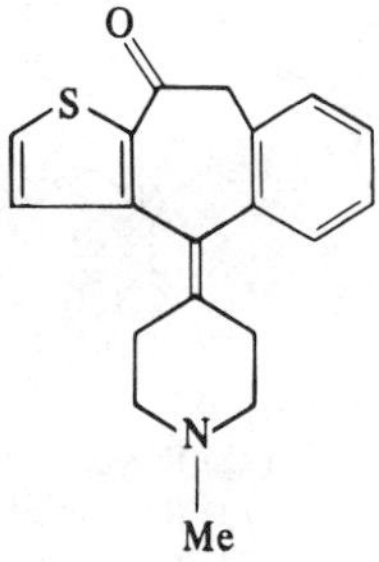

(71)

also appears to occur during work-up.[408] Tiaramide (4-{[5-chloro-2-oxo-3(2*H*)-benzothiazolyl]acetyl}-1-piperazine-ethanol) undergoes a variety of metabolic reactions including terminal *N*-sulphoconjugation following *N*-dealkylation as shown by FD spectra.[409] Extensive metabolism has also been reported for tolfenamic acid (2-carboxyl-3′-chloro-2′-methyldiphenylamine),[410] benzofenac (3-chloro-4-benzyloxyphenylacetic acid),[411] and 2-oxo-3-[4-(1-oxo-2-isoindol)-phenyl]butanamide[412] using GC/MS techniques. Direct-insertion methods have revealed mercapturic acid conjugates and other sulphur-containing metabolites of 2-acetamido-4-(chloromethyl)thiazole,[413] *N*-oxidation,[414] and glucuronide and sulphate[415] formation in tiaramide.

Diclofenac has been measured to 0.2 ng ml^{-1} in human plasma by GC/MS with an analogue standard in a method designed to measure blood levels following cutaneous application of a 1% cream. The drug was measured as its cyclic indolinone derivative.[416] A sensitive EI method for ketotifen and three metabolites in plasma has enabled the drug to be measured to 0.05 ng ml^{-1} and the hydroxy metabolite to 0.3 ng ml^{-1}.[417] GC/MS-based detection methods have also been developed for carprofen and zomepirac.[418]

Tricyclic Antidepressants. – Positive-ion FAB spectra have been used to characterize *N*-glucuronides of imipramine (72), amitriptyline (76), and chlorpromazine[419]

[408] P. H. Degen, A. Schweizer, and A. Sioufi, *J. Chromatogr.*, 1984, **290**, 33.
[409] L. J. Klunk, P. S. Riska, and D. E. Maynard, *Drug Metab. Dispos.*, 1982, **10**, 241.
[410] R. G. Khalifah, C. E. Hignite, P. J. Pentikainen, A. Penttila, and P. J. Neuvonen, *Eur. J. Drug Metab. Pharmacokinet.*, 1982, **7**, 269.
[411] I. Koruna, M. Ryska, M. Kuchar, J. Grimova, Z. Roubal, and O. Nemecek, *Biomed. Mass Spectrom.*, 1984, **11**, 121.
[412] K. Kigasawa, M. Hiiragi, K. Maruyama, M. Tanaka, Y. Ohsawa, and T. Hayashi, *Yakugaku Zasshi*, 1982, **102**, 866.
[413] J. J. Rafter and J. E. Bakke, *Drug Metab. Dispos.*, 1982, **10**, 654.
[414] H. Noguchi, K. Tada, and K. Iwasaki, *Xenobiotica*, 1982, **12**, 211.
[415] K. Iwasaki, T. Shiraga, K. Noda, K. Tada, and H. Noguchi, *Xenobiotica*, 1983, **13**, 273.
[416] H. Kadowaki, M. Shiino, I. Uemura, and K. Kobayashi, *J. Chromatogr.*, 1984, **308**, 329.
[417] C. Julien-Larose, M. Guerret, D. Lavene, and J. R. Kiechel, *Biomed. Mass Spectrom.*, 1983, **10**, 136.
[418] M. Dettwiler, S. Rippstein, and A. Jeger, *J. Chromatogr.*, 1982, **244**, 153.
[419] J. P. Lehman, C. Fenselau, and J. R. Depaulo, *Drug Metab. Dispos.*, 1983, **11**, 221.

(72) $R^1 = Me, R^2 = H$
(73) $R^1 = R^2 = H$
(74) $R^2 = H, R^2 = Cl$
(75) $R^1 = OH, R^2 = H$

(76) $R^1 = Me, R^2 = H$
(77) $R^1 = R^2 = H$
(78) $R^1 = Me, R^2 = H$
(79) $R^1 = H, R^2 = OH$

and the glucuronide of hydroxydesmethylimipramine.[419] Both GC/MS and direct-insertion methods have been used to show that desipramine (73), imipramine (72), desmethylchlorimipramine (74), and nortriptyline (77) are oxidatively deaminated in the rabbit to carboxylic acids,[420–422] desmethylimipramine is additionally metabolized to *N*-hydroxydesmethylimipramine (75),[421] and desmethylchlorimipramine undergoes hydroxylation and demethylation.[422] Similar hydroxy metabolites have been identified for oxaprotiline in man,[423] and FD has been used to characterize their *O*-glucuronides.[423]

A selected-ion monitoring method using analogue standards has been published for amitriptyline (76) and its metabolites nortriptyline (77), 10-hydroxyamitriptyline (78), and 10-hydroxynortriptyline (79); detection limits for the metabolites were 2, 2, and 5 ng ml^{-1}, respectively.[424] A similar EI assay for doxepin and desmethyldoxepin gave a detection limit better than 1 ng ml^{-1}.[425] GC/MS studies have shown good correlation between nortriptyline and debrisoquine hydroxylation in human liver,[426] and they have been used to determine *Z*-10-hydroxynortriptyline in the same tissue.[427] Pharmacokinetic studies have been reported for carbamazepine-10,11-epoxide,[428,429] imipramine,[430] and desipramine.[431] Clinical studies using GC/MS quantification have shown decreased levels of amitriptyline and its metabolites but increased levels of conjugates

[420] A. H. Beckett and A. J. Hutt, *J. Pharm. Pharmacol.*, 1982, **34**, 614.
[421] A. H. Beckett, A. J. Hutt, and G. E. Navas, *Xenobiotica*, 1983, **13**, 391.
[422] A. J. Hutt, G. E. Navas, and A. H. Beckett, *Xenobiotica*, 1982, **12**, 645.
[423] W. Dieterle, J. W. Faigle, H.-P. Kriemler, and T. Winkler, *Xenobiotica*, 1984, **14**, 311.
[424] R. Ishida, T. Ozaki, H. Uchida, and T. Irikura, *J. Chromatogr.*, 1984, **305**, 73.
[425] T. P. Davis, S. K. Veggeberg, S. R. Hameroff, and K. L. Watts, *J. Chromatogr.*, 1983, **273**, 436.
[426] C. von Bahr, C. Birgersson, A. Blanck, M. Goransson, B. Mellstrom, and K. Nilsell, *Life Sci.*, 1983, **33**, 631.
[427] B. Mellstrom, L. Bertilsson, C. Birgersson, M. Goransson, and C. von Bahr, *Drug Metab. Dispos.*, 1983, **11**, 115.
[428] T. Tomson, G. Tybring, and L. Bertilsson, *Clin. Pharmacol. Ther.*, 1983, **33**, 58.
[429] P. J. Wedlund, I. H. Patel, and R. H. Levy, *J. Pharmacokinet. Biopharm.*, 1982, **10**, 427.
[430] E. D. Peselow, S. I. Deutsch, and R. R. Fieve, *Res. Commun. Psychol. Psychiatr. Behav.*, 1983, **8**, 75.
[431] M. N. Musa, *Res. Commun. Psychol. Psychiatr. Behav.*, 1983, **8**, 61.

in patients with renal failure,[432] nortriptyline formation after intramuscular amitriptyline,[433] and lower levels of this metabolite in alcoholic patients than in normal controls.[434]

Phenothiazines. – Detailed mechanisms of phenothiazine fragmentation have been published,[435] piperazine-ring cleavage has been observed in EI spectra,[436] fragmentation of new 2-substituted 10-*N*-(aminoacyl)phenothiazines has shown different mechanisms for dechlorination and defluorination under EI conditions,[437] and the Cs^+ SIMS spectra of promazines have been shown to exhibit prominent $[M + H]^+$ ions.[438] Both *N*- and *S*-oxides of fluphenazine (80) give $M^{+\bullet}$ or pseudo-molecular ions in their FAB and DCI spectra.[439] Phenothiazine *S*-oxide metabolites decompose rapidly above 300 °C to the parent drug in GLC injector ports, but decomposition is only about 2% in the 250–270 °C range.[440] Desmethyl and *N*-oxide metabolites of chlorpromazine (81) have been found by GC/MS in guinea-pig liver and intestine;[441] the *N*-oxide metabolite was measured as its thermolytic-decomposition product, *N*-allyl-2-chlorophenothiazine (82). A selected-ion detection method for chlorpromazine, using prochlorperazine as the internal standard and sensitive to 0.25 ng ml^{-1}, has enabled single-dose kinetics

(80) $R^1 = (CH_2)_3$—N(piperazine)N—$(CH_2)_2$—OH, $R^2 = CF_3$

(81) $R^1 = (CH_2)_3$—NMe_2, $R^2 = Cl$

(82) $R^1 = CH_2$—CH=CH_2, $R^2 = Cl$

(83) $R^1 = CH_2$—CH(Me)—CH_2—NMe_2, $R^2 = OMe$

[432] M. Sandoz, S. Vandel, B. Vandel, B. Bonin, B. Hory, Y. St. Hillier, and R. Volmat, *Eur. J. Clin. Pharmacol.*, 1984, **26**, 227.
[433] B. Mellstrom, G. Alvan, L. Bertilsson, W. Z. Potter, J. Sawe, and F. Sjoqvist, *Clin. Pharmacol. Ther.*, 1982, **32**, 664.
[434] M. Sandoz, S. Vandel, B. Vandel, B. Bonin, G. Allers, and R. Volmat, *Eur. J. Clin. Pharmacol.*, 1983, **24**, 615.
[435] A. Hallberg, I. Al-Showaier, and A. R. Martin, *J. Heterocycl. Chem.*, 1984, **21**, 841.
[436] H. U. Shetty, E. M. Hawes, and K. K. Midha, *Biomed. Mass Spectrom.*, 1983, **10**, 601.
[437] S. Morosawa, S. Kamal, P. C. Dandiya, and H. L. Sharma, *Org. Mass Spectrom.*, 1982, **17**, 309.
[438] D. V. Dung, J. Marien, E. de Pauw, and J. Decuyper, *Org. Mass Spectrom.*, 1984, **19**, 276.
[439] P. Brooks and W. F. Heyes, *Biomed. Mass Spectrom.*, 1982, **9**, 522.
[440] K. Hall, P. K. F. Yeung, and K. K. Midha, *J. Chromatogr.*, 1982, **231**, 200.
[441] F. Hartmann, L. D. Gruenke, J. C. Craig, and D. M. Bissell, *Drug Metab. Dispos.*, 1983, **11**, 244.

of the drug to be measured.[442] A similar method for trifluorperazine, sensitive to 0.078 ng ml^{-1}, has been used to show bioequivalence of two tablet formulations[443] and to measure a half-life of 12.5 $\pm$ 1.4 hours in humans.[444] Fluphenazine has been measured to the same concentration using high-resolution (2000) monitoring of *m/z* 406.1563,[445] and the method has enabled the drug to be measured in human plasma for 32 hours after a single 5 mg dose.[446] A resolution of 5000 has produced an assay sensitive enough to detect 3 metabolites of levomepromazine (83) at the 0.1 ng ml^{-1} level in the blood of psychiatric patients.[447]

Antibiotics. – The newer ionization techniques have enabled a considerable number of studies to be carried out on this group of drugs. FAB spectra have been recorded for β-lactam antibiotics,[448,449] gentamycin,[450] azthreonam and its salts,[451] and novel triostins[452] and have been shown to give results superior to those obtained from techniques such as DCI and FD for anthracyclins.[453] However, in a comparison of positive- and negative-ion FAB, DCI, and ^{252}Cf ionization, the latter method was reported to give the best reproducibility.[454] Other papers reporting DCI and ^{252}Cf spectra have also appeared.[455,456]

In-beam EI ionization of penicillins yields $M^{+\bullet}$ or $[M + 1]^+$ ions even though no corresponding ions are seen on direct insertion.[457] However, the addition of ammonium salts to the probe has been reported to yield $M^{+\bullet}$ ions under otherwise normal conditions.[458] Under EI conditions, β-lactam antibiotics yield abundant fragment ions by lactam ring cleavage.[459] Collisional activation of these ions and examination of the fragments yield information on ring size, substituent type, and substituent position, but not on stereochemistry.[459]

442 G. McKay, K. Hall, J. K. Cooper, E. M. Hawes, and K. K. Midha, *J. Chromatogr.*, 1982, **232**, 275.
443 K. K. Midha, E. M. Hawes, E. D. Korchinski, J. W. Hubbard, G. McKay, J. K. Cooper, and R. M. H. Roscoe, *Biopharm. Drug Dispos.*, 1984, **5**, 25.
444 K. K. Midha, E. D. Korchinski, R. K. Verbeeck, R. M. H. Roscoe, E. M. Hawes, J. K. Cooper, and G. McKay, *Br. J. Clin. Pharmacol.*, 1983, **15**, 380.
445 G. McKay, K. Hall, R. Edom, E. M. Hawes, and K. K. Midha, *Biomed. Mass Spectrom.*, 1983, **10**, 550.
446 K. K. Midha, G. McKay, R. Edom, E. D. Korchinski, E. M. Hawes, and K. Hall, *Eur. J. Clin. Pharmacol.*, 1983, **25**, 709.
447 S. G. Dahl, H. Johnsen, and C. R. Lee, *Biomed. Mass Spectrom.*, 1982, **9**, 534.
448 J. L. Gower, *Int. J. Mass Spectrom. Ion Phys.*, 1983, **46**, 431.
449 J. L. Gower, G. D. Risbridger, and M. J. Redrup, *J. Antibiot.*, 1984, **37**, 33.
450 M. A. Baldwin, D. M. Carter, and K. J. Welham, *Org. Mass Spectrom.*, 1983, **18**, 176.
451 A. I. Cohen, P. T. Funke, and B. N. Green, *J. Pharm. Sci.*, 1982, **71**, 1065.
452 S. Santikarn, S. J. Hammond, D. H. Williams, A. Cornish, and M. J. Waring, *J. Antibiot.*, 1983, **36**, 362.
453 B. Gioia, E. Arlandini, and A. Vigevani, *Biomed. Mass Spectrom.*, 1984, **11**, 35.
454 L. David, S. Della Negra, D. Fraisse, G. Jeminet, I. Lorthiois, Y. Le Beyec, and J. C. Tabet, *Int. J. Mass Spectrom. Ion Phys.*, 1983, **46**, 391.
455 R. G. Smith, *Anal. Chem.*, 1982, **54**, 2006.
456 H. M. Fales, *Int. J. Mass Spectrom. Ion Phys.*, 1983, **53**, 59.
457 M. Ohashi, R. P. Barron, and W. R. Benson, *J. Pharm. Sci.*, 1983, **72**, 508.
458 A. K. Bose, B. N. Pramanik, and P. L. Bartner, *J. Org. Chem.*, 1982, **47**, 4008.
459 M. P. Barbalas, F. W. McLafferty, and J. L. Occolowitz, *Biomed. Mass Spectrom.*, 1983, **10**, 258.

Spectra of β-lactam antibiotics have also been recorded from an LC/MS system employing a microbore column.[460] Emitter CI spectra of aminoglycoside antibiotics have been shown to contain $[M + H]^+$ and diagnostic fragment ions.[461]

Metabolic studies on tetracycline antibiotics using a variety of methods for sample introduction have shown hydroxylation and demethylation of minocycline[462] and doxycycline in several species,[463] the formation of eleven metabolites including the previously unreported bisanhydroaklavinic acid and its glucuronide from aclacinomycin in man,[464] and the formation of four metabolites including two polar conjugates of marcallomycin in the mouse.[465] Nine metabolites of cyclosporin have been shown by FD to retain the intact oligopeptide ring,[466] the macrolide antibiotic rosaramicin has been shown by FAB techniques to give 20-bisureidorosaramicin in man,[467] and deacetylation has been detected for 9,3″-diacetylmidecamycin in man.[468]

Nitroglycerine. – Sensitive analytical methods are required to measure the low levels of nitroglycerine encountered in humans. The use of ^{15}N-containing standards under EI conditions has enabled 0.05 ng ml^{-1} to be measured in human plasma,[469] and the method has been used for pharmacokinetic studies of a new transdermal preparation[470] and to show wide interindividual variation in blood levels of the drug after intravenous infusion.[471] Increased sensitivity to the femtogram ml^{-1} level is provided by negative-ion CI,[472,473] with the latter method[473] enabling a drug level of 0.065 ng ml^{-1} to be measured at 48 hours in bioavailability studies.[474]

460 M. A. McDowall, D. E. Games, and J. L. Gower, *Int. J. Mass Spectrom. Ion Phys.*, 1983, **48**, 157.

461 N. Takeda, K. Harada, M. Suzuki, A. Tatematsu, and T. Kubodera, *Org. Mass Spectrom.*, 1982, **17**, 247.

462 H. J. C. F. Nelis and A. P. De Leenheer, *Drug Metab. Dispos.*, 1982, **10**, 142.

463 R. Bocker, *J. Chromatogr.*, 1983, **274**, 255.

464 M. J. Egorin, P. A. Andrews, H. Nakazawa, and N. R. Bachur, *Drug Metab. Dispos.*, 1983, **11**, 167.

465 P. Dodion, M. J. Egorin, J. M. Tamburini, C. E. Riggs, jun., and N. R. Bachur, *Drug Metab. Dispos.*, 1984, **12**, 209.

466 G. Maurer, H. R. Loosli, E. Schreier, and B. Keller, *Drug Metab. Dispos.*, 1984, **12**, 120.

467 C. Lin, M. S. Puar, D. Schuessler, B. N. Prananik, and S. Symchowicz, *Drug Metab. Dispos.*, 1984, **12**, 51.

468 T. Shomura, S. Someya, K. Umemura, M. Nishio, and S. Murata, *Yakugaku Zasshi*, 1982, **102**, 781.

469 A. Gerardin, D. Gaudry, and D. Wantiez, *Biomed. Mass Spectrom.*, 1982, **9**, 333.

470 P. Muller, P. R. Imhof, F. Burkart, L.-C. Chu, and A. Gerardin, *Eur. J. Clin. Pharmacol.*, 1982, **22**, 473.

471 P. R. Imhof, A. Sieber, J. Hodler, P. Muller, B. Ott, P. Fankhauser, L.-C. Chu, and A. Gerardin, *Eur. J. Clin. Pharmacol.*, 1982, **23**, 99.

472 D. Gaudry, A. Gerardin, C. Briand, and D. Wantiez, *Biomed. Mass Spectrom.*, 1984, **11**, 276.

473 J. A. Settlage, W. Gielsdorf, and H. Jaeger, *J. High Resolut. Chromatogr.*, 1983, **6**, 68.

474 W. Gielsdorf, E. Schlegel, J. A. Settlage, H. Jaeger, and E. Fink, *Arzneim.-Forsch.*, 1983, **33**, 1677.

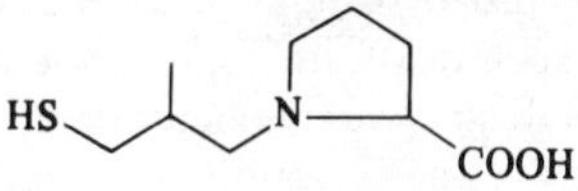

HS–CH2–CH(CH3)–CH2–N(pyrrolidine)–COOH

(84)

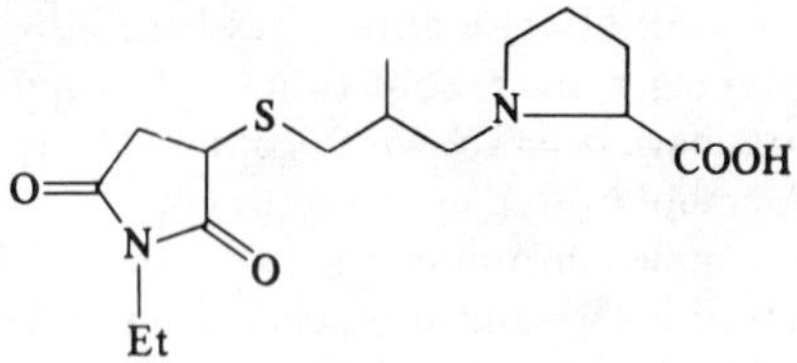

(85)

Captopril. – A brief account of the EI fragmentation of captopril (84) has appeared.[475] Two quantitative assays have been published;[476,477] both employed analogue standards, and one[477] was also capable of measuring three sulphur-conjugate metabolites to levels in the 10–25 ng ml^{-1} range. Bioavailability studies have also been carried out with a stable-isotope-labelled analogue, $[^2H_7]$-captopril, being given intravenously and with the unlabelled drug being administered by mouth.[476] In these and other studies the SH group in captopril was protected as its stable *N*-ethylmaleimide adduct (85).[478,479] GC/MS has also been used to show a longer half-life for the drug in patients with congestive heart failure than in normal subjects.[480]

Antimalarials. – A major mammalian and microbial metabolite of primaquine has been identified by direct-insertion and GC/MS techniques as the deaminated compound 8-(3-carboxy-1-methylpropylamino)-6-methoxyquinoline,[481,482] and a toxic methaemoglobin-forming metabolite has been identified as a substituted quinoneimine.[483] Reaction of chloroquine with chloroformates has enabled the drug to be measured in the presence of interfering desethyl metabolites; chloroquine gives 4-(2-methyl-1-pyrrolidyl)-7-chloroquinoline whereas the metabolites give carbamates.[484]

Valproic Acid. – Forty-eight metabolites of valproic acid (dipropylacetic acid) have been identified by GC/MS in rats;[485] glucuronide formation was a major route. pH-Dependent isomerization of these compounds to β-glucuronidase-resistant forms has been reported.[486] The use of $[^2H_6]$valproic acid has enabled

[475] V. Caplar, S. Rendic, F. Kajfez, H. Hofman, J. Kuftinec, and N. Blazevic, *Acta Pharm. Jugosl.*, 1982, **32**, 125.
[476] A. I. Cohen, R. G. Devlin, E. Ivashkiv, P. T. Funke, and T. McCormick, *J. Pharm. Sci.*, 1982, **71**, 1251.
[477] O. H. Drummer, B. Jarrott, and W. J. Louis, *J. Chromatogr.*, 1984, **305**, 83.
[478] O. H. Drummer, P. J. Worland, and B. Jarrott, *Biochem. Pharmacol.*, 1983, **32**, 1563.
[479] M. S. Bathala, S. H. Weinstein, F. S. Meeker, jun., S. M. Singhvi, and B. H. Migdalof, *J. Pharm. Sci.*, 1984, **73**, 340.
[480] R. J. Cody, M. L. Schaer, A. B. Covit, K. Pondolfino, and G. Williams, *Clin. Pharmacol. Ther.*, 1982, **32**, 721.
[481] C. D. Hufford, A. M. Clark, I. N. Quinones, J. K. Baker, and J. D. McChesney, *J. Pharm. Sci.*, 1983, **72**, 92.
[482] J. K. Baker, *J. Chromatogr.*, 1982, **230**, 69.
[483] A. Strother, R. Allahyari, J. Buchholz, I. M. Fraser, and B. E. Tilton, *Drug Metab. Dispos.*, 1984, **12**, 35.
[484] J. O. Kuye, M. J. Wilson, and T. Walle, *J. Chromatogr.*, 1983, **272**, 307.
[485] G. R. Granneman, S. I. Wang, J. M. Machinist, and J. W. Kesterson, *Xenobiotica*, 1984, **14**, 375.
[486] R. G. Dickinson, W. H. Hooper, and M. J. Eadie, *Drug Metab. Dispos.*, 1984, **12**, 247.

valproic acid metabolites to be identified in human serum in the presence of endogenous acids[487] and a half-life of 13.5 hours to be recorded in a pulse-label experiment.[488] Biphasic kinetics with the drug showing half-lives of 10.2 and 27.2 hours have been found in a uraemic epileptic child.[489] Of eight metabolites measured in plasma and brain from dogs and rats, only one, 2-propyl-2-pentenoic acid, was found in brain.[490] The concentration of this metabolite exceeded that of the parent drug in mice after withdrawal of the drug in an infusion experiment.[491]

Miscellaneous Metabolic Studies Using GC/MS. – The identification of the 5-hydroxy metabolite of tienilic acid (86) has been reported to be the first example of aromatic hydroxylation of a drug containing a thiophene ring.[492] The metabolite gave a complex mixture of products on reaction with diazomethane as the result of tautomerization (Scheme 14). Thioether metabolites of the imidazole-containing drug thiamazole have been identified in man and rat and are assumed to be formed by nucleophilic attack at C-5 of the imidazole ring following *N*-oxidation.[493] New metabolites of chloramphenicol[494–496] include an aldehyde, *p*-nitrobenzyl alcohol, and a dechloro metabolite that

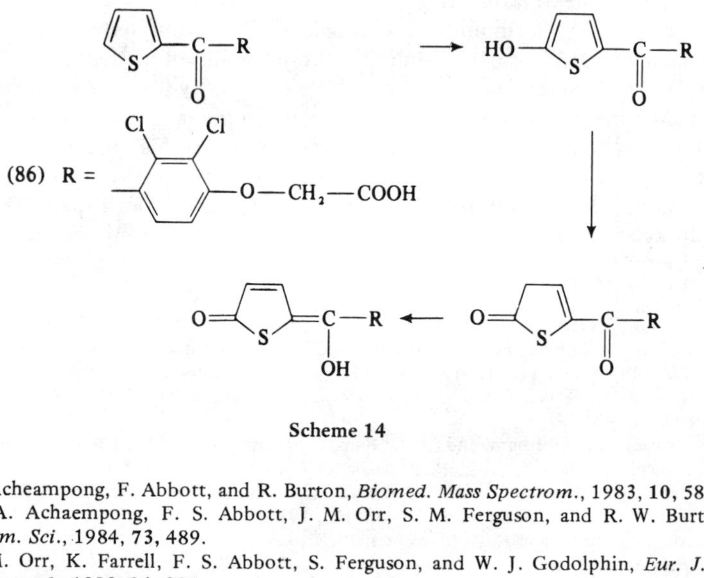

(86) R = (2,3-dichloro-4-(carboxymethoxy)phenyl: Cl, Cl, —O—CH_2—COOH)

Scheme 14

[487] A. Acheampong, F. Abbott, and R. Burton, *Biomed. Mass Spectrom.*, 1983, **10**, 586.
[488] A. A. Achaempong, F. S. Abbott, J. M. Orr, S. M. Ferguson, and R. W. Burton, *J. Pharm. Sci.*, 1984, **73**, 489.
[489] J. M. Orr, K. Farrell, F. S. Abbott, S. Ferguson, and W. J. Godolphin, *Eur. J. Clin. Pharmacol.*, 1983, **24**, 387.
[490] W. Loscher and H. Nau, *J. Pharmacol. Exp. Ther.*, 1983, **226**, 845.
[491] H. Nau and R. Zierer, *Biopharm. Drug Dispos.*, 1982, **3**, 317.
[492] D. Mansuy, P. M. Dansette, C. Foures, M. Jaouen, G. Moinet, and N. Bayer, *Biochem. Pharmacol.*, 1984, **33**, 1429.
[493] R. Twele, W. Kern, and G. Spiteller, *Xenobiotica*, 1983, **13**, 661.
[494] P. L. Morris, T. R. Burke, jun., J. W. George, and L. R. Pohl, *Drug. Metab. Dispos.*, 1982, **10**, 439.
[495] G. F. Bories, J. C. Peleran, J. M. Wal, and D. E. Corpet, *Drug Metab. Dispos.*, 1983, **11**, 249.
[496] P. L. Morris, T. R. Burke, jun., and L. R. Pohl, *Drug Metab. Dispos.*, 1983, **11**, 126.

binds to tissues under anaerobic conditions.[496] 1-*m*-Chlorophenylpiperazine has been identified as an active metabolite produced by dealkylation of the anti-depressants trazodone, etoperidone, and mepiprazole.[497] Bis-dealkylation of the morpholino nitrogen in minaprine leads to loss of the ring with the formation of an amine.[498] Ring cleavage of clavulanic acid gives 1-amino-4-hydroxybutan-2-one as the major metabolite in rat and dog.[499] Differentiation between aliphatic and aromatic *N*-oxides of niaprazine has been achieved; the aliphatic oxides were specifically reduced with SO_2 to compounds giving diagnostic ions at $[M - 16]^+$ and $[M - 18]^+$, whereas the aromatic oxides were reduced by $TiCl_3$ to compounds giving ions at $[M - 16]^+$ and $[M - 17]^+$.[500]

Impurities in tryptamines attributed to an 'inept chemist using inactive lithium aluminium hydride in wet solvents' have yielded information on the source of the illicit drugs,[501] and similar uses have been made of impurities present in cocaine.[502] Benzoyltropeine has been identified in samples of street heroin originating from Florence.[503] Methyl 4-(3-pyridyl)butyrate is a product of cocaine decomposition during smoking,[504] and decomposition of cocaine to benzoylecgonine in stored blood samples has been determined to be as much as 30% over 36 days.[505]

Cleavage of the thiazole ring gives major metabolites of clomethiazole.[506] Sulphur-containing metabolites of afloqualone, formed by defluorination, have been characterized,[507,508] and the drug is also metabolized by hydroxylation and *N*-acetylation.[509] Stable-isotope tracers have enabled hydroxy and *N*-demethylated metabolites of fentanyl to be characterized in the rat[510] and have been used to differentiate endogenous and metabolically produced hippuric acid in studies of the metabolism of LU 2443 (didehydro-4-methyl-5-phenyl-1,3,4-thiadiazolidin-2-thione).[511] Multiple metabolic routes along conventional pathways such as hydroxylation have been characterized for sobrerol,[512,513] procyclidine,[514]

[497] M. H. Fong, S. Garattini, and S. Caccia, *J. Pharm. Pharmacol.*, 1982, **34**, 674.
[498] M. H. Fong, A. Abbiati, E. Benfenati, and S. Caccia, *J. Chromatogr.*, 1983, **259**, 141.
[499] G. C. Bolton, G. D. Allen, C. W. Filer, and D. J. Jeffery, *Xenobiotica*, 1984, **14**, 483.
[500] M. Sanjuan, V. Rovei, J. Dow, and M. Strolin Benedetti, *Int. J. Mass Spectrom. Ion Phys.*, 1983, **48**, 93.
[501] J. S. Cowie, A. L. Holtham, and L. V. Jones, *J. Forensic Sci.*, 1982, **27**, 527.
[502] F. T. Noggle, jun. and C. R. Clark, *J. Assoc. Off. Anal. Chem.*, 1982, **65**, 756.
[503] F. Mari, E. Bertol, and M. Tosti, *Bull. Narcotics*, 1984, **36(i)**, 59.
[504] M. Novak and C. A. Salemink, *Bull. Narcotics*, 1984, **36(ii)**, 79.
[505] Y. Liu, R. D. Budd, and E. C. Griesemer, *J. Chromatogr.*, 1982, **248**, 318.
[506] A. Grupe and G. Spiteller, *J. Chromatogr.*, 1982, **230**, 335.
[507] M. Otsuka, S. Furuuchi, and S. Harigaya, *Chem. Pharm. Bull.*, 1983, **31**, 2799.
[508] M. Otsuka, T. Kurozumi, S. Furuuchi, S. Usuki, K. Kotera, and S. Harigaya, *Chem. Pharm. Bull.*, 1983, **31**, 2438.
[509] S. Furuuchi, M. Otsuka, Y. Miura, and S. Harigaya, *Drug Metab. Dispos.*, 1983, **11**, 371.
[510] T. Goromaru, H. Matsuura, T. Furuta, S. Baba, N. Yoshimura, T. Miyawaki, and T. Sameshima, *Drug Metab. Dispos.*, 1982, **10**, 542.
[511] R. Neidlein and Th. Eder, *Arzneim.-Forsch.*, 1982, **32**, 1292.
[512] P. C. Braga, L. De Angelis, R. Bossi, F. Scaglione, G. Scarpazza, L. Allegra, and F. Fraschini, *Eur. J. Clin. Pharmacol.*, 1983, **24**, 209.
[513] P. Ventura, M. Schiavi, and S. Serafini, *Xenobiotica*, 1983, **13**, 139.
[514] G. Paeme, R. Grimee, and A. Vercruysse, *Arch. Int. Pharmacodyn.*, 1982, **260**, 291.

9-deoxy-16,16-dimethyl-9-methylene prostaglandin E_2,[515] ciglitazone {5-[4-(1-methylcyclohexylmethoxy)benzyl]-2,4-thiazolidinedione},[516] chlorzoxazone,[517] nitrendipine,[518] trithiozine,[519] nitromethaqualone,[520] flunarizine,[521] gliclazide,[522] terfenadine,[523] quinfamide,[524] pargyline,[525] substituted benzophenones related to the benzodiazepines,[526,527] and the antiprotozoal agent 3a,4,5,6,7,7a-hexahydro-3-(1-methyl-5-nitro-1*H*-imidazol-2-yl)-1,2-benzisoxazole.[528] 1-Arylpyrazines are formed from several arylpyrazine-substituted drugs such as oxypertine and zolertine.[529]

Miscellaneous Metabolic Studies Using Direct-insertion Techniques. – Selective hydroxylation of the heterocyclic ring of phenazopyridine to give 2,6-diamino-5-hydroxy-3-(phenylazo)pyridine was thought to be the first report of this type of reaction.[530] Another group of metabolic reactions arousing recent interest is conjugation reactions with lipids. As an example of this, the antihypolipidaemic drug 1-(carboxyphenoxy)-10-(4-chlorophenoxy)decane has been shown to form cholesterol conjugates.[531] *N*-Dealkylation is the major metabolic route exhibited by a series of eleven *N*-alkylated diphenylpiperazines[532] and also leads to major metabolites of fendiline in man[533] and of oxatomide in several species.[534]

[515] S. Steffenrud, *Drug Metab. Dispos.*, 1983, **11**, 255.
[516] K. Yoshida, H. Torii, Y. Sugiyama, T. Fujita, and S. Tanayama, *Xenobiotica*, 1984, **14**, 249.
[517] R. Twele and G. Spiteller, *Arzneim.-Forsch.*, 1982, **32**, 759.
[518] H. Meyer, D. Scherling, and W. Karl, *Arzneim.-Forsch.*, 1983, **33**, 1528.
[519] A. G. Renwick, J. L. Pettet, B. Gruchy, and D. L. Corina, *Xenobiotica*, 1982, **12**, 329.
[520] M. Van Boven and P. Daenens, *J. Pharm. Sci.*, 1982, **71**, 1152.
[521] W. Meuldermans, J. Hendrickx, R. Hurkmans, E. Swysen, R. Woestenborghs, W. Lauwers, and J. Heykants, *Arzneim.-Forsch.*, 1983, **33**, 1142.
[522] H. Miyazaki, T. Fujii, K. Yoshida, S. Arakawa, H. Furukawa, H. Suzuki, A. Kagemoto, M. Hashimoto, and N. Tamaki, *Eur. J. Drug Metab. Pharmacokinet.*, 1983, **8**, 117.
[523] D. A. Garteiz, R. H. Hook, B. J. Walker, and R. A. Okerholm, *Arzneim.-Forsch.*, 1982, **32**, 1185.
[524] J. F. Baker, P. E. O'Melia, D. P. Benziger, S. D. Clemans, and J. Edelson, *Arch. Int. Pharmacodyn.*, 1982, **258**, 29.
[525] A. M. Weli, N.-O. Ahnfelt, and B. Lindeke, *J. Pharm. Pharmacol.*, 1982, **34**, 771.
[526] M. Ballabio, S. Caccia, S. Garattini, G. Guiso, and R. Reginato, *Arzneim.-Forsch.*, 1983, **33**, 959.
[527] M. Fujimoto, S. Hashimoto, S. Takahashi, K. Hirose, H. Hatakeyama, and T. Okabayashi, *Biochem. Pharmacol.*, 1984, **33**, 1645.
[528] W. J. A. VandenHeuvel, H. Skeggs, B. H. Arison, and P. G. Wislocki, *J. Pharm. Sci.*, 1983, **72**, 782.
[529] S. Caccia, A. Notarnicola, M. H. Fong, and E. Benfenati, *J. Chromatogr.*, 1984, **283**, 211.
[530] K. Bailey, B. H. Thomas, M. Vezina, L. W. Whitehouse, W. Zeitz, and G. Solomonraj, *Drug Metab. Dispos.*, 1983, **11**, 277.
[531] R. Fears, K. H. Baggaley, P. Walker, and R. M. Hindley, *Xenobiotica*, 1982, **12**, 427.
[532] W. Voelter and T. Kronbach, *J. Chromatogr.*, 1984, **290**, 1.
[533] H. F. Benthe and R. A. Thieme, *Biomed. Mass Spectrom.*, 1983, **10**, 65.
[534] W. Meuldermans, J. Hendricks, F. Knaeps, W. Lauwers, J. Heykants, and J. M. Grindel, *Xenobiotica*, 1984, **14**, 445.

Multiple metabolic routes along conventional pathways have been reported for citalopram,[535] fluvoxamine,[536] oltipraz,[537] nimodipine,[538] febarbamate,[539] and supidimide [3-(2,3-dihydro-1,1-dioxido-3-oxo-1,2-benzisothiazole-2-yl)-2-oxopiperidine].[540] In the case of supidimide,[540] metabolic cleavage of both heterocyclic rings occurred in the rat. Oxidation by introduction of a ketone group adjacent to heterocyclic nitrogen occurs as a major metabolic route for sulpipide in primates[541] and for niridazole in rats and mice.[542] The isotope-doublet technique has been used in metabolic studies of the anthelmintic mebendazole;[543] cleavage of the amide bond and reduction of the ketone were detected. The formation of cotinine from nicotine involves the nicotine-$\Delta^{1'(5')}$-iminium ion.[544] Other metabolic routes identified by direct-insertion techniques include hydroxylation of the methyl group of the hypolipidaemic drug sultosilic acid and oxidation of the alcohol to a carboxylic acid,[545] sulphoxide formation from the new cardiotonic agent 1,3-dihydro-4-methyl-5-[4-(methylthio)benzoyl]-2*H*-imidazol-2-one,[546] and *N*-de-ethylation of amiodarone.[547] The mass-spectral characteristics of amiodarone have been published.[548]

FD techniques have been used to show excretion of the sulphate conjugate of tiaramide in female but not male rats,[549] sulphoxide and *N*-glucuronide conjugates of cimetidine,[550,551] and both sulphate and glucuronide conjugates of phenazone.[552,553] Use of ammonia DCI techniques has led to the identification of a

535 E. Oyehaug, E. T. Ostensen, and B. Salvesen, *J. Chromatogr.*, 1984, **308**, 199.

536 H. M. Ruijten, H. De Bree, A. J. M. Borst, N. De Lange, P. M. Scherpenisse, W. R. Vincent, and L. C. Post, *Drug Metab. Dispos.*, 1984, **12**, 82.

537 A. Bieder, B. Decouvelaere, C. Gaillard, H. Depaire, D. Heusse, C. Ledoux, M. Lemar, J. P. Le Roy, L. Raynaud, C. Snozzi, and J. Gregoire, *Arzneim.-Forsch.*, 1983, **33**, 1289.

538 H. Meyer, E. Wehinger, F. Bossert, and D. Scherling, *Arzneim.-Forsch.*, 1983, **33**, 106.

539 J. Vachta and P. Gold-Aubert, *Eur. J. Drug Metab. Pharmacokinet.*, 1982, **7**, 147.

540 R. Becker, E. Frankus, I. Graudums, W. A. Gunzler, F.-Ch. Helm, and L. Flohe, *Arzneim.-Forsch.*, 1982, **32**, 1101.

541 J. J. Brennan, A. R. Imondi, D. G. Westmoreland, and M. J. Williamson, *J. Pharm. Sci.*, 1982, **71**, 1199.

542 B. A. Catto, C. I. Valencia, K. Hafez, E. H. Fairchild, and L. T. Webster, jun., *J. Pharm. Exp. Ther.*, 1984, **228**, 662.

543 R. J. Allan and T. R. Watson, *Eur. J. Drug Metab. Pharmacokinet.*, 1982, **7**, 131.

544 J. W. Gorrod and A. R. Hibberd, *Eur. J. Drug Metab. Pharmacokinet.*, 1982, **7**, 293.

545 S. G. Wood, D. Kirkpatrick, A. J. S. Jackson, D. R. Hawkins, W. H. Down, L. F. Chasseaud, and S. R. Biggs, *Xenobiotica*, 1982, **12**, 165.

546 K. Y. Chan, D. F. Ohlweiler, J. F. Lang, and R. A. Okerholm, *J. Chromatogr.*, 1984, **306**, 249.

547 R. J. Flanagan, G. C. A. Storey, D. W. Holt, and P. B. Farmer, *J. Pharm. Pharmacol.*, 1982, **34**, 638.

548 M. Bonati, F. Gaspari, V. D'Aranno, E. Benfenati, P. Neyroz, F. Galletti, and G. Tognoni, *J. Pharm. Sci.*, 1984, **73**, 829.

549 W. J. S. Lockley, B. Mead, A. P. Plumb, D. A. Smith, K. Iwasaki, A. Suzuki, K. Tada, and H. Noguchi, *Biochem. Pharmacol.*, 1982, **31**, 2433.

550 S. C. Mitchell, J. R. Idle, and R. L. Smith, *Drug Metab. Dispos.*, 1982, **10**, 289.

551 S. C. Mitchell, J. R. Idle, and R. L. Smith, *Xenobiotica*, 1982, **12**, 283.

552 J. Bottcher, H. Bassmann, R. Schuppel, and W. D. Lehmann, *Naunyn Schmiedberg's Arch. Pharmacol.*, 1982, **321**, 226.

553 W. D. Lehmann, J. Bottcher, H. Bassmann, R. Schuppel, and H. M. Schiebel, *Biomed. Mass Spectrom.*, 1982, **9**, 477.

hydroxylamine metabolite of procainamide in rat and man;[554] the compound was toxic and thought to be responsible for procainamide-induced lupus. The alcohol deterrent cyanamide is acetylated in several species, as determined by pulsed positive-ion/negative-ion CI.[555]

5 Miscellaneous Quantitative Methods

Assays for drugs alone employing deuteriated internal standards include those for the analgesic 3-phenoxy-*N*-methylmorphinan (detection limit 2.5 ng ml^{-1}),[556] the new neuroleptic (±)-3-ethyl-2,6-dimethyl-4,4a,5,6,7,8,8a,9-octahydro-4a,8a-*trans*-1*H*-pyrrolo[2,3-*g*]isoquinolin-4-one (2 ng ml^{-1}),[557] and warfarin (1 ng ml^{-1}).[558] Related methods that also measured metabolites are for ketobemidone and its *N*-demethylated metabolite,[559] isoniazid and four metabolites,[560] nicotine and cotinine in tissues,[561] and peptidoaminobenzophenone (2-*o*-chlorobenzoyl-4-chloro-*N*-methyl-*N'*-glycylglycinanilide) measured as its stable aminoquinalone degradation product to the 1 ng ml^{-1} level by negative-ion CI.[562,563] In an assay for phenazone and three metabolites, the deuteriated standards were obtained as metabolites from humans given [$^{2}H_{5}$]phenazone.[564] ^{13}C standards have been used for bepridil[565] and (−)-*threo*-chlorocitrate;[566] the latter assay employed positive-ion CI with methane as the reactant gas and gave a detection limit of 100 ng ml^{-1}.

Assays employing analogue standards and single-ion monitoring have been reported for melperone[567] in human plasma and for pinaverium bromide[568] in human serum. The pinaverium bromide assay used Raney nickel dequaternization of the drug to *N*-(6,6-dimethylbicyclo[3,1,1]-2-heptanylethoxyethyl)-perhydro-1,4-oxazine, which was measured to the 1 ng ml^{-1} level with a dehydro analogue as the standard. Dequaternization was also required for measurement

554 J. P. Uetrecht, B. J. Sweetman, R. L. Woosley, and J. A. Oates, *Drug Metab. Dispos.*, 1984, **12**, 77.

555 F. N. Shirota, H. T. Nagasawa, C. H. Kwon, and E. G. Demaster, *Drug Metab. Dispos.*, 1984, **12**, 337.

556 J. A. F. de Silva, J. Pao, and M. A. Brooks, *J. Chromatogr.*, 1982, **232**, 63.

557 B. H. Min, *J. Chromatogr.*, 1983, **277**, 340.

558 E. D. Bush, L. W. Low, and W. F. Trager, *Biomed. Mass Spectrom.*, 1983, **10**, 395.

559 U. Bondesson, P. Hartvig, L. Abrahamsson, and N.-O. Ahnfelt, *Biomed. Mass Spectrom.*, 1983, **10**, 283.

560 A. Noda, K.-Y. Hsu, Y. Aso, K. Matsuyama, and S. Iguchi, *J. Chromatogr.*, 1982, **230**, 345.

561 J. A. Thompson, M.-S. Ho, and D. R. Petersen, *J. Chromatogr.*, 1982, **231**, 53.

562 S. Hashimoto, E. Sakurai, M. Mizobuchi, S. Takahashi, K.-I. Yamamoto, and T. Momose, *Biomed. Mass Spectrom.*, 1982, **9**, 546.

563 S. Hashimoto, E. Sakurai, M. Mizobuchi, S. Takahashi, K. Yamamoto, and T. Momose, *Biomed. Mass Spectrom.*, 1984, **11**, 50.

564 B. K. Tang, H. Uchino, T. Inaba, and W. Kalow, *Biomed. Mass Spectrom.*, 1982, **9**, 425.

565 J. Vink, H. J. M. Van Hal, F. M. Kaspersen, and H. P. Wijnand, *Int. J. Mass Spectrom. Ion Phys.*, 1983, **48**, 217.

566 F. Rubio, F. De Grazia, B. J. Miwa, and W. A. Garland, *J. Chromatogr.*, 1982, **233**, 149.

567 K. Y. Chan and R. A. Okerholm, *J. Chromatogr.*, 1983, **274**, 121.

568 G. A. de Weerdt, R. P. Beke, H. G. Verdievel, F. Barbier, J. A. Jonckheere, and A. P. de Leenheer, *Biomed. Mass Spectrom.*, 1983, **10**, 162.

of the new antifibrillatory agent clofilium (4-chloro-*N*,*N*-diethyl-*N*-heptylbenzene butanaminium phosphate);[569] the methyl analogue was used as the standard and the drug was measured in the 25–1000 ng ml^{-1} range by monitoring two peaks. Multi-ion recording was also used for measurements of bencyclane to 0.1 ng ml^{-1} levels by positive-ion CI,[570] busulfan to 10 ng ml^{-1} by EI,[571] tolazoline[572] and γ-vinyl-γ-aminobutyric acid by EI,[573] and aprindine to 5 ng ml^{-1} by positive-ion CI in plasma,[574] all with analogue standards. An unrelated compound, cholestane, was used in an assay for the antiulcer agent geranylgeranylacetone, but the drug could nevertheless be measured down to 1 ng ml^{-1} with high precision.[575] Quantitative methods employing direct probe introduction have been reported for triquinol[576] and benzalkonium chloride.[577]

6 Miscellaneous Pharmacokinetic and Clinical Studies

Although HPLC now seems to be preferred for pharmacokinetic studies in many cases, GC/MS-based methods are frequently used when high sensitivity is required. Several examples have been given above, and a large number of others have been reported, particularly in the clinical literature. Some of these are outlined below. The rate of elimination of verapamil has been shown to increase with age in man[578] and in patients with atrial fibrillation.[579] No difference has been found in the half-lives of the two enantiomers of this drug[580] or its racemate or for four doses given at different times to chronically treated patients in a pulse-label experiment.[581] The drug did, however, show considerable intra-individual variation in its elimination characteristics.[581] Guanfacine has a half-life of 111.5 minutes in the rabbit, some 10–20 times less than that in humans,[582] and it shows a similar half-life in normal subjects and patients undergoing haemodialysis.[583] Steady-state levels in plasma have been measured at 4.1 ± 0.59 ng ml^{-1} after one year's treatment with 0.5–4 mg of the drug daily.[584]

[569] T. D. Lindstrom and R. L. Wolen, *J. Chromatogr.*, 1982, **233**, 175.
[570] W. Gielsdorf, J. A. Settlage, and H. Jaeger, *Arzneim.-Forsch.*, 1984, **34**, 290.
[571] H. Ehrsson and M. Hassan, *J. Pharm. Sci.*, 1983, **72**, 1203.
[572] R. M. Ward, M. J. Cooper, and B. L. Mirkin, *J. Chromatogr.*, 1982, **231**, 445.
[573] K. D. Haegele, J. Schoun, R. G. Alken, and N. D. Heubert, *J. Chromatogr.*, 1983, **274**, 103.
[574] T. Ito, H. Namekawa, and T. Kobari, *J. Chromatogr.*, 1983, **274**, 341.
[575] M. Tanaka, J. Hasegawa, J. Tsutsumi, and T. Fujita, *J. Chromatogr.*, 1982, **231**, 301.
[576] S. Popov, P. Demirev, N. Mollova, V. Nenov, A. Sidjimov, N. Marekov, R. Georgieva, and V. Ognyanova, *Int. J. Mass Spectrom. Ion Phys.*, 1983, **48**, 105.
[577] N. N. Daoud, P. A. Crooks, R. Speak, and P. Gilbert, *J. Pharm. Sci.*, 1983, **72**, 290.
[578] P. Anderson, U. Bondesson, C. Sylven, and H. Astrom, *Br. J. Clin. Pharmacol.*, 1982, **14**, 578P.
[579] P. Anderson, U. Bondesson, and C. Sylven, *Eur. J. Clin. Pharmacol.*, 1982, **23**, 49.
[580] M. Eichelbaum, G. Mikus, and B. Vogelgesang, *Br. J. Clin. Pharmacol.*, 1984, **17**, 453.
[581] M. Eichelbaum and A. Somogyi, *Eur. J. Clin. Pharmacol.*, 1984, **26**, 47.
[582] N. D. Barber and J. L. Reid, *J. Pharm. Pharmacol.*, 1983, **35**, 114.
[583] W. Kirch, H. Kohler, and T. Axthelm, *Eur. J. Drug Metab. Pharmacokinet.*, 1982, **7**, 277.
[584] T. Hedner, G. Nyberg, and T. Mellstrand, *Clin. Pharmacol. Ther.*, 1984, **35**, 604.

Bioavailability studies using conventional methods have shown bioequivalence between four preparations of indomethacin,[585] a mean value of 92% for terodiline,[586] and a reduction in systemic availability of the prostaglandin analogue trimoprostil in the presence of food.[587] Bioavailability for fast-release formulations of etilefrine was 35% whereas that for a sustained-release formulation was only 17%.[588] Substantial presystemic elimination was thought to result in a low bioavailability for methylphenidate,[589] and studies using the stable-isotope method have shown high absorption of methimazole.[590] Rectal administration of atropine in children results in a peak uptake at 15 minutes with a plasma concentration of 0.7 ng ml^{-1}.[591]

Comparatively long elimination half-lives have been recorded for several drugs, for example 63 hours for terodiline[586] and 39.4 hours for pemoline.[592] Debrisoquine hydroxylation has been studied using negative-ion CI, and the metabolite appears to be formed by a specific cytochrome P450 isozyme.[593] Hydralazine has been measured in the rat,[594] and in a pharmacokinetic study of isoniazid both hydrazine and acetylhydrazine have been detected.[595] Release of indomethacin from the pro-drugs oxametacin and acemetacin has been studied in dogs;[596] the concentration of the released drug was compared with that of intravenously administered $[^2H_4]$indomethacin using $[^2H_9]$indomethacin as the internal standard. Good correlation exists between the serum concentration of terbutaline and improvement of lung function in asthmatics,[597] and plasma concentrations above 20 ng ml^{-1} are associated with uterine contractions sufficient to cause second trimester abortions.[598] Measurements at 24 hours of the plasma concentration of dothiepin have been used to predict steady-state levels after 4 weeks of treatment,[599] and in a study with mefloquinine no pharmacokinetic

[585] J. A. Settlage, W. Gielsdorf, M. Nieder, and H. Jaeger, *Arzneim.-Forsch.*, 1983, **33**, 885.
[586] B. Karlen, K.-E. Andersson, G. Ekman, S. Stromberg, and U. Ulmsten, *Eur. J. Clin. Pharmacol.*, 1982, **23**, 267.
[587] R. J. Wills, *J. Clin. Pharmacol.*, 1984, **24**, 194.
[588] J. H. Hengstmann, R. Hengstmann, S. Schwonzen, and H. J. Dengler, *Eur. J. Clin. Pharmacol.*, 1982, **22**, 463.
[589] W. Wargin, K. Patrick, C. Kilts, C. T. Gualtieri, K. Ellington, R. A. Mueller, G. Kraemer, and G. R. Breese, *J. Pharmacol. Exp. Ther.*, 1983, **226**, 382.
[590] R. Jansson, P. A. Dahlberg, and B. Lindstrom, *Int. J. Clin. Pharmacol. Ther. Toxicol.*, 1983, **21**, 505.
[591] G. L. Olsson, A. Bejersten, H. Feychting, L. Palmer, and B.-M. Pettersson, *Anaesthesia*, 1983, **38**, 1179.
[592] O. J. Igwe and J. W. Blake, *Drug Metab. Dispos.*, 1983, **11**, 120.
[593] A. R. Boobis, S. Murray, G. C. Kahn, G.-M. Robertz, and D. S. Davies, *Mol. Pharmacol.*, 1983, **23**, 474.
[594] A. J. Streeter and J. A. Timbrell, *Drug Metab. Dispos.*, 1983, **11**, 184.
[595] I. W. Beever, I. A. Blair, and C. T. Dollery, *Br. J. Pharmacol.*, 1982, **77**, 488P.
[596] Y. Matsuki, T. Ito, M. Kojima, H. Katsumura, H. Ono, and T. Nambara, *Chem. Pharm. Bull.*, 1983, **31**, 2033.
[597] W. Van Den Berg, J. G. Leferink, W. Tabingh Suermondt, J. Kreukniet, R. A. A. Maes, R. Serra and P. L. B. Bruynzeel, *Int. J. Clin. Pharmacol. Ther. Toxicol.*, 1983, **21**, 24.
[598] K. Green, D. Vesterqvist, M. Bygdeman, N. J. Christenssen, and S. Bergstroem, *Prostaglandins*, 1982, **24**, 451.
[599] K. P. Maguire, T. R. Norman, I. McIntyre, and G. D. Burrows, *Clin. Pharmacokinet.*, 1983, **8**, 179.

changes were observed over a period of 21 weeks.[600] Pyridostigmine has been detected in breast milk at 36–113% of the concentration in maternal but not in the infant plasma, suggesting that drug treatment is no obstacle to breast feeding.[601] Pharmacokinetics of the anaesthetics vencuronium and pancuronium in man have been studied after thermal *N*-demethylation,[602] and both drugs have been shown to cross the placental barrier to give measurable concentrations in the foetus.[603]

[600] I. Mimica, W. Fry, G. Eckert, and D. E. Schwartz, *Chemotherapy*, 1983, **29**, 184.
[601] L.-I. Hardell, B. Lindstrom, G. Lonnerholm, and P. O. Osterman, *Br. J. Clin. Pharmacol.*, 1982, **14**, 565.
[602] R. Cronnelly, D. M. Fisher, R. D. Miller, P. Gencarelli, L. Nguyen-Gruenke, and N. Castagnoli, jun., *Anesthesiology*, 1983, **58**, 405.
[603] P. A. Dailey, D. M. Fisher, S. M. Shnider, C. L. Baysinger, Y. Shinohara, R. D. Miller, T. K. Abboud, and K. C. Kim, *Anesthesiology*, 1984, **60**, 569.

11

Metal-containing and Inorganic Compounds Investigated by Mass Spectrometry

BY J. CHARALAMBOUS

1 Introduction

This report reviews articles that appeared between June 1982 and June 1984 and concerned the mass-spectral behaviour of metallic and inorganic compounds. In the context of this report, elements such as boron, silicon, phosphorus, arsenic, sulphur, and selenium are considered as metals, and some of their compounds are included. Because of the breadth of the area covered, only selective references are given in each section.

During the period under review there has been a considerable increase in the application of new mass-spectral techniques to the study of metallic and inorganic compounds. The main thrust has been in two areas:

(i) The application of new techniques of ionization to the study of thermally labile compounds and/or species of low volatility. This has proved particularly useful in investigating large metal-containing compounds and especially those that are biologically important.

(ii) The combination of new methods of ionization with techniques for characterization of ion structure. This approach has opened new frontiers in the study of metallic compounds and will have considerable impact on various aspects of co-ordination chemistry and catalysis.

The format adopted in this chapter has been modified from that used previously in order to reflect these new areas of interest.

2 Main-group Organometallic Compounds

During the period under review mass spectrometry has been used extensively for the study of main-group organometallic species. Most applications have involved EI techniques and have been concerned simply with the determination of the molecular weights of the compounds under investigation. Such studies are not included in this report unless the type of compound investigated has received no attention previously.

Acyclic Compounds. – Ion-cyclotron-resonance (ICR) spectrometry has been used to investigate the gas-phase positive-ion chemistry of trimethylboron and trimethylaluminium.[1] The study of halide transfer to $[Me_2B]^+$ and $[Me_2Al]^+$,

[1] M. M. Kappes, J. S. Uppal, and R. H. Staley, *Organometallics*, 1982, **1**, 1303.

the principal ions produced by electron impact on Me_3B and Me_3Al, respectively, has enabled the determination of the halide affinities of these ions and other thermodynamic parameters. It has been shown that a variety of neutral Lewis bases (*e.g.* Me_2S or PhCN) condense with $[Me_2B]^+$ to give one-ligand complexes, but they give two-ligand complexes with $[Me_2Al]^+$. The relative order of ligand-binding energies has been established. ICR has also been used to examine the reaction of alkanol-alkoxide negative ions $[R^1O\cdots H\cdots OR^2]^-$ with alkoxysilanes, Me_3SiOR.[2] The reaction leads to both $[M+R^1O]^-$ and $[M+R^2O]^-$ ions of trigonal-bipyramidal geometry. Analogous reactions do not occur between Me_3SiX and $[R^1O\cdots H\cdots OR^2]^-$ when X = F, NHR, NR_2, $SiMe_3$, alkyl, allyl, propargyl, benzyl, or aryl, but $[R^1O\cdots H\cdots X]^-$ ions of low abundance are formed when X = OH, OCOMe, OCN, or SR. Cyclic ethers (1; $n = 2$–4) react with the alkanol-alkoxide negative ions to give ions of type (2).

The EI spectra of the aminoboranes (3; X = Cl or N_3) and of the iminoborane Bu^t—B=N—Bu^t have been reported,[3] and the spectra of dialkoxydimethylsilanes,[4] 2-thienylchloromethylsilanes,[5] and the silane derivatives $(EtO)_3SiCH{=}CHR$ [R = (4)][6] have been studied. Fragmentation pathways have been suggested for several hexa-alkyldisiloxanes.[7] The EI spectra of the tris(trimethylsilyl)silanes $(Me_3Si)_3SiX$ [X = H, Cl, $SiMe_3$, or $Si(SiMe_3)_3$][8] and of several tris(trimethyl-

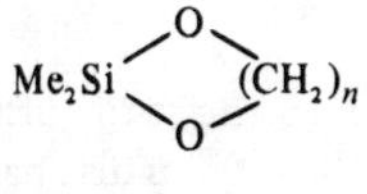

(1) 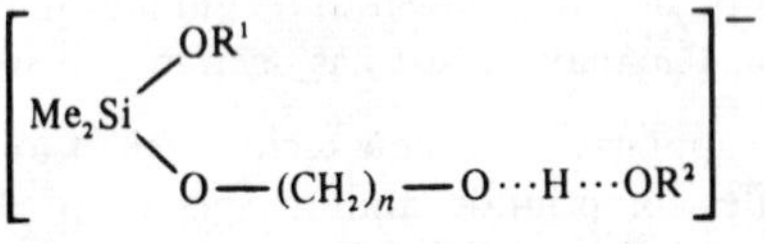 (2)

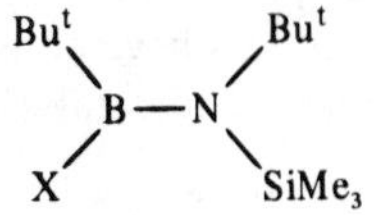

(3) 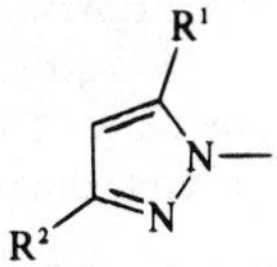 (4) $R^1 = H, R^2 = Me$
$R^1 = R^2 = Me$

[2] R. N. Hayes, J. H. Bowie, and G. Klass, *J. Chem. Soc., Perkin Trans. 2*, 1984, 1167.

[3] P. Paetzold, C. Plotho, G. Schmid, R. Boese, B. Schrader, D. Bougeard, U. Pfeiffer, R. Gleiter, and W. Schäfer, *Chem. Ber.*, 1984, **117**, 1089.

[4] R. S. Mosovirov, I. A. Borisova, L. B. Gazizova, E. P. Nedogrei, S. S. Zlotskii, R. A. Karakhanov, and D. L. Rakhankulov, *J. Gen. Chem. U.S.S.R. (Engl. Transl.)*, 1983, **53**, 1428.

[5] E. Lukevits, N. P. Erchak, V. F. Mtorykina, I. B. Mazheika, and S. Kh. Rozite, *J. Gen. Chem. U.S.S.R. (Engl. Transl.)*, 1983, **53**, 1857.

[6] V. D. Sheludyakov, V. I. Zhun, V. G. Lakhtin, V. N. Bochkarev, T. F. Slyusarenko, V. M. Mosova, A. V. Kisin, and N. V. Alekseev, *J. Gen. Chem. U.S.S.R. (Engl. Transl.)*, 1983, **53**, 2274.

[7] R. Pfefferkorn, H. F. Grützmacher, and D. Krick, *Int. J. Mass Spectrom. Ion Phys.*, 1983, **47**, 515.

[8] A. G. Brook, A. G. Harrison, and R. K. M. R. Kallury, *Org. Mass Spectrom.*, 1982, **17**, 360.

silyl)acylsilanes $(Me_3Si)_3Si—CO—R$ (R = *e.g.* Ph, C_6F_5, CMe_3, or 1-adamantyl)[9] have been studied. The mass spectra of the former show abundant odd-electron ions for which disilene (>Si=Si<) structures have been suggested. The spectra of the acyl compounds exhibit few molecular ions. When R = alkyl, fragmentation of $M^{+\bullet}$ leads to abundant ions $[M - R^{\bullet}]^+$ and $[M - RCO^{\bullet}]^+$. In contrast, in the spectra of aroyl compounds peaks for the analogous ions are absent or small, and the spectra are dominated by $[M - Me^{\bullet}]^+$ and $[M - Me^{\bullet} - C_3H_8OSi]^+$ peaks. The methane CI spectra of some tris(trimethylsilyl)acylsilanes have been reported.[9]

Mass-spectral studies of several *cis*- and *trans*-carboalkoxycyclopropylsilanes of type (5) have shown that under electron impact the main decomposition routes involve alkyl loss from silicon, trialkylsilyl-ion formation, and trialkoxysilane elimination.[10,11] Interestingly, and in contrast to carbon analogues, the silicon atom inhibits ring opening. The trialkoxysilane elimination is less significant for *cis* isomers than for the *trans* isomers. This unexpected inverse stereochemical effect has been explained in terms of silicon–oxygen bond formation, which is possible only in the *cis* isomer for geometric reasons. Differences are also apparent in the CI spectra, which show only two significant ions, $[M + H]^+$ and $[M - R^{\bullet}]^+$. The latter is more pronounced for the *cis* isomers, indicating that its formation involves neighbouring-group participation.[11]

Various boron compounds containing the trimethylsilyl group [*e.g.* (6)] have been characterized.[12] The mass spectra of several pentamethylcyclopentadienylsilanes and -germanes (7) have been reported.[13] Small molecular-ion peaks have been observed in the mass spectra of the alkyl- and aryl-(porphyrinato)gallium(III) complexes Ga(por)(R) ($porH_2$ = octaethylporphyrin or tetraphenylporphyrin, R = Me, Et, Bu^n, Ph, or *p*-MeC_6H_4).[14]

The arsenic compound (8),[15] two novel cryptands prepared from it,[15] and several organoarsenic catecholates (9)[16] have been characterized by mass spectrometry. A study of the LAMMA positive-ion mass spectra of the compounds Ph_3Met (Met = As, Sb, or Bi) has been reported.[17] The abundance of the Met^+

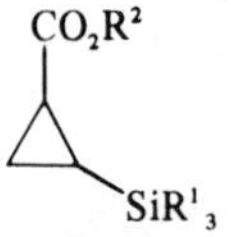

CO_2R^2 ... SiR^1_3 (5)   B—$C(SiMe_3)_3$ (6) $MetR_3$ (7) Met = Si or Ge, R = alkyl

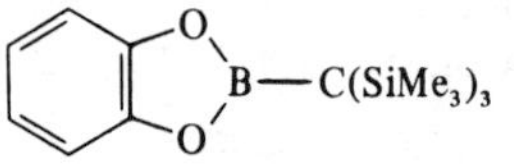

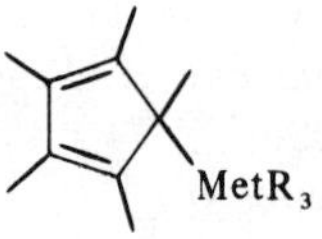

[9] A. G. Brook, R. K. M. R. Kallury, and V. C. Poon, *Organometallics*, 1982, **1**, 987.
[10] K. Vékey, J. Tamás, G. Czira, and I. E. Dalgy, *Org. Mass Spectrom.*, 1982, **17**, 620.
[11] K. Vékey and J. Tamás, *Int. J. Mass Spectrom. Ion Phys.*, 1983, **47**, 519.
[12] C. Eaborn, M. N. El-Kheli, N. Retta, and J. D. Smith, *J. Organomet. Chem.*, 1983, **249**, 23.
[13] P. Jutzi, H. Saleske, D. Bühl, and H. Grohe, *J. Organomet. Chem.*, 1983, **252**, 29.
[14] A. Coutsolelos and R. Guilard, *J. Organomet. Chem.*, 1983, **253**, 273.
[15] J. Ellerman, A. Veit, E. Lindner, and S. Hoehne, *J. Organomet. Chem.*, 1983, **252**, 153.
[16] R. H. Fish and R. S. Tannous, *Organometallics*, 1982, **1**, 1238.
[17] B. Ollman, K.-D. Kupka, and F. Hillenkamp, *Int. J. Mass Spectrom. Ion Phys.*, 1983, **47**, 31.

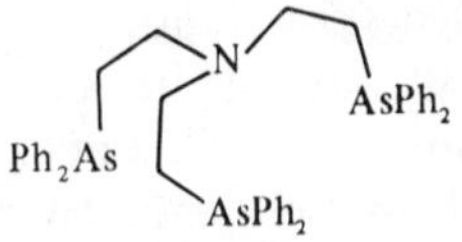

(8)

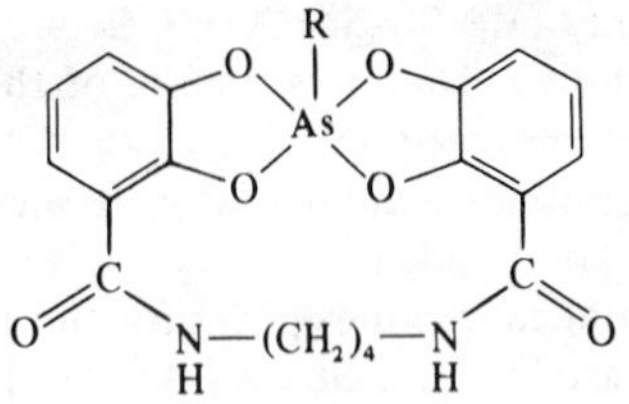

(9) R = Me or Ph

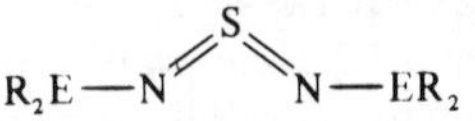

(10) E = P, As, Sb, or Bi, R = Bu
E = As, R = Me, c-C_6H_{11}, or Ph

ion and the extent of fragmentation increase with increasing size of the metal. The EI spectra of bis- and tris-(trimethylsilyl)methyl derivatives of antimony have been reported.[18] The compounds $(R_2CH)_3Sb$ and $(R_2CH)_2SbCl$ ($R = Me_3Si$) afford abundant molecular ions, but such ions are not present in the spectra of $(R_2CH)_4Sb_2$, $(R_3C)SbCl_2$, or $(R_2CH)SbCl_2$. Both dichloro compounds show abundant ions for which the stiba-alkene structure $[R_2C{=}SbCl]^+$ has been proposed.[18] In the EI spectrum of Ph_4Bi a small molecular-ion peak has been observed.[19] The spectra of $Bu^t{}_2BiBr$ and of the sulphur di-imide compounds (10) have been examined.[20] In the latter, the SN_2 unit shows considerable resistance to fragmentation.

In the mass spectra of a series of benzeneseleninic acids $XC_6H_4SeO_2H$ (X = *m*-Cl, *p*-Cl, *m*-Br, *p*-Br, *p*-Me, *m*-NO_2, or *p*-NO_2), besides very small peaks due to the molecular ions, several peaks at higher mass numbers and of greater size have been observed.[21] These are most probably due to $(XC_6H_4Se)_2$ and $(XC_6H_4)_2SeO$, which arise from thermal decomposition of the acids. In the esters $R^1SeCOOR^2$ ($R^1 = R^2 = Et$; $R^1 = Pr^n$, $R^2 = Me$ or Et), molecular ions fragmenting by loss of $R^1Se^{\bullet}$ have been observed.[22] It has been shown that the CID charge-reversal spectrum of $[MeSe]^-$ is identical with the CID mass spectrum of $[MeSe]^+$.[23] EI mass spectra of $(CF_3)_2Te$, $(CF_2Cl)_2Te$, and $(CF_3)_2TeF_2$ have been reported.[24]

Cyclic Compounds with Carbon Atoms in the Ring. – The EI mass spectra of the boron heterocycles (11)–(13) have been reported.[25] All compounds except

[18] H. J. Breunig, W. Kanig, and A. Soltani-Neshan, *Polyhedron*, 1983, **2**, 291.
[19] F. Calderazzo, A. Morvillo, G. Pelizzi, and R. Poli, *J. Chem. Soc., Chem. Commun.*, 1983, 507.
[20] M. Herberhold, W. Ehrenreich, and K. Guldner, *Chem. Ber.*, 1984, **117**, 1999.
[21] A. Benedetti, C. Preti, L. Tassi, and G. Tosi, *Aust. J. Chem.*, 1982, **35**, 1365.
[22] B. Sturm and G. Gattow, *Z. Anorg. Allg. Chem.*, 1983, **513**, 183.
[23] T. A. Lehman, M. M. Bursey, D. J. Harvan, J. R. Hass, D. Liotta, and L. Waykole, *Org. Mass Spectrom.*, 1982, **17**, 607.
[24] S. Herberg and D. Naumann, *Z. Anorg. Allg. Chem.*, 1982, **492**, 95.
[25] S. M. Van Der Kerk, P. H. M. Budzelaar, A. L. M. Van Eekeren, and G. J. M. Van Der Kerk, *Polyhedron*, 1984, **3**, 271.

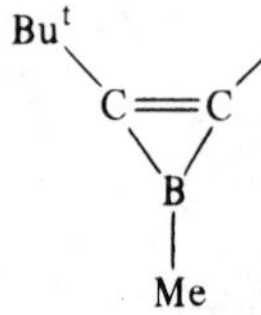

(11) 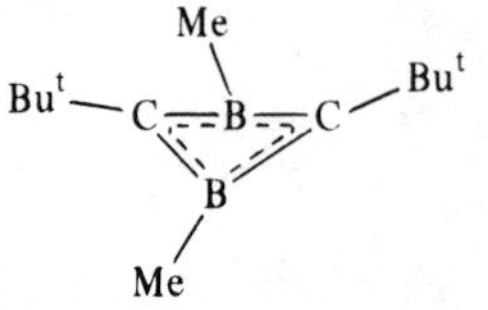 (12) 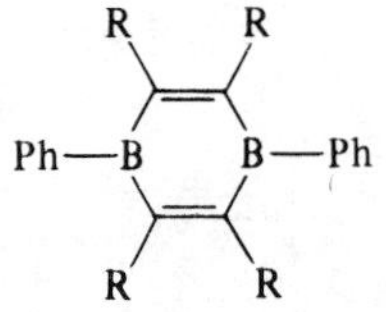 (13) R = alkyl

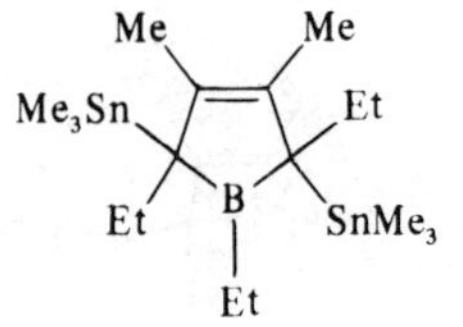

(14) 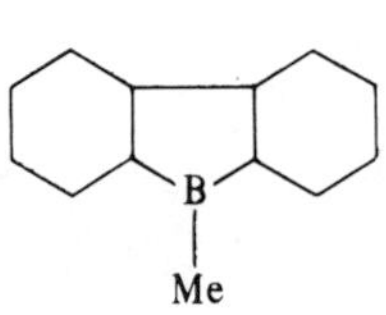 (15) 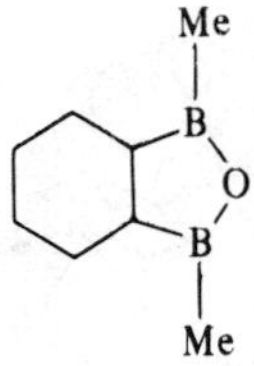 (16)

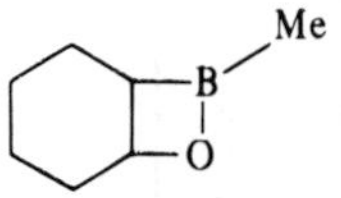

(17) 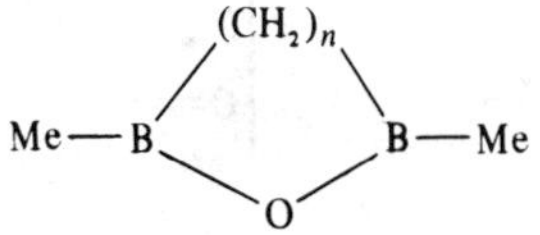 (18) $n = 2$ or 3 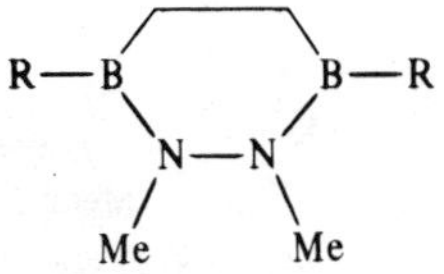 (19) R = Cl or Me

(12) show abundant molecular ions and generally fragment by loss of an alkyl group followed by loss of olefin and methane. In the case of compound (11), the BC_2 nucleus shows considerable resistance to fragmentation. The borolene (14) shows a very small molecular-ion peak, and the base peak corresponds to $[Me_3Sn]^+$.[26] For compound (15), both EI and (ammonia and methane) CI spectra have been reported.[27] In the former, abundant molecular ions are observed and the base peak corresponds to $[C_7H_{13}B]^+$. The CI spectra show peaks corresponding to $[M + NH_2]^+$ and $[M - H]^+$. The heterocycles (16) and (17) have been identified by mass spectrometry and their EI spectra reported.[27] Prominent molecular-ion peaks have been observed in the spectra of compounds (18) and (19),[28] and there is no evidence for dimerization in the case of the nitrogen-containing heterocycles. Derivatization of substituted benzeneboronic acids with 1,2- or 1,3-diols, resulting in the formation of boronate esters [*e.g.* (20)], has been utilized in the analysis of the acids by GC/MS.[29] Fragmentation pathways for the esters were investigated using isotope labelling and linked-scanning techniques. The spectra of a series of transannularly bonded com-

[26] B. Wrackmeyer, *Organometallics*, 1984, **3**, 1.
[27] S. M. Van Der Kerk, J. C. Roos-Venekamp, A. J. M. Van Beijnen, and G. J. M. Van Der Kerk, *Polyhedron*, 1983, **2**, 1337.
[28] W. Haubold, A. Gemmler, and U. Kraatz, *Z. Anorg. Allg. Chem.*, 1983, **507**, 222.
[29] C. Longstaff and M. E. Rose, *Org. Mass Spectrom.*, 1982, **17**, 508.

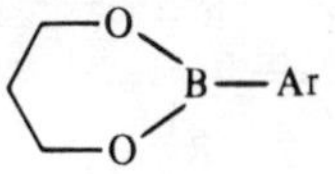

(20)

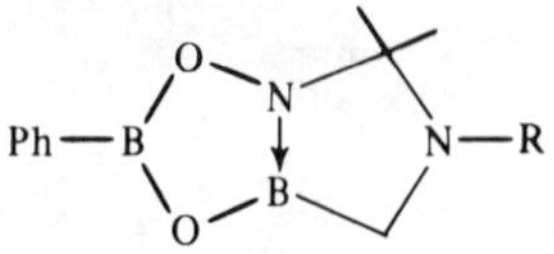

(21)

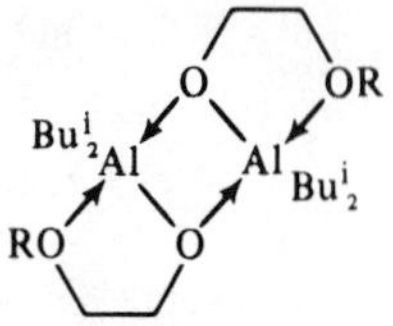

(22)

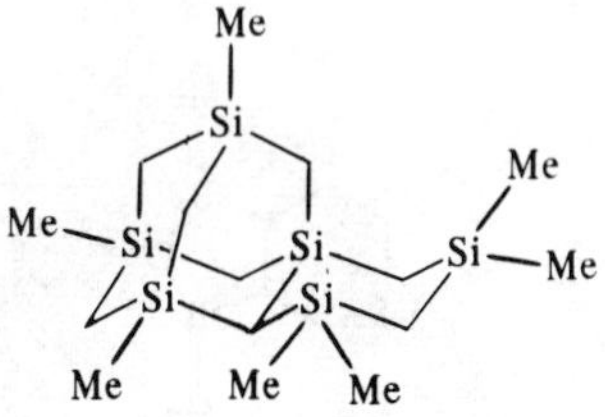

(23)

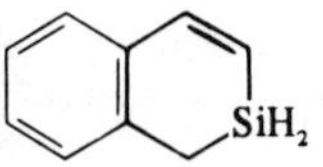

(24)

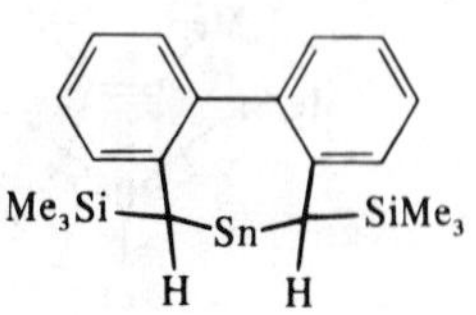

(25)

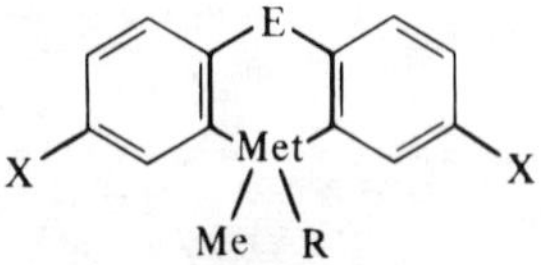

(26) Met = Si or Ge, E = O, NEt, or S
X = H, Br, or Me
R = Cl, Me, OEt, or aryl

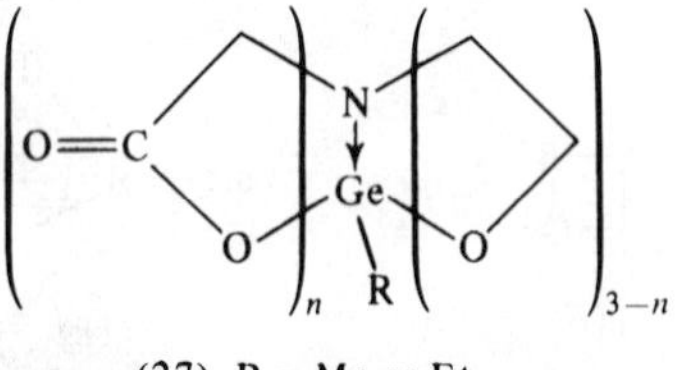

(27) R = Me or Et

pounds (21) have been reported.[30] All compounds show molecular-ion peaks, but these are fairly small. Mass-spectral studies of the products arising from the partial hydrolysis of Bu^i_3Al with monoalkyl ethers of ethylene glycol ($ROCH_2CH_2OH$) have shown that they have dimeric structures (22).[31]

Various silicon-containing cyclic compounds [*e.g.* (23) and (24)] have been characterized.[32–35] The spectra of silepines and stannepines [*e.g.* (25)] have been reported.[36] The effect of the metal, the heteroatom E, and the substituent X on the fragmentation behaviour of compounds of type (26) has been examined.[37] Studies of various germatranes (27) have shown that the ions of highest mass in their spectra correspond to $[M - CO_2]^{+\bullet}$.[38] In a study of heterocycle (28), three

[30] W. Kleigel, *J. Organomet. Chem.*, 1983, **253**, 9.
[31] A. I. Belokon and V. N. Bochkareve, *J. Gen. Chem. U.S.S.R. (Engl. Transl.)*, 1983, **53**, 83.
[32] G. Fritz and K. P. Wörns, *Z. Anorg. Allg. Chem.*, 1984, **512**, 103.
[33] G. Fritz and A. Wörsching, *Z. Anorg. Allg. Chem.*, 1984, **512**, 131.
[34] H. Vuper and T. J. Barton, *J. Chem. Soc., Chem. Commun.*, 1982, 1211.
[35] T. J. Barton and G. T. Burns, *Organometallics*, 1982, **1**, 1455.
[36] L. M. Engelhardt, W. Leung, C. L. Raston, P. Twiss, and A. H. White, *J. Chem. Soc., Dalton Trans.*, 1982, 321.
[37] A. A. Simonenko, V. N. Bocharev, I. E. Saratov, and V. O. Reikhfel'd, *J. Gen. Chem. U.S.S.R. (Engl. Transl.)*, 1982, **52**, 2110.
[38] T. K. Gay, N. Yu. Kjromova, D. A. Ivashchenko, S. N. Tandura, A. E. Chernyshev, V. N. Bochkarev, N. A. Minaeva, V. S. Nikitin, N. V. Alekseev, and V. F. Mironov, *J. Gen. Chem. U.S.S.R. (Engl. Transl.)*, 1983, **53**, 1187.

(28) 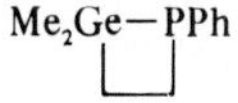 (29) 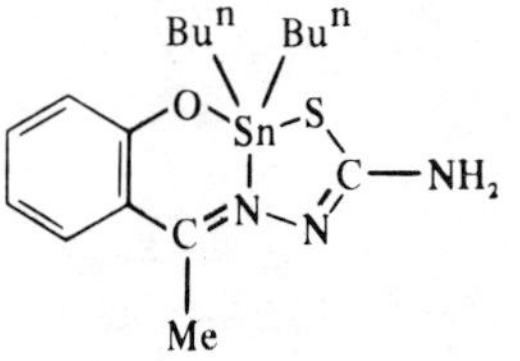 (30)

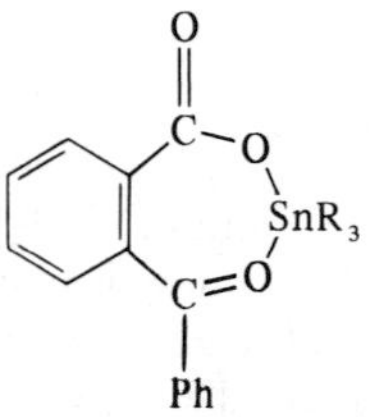

(31) 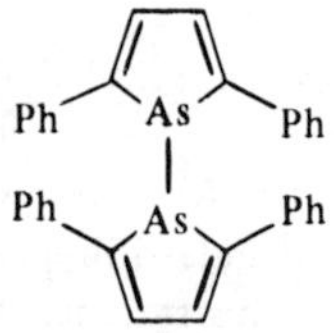 (32)

fragmentation pathways have been identified, namely loss of a methyl radical and rearrangement reactions involving loss of CH_2S and C_2H_4.[39] The molecular ion has been observed in the spectrum of the phosphagermetane (29).[40] Studies of the tin(IV) compounds (30)[41] and (31; R = Me, Et, Pr^n, Bu^n, Ph, or cyclohexyl)[42] have been reported. Molecular ions are observed only for (30) and (31; R = Ph), and the main fragmentation reactions involve cleavage of the Sn—C bond.

In the EI spectrum of the diarsole (32) a prominent molecular-ion peak and fragmentation reactions involving cleavage of the As—As bond and loss of phenylacetylene have been observed.[43] Studies of the spectra of the xanthates $PhMet(S_2COR)_2$ (Met = As, R = Me or Pr^i; Met = Sb, R = Pr^i)[44] and $Me_2Bi(S_2COR)$ (R = Me, Et, Pr^n, Pr^i, Bu^n, or Bu^i)[45] have shown that none of the compounds exhibits molecular ions and that ions arising from loss of the xanthate group or from metal–carbon cleavage are present. Similar behaviour has been observed for the bismuth carboxylates $RBi(O_2CCF_3)_2$ (R = Me or Ph) and $R_2Bi(O_2CCF_3)$ (R = Ph or p-MeC_6H_4).[46]

Cyclic Compounds without Carbon Atoms in the Ring. – The EI spectra of a large number of peralkylcyclosilanes $(R_2Si)_n$ ($n = 3$;[47,48] $n = 4$;[49,50] $n = 5$–7[50])

39 J. Barrau, G. Rima, and J. Satgé, *J. Organomet. Chem.*, 1983, **252**, C73.
40 J. Escudie, C. Couret, J. Satgé, and J. D. Andriamizaka, *Organometallics*, 1982, **1**, 1261.
41 A. Saxena and J. P. Tandon, *Polyhedron*, 1984, **3**, 681.
42 S. W. Ng and J. J. Zuckerman, *J. Organomet. Chem.*, 1983, **249**, 81.
43 G. Märkl and H. Hauptmann, *J. Organomet. Chem.*, 1983, **248**, 269.
44 R. K. Gupta, A. K. Rai, R. C. Mehrotva, and V. K. Jain, *Polyhedron*, 1984, **3**, 721.
45 M. Wieber and H. G. Rüdling, *Z. Anorg. Allg. Chem.*, 1983, **505**, 150.
46 G. B. Deacon, W. R. Jackson, and J. M. Pfeiffer, *Aust. J. Chem.*, 1984, **37**, 527.
47 S. Masumune, H. Tobita, and S. Murakami, *J. Am. Chem. Soc.*, 1983, **105**, 6524.
48 H. Watanabe, T. Okawa, M. Kato, and Y. Nagai, *J. Chem. Soc., Chem. Commun.*, 1983, 781.
49 B. J. Helmer and R. West, *Organometallics*, 1982, **1**, 1458.
50 H. Watanabe, T. Muraoka, M. Kageyama, K. Yoshizumi, and Y. Nagai, *Organometallics*, 1984, **3**, 141.

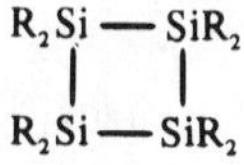

$R_2Si—SiR_2$
$R_2Si—SiR_2$

(33)

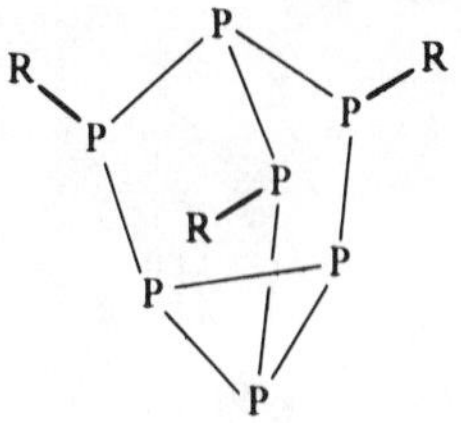

(34)

[*e.g.* (33)] and of octakis(trimethylsilyl)cyclotetrasilane[51] have been reported, and evidence for the formation of polycyclostibanes $(RSb)_n$ has been obtained from mass-spectral studies.[18] All the cyclosilanes investigated exhibit molecular ions. In contrast, the molecular-ion peak is absent from the spectrum of the cyclostannane $(PhSn)_6$.[52] The fragmentation of the peralkylcyclosilanes generally involves initial loss of either an alkyl substituent or an olefin from the alkyl substituent to leave an Si—H bond. In the tetra and penta systems the cyclic Si_{4-5} unit remains intact during fragmentation, whereas in the tri systems ring fragmentation occurs.[50] The mass spectra of geometrical isomers of $(Bu^tMeSi)_4$ are indistinguishable.[49] Molecular ions have been reported for the heptaphosphanortricyclene derivatives (34; R = Pr^i, $SiMe_3$, or $SnMe_3$), and mass spectrometry has been used to identify the compounds $P_7(SiMe_3)_{3-n}(SnMe_3)_n$.[53]

The spectra of the boron heterocycles (35) and (36),[3] of the cyclic siloxanes $(Bu^tMeSi)_4O_n$ (n = 1, 2, or 3) [*e.g.* (37)],[54] and of some silicon–nitrogen heterocycles such as (38)[55] have been reported. For the latter it has been suggested that fragmentation affords ions that retain the original ring structure and that the elimination of neutral species from the exocyclic chains leads to the formation of new rings. Mass spectrometry has shown that the novel cycloaluminadisilatriazane (39)[56] is dimeric in the vapour state and has indicated cyclic structures for the compounds $[Et_2MetPEt_2]_3$ (Met = Ga or In) (40).[57] The fragmentation behaviour of both compounds (39) and (40) involves preferential loss of exocyclic groups, suggesting considerable stability for the six-membered rings in these compounds. EI studies of some phosphorus–tin heterocycles [*e.g.* (41)],[58] of various phosphazene monomers [*e.g.* (42; $R^1 = Pr^n$, R^2 = Cl)],[59–61]

[51] Y. Chen and P. P. Gaspar, *Organometallics*, 1982, **1**, 1410.
[52] M. Dräger, B. Mathiasch, L. Ross, and M. Ross, *Z. Anorg. Allg. Chem.*, 1983, **505**, 99.
[53] G. Fritz, K. D. Hoppe, W. Hönle, D. Weber, C. Mujica, V. Manriquez, and H. G. Schnering, *J. Organomet. Chem.*, 1983, **249**, 63.
[54] B. J. Helmer and R. West, *Organometallics*, 1982, **1**, 1463.
[55] M. M. Il'in, A. Moskovkin, V. N. Tolanov, I. V. Miroschnichenko, V. N. Bochkarev, and A. E. Chernyshev, *J. Gen. Chem. U.S.S.R. (Engl. Transl.)*, 1983, **53**, 91.
[56] V. Wannagat, T. Blumenthal, D. J. Brauer, and H. Bürger, *J. Organomet. Chem.*, 1983, **249**, 33.
[57] F. Maury and G. Constant, *Polyhedron*, 1984, **3**, 581.
[58] M. Baudler and H. Suchomel, *Z. Anorg. Allg. Chem.*, 1983, **505**, 39.
[59] H. R. Allcock, P. R. Susko, and T. L. Evans, *Organometallics*, 1982, **1**, 1443.
[60] R. D. Minard and H. R. Allock, *Org. Mass Spectrom.*, 1982, **17**, 351.
[61] P. J. Harris and K. B. Williams, *Org. Mass Spectrom.*, 1984, **19**, 248; *Inorg. Chem.*, 1984, **23**, 1495.

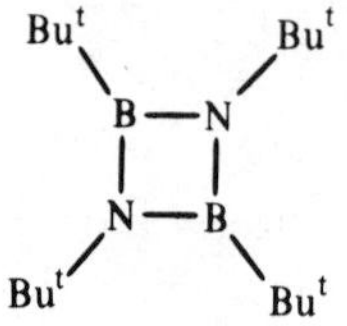

(35) 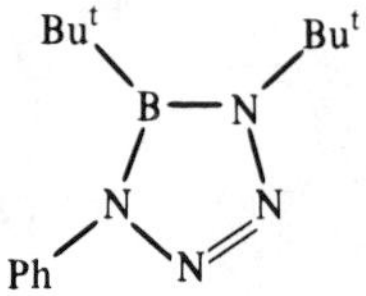 (36) 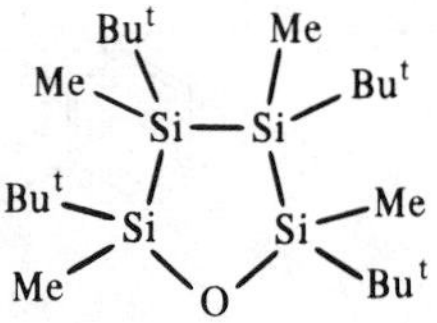 (37)

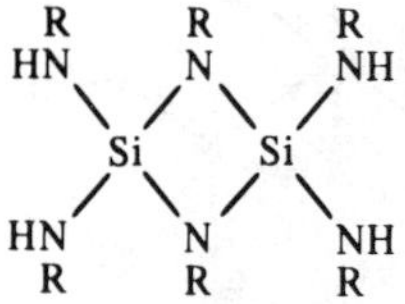

(38) $R = Me_3Si$ 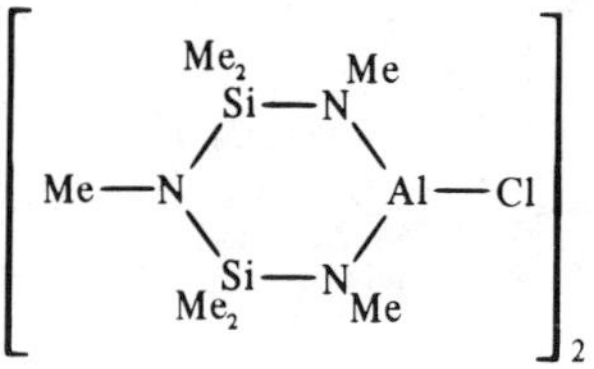 (39)

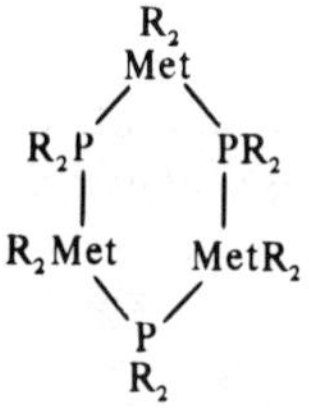

(40) 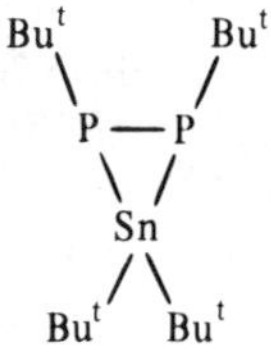 (41) 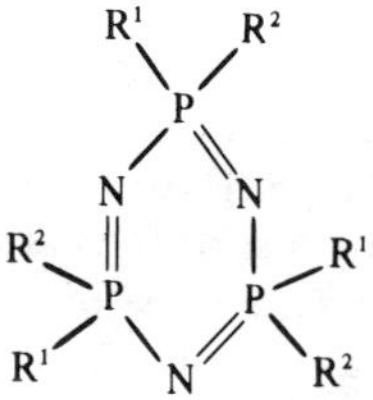 (42)

and of an interesting mass-spectrometrically induced polymerization of (42; $R^1 = R^2 = PhO$)[62] have been reported. Finally, the laser desorption (LD) spectrum of a phosphazene of high molecular weight has been described.[63]

3 Transition-metal Organometallic Compounds

Metal Carbonyl and Related Complexes. – Studies of the positive-ion EI spectra of many complexes containing carbonyl as well as other neutral ligands {*e.g.* $M(CO)_5L$ [M = Cr, L = Me_2NNS;[64] M = Cr or W, L = 1,8-naphthyridine or phthalazine;[65] M = Cr, Mo, or W, L = $AsMe_{3-n}(NMe_2)_n$;[66] M = Cr or W, L = MeNC or CF_3NC[67]], $M(CO)_4L$ [M = Cr, Mo, or W, L = (43; E = P or As);[68] M = Cr, L = $Me_2PCH_2PBu^t{}_2$ or $Ph_2PCH_2AsPh_2$[69]], (44),[70] and

[62] M. Glerio, G. Audisio, P. Traldi, S. Daolio, and E. Vecchi, *J. Chem. Soc., Chem. Commun.*, 1983, 1380.
[63] R. J. Cotter and J. C. Tabet, *Int. J. Mass Spectrom. Ion Phys.*, 1983, **53**, 151.
[64] H. W. Roesky, R. Emmert, W. Isenberg, M. Schmidt, and G. M. Sheldrick, *J. Chem. Soc., Dalton Trans.*, 1983, 183.
[65] K. R. Dixon, D. T. Eadie, and S. R. Stobart, *Inorg. Chem.*, 1982, **21**, 4318.
[66] F. Kober and M. Kerber, *Z. Anorg. Allg. Chem.*, 1983, **505**, 119.
[67] D. Lentz, *Chem. Ber.*, 1984, **117**, 415.

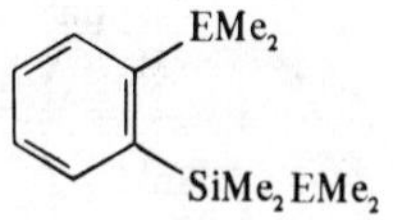

EMe_2 / $SiMe_2EMe_2$

(43)

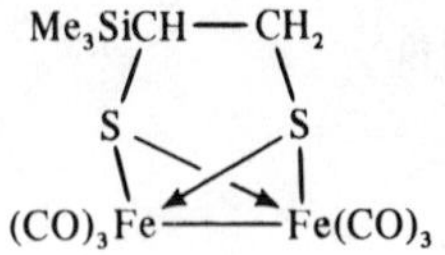

$Me_3SiCH—CH_2$, S, S, $(CO)_3Fe$, $Fe(CO)_3$

(44)

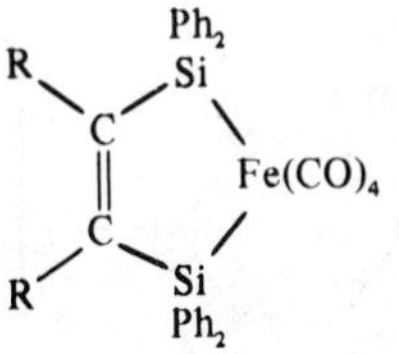

Ph_2 Si, R, C, C, R, Si Ph_2, $Fe(CO)_4$

(45)

OC, CO, C, O, Re, OC, X, N, R, H N—R, C—R

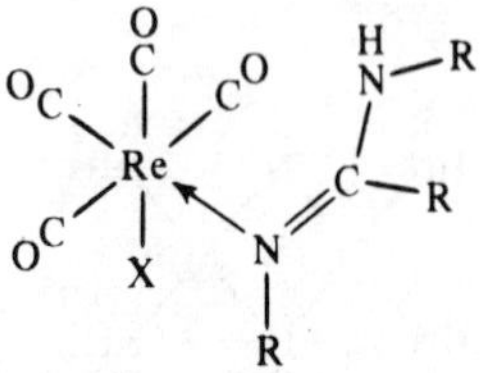

(46) R = H, alkyl, or aryl
X = Cl or Br

$Re_2(CO)_{10-n}(PPh_3)_n$ ($n = 2$, 3, or 4)[71]} have provided further examples for the preferential loss of carbonyl ligands from the molecular ions of complexes of this type. However, the molecular ion of $W(CO)_4[(Ph_2P)_2NCH_2CH_2OH]$ fragments by loss of H_2O as well as of CO.[72]

Preferential loss of carbonyl ligands from the molecular ion is also shown by ruthenium[73] and iron[74] carbonyls containing one-electron ligands, such as $Ru(CO)_3(dipy)F$, $Ru(CO)_3(SbPh_3)_2F$, $Fe_2(CO)_8(SiPh_2)_2$, $Fe_2(CO)_7(SiMePh)_2$, $Fe_2(CO)_6(PPh_3)_2(SiPh_2)_2$, and the metallocycles (45; R = Me or Et). In contrast, in the spectra of complexes of type (46), no molecular ion is observed; the peak of highest mass corresponds to $[M-X^{\bullet}]^+$, and when X = Br the ion $[Re(CO)_4Br]^+$ is also present in some cases.[75] Similarly, the fragmentation of $Fe(CO)_4(SnMe_3)_2$ involves preferential loss of methyl radical.[76]

Several mass-spectral studies of metal carbonyls and related compounds involving the use of other ionization techniques have been reported. Field desorption has been found useful in characterizing the complexes (47; $L^1 = L^2 = CO$; $L^1 = CO$, $L^2 = py$; $L^1 = L^2 = py$; $L^1 = L^2 = PPh_3$).[77] For the complex *fac*-$[Re(CO)_3(dipy)Cl]$ it has been reported that its FAB mass spectrum contains no ion with all the ligands intact and that the base peak corresponds to

[68] P. Aslanidis and J. Grobe, *J. Organomet. Chem.*, 1983, **249**, 103.
[69] L. Weber and D. Wewers, *Chem. Ber.*, 1984, **117**, 1103.
[70] E. A. Chernyshev, O. V. Kuz'min, A. V. Lebedev, A. I. Gusev, M. G. Los', N. V. Alekseev, N. S. Nametkin, V. D. Tyurin, A. M. Krapivin, N. A. Kubasova, and V. G. Zaikin, *J. Organomet. Chem.*, 1983, **252**, 143.
[71] S. W. Lee, L. F. Wang, and C. P. Cheng, *J. Organomet. Chem.*, 1983, **248**, 189.
[72] H. Fick and W. Beck, *J. Organomet. Chem.*, 1983, **252**, 83.
[73] E. Horn and M. R. Snow, *Aust. J. Chem.*, 1984, **37**, 35.
[74] F. H. Carré and J. J. E. Moreau, *Inorg. Chem.*, 1982, **21**, 3099.
[75] J. A. Clark and M. Kilner, *J. Chem. Soc., Dalton Trans.*, 1984, 389.
[76] H. J. Breunig, *Polyhedron*, 1984, **3**, 757.
[77] P. O. Nubel and T. L. Brown, *Organometallics*, 1984, **3**, 29.

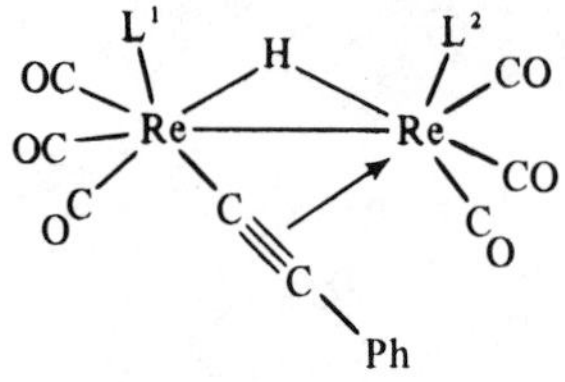

L¹ H L² OC CO Re Re OC CO C C OC C O Ph

(47)

$[M - Cl^{\bullet}]^{+}$.[78] No other details of the spectrum have been provided, but mechanisms for the formation of $[M - Cl^{\bullet}]^{+}$ ions have been considered. Similarly, for the complex *cis,trans*-$[Mo(CO)_2(PBu^n{}_3)_2(dipy)]$ loss of hydrogen and formation of the ions $[M - 1]^{+}$ and $[M - PBu^n{}_3 - 1]^{+}$ have been reported.[79] In the FAB spectra of $[Ir(CO)(PPh_3)_2Cl]$ and its rhodium analogue, molecular ions have been observed together with ions arising from loss of CO, PPh_3, or $Cl^{\bullet}$. However, the molecular ion was not represented in the spectrum of $[Rh(PPh_3)_3Cl]$.[80] An extensive study of the SIMS spectra of compounds containing cationic complexes of chromium and molybdenum {*e.g.* $[Cr(NO)(CNR)_{5-n}(L)_n](PF_6)$ ($n = 0$ or 1, L = Cl or amine), $[Mo(NO)(CNR)_5](PF_6)$, and $[Mo(CNR)_7](PF_6)_2$} has demonstrated the usefulness of this technique for the study of certain types of complex.[81] It has been shown that informative spectra, exhibiting the molecular cation and a characteristic fragmentation pattern, can be readily obtained for compounds that contain either stable monocations or stable dications that may be readily reduced to the corresponding monocation.

Complexes Containing Hydrocarbon Ligands. – Molecular ions have been observed in the EI spectra of the metallocycles (48; Met = Ti, Zr, Hf, or Nb).[82] The main fragmentation route leads to $[M - C_8H_6(SiMe_3)_2]^{+\bullet}$ for the titanium and niobium complexes, to $[M - C_8H_8(SiMe_3)_2]^{+\bullet}$ for the zirconium and hafnium analogues, and to the respective organic fragment peaks. This difference in fragmentation behaviour may reflect the readier availability of lower oxidation states for titanium and niobium, thus allowing for reductive elimination. The spectra of several η^1, η^2- and η^3-cyclo-octenyl and η^3-allyl-β-diketonate nickel complexes [*e.g.* (49)] have been reported.[83] Molecular ions are present in all spectra, but generally these are not abundant and are accompanied by one or two metal-containing fragment ions. In the η^3-allyl complexes $(\eta^3\text{-}C_3H_4R^1)$-$Met(CO)_3(CNR^2)$ (Met = Mn or Re, R^1 = H or Me, R^2 = Me or $SnMe_3$)[84] and the η^1,η^2-azaborolinyl complexes (50; Met = Mn, L = CO; Met = Mo, L = $\eta^3\text{-}C_3H_5$)[85] molecular ions that fragmented by loss of CO have been observed.

[78] R. L. Cerny, B. P. Sullivan, M. M. Bursey, and T. J. Meyer, *Anal. Chem.*, 1983, **55**, 1954.
[79] J. A. Conner and C. Overton, *J. Chem. Soc., Dalton Trans.*, 1982, 2397.
[80] T. R. Sharp, M. R. White, J. F. Davis, and P. J. Stang, *Org. Mass Spectrom.*, 1984, **19**, 107.
[81] J. L. Pierce, D. E. Wigley, and R. A. Walton, *Organometallics*, 1982, **1**, 1328.
[82] M. F. Lappert, C. L. Raston, B. W. Shelton, and A. H. White, *J. Chem. Soc., Dalton Trans.*, 1984, 893.
[83] W. Keim, A. Behr, and G. Kraus, *J. Organomet. Chem.*, 1983, **251**, 377.
[84] M. Moll, H. Behrens, H.-J. Seibold, and P. Merbach, *J. Organomet. Chem.*, 1983, **248**, 329.
[85] G. Schmid, V. Höhner, D. Kampmann, F. Schmidt, D. Bläser, and R. Boese, *Chem. Ber.*, 1984, **117**, 672.

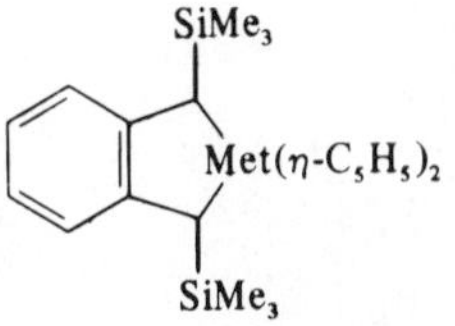

SiMe3 Met(η-C_5H_5)$_2$ SiMe3 (48)

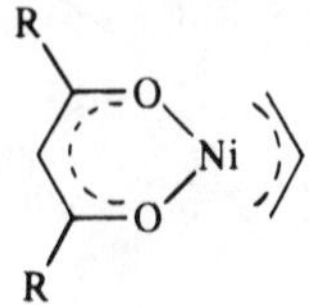

R O Ni O R (49)

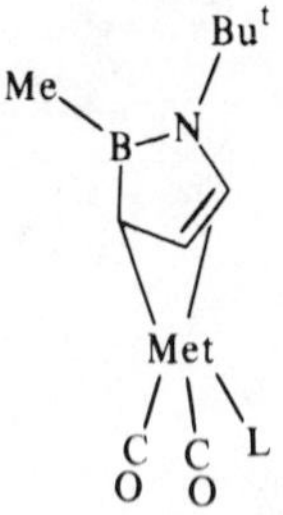

Me B N Bu^t Met C O C O L (50)

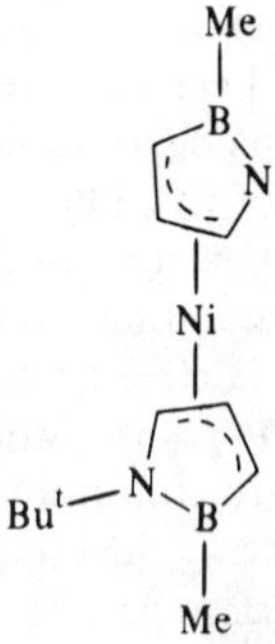

Me B N Bu^t Ni N Bu^t B Me (51)

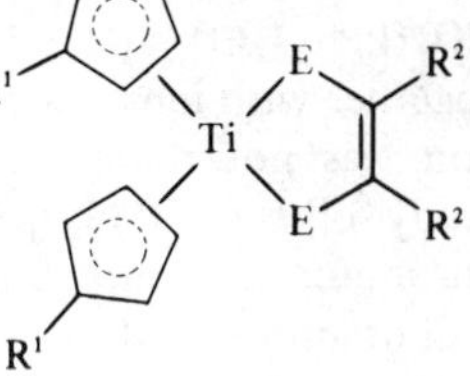

R^1 Ti E E R^2 R^2 R^1 (52)

The EI spectra of various (η^3-cycloalkenyl)(η^5-C_5H_5)Ni complexes,[86] the η^3-azaborolinyl complex (51),[87] (η^5-C_5H_5)$_2$Sc(BH_4),[88] (η^5-$C_5H_4R^1$)(η^5-$C_5H_5R^2$)-Nb(O)Cl [R^1 = H, R^2 = Me; R^1 = H, R^2 = Bu^t; R^1 = Me, R^2 = Bu^t; R^1 = Bu^t, R^2 = Ph(Me)CH],[89] (η^5-C_5H_5)$_2$W(R^1)(OR^2) (R^1 = Me, R^2 = Ph, *p*-MeC_6H_4, *p*-$MeOC_6H_4$, or $PhCH_2$; R^1 = *p*-MeC_6H_4, R^2 = Me),[90] the titanium complexes (52; E = S, R^1 = H or Me, R^2 = $MeCO_2$; E = Se, R^1 = H, R^2 = $MeCO_2$ or F),[91] and the metallocycloalkanes (53; Met = Mo or W, n = 0, 1, or 2)[92] have been reported. Whereas the fragmentation of the scandium, niobium, and titanium complexes involves loss of intact cyclopentadienyl groups from their respective molecular ions, no such reaction is shown by the molybdenum or tungsten complexes. Molecular ions have been observed in the EI spectra of several ferrocenophanes,[93,94] such as (54)–(56), and of the oligomeric ferrocenophanes

86 H. Lehmkuhl, A. Rufinska, C. Naydowski, and R. Mynott, *Chem. Ber.*, 1984, **117**, 376.
87 G. Schmid, D. Kampmann, U. Höhner, D. Bläser, and R. Boese, *Chem. Ber.*, 1984, **105**, 1052.
88 M. Mancine, P. Bougeard, R. C. Burns, M. Mlekuz, B. G. Sayer, J. I. A. Thompson, and M. J. McGlinchey, *Inorg. Chem.*, 1984, **23**, 1072.
89 R. Broussier, J. D. Olivier, and B. Gautheron, *J. Organomet. Chem.*, 1983, **251**, 307.
90 M. Canestrari and M. L. H. Green, *Polyhedron*, 1982, **1**, 629.
91 C. M. Bolinger and T. B. Rauchfuss, *Inorg. Chem.*, 1982, **21**, 3947.
92 E. Lindner, E. V. Küster, W. Hiller, and R. Fawzi, *Chem. Ber.*, 1984, **117**, 127.
93 D. Seyferth and H. P. Withers, jun., *Organometallics*, 1982, **1**, 1275.
94 J. Mirek, S. Rachwal, and B. Kawalek, *J. Organomet. Chem.*, 1983, **248**, 107.

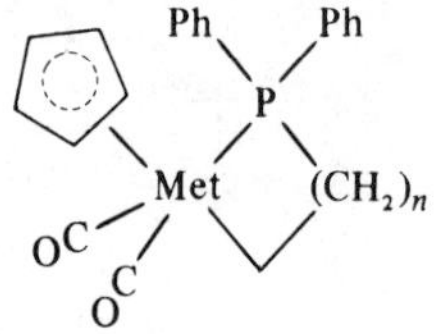

(53) 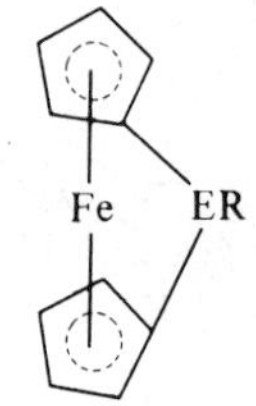 (54) 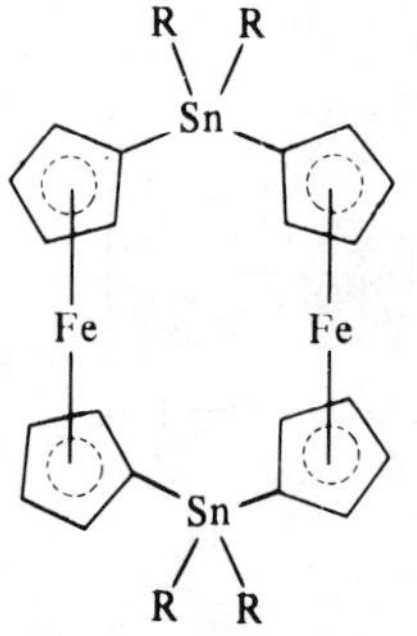 (55)

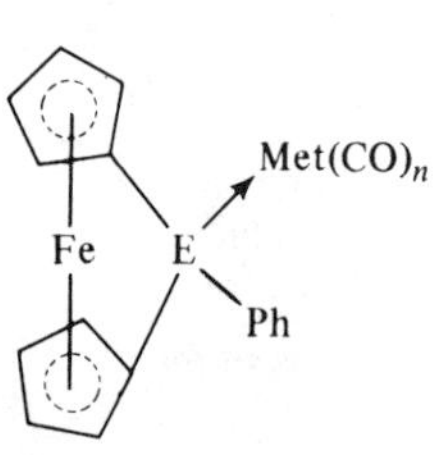

(56) 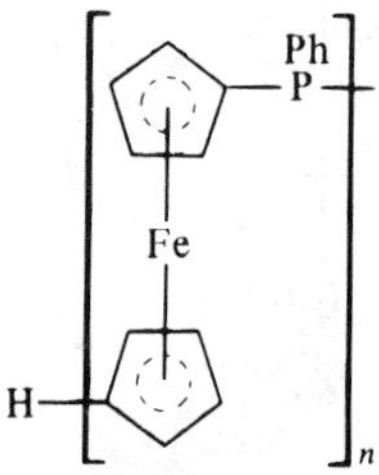 (57) $n = 2$–4 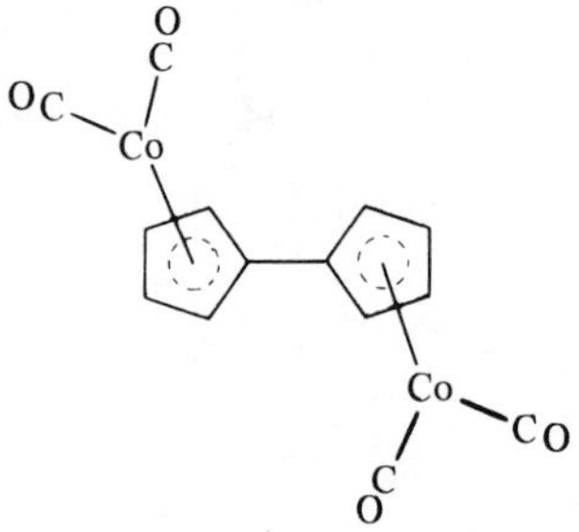 (58)

(57).[95] Of special interest is the fragmentation behaviour of compounds of type (56), whose molecular ions decompose by successive loss of CO followed by loss of the metal atom. The spectra of all the fulvalene complexes $(\eta^5,\eta^5\text{-}C_{10}H_8)(CO)_6Met_2$ (Met = Cr or W) and $(\eta^5,\eta^5\text{-}C_{10}H_8)(CO)_5MoRu$ [*e.g.* (58) and (59)] show molecular ions, but these are more prominent in the spectra of the complexes incorporating metals of the second or third transition series.[96] The EI spectra of the *exo*-alkyl-η^5-cyclohexadienyl complexes (60; R = Me, Et, Pr^n, Bu^n, Bu^t, or Ph)[97] and the aroyl-substituted bis-(η^6-arene)chromium complexes (61; *e.g.* $R^1 = H$, $R^2 = Ph$; $R^1 = PhCO$, $R^2 = Ph$)[98] have been reported, and several $Fe(\eta^6\text{-arene})(POMe_3)_2$ and $Fe(\eta^6\text{-arene})(\eta^4\text{-diene})$ complexes have been characterized by means of high-resolution mass spectra.[99]

FD mass spectrometry has proved useful in determining the molecular weights of several thermally labile complexes such as $(\eta^5\text{-}C_5H_5)Met(CO)_2\text{-}[PPh_2(CH_2)_nCl]$ (Met = Mo or W, $n = 1$–3),[92] (62),[92] and (63).[100–101] In the case of the ferracyclic complex (63), as well as other related compounds, com-

[95] H. P. Withers, jun., D. Seyferth, J. D. Fellmann, P. E. Garron, and S. Martin, *Organometallics*, 1982, **1**, 1283.
[96] K. P. C. Vollhardt and T. W. Weidman, *Organometallics*, 1984, **3**, 82.
[97] H. Werner, R. Werner, and C. Burschka, *Chem. Ber.*, 1984, **117**, 152.
[98] C. Elschenbroich, E. Bilger, J. Heck, F. Stohler, and J. Heinzer, *Chem. Ber.*, 1984, **117**, 23.
[99] S. D. Ittel and C. A. Tolman, *Organometallics*, 1982, **1**, 1432.
[100] N. Bild, E. R. F. Gesing, C. Quiquerez, and A. Wehrli, *J. Chem. Soc., Chem. Commun.*, 1983, 172.
[101] N. Bild and E. R. F. Gesing, *J. Organomet. Chem.*, 1983, **248**, 85.

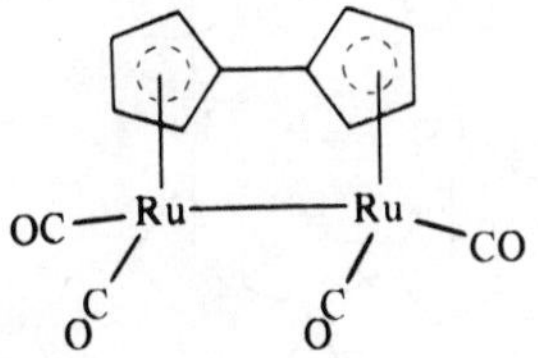

(59) 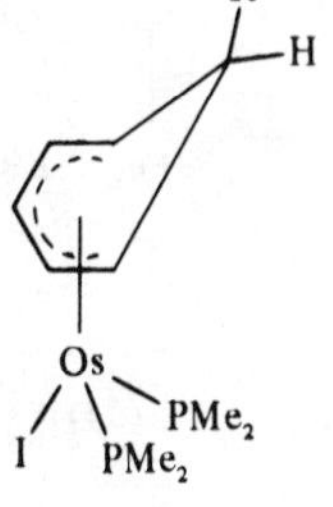 (60) 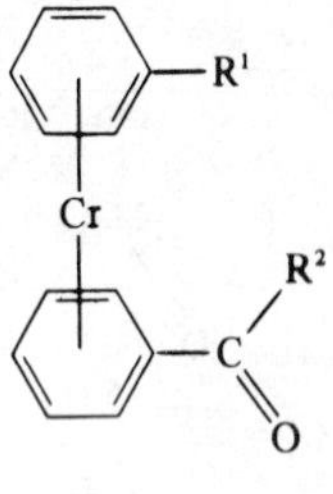 (61)

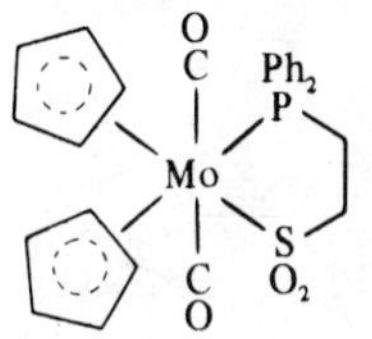

(62) 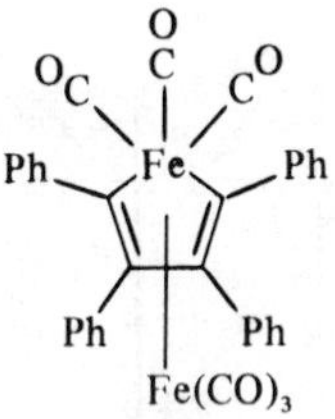 (63) 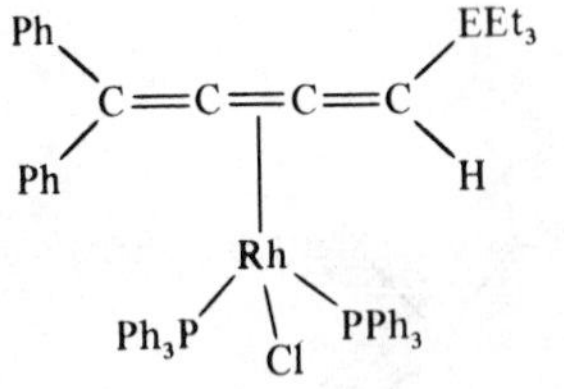 (64) E = Si, Ge, or Sn

parative EI studies have also been reported.[100–101] These have shown that the application of routine EI techniques to such thermally labile compounds provides little or no information regarding their composition. However, it has been established that, by lowering the ion-source and sample-inlet temperatures to 40–50 °C below the melting/decomposition temperature of the compound under investigation and by using an 'in-beam' mode of sample introduction, moderately abundant molecular ions can be obtained. FD mass spectra of several cationic η^3-allylic complexes of palladium and nickel {*e.g.* [(η^3-C_3H_3Me)-$Pd(PPh_3)_2$](PF_6)} have been reported.[102] These spectra can be obtained easily and are characterized by the cation being abundant and there being very few fragments. In contrast, the FAB spectra of these complexes are difficult to produce and exhibit relatively few intact cations and several fragment ions. Few molecular ions and many fragment ions have also been observed in the FAB spectra of various cumulene complexes [*e.g.* (64)] and in the spectrum of (η^2-C_2H_4)$Pt(PPh_3)_2$.[80] In these compounds the fragment ions arise mainly by cleavage of metal–ligand bonds. Comparative studies of the methane CI and EI mass spectra of several hydrocarbon complexes such as (η^5-C_5H_5)$_2$Cr, (η^5-C_5H_5)$Mn(CO)_3$, and (η^5-C_6H_6)$Cr(CO)_3$ have been reported.[103] Generally, the CI spectra show abundant ions in the molecular-ion region ($[M + H]^+$ and $M^{+\bullet}$) and few or no fragment ions. The formation of $M^{+\bullet}$ ions by charge transfer is more important than the formation of $[M + H]^+$ by protonation, reflecting the

[102] I. Thatchenko, D. Neibecker, D. Fraisse, F. Gomez, and D. F. Barofsky, *Int. J. Mass Spectrom. Ion Phys.*, 1983, **46**, 499.
[103] J. Müller, E. Baumgartner, and C. Hänsch, *Int. J. Mass Spectrom. Ion Phys.*, 1983, **47**, 523.

low IEs of the complexes. The $M^{+\bullet}$ and $[M + H]^+$ ions show different fragmentation pathways, indicating that, as with organic compounds, complementary information can be obtained from CI and EI spectra of organometallic complexes. For complexes of type (η^3-arene)$Cr(CO)_3$ [arene = *e.g.* PhCOR or 1,3,5-$(Me)_3C_6H_3$], negative-ion CI spectra, using CH_4 or NH_3 as reactant gas, have been reported. The complexes containing the PhCO group show abundant $M^{-\bullet}$ ions, whereas complexes containing alkyl-substituted arenes exhibit $[M - H]^-$ as the base peak as well as $[M + H]^-$ ions.[104]

Transition-metal Cluster Compounds. – The EI spectra of a wide range of clusters have been reported,[105–109] and the use of field desorption for the characterization of such compounds continues.[109–112] Applications of DCI[113] and FAB[114] mass spectrometry to the study of clusters have been noted, and the value of mass spectrometry in characterizing carbido clusters has been briefly discussed.[115]

4 Chelate, Macrocyclic, and Other Complexes

Considerable interest continues in the mass-spectral behaviour of metal chelates, macrocyclic complexes, and other types of metal complex. Most studies have been concerned with EI spectra, but extensive use of other ionization techniques (CI, FAB, FD, LD, EHI, ^{272}Cf plasma desorption) has also been made. The application of some of the latter techniques has produced very useful results and has demonstrated their value in handling higher-mass, unstable, or charged complexes.

Metal Chelates. – *Metal β-Diketonates*. The EI spectra of several chelates derived from 1,1,1-trichloro-2,4-pentanedione (tcacH) of types $Met(tcac)_2$ (Met = Cu, Co, or Mn) and $Met(tcac)_3$ (Met = Al, Cr, or Fe) have been examined and contrasted with analogous complexes derived from fluorinated or non-halogenated ligands.[116] Transfer of chlorine from ligand to metal accompanied by elimination of CO or other neutral even-electron fragments is an important type of reaction

104 G. A. Vaglio, P. Volpe, and L. Operti, *Org. Mass Spectrom.*, 1982, **17**, 617.
105 D. E. Samkoff, J. R. Shapley, M. R. Churchill, and H. J. Wasserman, *Inorg. Chem.*, 1984, **23**, 397.
106 A. C. Willis, G. N. Buuren, R. K. Pomeroy, and F. W. B. Einstein, *Inorg. Chem.*, 1983, **22**, 1162.
107 S. P. Foster, K. M. Mazckey, and B. K. Nicholson, *J. Chem. Soc., Chem. Commun.*, 1982, 1156.
108 D. Seyferth and H. P. Withers, *Organometallics*, 1982, **1**, 1294.
109 M. R. Churchill, C. Bueno, J. T. Park, and J. R. Shapley, *Inorg. Chem.*, 1984, **23**, 1017.
110 M. Mlekuz, P. Bougeard, M. J. McGlinchey, and G. Jaouen, *J. Organomet. Chem.*, 1983, **253**, 117.
111 J. B. Keister and J. R. Shapley, *Inorg. Chem.*, 1982, **21**, 3304.
112 J. T. Park, J. R. Shapley, M. R. Churchill, and C. Bueno, *Inorg. Chem.*, 1983, **22**, 1579.
113 C. Brunnée, *Int. J. Mass Spectrom. Ion Phys.*, 1982, **45**, 51.
114 M. R. White and P. J. Stang, *Organometallics*, 1983, **2**, 1382.
115 B. F. G. Johnson, J. Lewis, W. J. H. Nelson, J. N. Nicholls, and M. D. Vargas, *J. Organomet. Chem.*, 1983, **249**, 255.
116 M. L. Morris and R. D. Koob, *Org. Mass Spectrom.*, 1982, **17**, 503.

in all complexes. This rearrangement was rationalized in terms of the hard/soft acid/base concept.[116] The complexes show internal redox reactions analogous to those shown by the complexes derived from fluorinated or non-halogenated ligands, but the importance of these reactions was reported to be reduced by the facility of the chlorine rearrangement. EI spectra of some palladium(II) chelates of fluorinated diketones[117] and of tris(acetylacetonato)holmium(III)[118] have been reported. Complex mixtures of geometrical isomers of tris-(β-diketonato)chromium(III) have been investigated using combined GC/MS.[119]

The SIMS spectra of the acetylacetonato complexes $Met(acac)_2$ (Met = Ni or Cu), $Met(acac)_3$ (Met = Fe, Cr, Mn, or Co), $VO(acac)_2$, and $MoO_2(acac)_2$ as well as of the related complexes $Cr(tfac)_3$ (tfacH = CF_3COCH_2COMe) and $Cr(btac)_3$ (btacH = $PhCOCH_2COCF_3$) have been reported.[120] Unlike the EI mass spectra of the tris-chelates, which show molecular ions of type $[Met(\beta\text{-dike})_3]^{+\cdot}$, their SIMS spectra do not contain this form of molecular ion. A feature of the SIMS spectra of the tris-chelates is the formation of dimetallic ions, including cationization of $Met(\beta\text{-dike})_3$ to give the structurally informative secondary ions $[C + Met(\beta\text{-dike})_3]^+$ (C = cationizing agent). In addition to cationization by sodium (from the NaCl matrix) and silver (from the silver support), self-cationization and cationization by another first-row transition metal have also been observed. The spectra of both the nickel and copper bis-chelates show metal cluster ions $[Met_n(acac)_m]^+$ up to $n = 3$. The observation of these cluster ions in the vapour phase shows no correlation with the degree of association of the neutral complexes in the solid state. The spectra of $VO(acac)_2$ and $MoO_2(acac)_2$ show few molecular ions and fragment ions arising from loss of oxygen or the acac ligand. The behaviour of these complexes has been rationalized in terms of facile cationization at the nucleophilic terminal oxygen atom to yield a species, not necessarily observed, which is the precursor of fragment ions of lower mass.

Laser-desorption spectra of $Co(acac)_3$ and $Fe(acac)_3$ supported on silver foils contain dimetallic ions, including those that incorporate silver from the foil, *e.g.* $[Met_2(acac)_3]^+$ and $[AgMet(acac)_2]^+$.[120] In the LD spectra, as in SIMS, no intact molecular ions are observed; however, many of the same dimetallic cluster ions are formed.

Metal Oximates. The positive-ion EI spectra of several nickel,[121,122] palladium,[121] and platinum[121] *vic*-dioximato complexes $MetL_2$(LH = *vic*-dioxime) (65) have been studied. In all cases the molecular ions are very stable and in most cases correspond to the base peak. Usually, the stabilities of the molecular ions are in the sequence Pt > Ni > Pd. The main decomposition routes of the molecular ions involve loss of the fragments (L − H), $L^\cdot$, or (L + H). The negative-ion spectra of some nickel complexes (65; $R^1 = R^2 = H$; $R^1 = H$, $R^2 = Me$;

[117] M. Das, *Inorg. Chim. Acta*, 1983, **76**, L111.
[118] N. V. Zakurin, Yu. S. Nekrasov, M. D. Reshetova, and A. Yu. Vasil'kov, *Inorg. Chem. Acta*, 1983, **76**, L161.
[119] R. E. Sievers and K. C. Brooks, *Int. J. Mass Spectrom. Ion Phys.*, 1983, **47**, 527.
[120] J. L. Pierce, K. L. Busch, R. G. Cooks, and R. A. Walton, *Inorg. Chem.*, 1982, **21**, 2597.
[121] J. B. Westmore and D. K. C. Fung, *Inorg. Chem.*, 1983, **22**, 902.

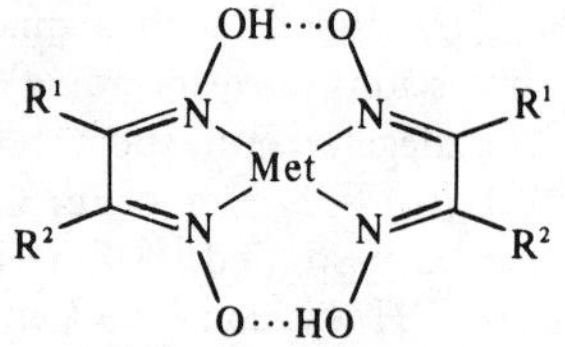

(65) 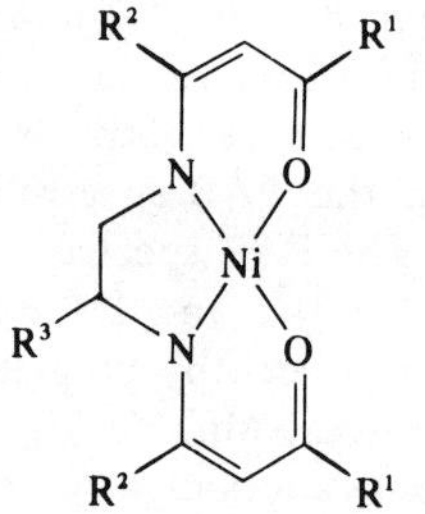 (66) 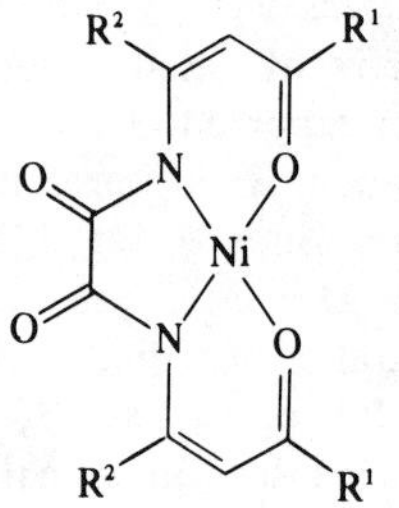 (67)

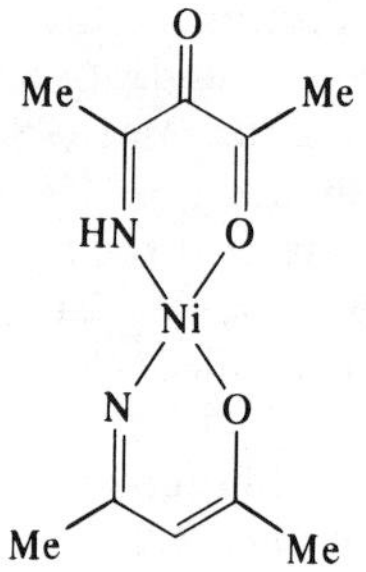

(68)

$R^1 = R^2 = Me$; $R^1 = R^2 = Ph$) have been reported.[122] In all cases, except when $R^1 = R^2 = H$, the spectra show molecular ions that fragment mainly by loss of $OH^{\bullet}$ and H_2O. The spectra of cadmium(II) and lead(II) salicylaldoximates have been studied.[123]

Schiff-base Chelates. Ligand fragmentation in nickel(II) complexes of Schiff bases derived from 1,1,1-trifluoro-2,4-pentanedione and 1,2-diaminoethane, 1,3-diaminopropane, 1,4-diaminobutane, or 1,2-diaminobenzene shows little change when compared to the fragmentation of the uncomplexed ligands.[124] In contrast, the copper(II) complexes exhibit additional fragmentation pathways that have been attributed to a reduction of the copper co-ordination centre in the complex. Mass-spectral methods have proved useful for characterizing both various nickel(II) complexes (66) derived from tetradentate Schiff bases and products arising from their aerial or chemical oxidation, *e.g.* (67).[125] The EI spectra of both oxidized and parent compounds are characterized by prominent molecular-ion peaks. The isobutane CI spectra generally show prominent $[M + H]^+$ peaks, with smaller peaks also at $[M + 39]^+$ and $[M + 57]^+$. The structure of (68), the product arising from the hydrogen peroxide oxidation of (66; $R^1 = R^2 = Me$, $R^3 = H$), was deduced from its fragmentation behaviour.

[122] J. Charalambous, G. Soobramanien, A. D. Stylianou, G. Manini, L. Operti, and G. A. Vaglio, *Org. Mass Spectrom.*, 1983, **18**, 406.
[123] P. Lumme and P. Knuttila, *J. Therm. Anal.*, 1982, **25**, 139.
[124] M. L. Morris and R. D. Koob, *Org. Mass Spectrom.*, 1983, **18**, 305.
[125] S. Dilli, A. M. Maitra, and E. Patsalides, *Inorg. Chem.*, 1982, **21**, 2832.

Chelates with Metal–Sulphur Bonds. A comparative study of the FD and FAB spectra of the diethyldithiocarbamate complex MoO_2L_2 ($LH = Et_2NCS_2H$) has been reported.[78] The FD spectrum shows only the molecular ion. In contrast, this ion is not present in the FAB spectrum where the ion of highest mass corresponds to $[M-O]^+$. The FAB spectrum also shows ions corresponding to $[M-O-O]^+$, $[M-O-NEt_2]^+$, and $[M-L]^+$. The negative-ion spectra of several other diethyldithiocarbamato complexes $MetL_n$ ($n = 2$, Met = Fe, Co, Ni, Pd, or Cu; $n = 3$, Met = Cr, Mn, Fe, or Co) have been studied.[126] For the tris-chelates, only small abundances of $M^{-\bullet}$ are observed. This is consistent with the electron entering a metal-based orbital to give a reduced $[Met^{II}L_3]^-$ species, which is the precursor of the observed $[Met^{II}L_2]^-$ and $[L]^-$ product ions. For the bis-chelates, a much higher proportion of the total ion current is carried by $M^{-\bullet}$ with $[L]^-$ being the next most abundant species. The positive-ion spectrum of the diethyldithiocarbamato complex $MoL_3 \cdot SOCl_2$ shows the molecular ion, as well as fragment ions corresponding to $[M-Cl^\bullet]^+$ and $[M-S-Cl^\bullet]^+$.[127] This most unusual behaviour suggests that the neutral $SOCl_2$ ligand is bound very strongly to the metal. The positive-ion EI spectrum of bis(benzyldithiocarbamato)platinum(II) has been studied and a fragmentation scheme for its decomposition suggested.[128]

The EI spectra of the metal(II) and metal(III) dithiophosphinato complexes $Met(Et_2PS_2)_n$ (Met = Ni, Pd, or Pt, $n = 2$; Met = Sb, Bi, In, Cr, Rh, or Ir, $n = 3$) are dominated by metal-containing ions and, with the exception of the Sb, Bi, and In compounds, show abundant molecular ions.[129] These features have been attributed to the π-acceptor character of the metal towards the dithiophosphinato ligand and the consequent multiple metal–ligand bonding. Metal-containing ions also dominate the spectra of the metal(I) complexes. The thallium(I) complex, which is monomeric in the solid state, shows only monometallic ions in its spectrum: $[R_2PS_2Tl]^+$ and the bare-metal ion. In contrast, the gold(I) and copper(I) complexes, which are respectively dimeric and tetrameric in solution, show only dimetallic ions in their spectra. In the tetrachloromethane negative-ion CI spectrum of bis(dipropyldithiophosphato)zinc(II) the base peak corresponds to the ligand ion and the only other peak is due to the molecule plus a chloride ion, $[M+Cl]^-$.[130] This ionization technique has also been used for the analysis of dithiophosphato complexes in lubricating hydrocarbon oils. The EI spectrum of bis(thiosalicyldehydro)iron(II) has been reported.[131]

Miscellaneous Neutral Chelates. Both EI and FD mass spectra of the nickel(II) camphorato complex (69) have been reported.[132] Whereas the FD spectrum shows peaks due to the intact dimer, the EI spectrum shows several mono-

[126] I. K. Gregor and M. Guilhaus, *Org. Mass Spectrom.*, 1982, **17**, 575.
[127] H. Sugimoto, T. Higashi, A. Maeda, M. Mori, H. Masuda, and T. Taga, *J. Chem Soc., Chem. Commun.*, 1982, 1234.
[128] G. A. Katsoulos, G. E. Manoussakis, and C. A. Tsipis, *Polyhedron*, 1984, **3**, 735.
[129] S. Heinz, H. Keck, and W. Kuchen, *Org. Mass Spectrom.*, 1984, **19**, 82.
[130] R. P. Morgan, C. A. Gilchrist, K. R. Jennings, and I. K. Gregor, *Int. J. Mass Spectrom. Ion Phys.*, 1983, **46**, 309.
[131] P. J. Marini, K. S. Murray, and B. O. West, *J. Chem. Soc., Dalton Trans.*, 1983, 143.
[132] V. Schurig and W. Bürkle, *J. Am. Chem. Soc.*, 1982, **104**, 7573.

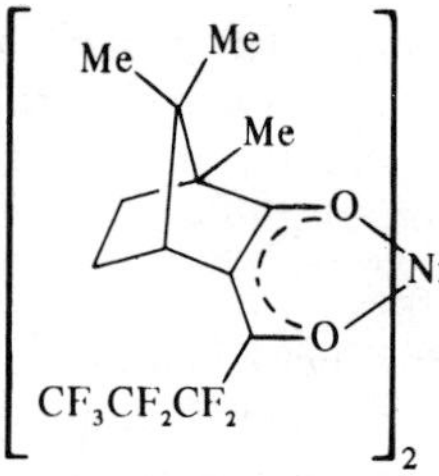

Me Me Me O Ni O $CF_3CF_2CF_2$]$_2$

(69)

metallic ions and only traces of dimetallic species. Several nickel(II) chelates involved as intermediates in the formation of nickel phthalocyaninate from phthalonitrile have been identified using mass spectrometry.[133] The EI spectra of some 2-hydroxynaphthalene-1-azo-2′-pyridine *N*-oxide metal chelates have been reported,[134] and the thermal decomposition of bis-(4-chloroacetoacetanilido)-copper(II) adducts with Lewis bases has been investigated by TGA/MS.[135]

Cationic Chelates. The FAB spectrum of *trans,cis*-[Re(dipy)(PMe_2Ph)Cl_2](PF_6), which contains the chelating ligand 2,2′-dipyridine, shows the complete cation and several fragment ions.[78] The spectrum of the complex *trans,cis*-[Re(dipy)-(PMe_2Ph)$_2$(CO)$_2$](PF_6) is simpler and consists of two major peaks: the base peak for the complete cation and the fragment ion [Re(dipy)(PMe_2Ph)(CO)$_2$]$^+$. The absence of peaks due to successive losses of the carbonyl groups from the complete cation prior to loss of other ligands is noteworthy. This feature contrasts with the EI behaviour of neutral complexes containing similar ligands, where the carbonyl groups are lost before elimination of any other ligands commences. MS/MS studies combined with FAB ionization of technetium and iron complexes of the type [Met(chel)$_2$$Cl_2$]($ClO_4$) [chel = *e.g.* 1,2-bis(dialkylphosphino)ethane and *o*-phenylenebis(dimethylarsine)] have shown that the stability and mode of fragmentation of the cation [Met(chel)$_2$$Cl_2$]$^+$ are determined by the size of the groups attached to the donor atoms of the chelating ligand and the metal.[136] Both the FD and FAB spectra of [Ag(dipy)$_2$](ClO_4) show two main peaks corresponding to [Ag(dipy)$_2$]$^+$ and [Ag(dipy)]$^+$. The fragmentation in the FD spectrum is unusual.[136]

The EHI spectrum of [Ru(dipy)$_3$]Cl_2, obtained by using glycerol (G) as solvent and sodium chloride as supporting electrolyte, shows only one peak corresponding to a metal-containing ion, [Ru(dipy)$_3$]$^{2+}$, and a very small peak for protonated dipyridine ions. In contrast, in the corresponding spectrum of [Cr(dipy)$_3$]Cl_2 the protonated dipyridine ions are relatively abundant and, more significantly, several metal-containing ions are present. These include the chromium(III) ions [Cr(dipy)$_2$Cl(G − H)]$^+$ and [Cr(dipy)$_2$$Cl_2$]$^+$, whose presence

[133] C. H. Yang and C. T. Chang, *J. Chem. Soc., Dalton Trans.*, 1982, 2539.
[134] N. Koprivanac, J. Jovanović-Kolar, and V. Kramer, *Int. J. Mass Spectrom. Ion Phys.*, 1983, **47**, 531.
[135] K. H. Ohrbach, G. Radhoff, and A. Kettrup, *Int. J. Mass Spectrom. Ion Phys.*, 1983, **47**, 59.
[136] S. E. Unger, *Anal. Chem.*, 1984, **56**, 363.

has been accounted for in terms of ligand exchange, with Cl^- and solvent, and oxidation of the chromium(II) complex in glycerol. The behaviour of these complexes reflects their solution chemistry and is indicative of the potential of EHI mass spectrometry as a probe of the solution chemistry of ion–ligand interactions.[137]

Macrocycles. – The mass-spectral behaviour of the biologically important corrins has received considerable attention.[138–144] A study of vitamin B_{12} involving the combined use of caesium desorption ionization and FTMS has been reported.[138] The application of FAB to these compounds has been briefly reviewed.[139] *o*-Nitrophenyl octyl ether has been found to be a useful matrix for FAB studies of corrins; with this matrix molecular cations are produced and hydrogenation reactions are suppressed.[144] The same matrix also proved to be useful for the analysis of iron porphyrin complexes.[144] The EI spectra of the iron(III) porphyrin carboxylates $Fe(por)(RCO_2)$ ($porH_2$ = octaethylporphyrin or tetraphenylporphyrin, R = Me, Et, or Ph) have been reported.[145] The base peak corresponds to $[(por)Fe]^+$ or $[(por)FeH]^{+\bullet}$ resulting from the fission of the metal–carboxylate bond without previous fragmentation of the carboxylate group. The LD spectrum of a cobalt(II) porphyrin dimer has been reported.[62]

Californium-252 plasma-desorption mass spectrometry has been further established as a method suitable for the study of the thermally unstable and chemically fragile chlorophylls. Both positive- and negative-ion spectra of chlorophyll α have been described, and fragmentation paths have been suggested.[146–148] This technique has also been applied to the characterization of the products of the chlorophyll allomerization reaction and the study of chlorophyll aggregates.[149]

A study of the temperature dependence of the spectra of several metal(II) complexes of phthalocyanine has been reported.[150] In the range 600–670 K the relative ion abundances are independent of residence time in the ion-source region. Interestingly, the relative abundance of the doubly charged molecular ions decreases as the temperature is raised. This is not the result of increased

[137] K. W. S. Chan and K. D. Cook, *J. Am. Chem. Soc.*, 1982, **104**, 5031.
[138] M. E. Castro and D. H. Russell, *Anal. Chem.*, 1984, **56**, 578.
[139] J. M. Miller, *J. Organomet. Chem.*, 1983, **249**, 299.
[140] J. Meili and J. Seibl, *Int. J. Mass Spectrom. Ion Phys.*, 1983, **46**, 367.
[141] H. Schwarz, K. Eckart, and L. C. E. Taylor, *Org. Mass Spectrom.*, 1982, **17**, 458.
[142] P. Schulthess, D. Ammonn, W. Simon, C. Caderas, R. Stepánek, and B. Kräutler, *Helv. Chim. Acta*, 1984, **67**, 1026.
[143] B. Kräutler, *Helv. Chim. Acta*, 1984, **67**, 1053.
[144] J. Meili and J. Seibl, Proceedings of the 31st Conference of Mass Spectrometry and Allied Topics, Boston, MA, 8–13 May, 1983, p. 294.
[145] H. Oumous, C. Lecomte, J. Protas, P. Cocolios, and R. Guilard, *Polyhedron*, 1984, **3**, 651.
[146] R. J. Cotter and J. Tabet, *Int. J. Mass Spectrom. Ion Phys.*, 1983, **53**, 151.
[147] B. T. Chait and F. H. Field, *J. Am. Chem. Soc.*, 1982, **104**, 5519.
[148] B. Chai and F. H. Field, *J. Am. Chem. Soc.*, 1984, **106**, 1931.
[149] J. E. Hunt, P. M. Schaber, T. J. Michalski, R. C. Dougherty, and J. J. Katz, *Int. J. Mass Spectrom. Ion Phys.*, 1983, **53**, 45.
[150] S. M. Schildcourt, *J. Am. Chem. Soc.*, 1983, **105**, 3852.

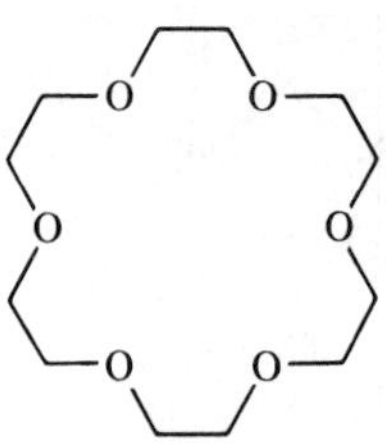

(70) 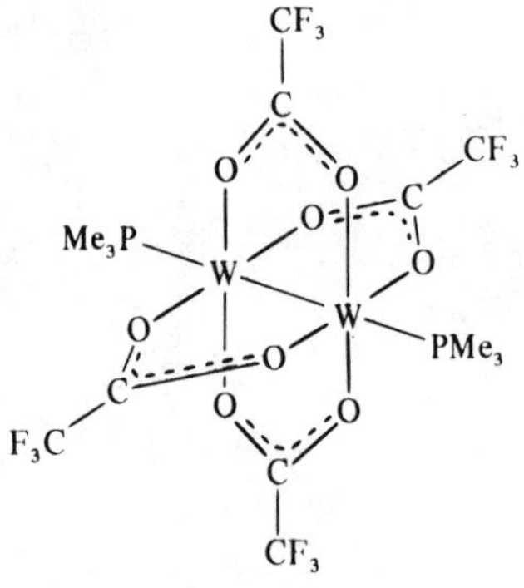 (71) 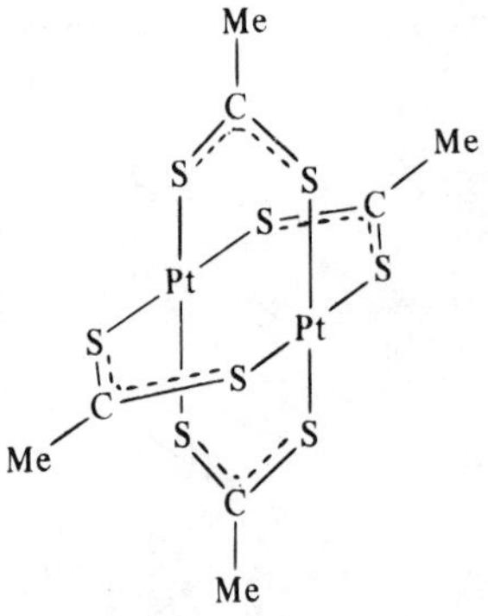 (72)

fragmentation of $M^{+\bullet}$ and has been attributed to thermal enhancement of single ionization by vibrationally induced autoionization.

Complex formation between metal cations and macrocyclic ligands has been observed by obtaining the FAB mass spectra of solutions of the ligands and metal salts in aqueous glycerol.[151] The spectra obtained in this way show abundant ions of the type $[\text{ligand} + \text{Met}^{n+} + \text{A}^{(n-1)-}]^+$, where A is the anion. For example, solutions of lead(II) chloride and the crown ether (70) yield ions corresponding to $[\text{crown} + \text{PbCl}]^+$. This approach provides convenient means for analysing traces of metals[152] and for investigating rapidly and semi-quantitatively the complexing ability of macrocylic ligands in solution.[153]

Carboxylates and Related Complexes. – The EI spectra of the dimeric complexes (71),[154] $Mo_2(OPr^i)_6(PhCO_2)_2$,[155] and $Pt_2(MeCS_2)_4$ (72),[156] which contain bridging carboxylato or thiocarboxylato ligands, have been reported. In all cases the spectra show abundant molecular ions corresponding to the dimeric unit, as well as other dimetallic ions. Molecular ions corresponding to the dimer unit have also been observed in the FD spectra of the structurally analogous compounds (73).[157] In contrast, in the EI spectrum of the dimeric amidino complex (74) the molecular ion is absent and the peak with highest mass corresponds to $[Cu_2L_2]^+$.[158] Similar behaviour is shown by other amidino complexes of copper(II).

A study of the EI spectra of zinc(II), magnesium(II), cobalt(II), and manganese(II) acetates indicates that these compounds undergo pyrolysis when introduced into the ion source *via* the direct-insertion probe to give the tetrametallic basic acetates $Met_4O(MeCO_2)_6$.[159] This is indicated by the presence of ions such

[151] R. A. W. Johnstone and I. A. S. Lewis, *Int. J. Mass Spectrom. Ion Phys.*, 1983, **46**, 451.
[152] R. A. W. Johnstone, I. A. S. Lewis, and M. E. Rose, *Tetrahedron*, 1983, **39**, 1597.
[153] R. A. W. Johnstone and M. E. Rose, *J. Chem. Soc., Chem. Commun.*, 1983, 1268.
[154] D. J. Santure, J. C. Huffman, and A. P. Sattelberger, *Inorg. Chem.*, 1984, **23**, 938.
[155] M. H. Chisholm, J. C. Huffman, and C. C. Kirkpatrick, *Inorg. Chem.*, 1983, **22**, 1704.
[156] C. Berlitto, A. Flamini, L. Gastaldi, and L. Scaramuzza, *Inorg. Chem.*, 1983, **22**, 444.
[157] H. P. M. M. Ambrosius, F. A. Cotton, L. R. Falvello, H. T. J. M. Hintzen, T. J. Melton, W. Schwotzer, M. Tomas, and J. G. M. Linden, *Inorg. Chem.*, 1984, **23**, 1611.
[158] M. Kilner and A. Pietrzykowski, *Polyhedron*, 1983, **2**, 1379.

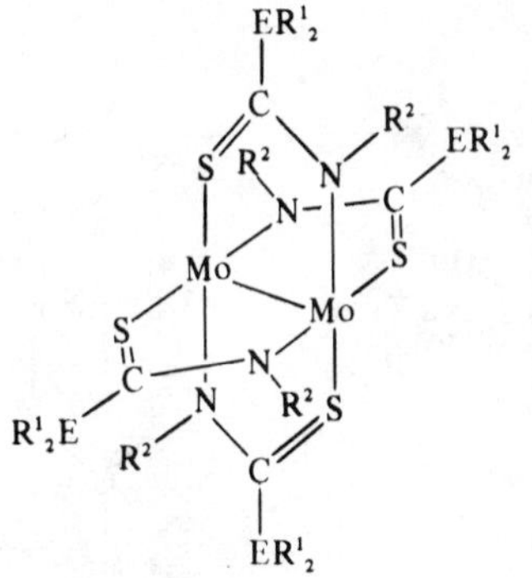

(73) $E = P, R^1 = Ph, R^2 = Me$ or Ph

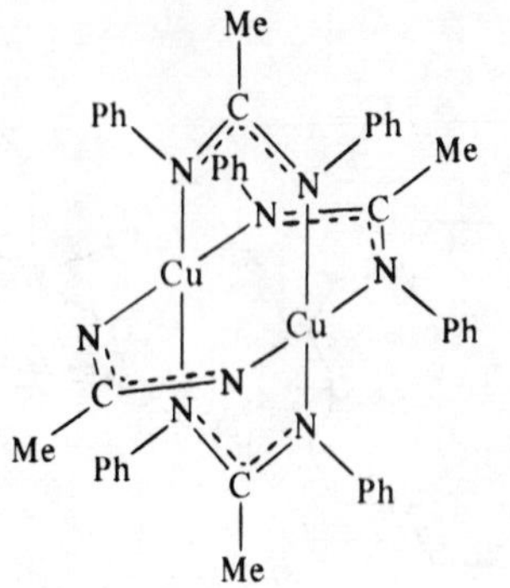

(74)

as $[Met_4O(MeCO_2)_n]^+$ ($n = 6$ or 5, Met = Co or Mn; $n = 5$, Met = Mg or Zn) in the spectra. The fragmentation behaviour of the structurally related compound $Be_4O(NO_3)_6$ has been discussed.[160]

Alkoxides and Related Compounds. – Prominent ions in the EI mass spectra of the molybdenum(VI) oxo alkoxides $MoO(OR)_4$ and $MoO_2(OR)_2$ ($R = Pr^i$, Bu^t, or Bu^tCH_2) have been reported, and the fragmentation behaviour of these compounds has been examined using ^2H-labelling.[161] The pyridine and dipyridine adducts $MoO_2(OR)_2(py)_2$ and $MoO_2(OR)_2(dipy)$ lose the neutral ligands in the vapour state and give spectra identical to that of $MoO_2(OR)_2$. On the other hand, the pyridine adduct of $TiCl_2(OPh)_2$ retains the Lewis base in the vapour state and shows a molecular ion corresponding to $[TiCl_2(OPh)_2(py)_2]^{+\cdot}$, as well as a fragment ion arising from loss of one pyridine molecule from $M^{+\cdot}$.[162] Fragmentation studies have been reported for iron(III) isopropoxide.[163] The spectra of titanium(IV) *F*-t-butoxide,[164] $Te(SCF_3)_2$,[165] and the dimetallic alkoxide $AlZr(OPr^i)$[166] have been examined. Mass spectrometry has been employed for the characterization of the products arising from the alcoholysis of metal aluminohydrides.[167] The products $[MetAlH_{4-n}(OR)_n]$ (Met = Li or Na, $R = Et$ or Bu^t) are dimeric in the vapour state.

[159] G. L. Marshall, *Org. Mass Spectrom.*, 1983, **18**, 168.
[160] V. A. Sipachev, N. I. Tuseev, Y. K. Nekrasov, and R. F. Galimzyanov, *Polyhedron*, 1982, **1**, 820.
[161] M. H. Chisholm, K. Folting, J. C. Huffman, and C. C. Kirkpatrick, *Inorg. Chem.*, 1984, **23**, 1021.
[162] K. C. Malhotra, K. C. Mahajan, and S. C. Chaudhry, *Polyhedron*, 1984, **3**, 125.
[163] N. Ya. Turova, T. V. Rogova, N. I. Kozlova, and A. I. Zhirov, *Sov. J. Coord. Chem. (Engl. Transl.)*, 1983, **9**, 708.
[164] J. M. Canich, G. L. Gard, and J. M. Shreeve, *Inorg. Chem.*, 1984, **23**, 441.
[165] N. Ya. Turova, S. I. Kucheiko, and N. I. Kozlova, *Sov. J. Coord. Chem. (Engl. Transl.)*, 1983, **9**, 513.
[166] A. I. Belokon', V. N. Bochkarev, V. V. Gavrilenko, T. D. Danina, and G. I. Belik, *J. Gen. Chem. U.S.S.R. (Engl. Transl.)*, 1983, **53**, 1843.
[167] T. R. Bierschenk and R. J. Lagow, *Inorg. Chem.*, 1983, **22**, 359.

5 Metal Cluster Ions and Miscellaneous Inorganic Compounds

During the period under review considerable attention has been given to the study of metal and metal-containing cluster ions. The production and stability of alkali-metal halide cluster ions $[M(MX)_n]^+$ and $[X(MX)_n]^-$ have been investigated using SIMS,[168–171] FAB,[172,173] and LD[62] techniques, and the gas-phase decomposition of such ions has been examined. All studies support previous findings regarding the presence of certain particularly stable cluster-ion species in the spectra, and structural proposals have been made for such species. In the case of small cluster ions, results of SIMS studies have been presented as evidence of a mechanism for the formation of gas-phase cluster ions.[170] The mass spectra of aluminium bromide[174] and copper halide[175] clusters have been reported. These spectra have been obtained by EI ionization of neutral clusters generated by quenching the metal halide vapour in helium gas. In the spectrum of aluminium bromide, cluster ions of type $[(AlBr_3)_n]^{+\cdot}$ containing up to 30 atoms are present, and these are particularly abundant when $n = 2, 4$, or 6. The copper halide spectra show a pronounced peak corresponding to $[Cu_{14}X_{13}]^+$ ($X = Cl$ or Br).[175] Applications of mass spectrometry to the study of metal clusters in the gas phase, concerned mainly with the determination of cluster distribution[176–178] and oxidation reactions[179,180] of clusters, have been described.

Several boranes[181–185] and osma- and irida-boranes[186] have been characterized by mass spectrometry. In the case of the diazo borane $1,10\text{-}B_{10}H_8(N_2)_2$,

168 T. M. Barlak, J. E. Campana, J. R. Wyatt, B. I. Dunlap, and R. J. Colton, *Int. J. Mass Spectrom. Ion Phys.*, 1983, **46**, 523.
169 K. G. Standing, R. Beavis, W. Ens, and B. Schueler, *Int. J. Mass Spectrom. Ion Phys.*, 1983, **53**, 125.
170 X. B. Cox, R. W. Linton, and M. M. Bursey, *Int. J. Mass Spectrom. Ion Phys.*, 1984, **55**, 281.
171 T. M. Barlak, J. E. Campana, J. R. Wyatt, and R. J. Colton, *J. Phys. Chem.*, 1983, **87**, 3441.
172 J. E. Campana and B. N. Green, *J. Am. Chem. Soc.*, 1984, **106**, 531.
173 B. I. Dunlap, J. E. Campana, B. N. Green, and R. H. Bateman, *J. Vac. Sci. Technol.*, 1983, **1**, 432.
174 T. P. Martin and J. Diefenbach, *J. Am. Chem. Soc.*, 1984, **106**, 623.
175 T. P. Martin and A. Kakizaki, *J. Chem. Phys.*, 1984, **80**, 3956.
176 S. J. Riley, E. K. Parks, C. R. Mao, L. G. Pobo, and S. Wexler, *J. Phys. Chem.*, 1982, **86**, 3911.
177 D. E. Powers, S. G. Handen, M. E. Geusic, A. C. Puiu, J. B. Hopkins, T. G. Dietz, M. A. Duncan, P. R. R. Langridge-Smith, and R. E. Smalley, *J. Phys. Chem.*, 1982, **86**, 2556.
178 K. Ervin, S. K. Loh, N. Aristov, and P. B. Armentrout, *J. Phys. Chem.*, 1983, **87**, 3593.
179 S. J. Riley, E. K. Parks, G. C. Nieman, L. G. Pobo, and S. Wexler, *J. Chem. Phys.*, 1984, **80**, 1360.
180 P. B. Armentrout, S. K. Loh, and K. M. Ervin, *J. Am. Chem. Soc.*, 1984, **106**, 1161.
181 T. Whelan, P. Brint, T. R. Spalding, W. S. McDonald, and D. R. Lloyd, *J. Chem. Soc., Dalton Trans.*, 1982, 2469.
182 M. A. Nelson, M. Kameda, S. A. Snow, and G. Kodama, *Inorg. Chem.*, 1982, **21**, 2898.
183 D. F. Gaines, J. A. Heppert, D. E. Coons, and M. W. Jorgenson, *Inorg. Chem.*, 1982, **21**, 3662.
184 J. A. Heppert, M. A. Kulzick, and D. F. Gaines, *Inorg. Chem.*, 1984, **23**, 14.
185 J. A. Heppert and D. F. Gaines, *Inorg. Chem.*, 1982, **21**, 4117.
186 J. Bould, N. N. Greenwood, and J. D. Kennedy, *J. Organomet. Chem.*, 1983, **249**, 11.

successive losses of N_2 molecules from the molecular ion followed by loss of H_2 have been observed;[181] for the borane $(\mu\text{-}Cl_2B)B_5H_8$, facile loss of the μ-boryl substituent has been reported.[182] Polyhedral boron chlorides of type B_nCl_n ($n = 9$–12) as well as compounds such as $H_2B_9Cl_7$ and HB_9Cl_8 have been identified by mass spectrometry.[187] In addition, the existence of other such compounds ($n = 13$–20) has been indicated by this technique. For all compounds with $n \leqslant 14$, molecular ions have been observed along with ions arising from loss of BCl_3 groups. High- and low-eV EI spectra of diboron tetraiodide have been reported. At 80 eV the base peak corresponds to $[B_2I_3]^+$ but the molecular ion is prominent and all other diboron-containing ions are present.[188]

The EI mass spectra of several derivatives of sulphur hexafluoride of type YSF_5 (Y = CF_3CH_2,[189] CF_3CFD,[189] CF_3CFCl,[189] $NSCl_2$,[190] NSFCl,[190] $NSeCl_2$,[191] $NPCl_3$,[191] and $EtCO_2$[192]) have been reported. Molecular ions are present only when Y = $NSeCl_2$, $NPCl_3$, or $EtCO_2$, and in all cases the main fragmentation pathways involve cleavage of the Y—S bond and/or loss of fluorine. When Y = $NPCl_3$ the presence of abundant ions such as $[F_2PCl_2]^+$ and $[FPCl_3]^+$ is indicative of reactions involving F migration to phosphorus. For the related compound $MeCH{=}SF_4$, an abundant molecular ion has been reported.[192] In contrast, the tellurium hexafluoride derivatives $YTeF_5$ (Y = OH, OF, or OCl) exhibit very small molecular-ion peaks along with base peaks due to $[TeF_5]^+$.[193] The EI spectra of various diazasulphanes of type (*75*; $n = 1$, 3, or 4) have been reported.[194] Molecular ions are present when $n = 1$ and $n = 4$ ($R = C_2F_5$). In the FD spectrum of (*76*) the molecular ion along with an abundant ion corresponding to $[M/2]^+$ has been observed.[195] The positive-ion LAMMA spectra of

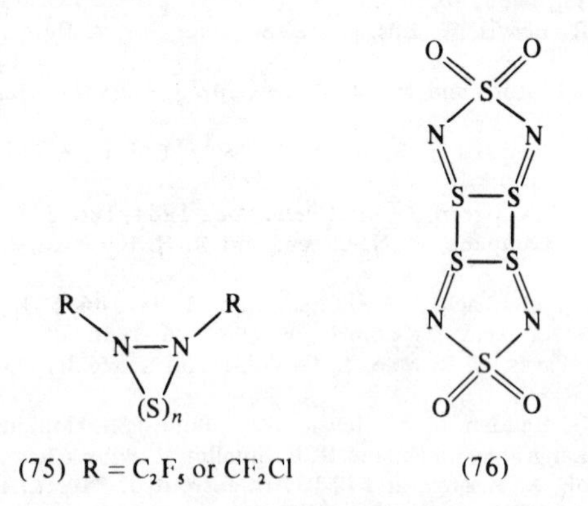

(*75*) $R = C_2F_5$ or CF_2Cl (*76*)

[187] D. A. Saulys, N. A. Kutz, and J. A. Morrison, *Inorg. Chem.*, 1983, **22**, 1821.
[188] W. Haubold and P. Jacob, *Z. Anorg. Allg. Chem.*, 1983, **507**, 231.
[189] H. F. Efner, R. Kirk, R. E. Noftle, and M. Uhrig, *Polyhedron*, 1982, **1**, 723.
[190] J. S. Thrasher, N. S. Hosmane, D. E. Maurer, and A. F. Clifford, *Inorg. Chem.*, 1982, **21**, 2506.
[191] J. S. Thrasher and K. Seppelt, *Z. Anorg. Allg. Chem.*, 1983, **507**, 7.
[192] B. Pötter and K. Seppelt, *Inorg. Chem.*, 1982, **21**, 3147.
[193] C. J. Schack, W. W. Wilson, and K. O. Christe, *Inorg. Chem.*, 1983, **22**, 18.
[194] R. C. Kumar and J. M. Shreeve, *Inorg. Chem.*, 1984, **23**, 238.
[195] H. W. Roesky, R. Emmert, and T. Gries, *Chem. Ber.*, 1984, **117**, 404.

Na_2SO_4, Na_2SO_3, and $Na_2S_2O_3$ have been reported, and the use of this technique for the analysis of alkali sulphoxy anions has been reported.[196] An interesting use of SIMS allowing the differentiation between phosphine (R_3P) and phosphonium ion ($[R_3PH]^+$) moieties present as ligands or counter-ions in metal complexes has been described.[197]

6 Reactions of Gaseous Metal or Metal-containing Ions with Organic Compounds

The study of gas-phase reactions of metal or metal-containing ions with a variety of organic compounds has attracted considerable attention. Such studies can provide thermodynamic and kinetic information about some of the basic processes of co-ordination – and in particular of organometallic – chemistry. Furthermore, they can help in elucidating important aspects of various catalytic processes. Consequently, they are expected to attract increasing interest not only from mass spectrometrists but also from many co-ordination chemists. The reactions have been investigated mainly by ICR, but ion-beam, flowing-afterglow, and spark-source techniques have also been employed.

Several studies concerned with the reactions of Ti^+,[198] Fe^+,[198–202] Co^+,[202] Ni^+,[201–204] or Rh^+[201,205] with various alkanes have been reported. A detailed discussion of ICR results is given in Chapter 6, but here it is noted that reactions of metal ions with propanes or larger alkanes generally involve exothermic processes and lead to ions of type $[MetC_nH_{2n}]^+$. For some comparable systems it has been established that product-ion distributions are independent of the method used to generate the Met^+ ion.[200] The formation of the $[MetC_nH_{2n}]^+$ ions involves elimination of a hydrogen molecule (Scheme 1) or of a small

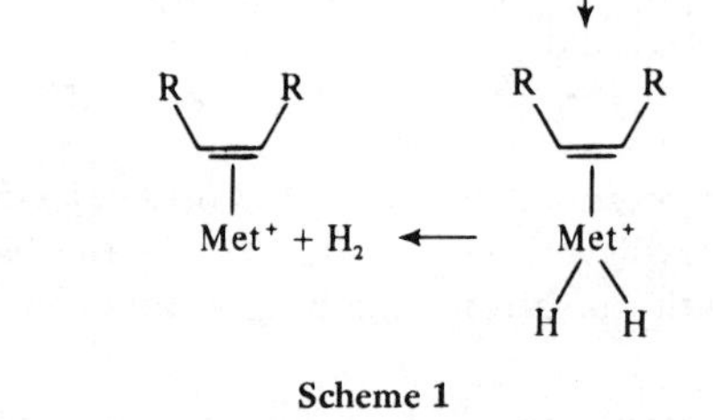

$$Met^+ + RCH_2CH_2R \longrightarrow R{-}CH(Met^+{-}H)CH_2R \longrightarrow (RCH{=}CHR)Met^+(H)(H) \longrightarrow (RCH{=}CHR)Met^+ + H_2$$

Scheme 1

[196] F. J. Bruynseels and R. E. Grieken, *Anal. Chem.*, 1984, **56**, 871.
[197] J. L. Pierce, D. DeMarco, and R. A. Walton, *Inorg. Chem.*, 1983, **22**, 9.
[198] G. D. Byrd, R. C. Burnier, and B. S. Freiser, *J. Am. Chem. Soc.*, 1982, **104**, 3565.
[199] J. Wronka and D. P. Ridge, *J. Am. Chem. Soc.*, 1984, **106**, 67.
[200] B. S. Larsen and D. P. Ridge, *J. Am. Chem. Soc.*, 1984, **106**, 1912.

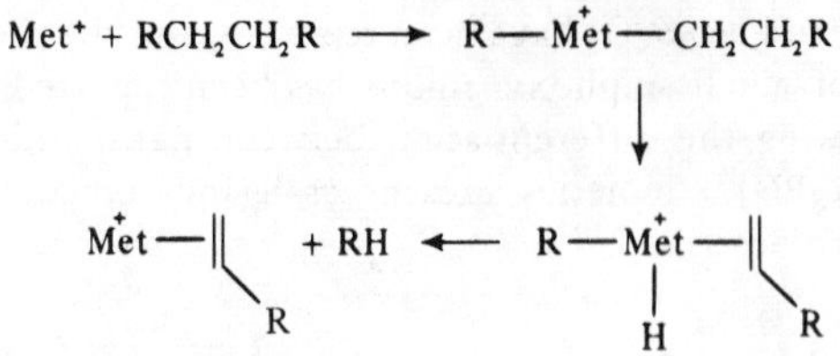

$$Met^+ + RCH_2CH_2R \longrightarrow R{-}\overset{+}{Met}{-}CH_2CH_2R$$

$$\downarrow$$

$$\overset{+}{Met}{-}(CH_2{=}CHR) + RH \longleftarrow R{-}\overset{+}{Met}(H){-}(CH_2{=}CHR)$$

Scheme 2

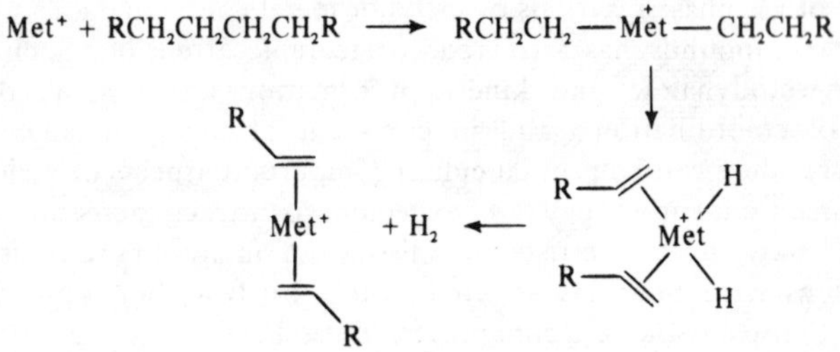

$$Met^+ + RCH_2CH_2CH_2CH_2R \longrightarrow RCH_2CH_2{-}\overset{+}{Met}{-}CH_2CH_2R$$

$$\downarrow$$

$$Met^+(CH_2{=}CHR)_2 + H_2 \longleftarrow (RCH{=}CH_2)_2\overset{+}{Met}H_2$$

Scheme 3

alkane (Scheme 2). In the systems involving Fe^+, Co^+, or Ni^+ the alkane- and H_2-elimination reactions are competing processes, but in the systems involving Ti^+ or Rh^+ H_2 elimination is dominant. It has been suggested that in systems exhibiting alkane cleavage the first step involves oxidative addition of the metal ion across a C—C bond. For systems exhibiting dehydrogenation it has been postulated that the first step involves oxidative addition to a C—H bond or, in the case of reactions involving Ni^+ and alkanes larger than propane, oxidative addition to an internal C—C bond (Scheme 3). In the latter case the addition step is followed by two hydrogen transfers.[201] The structures of various ions arising from the reactions of Met^+ with the alkanes have been investigated using CID methods. In reactions involving Ni^+ or Fe^+ and methane or ethane, only endothermic processes have been observed. The reactions with methane give $[MetH]^+$ as the major product ion, whereas the reactions with ethane yield two major products, $[MetH]^+$ and $[MetMe]^+$. From the latter reactions the bond-dissociation energies $D^0(Fe^+{-}Me) = 284 \pm 20$ kJ mol^{-1} and $D^0(Ni^+{-}Me) = 198 \pm 20$ kJ mol^{-1} have been obtained.[201]

ICR studies of the reactions of Co^+ and $[Co(CO)_n]^+$ ions with mono- and di-substituted n-butanes have been reported.[206] For the halides Bu^nX (X = F, Cl, or Br) reaction with Co^+ affords the butyl cation and the metal–butadiene

[201] L. F. Halle, P. B. Armentrout, and J. L. Beauchamp, *Organometallics*, 1982, **1**, 963.
[202] D. B. Jacobson and B. S. Freiser, *J. Am. Chem. Soc.*, 1983, **105**, 5197.
[203] L. F. Halle, R. Houriet, M. M. Kappes, R. H. Staley, and J. L. Beauchamp, *J. Am. Chem. Soc.*, 1982, **104**, 6293.
[204] D. B. Jacobson and B. S. Freiser, *J. Am. Chem. Soc.*, 1983, **105**, 736.
[205] G. D. Byrd and B. S. Freiser, *J. Am. Chem. Soc.*, 1982, **104**, 5944.
[206] A. Tsarbopoulos and J. Allison, *Organometallics*, 1984, **3**, 86.

complex $[Co(C_4H_6)]^+$. The latter ion also results from reactions of Co^+ with 4-halo-1-butanols, but the bifunctional compounds show a richer behaviour towards the metal ion than their monofunctional analogues. A model predicting the relative strengths of metal–ligand interactions in such systems was proposed.[206] Studies of reactions of spark-generated Co^+ ions with acetaldehyde, ethanol, methyl ether, and acetone have shown that most of the product ions formed are those expected by analogy with cationization by ground-state species.[207] However, these products are sometimes accompanied by other ions whose formation is likely to involve excited Co^+ ions. The Co^+ ion reacts with primary and secondary amines by insertion into the N—H bond.[208] Reactions involving the alkyl group are also observed when the alkyl chain is large, but insertions into the C—N bond are not generally favoured. In reactions between $[Fe(CO)_4]^-$ and various perhalogenated methanes, the main products arise by halogen transfer processes, but in some cases product ions of oxidative addition have also been observed.[209]

7 Knudsen-cell Mass Spectrometry

New developments in the study of vaporization processes and of chemical equilibria at high temperatures by Knudsen-cell mass spectrometry have been discussed.[210] Knudsen-cell techniques have been employed in the following: the determination of the spectra of the chlorides Pd_6Cl_{12}, Pt_6Cl_{12}, and $Pd_nPt_{6-n}Cl_{12}$,[211] the measurement of relative abundances and appearance energies of positive ions in the mass spectrum of POF_3,[212] the characterization of AsOCl,[213] SbOCl,[213] PSCl,[214] and PO_2Cl,[215] as well as the determination of the heats of formation of these species, the study of the thermal decomposition of scandium, yttrium, and lanthanum perrhenates,[216] the determination of the vapour pressures of CsF and CsCl,[217] and studies of the vapour-phase composition over alkali halide/alkaline-earth halide mixtures,[218] As_2S_3,[219] and As_2O_3.[220] Knudsen-cell techniques have also been employed in investigations concerned with organometallic compounds, such as in the determination of the enthalpy of

207 E. S. Ackerman, W. L. Grady, and M. M. Bursey, *Int. J. Mass Spectrom. Ion Phys.*, 1984, **55**, 275.
208 B. D. Radecki and J. Allison, *J. Am. Chem. Soc.*, 1984, **106**, 946.
209 R. N. McDonald, P. L. Schell, and W. D. McGhee, *Organometallics*, 1984, **3**, 182.
210 J. Drowart, *Int. J. Mass Spectrom. Ion Phys.*, 1982, **45**, 243.
211 H. Schäfer, U. Wiese, and H. Rabeneck, *Z. Anorg. Allg. Chem.*, 1984, **513**, 157.
212 O. Neskovic, M. Miletic, M. Veljkovic, D. Golobocantin, and K. F. Zmbov, *Int. J. Mass Spectrom. Ion Phys.*, 1983, **47**, 141.
213 M. Binnewies, *Z. Anorg. Allg. Chem.*, 1983, **505**, 32.
214 M. Binnewies, *Z. Anorg. Allg. Chem.*, 1983, **507**, 66.
215 M. Binnewies, *Z. Anorg. Allg. Chem.*, 1983, **507**, 77.
216 K. V. Ovchinnikov, E. N. Nikolaev, and G. A. Semenov, *J. Gen. Chem. U.S.S.R. (Engl. Transl.)*, 1983, **53**, 852.
217 H. Kawano, T. Kenpo, and Y. Hidaka, *Bull. Chem. Soc. Jpn.*, 1984, **57**, 581.
218 H. H. Emons, W. Horlbeck, and D. Kiessling, *Z. Anorg. Allg. Chem.*, 1983, **507**, 142.
219 K. H. Lau, R. D. Brittain, and D. L. Hildenbrand, *J. Phys. Chem.*, 1982, **86**, 4429.
220 R. D. Brittain, K. H. Lau, and D. L. Hildenbrand, *J. Phys. Chem.*, 1982, **86**, 5072.

sublimation of bis-(2,2-dimethylpropyl)magnesium[221] and in the study of the mass spectrum of methyl-lithium.[222,223]

8 Inductively Coupled Plasma/Mass Spectrometry

Most of the mass-spectrometric techniques described in this chapter have been used for qualitative and quantitative elemental analysis, particularly spark-source mass spectrometry, SIMS, thermal ionization, and various laser methods. Recent progress in such mass-spectrometric inorganic analysis has been reviewed well.[224] For routine determination of elements in aqueous solution, the relatively new and additional technique of inductively coupled plasma/mass spectrometry (ICP/MS) shows considerable potential.[225,226] Samples are usually introduced into a standard plasma source (as for emission spectrometry) through a conventional nebulizer, but other sample-inlet systems suitable for a plasma source are also applicable (*e.g.* flow injection[227] and laser ablation of solid samples[226] as used with the emission source[228]). Dissociation, atomization, and ionization of the sample take place in the tail flame, from which an ionized gas sample is extracted and introduced into the vacuum system of a quadrupole mass spectrometer. The plasma torch is positioned horizontally to facilitate the ICP/MS coupling. Instrumental aspects have been described in detail.[229–231] The chief attributes of the technique are good sensitivity and selectivity over a very wide range of elements (typical detection limits are 0.01–1 ng ml^{-1}), wide dynamic range (up to 10^7 ng ml^{-1}), simple and fast operation (about one sample every 2 min), a large degree of freedom from matrix effects, and simple spectra, predominantly of singly charged elemental ions (molecular ions and analyte oxides appear to contribute little to the data obtained). Real applications of the technique are only just beginning, but with the advent of two commercial ICP/MS instruments[231,232] rapid, quantitative, sensitive, and multi-element analyses should soon appear in the literature and lead to assessment of the true power of this promising method.

[221] O. S. Akkerman, G. Schat, E. A. I. M. Evers, and F. Bickelhaupt, *Recl. Trav. Chim. Pays-Bas*, 1983, **102**, 109.
[222] J. W. Chinn, jun. and R. J. Lagow, *Organometallics*, 1984, **3**, 75.
[223] J. W. Chinn, jun. and R. J. Lagow, *J. Am. Chem. Soc.*, 1984, **106**, 3694.
[224] F. Adams, *Spectrochim. Acta, Part B*, 1983, **38**, 1379.
[225] A. R. Date and A. L. Gray, *Spectrochim. Acta, Part B*, 1983, **38**, 29.
[226] A. R. Date, *Trends Anal. Chem.*, 1983, **2**, 225.
[227] R. S. Houk and J. J. Thompson, *Biomed. Mass Spectrom.*, 1983, **10**, 107.
[228] T. Ishizuka and Y. Uwamino, *Spectrochim. Acta, Part B*, 1983, **38**, 519.
[229] A. L. Gray and A. R. Date, *Analyst (London)*, 1983, **108**, 1033.
[230] A. R. Date and A. L. Gray, *Analyst (London)*, 1983, **108**, 159.
[231] D. Douglas, G. Rosenblatt, and E. Quan, *Q. J. Plasma Spectrosc.*, 1983, **3**, 140.
[232] J. Cantle, *Lab. Pract.*, 1983, **32** (June), 31.